André Balogh · Leonid Ksanfomality ·
Rudolf von Steiger
Editors

Mercury

Foreword by André Balogh, Leonid Ksanfomality and
Rudolf von Steiger

Previously published in *Space Science Reviews* Volume 132,
Issues 2–4, 2007

André Balogh
International Space Science
Institute (ISSI),
Bern, Switzerland

Leonid Ksanfomality
Space Research Institute (IKI),
Moscow, Russia

Rudolf von Steiger
International Space Science
Institute (ISSI),
Bern, Switzerland

Cover illustration: Three planets after sunset over Paranal Observatory, Chile. The lower planet is Mercury, the brightest at the centre is Venus, the one on the left is Saturn. Image courtesy and © Stéphane Guisard.

All rights reserved.

Library of Congress Control Number: 2008920289

ISBN 978-0-387-77538-8   ISBN 978-0-387-77539-5 (eBook)

Printed on acid-free paper.

© 2008 Springer Science+Business Media, BV

No part of this work may be reproduced, stored in a retrieval system, or transmitted in any form or by any means, electronic, mechanical, photocopying, microfilming, recording or otherwise, without the written permission from the Publisher, with the exception of any material supplied specifically for the purpose of being entered and executed on a computer system, for the exclusive use by the purchaser of the work.

1

springer.com

Contents

Introduction
A. Balogh · L. Ksanfomality · R. von Steiger   1

The Origin of Mercury
W. Benz · A. Anic · J. Horner · J.A. Whitby   7

Mercury's Interior Structure, Rotation, and Tides
T. Van Hoolst · F. Sohl · I. Holin · O. Verhoeven · V. Dehant · T. Spohn   21

Interior Evolution of Mercury
D. Breuer · S.A. Hauck II · M. Buske · M. Pauer · T. Spohn   47

The Origin of Mercury's Internal Magnetic Field
J. Wicht · M. Mandea · F. Takahashi · U.R. Christensen · M. Matsushima · B. Langlais   79

The Surface of Mercury as Seen by Mariner 10
G. Cremonese · A. Sprague · J. Warell · N. Thomas · L. Ksamfomality   109

Radar Imaging of Mercury
J.K. Harmon   125

Earth-Based Visible and Near-IR Imaging of Mercury
L. Ksanfomality · J. Harmon · E. Petrova · N. Thomas · I. Veselovsky · J. Warell   169

Mercury's Surface Composition and Character as Measured by Ground-Based Observations
A. Sprague · J. Warell · G. Cremonese · Y. Langevin · J. Helbert · P. Wurz · I. Veselovsky · S. Orsini · A. Milillo   217

Processes that Promote and Deplete the Exosphere of Mercury
R. Killen · G. Cremonese · H. Lammer · S. Orsini · A.E. Potter · A.L. Sprague · P. Wurz · M.L. Khodachenko · H.I.M. Lichtenegger · A. Milillo · A. Mura   251

Electromagnetic Induction Effects and Dynamo Action in the Hermean System
K.-H. Glassmeier · J. Grosser · U. Auster · D. Constantinescu · Y. Narita · S. Stellmach   329

Hermean Magnetosphere-Solar Wind Interaction
M. Fujimoto · W. Baumjohann · K. Kabin · R. Nakamura · J.A. Slavin · N. Terada · L. Zelenyi   347

Magnetosphere–Exosphere–Surface Coupling at Mercury
S. Orsini · L.G. Blomberg · D. Delcourt · R. Grard · S. Massetti · K. Seki ·
J. Slavin **369**

Plasma Waves in the Hermean Magnetosphere
L.G. Blomberg · J.A. Cumnock · K.-H. Glassmeier · R.A. Treumann **393**

Particle Acceleration in Mercury's Magnetosphere
L. Zelenyi · M. Oka · H. Malova · M. Fujimoto · D. Delcourt · W. Baumjohann **411**

Missions to Mercury
A. Balogh · R. Grard · S.C. Solomon · R. Schulz · Y. Langevin · Y. Kasaba ·
M. Fujimoto **429**

Introduction

**André Balogh · Leonid Ksanfomality ·
Rudolf von Steiger**

Originally published in the journal Space Science Reviews, Volume 132, Nos 2–4.
DOI: 10.1007/s11214-007-9293-0 © Springer Science+Business Media B.V. 2007

It is remarkable that Mercury, the messenger of the gods in ancient mythology, was discovered and identified in antique times. Unlike Venus and Mars, or even Jupiter and Saturn, Mercury is extremely difficult to observe by the unaided eye. It is, in fact, also very difficult to observe with astronomical telescopes, because of its proximity to the Sun and its small size. This means that, while Venus is a proud and very bright evening and morning "star", Mercury is, at its greatest visual elongation from the Sun, a very faint "twilight" or "daybreak" star. Despite that, the ancient Egyptian, Babylonian, Greek and other astronomers in the classical world recognised it as a wanderer against the starry background of the sky, noting its swift motion, and thus finding it an appropriate role in their mythology. For the Greeks, the god Hermes represented the planet and for the Romans, the god (and the planet) became Mercurius, now in English Mercury. Both appellations survive in planetary sciences: it is equally possible to speak of the Hermean magnetic field or the magnetic field of Mercury.

While there are fleets of increasingly sophisticated spacecraft targeting Mars—and Venus, Jupiter and Saturn remain of continuing interest to planetary scientists and space agencies—Mercury has suffered from being very difficult to reach and very difficult to observe from Earth. However, since the visits of the pioneering Mariner 10 spacecraft, there has been a steady effort to achieve a better understanding of Mercury. This has been done (a) by a sustained programme of ground-based observations by a dedicated set of planetary scientists, (b) by very fully exploiting the so-far unique Mariner 10 archive of observations

A. Balogh (✉)
International Space Science Institute (ISSI), Hallerstrasse 6, Bern 3012, Switzerland
e-mail: balogh@issibern.ch

L. Ksanfomality
Space Research Institute, Russian Academy of Sciences, Profsoyuznaya St. 84/32,
Moscow 117810, Russia

R. von Steiger
International Space Science Institute (ISSI), Hallerstrasse 6 Bern 3012, Switzerland

A nineteenth century illustration imagining the ancient Egyptians admiring Mercury at sunset. (After an engraving of Régnier Barbant in G. Flammarion, *Astronomie Populaire*, 1881, discovered by Réjean Grard during the long, drawn-out study phase of what became the BepiColombo mission)

and (c) by theoretical and modelling research to understand the evolution and current state of the planet and its known peculiarities, in particular its high density and its planetary magnetic field. It is generally recognized that understanding Mercury is one of the keys to understanding the origin and formation of the solar system.

The Workshop on Mercury held at the International Space Science Institute in Bern, on June 26–30, 2006, gathered a group of scientists who have dedicated a significant part of their career to this planet. The main objective was to review the achievements of the past 30 years and more in Mercury research, at the dawn of a new phase for the scientific investigations of Mercury by the forthcoming MESSENGER and BepiColombo missions. This volume is the result of the collaborations established at the Workshop.

The images taken by Mariner 10 almost 35 year ago have been extensively studied, although even now new and innovative image processing (not available to the first researchers after Mariner 10) have yielded new understanding of the surface features, as summarised by Gabriele Gremonese and his co-workers in this volume. However, in the absence of new space-based observations since Mariner 10, there have been remarkable achievements by ground-based observers. The continuously improved radar imaging of Mercury, particularly by the Goldstone radar and most importantly by the Arecibo giant dish (as described in this volume by John Harmon) has produced a range of observations which have provided major discoveries. Most prominent among these is the identification of radar-bright deposits in (mostly) near polar craters, indicating the possible presence of water ice where permanent shadowing by the crater walls may have preserved accumulated ice deposits over time. Further evidence for that is that the lesser shading of mid-latitude craters provides radar images of deposits in the expected shaded portion of such craters. In addition to this discovery, radar images of remarkable resolution and clarity have been obtained on the hemisphere not imaged by Mariner 10. A new light has also been shed on several surface features that had been imaged by Mariner 10. Our knowledge of Mercury has greatly benefited from the radar observations; it is likely that further radar observations may again be used as complementary information when the new missions, MESSENGER in the first place, will provide images of the so-far unseen side of Mercury.

Visual observations of Mercury from the ground are very difficult. Evidence for that comes from observations and drawings made by prominent planetary astronomers before the Mariner 10 mission that were found to be largely unrepresentative of the real surface features. New technologies in imaging and image processing have overcome some of those early difficulties. Interestingly, the ground-based visual observations have brought additional information on aspects of the nominally much higher resolution observations by Mariner 10 and also, of course, on intriguing features on Mercury's so-far unknown hemisphere as described by Leonid Ksanfomality and his co-authors in this volume. It seems that it will be impossible to better the ground-based images reported on in that paper, but experience shows that ground-based images may well remain useful in the future.

The composition and texture of Mercury's surface are important unknowns in efforts to determine the history and evolution of the planet. Spectral imaging of Mercury at visible and near-infrared wavelengths from Earth faces, of course, the same technical problems and may be even more difficult than the imaging of the visible features. Nevertheless, as reported by Ann Sprague and co-workers in this volume, notable progress has been made in multispectral imaging and in the analysis, interpretation and modelling of the observations. Both known and unknown hemispheres seem to show the silicate-dominated, heavily cratered surface characteristics, with a probably very important part influenced in its details by space-weathering, that is the direct impact of particulate matter as well as atomic particles in the solar wind and cosmic rays.

Moving away from the surface, Mercury has no atmosphere but only a tenuous, almost certainly highly variable exosphere. While the expected atomic hydrogen, helium and atomic oxygen components were observed from close-by during the Mariner 10 flybys of Mercury, the quantitative results have remained in some doubt and will be clarified with the arrival of the orbiters around the planet. As recounted in detail by Rosemary Killen and her co-authors (this volume), real progress has been made in the past 15 to 20 years from the observation of sodium and potassium emission lines observed from the ground. Although the spatial resolution of the observations is rather coarse when compared to even the visible or near-infrared observations from Earth, the Na lines have shown a remarkable variability in intensity and in location. There are strong presumptions that the emissions are modulated by activity in the magnetosphere of Mercury. Stefano Orsini and co-workers in this volume explore the connections in the obviously complex and time variable surface–exosphere–magnetosphere system.

Mercury's magnetosphere remains a fascinating subject and it is very important that new observations from the forthcoming orbiters extend very significantly the seemingly very small, yet highly productive data archive from two flybys of Mariner 10. In all, the total amount of data corresponds to about 45 minutes from the first and third Mariner 10 flybys. Although no new data can be obtained without visiting Mercury again, the data we possess have been subjected to many imaginative interpretations. On the one hand, there is an Earth-like aspect, but with the planet filling a much greater proportion of the volume of the magnetosphere than is the case at the Earth. On the other hand, the small planetary magnetic field, the higher variability (in absolute terms) of the dynamic effects of the solar wind produce very short time and spatial scales for magnetospheric phenomena than those with which we are familiar at Earth. The absence of an ionosphere is a major difference between Mercury and Earth, and the role played by the surface and exosphere in closing current systems remains unknown. Reviews in this volume by Matsumi Fujimoto, Lev Zelenyi, Lars Blomberg and their respective co-authors provide a good progress report of the numerous analyses and interpretations of the Mariner 10 data.

Although Mercury's comparatively high density was known before Mariner 10—implying a large, iron-dominated core—the discovery of the planetary scale magnetic field was completely unexpected. This discovery justified totally the inclusion of a magnetometer in the space probe's payload despite expert opinion that insisted on the frozen state of

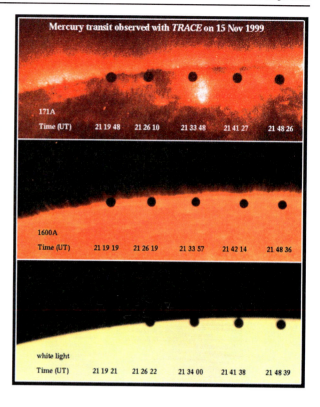

A time-elapsed series of the Mercury transit across the solar disk on November 15, 1999, observed by the TRACE spacecraft in three wavelengths, corresponding to (from top to bottom) the hot lower corona, the chromosphere and the photosphere. The transit is a very good illustration of the difference in scales between the Sun and the planet. Observations of Mercury transit have historically helped with the timing of Mercury's orbit and also to discover general relativity effects in its orbital period. (With acknowledgement to the TRACE team, Lockheed Martin Solar and Astrophysics Laboratory.)

Mercury's core and therefore the certain absence of a planetary dynamo. Clearly, we need considerably more data on the planetary magnetic field that can only come from the orbiters. Yet the simple existence and small magnitude of the magnetic field has continued to provide very big challenges to understanding the evolution and the current state of Mercury's interior, and constructing a planetary dynamo based on the interior modelling. The qualitative and quantitative constraints on the interior and the best current theories and models of the evolution and interior structure are reviewed in this volume by Tim van Hoolst and Doris Breuer and their respective co-authors. The increasingly sophisticated considerations that have become necessary to account for the relatively simple factual observations of the density and the magnetic field have now yielded an understanding that could be, but perhaps will not be challenged, although certainly refined by the observations to come.

Identifying the origin of the magnetic field remains a source of considerable difficulty. Not only is the internal structure of Mercury challenged by the existence of a planetary scale magnetic field, but what we know of the formation and functioning of planetary dynamos in general. The Mariner 10 data set is simply not sufficient to characterise and constrain an internal dynamo, or possible alternative sources in any detail. Nevertheless, over the past few years many of the conceptual and modelling difficulties have been thoroughly researched and many of those clarified. The status of our understanding of Mercury's internal magnetic field, including the very latest and significant developments, are described by Johannes Wicht and his co-authors in this volume. On this topic, perhaps more than any other related to Mercury, the new observations to come from the forthcoming orbiters are absolutely vital. Magnetic field measurements may well provide the most important information on the planet's interior as well as its environment. Therein lies a significant difficulty. The magnetic field that will be measured by the orbiters is a complex, nonlinearly interacting sum of the field internal to the planet and that generated by the highly variable currents due to the interaction of the solar wind with the planetary magnetic field. One representative aspect of this interplay between internal and external effects is the component of the magnetic field due to the induction field in the conductive core generated by external currents, as discussed in this volume by Karl-Heinz Glassmeier and his co-authors. Estimates of the contributions by the internal and external sources show that at any time and any place in the orbit around Mercury the contributions are likely to be of comparable magnitude, yet also variable, thus making the careful measurement and analysis of the magnetic field one of the most important objectives of the two orbiter missions.

There have been many proposals since Mariner 10 for the necessary space missions to Mercury, as described in this volume by André Balogh and his co-authors. It is quite remarkable that, after a lull of 35 years, two major space probes are targeting the planet Mercury. The first of these, NASA's MESSENGER, will make a first flyby of the planet in January 2008, before another flyby in 2008, then one in 2009, before finally inserted into orbit around Mercury in 2011. The joint, more ambitious, two-spacecraft BepiColombo mission by the European Space Agency and Japan's Institute of Space and Astronautical Sciences (a part of JAXA) will be launched in 2013 and will reach Mercury orbit in 2019. These two space missions will satisfy the lively curiosity of the two generations of planetary scientists who have worked, since the pioneering days of NASA's Mariner 10 mission in 1974–1975, on discovering the properties and peculiarities of this, the closest planet to the Sun and at least a distant cousin of the Earth in the family of terrestrial planets.

There is, however, a fundamental question that may not be answered, at least not simply or directly so, by the forthcoming space missions. That question is the origin of Mercury. As a member of the family of the four terrestrial planets (and the Moon), the formation of Mercury is an important aspect of our understanding of the terrestrial planets, their origin, formation and evolution. Willy Benz and his co-authors revisit, in this volume, this fundamental question and set out the likely scenarios that lead to the planet Mercury as we know it, in the orbit that we also know. It seems quite likely that neither planet nor its orbit is "original". This question is expected to remain on the agenda for the future, not least in an era when planetary systems around other stars increasingly become objects of close scrutiny.

The Editors are very happy to thank all those who have contributed to this volume and to the workshop. First of all, we thank the authors for their integrated approach to distil the presentations and discussions in the Workshop, in particular those who accepted the responsibility to coordinate the articles in this volume. All the papers were peer reviewed by referees, and we thank the reviewers for their helpful and critical reports. We also thank the directorate of ISSI for selecting Mercury as the topic for the workshop, and the advice of ISSI's Science Committee on this subject. We thank the staff of ISSI, in particular Roger Bonnet for his support of the workshop, and also Brigitte Fasler, Andrea Fischer, Vittorio Manno, Saliba F. Saliba, Irmela Schweizer and Silvia Wenger for their help, patience and good humour to provide a productive environment for the workshop.

The Origin of Mercury

W. Benz · A. Anic · J. Horner · J.A. Whitby

Originally published in the journal Space Science Reviews, Volume 132, Nos 2–4.
DOI: 10.1007/s11214-007-9284-1 © Springer Science+Business Media B.V. 2007

Abstract Mercury's unusually high mean density has always been attributed to special circumstances that occurred during the formation of the planet or shortly thereafter, and due to the planet's close proximity to the Sun. The nature of these special circumstances is still being debated and several scenarios, all proposed more than 20 years ago, have been suggested. In all scenarios, the high mean density is the result of severe fractionation occurring between silicates and iron. It is the origin of this fractionation that is at the centre of the debate: is it due to differences in condensation temperature and/or in material characteristics (e.g. density, strength)? Is it because of mantle evaporation due to the close proximity to the Sun? Or is it due to the blasting off of the mantle during a giant impact?

In this paper we investigate, in some detail, the fractionation induced by a giant impact on a proto-Mercury having roughly chondritic elemental abundances. We have extended the previous work on this hypothesis in two significant directions. First, we have considerably increased the resolution of the simulation of the collision itself. Second, we have addressed the fate of the ejecta following the impact by computing the expected reaccretion timescale and comparing it to the removal timescale from gravitational interactions with other planets (essentially Venus) and the Poynting–Robertson effect. To compute the latter, we have determined the expected size distribution of the condensates formed during the cooling of the expanding vapor cloud generated by the impact.

We find that, even though some ejected material will be reaccreted, the removal of the mantle of proto-Mercury following a giant impact can indeed lead to the required long-term fractionation between silicates and iron and therefore account for the anomalously high mean density of the planet. Detailed coupled dynamical–chemical modeling of this formation mechanism should be carried out in such a way as to allow explicit testing of the giant impact hypothesis by forthcoming space missions (e.g. MESSENGER and BepiColombo).

Keywords Mercury: origin · Planets: formation · Numerics: simulation

---

W. Benz (✉) · A. Anic · J. Horner · J.A. Whitby
Physikalisches Institut, University of Bern, Sidlerstrasse 5, 3012 Bern, Switzerland
e-mail: wbenz@space.unibe.ch

J. Horner
Astronomy Group, The Open University, Walton Hall, Milton Keynes MK7 6AA, UK

1 Introduction

The density of Mercury, mean density 5.43 g/cm^3 (Anderson et al. 1987), uncompressed mean density $\sim$5.3 g/cm^3 (Cameron et al. 1988), is anomalously high. For comparison we note that the uncompressed mean density of the Earth is just $\sim$4.45 g/cm^3 (Lewis 1972). From this Urey (1951, 1952) noted that Mercury must have an iron-to-silicate ratio much larger than that of any other terrestrial planet. The silicate-to-iron mass ratio is usually estimated to lie in the range from 30 : 70 to 34 : 66 or roughly 0.5. Harder and Schubert (2001) argued that the presence of sulfur in the core could lead to even smaller ratios and that a planet entirely made of FeS could not be excluded. All of these ratios are many times smaller than those of any of the other terrestrial planets or the Moon.

A variety of hypotheses have been suggested to account for the anomalously high mean density of Mercury. In all cases, the close proximity of Mercury to the Sun plays a crucial role and all theories invoke processes that result in some level of fractionation between iron and silicates during the very early phases of the solar system in order to explain this strange mean density. Amazingly, all these scenarios and ideas date back some 20 years or more. As far as the origin of the planet is concerned, very little new work has been carried out during the last two decades. In our opinion, this reflects more the lack of new relevant data than a lack of interest in the origin of this end member of the solar system. Although new ground-based observations of Mercury have been made since the Mariner 10 mission (Sprague et al. 2007), these have not yielded a consensus on the detailed geochemical and geophysical parameters necessary to distinguish between models of Mercury's formation. It is clear that with the two new space missions dedicated to study Mercury in unprecedented detail (NASA's MESSENGER and ESA's BepiColombo; see e.g. Balogh et al. (2007)), this situation is about to change drastically. It is therefore critical to revisit the problem of the origin of Mercury and to work out models that make testable predictions in order to prepare the necessary framework in which to discuss the measurements the two future missions will be able to carry out.

Mercury formation models which have been proposed to account for this anomaly can be classified into two broad categories according to the time at which the fractionation occurs. In the first category, we find models that explain the anomalous composition of the planet as a result of fractionation that occurred during the formation of the planet proper. The second set of models encompasses those for which the planet forms first with roughly chondritic abundances and fractionates shortly thereafter. We shall briefly review these two categories in Sect. 2.

Studying Mercury's origin involves studying the dynamics and chemistry of the protoplanetary nebula in close proximity to the star. Since planets grow through collisions, the study of the formation of Mercury is also an investigation of these processes in a region where these collisions are particularly violent. Although the details of the study may be specific to just this planet, it holds implications for the formation of rocky planets (or the cores of giant planets) in general, and may provide a means for choosing between different theories.

2 The Formation of Mercury: Scenarios and Ideas

In this section we briefly recall the different scenarios that have been proposed to explain Mercury's anomalous composition. In all the currently available scenarios, the main point is to achieve enough chemical fractionation to account for the high density of the planet. Not surprisingly, all these scenarios take place very early on in the history of the solar system either as an ongoing process during the formation of the planets, or during the late stages of accretion or shortly thereafter. They all rely in some way on the peculiar position of the planet, namely its close proximity to the Sun.

2.1 Fractionation During Formation

In this class of models, the anomalous density of Mercury results from fractionation occurring during the formation process itself. In its simplest form, fractionation is obtained as a result of an equilibrium condensation process in a proto-planetary nebula in which the temperature is a monotonic function of the distance to the Sun (Lewis 1972, 1974; Barshay and Lewis 1976; Fegley and Lewis 1980). Such models predict that the condensates formed at Mercury's distance were both extremely chemically reduced and extremely poor in volatiles and FeO. Metallic iron would be partially condensed while refractory minerals rich in calcium, aluminum, titanium and rare earths would be fully condensed. The bulk average density of the condensed material would therefore be much higher in the Mercury region than in the formation regions of the other terrestrial planets hence explaining the high mean density.

Although such simple models of the chemical behavior of the solid material in the early solar nebula successfully predict some of the most general compositional trends of solar system bodies, it was recognized by Goettel and Barshay (1978) and later by Lewis (1988), that this mechanism cannot explain the anomalous density of Mercury. The main reason for the failure of this model is the relatively small difference in the condensation temperature of core and mantle material. This implies also a close spatial proximity in the nebula while the area over which the material must be collected to actually bring a planetary mass together, is much larger. The high-density material is simply diluted with lower density material. Lewis (1988, and references therein) showed that this results in a maximum core mass fraction of about 36% as compared to the 70% for the actual planet. Hence, simple condensation–accretion models fail to explain the mean density of Mercury.

To circumvent these difficulties, various additional fractionation mechanisms operating during, or even before, the start of planetary accretion have been proposed. While some combination of these mechanisms based on microscopic differences between silicates and iron (ferromagnetism, strength, etc.) may possibly lead to higher mean densities, there exist no compelling reasons why these mechanisms should have been more active at Mercury than other places in the solar system (see Weidenschilling 1978 for a detailed discussion). Weidenschilling (1978), on the other hand, proposed that the additional fractionation results from a combination of gravitational and drag forces. As the early condensates orbited the Sun, immersed in a gaseous disk, they felt a drag force that depends in a complex fashion upon the size and shape of the condensed particles and upon the structure of the nebula. As a result of this drag force, orbiting bodies lose angular momentum and spiral inward. In a simple quantitative model, Weidenschilling (1978) showed that the rate of orbital decay is slower for larger and/or denser bodies. With suitable but reasonable assumptions for the initial conditions, Weidenschilling (1978) showed that the fractionation required to produce iron-rich planets can be achieved.

Following the three-dimensional dynamics of a dusty gas over periods of time vastly exceeding a dynamical timescale is a complicated problem, especially since the dynamics of the gas itself is still up for debate. For example, the origin of the turbulence, the existence of instabilities, the presence of spiral waves, among others, are still unclear. Hence, short of a better understanding of the dynamics of this multicomponent fluid, it is difficult to assess to what extent models based on fractionation occurring in a laminar nebula before and during planet formation are realistic.

2.2 Fractionation after Formation

Cameron (1985) proposed that, during the early evolution of the solar nebula, temperatures at the position of Mercury were probably in the range 2,500–3,000 K. If a proto-Mercury existed at the time, partial volatilization of the mantle would occur thus creating a heavy silicate atmosphere which could over time be removed by a strong solar wind. Fegley and Cameron (1987) computed the expected bulk chemical composition of the mantle as a function of evaporated fraction using both ideal and nonideal magma chemistry. They showed that starting with a proto-Mercury of chondritic abundance (2.25 times the mass of the present day planet) Mercury's mean density can be obtained after 70–80% of the mantle has evaporated. At this point, the remaining mantle is depleted in the alkalis, FeO and SiO_2, but enriched in CaO, MgO, Al_2O_3 and TiO_2 relative to chondritic material. Fegley and Cameron (1987) argued that this anomalous composition represents a unique signature of this formation scenario that could eventually be measured by a dedicated spacecraft mission.

This scenario has the clear advantage of having its consequences calculated in enough detail to allow potentially explicit testing. However, it also suffers from a number of difficulties. For example, it is not clear whether high enough temperatures can be reached and maintained for long enough in the solar nebula *after* a suitable proto-Mercury has been formed. Furthermore, as already identified by Cameron (1985) himself, the solar wind may not be efficient enough to remove the heavy silicate atmosphere thus preventing a significant evaporation of the mantle.

In another scenario to explain Mercury's anomalous density, the removal of a large fraction of the silicate mantle from the originally more massive proto-Mercury is achieved following one (or possibly more) giant impacts (Smith 1979; Benz et al. 1988; Cameron et al. 1988). In this hypothesis, a roughly chondritic Mercury (2.25 times the mass of present day Mercury) is hit by a sizable projectile (about 1/6 its mass in the calculations by Benz et al. 1988) at relatively high velocity. Such an impact results in the loss of a large fraction of the mantle leaving behind essentially a bare iron core (see Sect. 3). The existence of large projectiles was first suggested by Wetherill (1986), who realized that terrestrial proto-planets probably suffered collisions with bodies of comparable mass during the final stages of their formation. He also proposed that the high relative velocities in Mercury's formation region could lead to particularly disruptive collisions making the formation of Mercury unique among the terrestrial planets.

Simulations (Benz et al. 1988; Cameron et al. 1988) have shown that the required removal of the silicate mantle can be achieved by a giant impact. However, the question of the long-term fate of the material ejected from such an impact has never been properly investigated. Indeed, most of the ejected material following the impact is still orbiting the Sun on Mercury crossing orbits and will therefore eventually collide with the planet and be reaccreted over time unless it is removed by some other processes. If a significant fraction should indeed be reaccreted, the fractionation obtained as a result of the collision would only be short lived and therefore would not explain the present-day mean density of the planet.

Both gravitational scattering and the Poynting–Robertson effect have been invoked as possible ejecta-removal mechanisms. However, the former is found to remove only a very small amount of material (see Sect. 4.1) and the efficiency of the second depends on the size distribution of the ejecta. Simple condensation models based on equilibrium thermodynamics (Anic 2006) show that the expanding vapour cloud following the impact would lead to the formation of small-sized condensates (see Sect. 4.1). These small-sized condensates can readily be affected by nongravitational forces such as those originating from the Poynting–Robertson effect. Hence, from a dynamical point of view, the giant impact scenario as proposed by Benz et al. (1988) and Cameron et al. (1988) appears to be possible. It remains to be determined, however, whether the chemical signature of such a giant impact is compatible with the bulk chemistry of Mercury.

3 Simulations of Giant Impacts

3.1 Initial Conditions

As the target body in our collision simulations, we adopt a proto-Mercury that has roughly chondritic abundance. We built such a proto-Mercury by increasing the mass of the silicate mantle of the planet until the present-day core mass represents only about 1/3 of the planet's total mass. With this structure, the total mass of proto-Mercury amounts to 2.25 times the mass of present-day Mercury. Its internal structure is computed using the usual equations assuming an adiabatic temperature profile (Spohn et al. 2001). For the equation of state (EOS) in our calculations we use ANEOS (Thompson and Lauson 1984). This analytical EOS relates temperature and density to pressure, and describes mixed states (liquid–vapour, solid–vapour) using the Helmholtz free energy potential. The equation requires 24 coefficients for a given material which, for the most part, can be derived from laboratory experiments. We assume that the mantles of the projectile and the target consist of dunite (a rock consisting largely of forsteritic olivine Mg_2SiO_4) which has similar bulk properties to mantle rock. The table of parameters for dunite was given by Benz et al. (1989). The core of the planet is assumed to consist of pure iron.

Finally, we must specify boundary conditions and in particular the value of the temperature at the surface of the planet and the projectile. Since this value is not known at the time of formation of the planet, we use two different values which should bracket the possibilities reasonably well. In one case, we use the present-day mean surface temperature of 452 K and for the other we consider a much hotter body with a surface temperature of 2,000 K. We shall refer to these two models in the text as *cold* and *hot*.

3.2 Numerics and Model Assumptions

Following Benz et al. (1988), we use a 3D Smooth Particle Hydrodynamics (SPH) code to simulate the impacts. SPH is a Lagrangian method in which the motion of the mass elements (particles) is followed over time. Given that SPH has already been described many times in the literature and that we use a fairly standard implementation of the method, we refer the reader to reviews by Benz (1990) and Monaghan (1992) for further detailed explanation of the method. In the present work, we use the version of SPH described by Benz (1990), with only a small number of modifications. The major change is the use of individual artificial viscosity coefficients that vary over time using the shock detection algorithm proposed by Morris and Monaghan (1997), which minimises the viscosity outside shocks.

In all cases, we neglected the strength of the material. This assumption is reasonable given the size of the bodies involved for which self-gravity and pressure gradients are the dominating forces. Self-gravity is computed using a hierarchical binary tree as discussed by Benz (1990). We also neglect radiative losses during the impact (cooling due to adiabatic expansion is included). The main reason for this is to avoid the considerable additional numerical work that would be needed to compute such radiative losses. From a physical point of view, the assumption is justified by the fact that the simulations proper extend only over a relatively short amount of real time during which the radiative losses should remain small. We investigated simple models of radiative cooling as part of the condensation calculations presented in Sect. 4.1.

The simulations were carried out until the ultimate fate of the material could be reliably determined. At that time, we identified the material having being lost by the planet using the same iterative procedure (based on binding energy) as described by Michel et al. (2002). For the material remaining gravitationally bound, we compute the fractions of silicate and iron in order to determine the rock-to-iron (R/I) ratio.

3.3 Results

We performed a number of simulations of giant collisions with different projectile masses, impact velocity and impact geometries in order to find collisions that lead to a suitable fractionation. However, we did not carry out an extensive search to find all the possible initial conditions leading to the desired result. Hence, we cannot compute the actual probability of such an event. However, we note that success (see Table 1) does not involve exceptional geometries or mass ratios. On the other hand, the velocity at which the two large bodies must collide in order to ensure almost complete mantle loss is relatively high. Such high relative velocities are much more likely to occur in the inner regions of the solar system

Table 1 Simulations involving the "cold" (runs 1–12) and "hot" (runs 13–17) proto-Mercury

Run	b [R]	v_{rel} [km/s]	m_{imp}/m_{tar}	N_p	N_t	R/I	M_f
1	0.7	30	0.1	11'326	113'129	1.32	1.73
2	0.5	27	0.1	15'543	155'527	1.10	1.50
3	0	20	0.1	11'326	113'129	1.25	1.63
4	0.5	26	0.167	23'556	141'392	0.78	1.18
5	0.53	28	0.167	23'556	141'392	0.59	1.01
6	0	20	0.167	23'556	141'392	0.61	0.92
7	0	26	0.167	23'556	141'392	0.11	0.15
8	0.7	26	0.2	49'485	247'423	1.11	1.52
9	0.7	28	0.2	28'269	141'392	0.94	1.38
10	0.6	30	0.2	27'411	137'200	0.51	0.94
11	0.5	28	0.2	27'411	137'200	0.50	0.86
12	0	28	0.2	27'411	137'200	–	0
13	0.47	23	0.167	21'205	127'249	0.87	1.27
14	0.5	24	0.167	21'205	127'249	0.86	1.26
15	0	19	0.2	28'269	141'392	0.71	1.01
16	0.46	25.5	0.2	28'269	141'392	0.51	0.87
17	0.5	28	0.2	28'269	141'392	0.49	0.85

N_p is the number of particles in the projectile and N_t in the target. The impact parameter b is given in units of target radius and the relative velocity is given in km/s. R/I is the silicate to iron ratio in the surviving planet and M_f is its final mass in units of present-day Mercury mass. The planet in run 12 disintegrated where the Keplerian velocities are already large. Hence, extreme collisional fractionation of full-grown planets can, from a theoretical point of view, involve only planets orbiting deep in the potential well. This is consistent with the fact that Mercury is the only planet in the solar system with such a high mean density.

The initial conditions for the simulations performed and the final characteristics of the surviving planets are given in Table 1. Note that some of the runs are very similar to those performed by Benz et al. (1988) but with a considerable increase in the number of particles used (typically 20 to 50 times more).

In the various cases listed in this table, the collisions leading to a final mass of $M_f \approx 1$ (in units of present day Mercury mass) and a silicate to iron mass ratio $R/I = 0.4$–0.6 can be considered as successful in the sense that they reproduce the bulk characteristics of present-day Mercury. In fact, depending upon the subsequent reaccretion of a fraction of the silicate mantle (see Sect. 4), simulations with R/I less than the present-day value should be considered as successful.

Note that, in order to remove a sizable fraction of the silicate mantle, the collision speed must be quite high, especially in the case of an off-axis collision for which the strength of the shock is significantly weaker (all other parameters being equal). Statistically, the most probable collisions are those with $b = 0.7 R_{\text{proto-Mercury}}$ (Shoemaker 1962) where the impact parameter b is defined as the distance from the centre of the target to the centre of the impactor along a line normal to their relative velocity (b is thus zero for a head-on collision and $R_{\text{proto-Mercury}} + R_{\text{impactor}}$ for a grazing collision). However, for realistic relative velocities and reasonable-sized projectiles, these dynamically most probable collisions seem not to result in a large enough loss of mantle material.

Overall, the simulations that appear to yield potentially satisfactory results are runs 6, 10, 11 in the case of a "cold" proto-Mercury and runs 16, 17 for the "hot" progenitor. Hence, as far as the initial blasting off of the mantle is considered, the thermal state of the progenitor does not appear to play a major role. Collisions involving "hot" bodies are not overwhelmingly more disruptive that those involving "cold" ones. In fact, similar results can be obtained by relatively small changes in collision characteristics. To illustrate a typical collision, Fig. 1 shows a set of four snapshots illustrating run 11.

We note how severe this collision actually is. The planet is nearly destroyed in the process and it is actually gravitational reaccumulation that brings the core of the planet back together. Such nearly destructive collisions are required if most of the mantle of a roughly chondritic proto-Mercury is to be removed. This also shows that destroying *large* bodies by means of collisions is not so easy and requires large impactors and high velocities. We argue that this implies that such events can only occur in regions near the star where the collision velocities can be high enough. If this is correct, it could explain why only Mercury fractionated to such an extent even though all the other terrestrial planets also experienced giant collisions during their formation. This makes Mercury particularly important for the study of terrestrial planet formation. We also note that as a result of the severity of the impact, all the material reaches high temperatures and thus the assumption made by Harder and Schubert (2001), that a Mercury formed by means of a giant impact could have a volatile rich composition and lose more iron than sulfur during the collision, seems unlikely to be true.

Finally, we point out that our run 5 had almost identical initial conditions to run 13 by Benz et al. (1988) and that the outcomes are very similar even though in this work we have been able to use about 50 times as many particles!

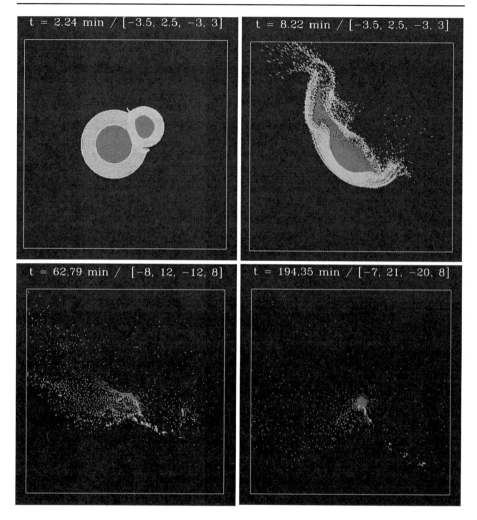

Fig. 1 Snapshots from the evolution of run 11. Particles in a slice running through the central plane are plotted. Velocity vectors are normalized to the maximum vector in each figure and plotted at particle locations. Iron is shown in *dark grey*, whilst *light grey* represents silicates. The time after first contact (in minutes), along with the coordinates of the quadrant given in units of target radius, is given above each snapshot

4 The Fate of the Ejecta

As mentioned in Sect. 1, a giant impact removes most of the rocky mantle is not sufficient to explain the present-day bulk composition of Mercury. It is also necessary to demonstrate that the overwhelming part of the ejected matter is not reaccreted by the planet over time. For this to happen, it needs to be removed from Mercury crossing orbits before it collides with the planet again. In order to address these issues, we first computed the size distribution of the ejected matter (Sect. 4.1) and then compared the timescale required by the Poynting–Robertson effect to remove the ejecta (Sect. 4.2) with the timescale until collision with the planet (Sect. 4.3). We also investigated how effective gravitational torques exerted by other planets can be in ejecting the material (Sect. 4.3).

4.1 Size Distribution of the Ejecta

To compute the final size distribution of the ejected matter it is necessary to follow its thermodynamical evolution. This is conveniently done by using a $T-\rho$ diagram such as that sketched in Fig. 2. In our calculations we assume equilibrium thermodynamics, neglecting all rate-dependent effects. To check the importance of the equation of state, we computed the cooling curves using both a perfect gas EOS and ANEOS. For simplicity, but partially justified by the short duration of the impact, we also omit radiative losses and assume that the internal energy of the hot gas is entirely transformed into the kinetic energy of the expansion. Finally, in following the ejected matter, we treat each SPH particle as an independent piece of material ignoring the potential interactions (heat exchange, collisions, etc.) between the expanding particles. The overall size distribution is obtained by summing up the distributions obtained for all ejected SPH particles.

Upon being struck by a very large, fast-moving body, a large fraction of the target material is compressed to extreme pressures at which both metallic and siliceous liquids exhibit characteristics of a supercritical fluid ("hot vapour" in what follows). The path followed by the material during this compression phase is shown by ① in Fig. 2. During the subsequent very rapid decompression of the compressed liquid and the expansion of the hot vapour the matter undergoes a phase transition from either the liquid or the vapour side of the vapour-liquid dome (respectively ② and ③ in Fig. 2). Depending upon the cooling path taken by the hot vapour, we use two different approaches to compute the size distribution of the condensates following the phase transition.

In the case that the transition occurs along path ③, we use the homogenous condensation model of Raizer (1960). In this model, when the expanding vapour cloud crosses the vapour–liquid boundary given by the Clausius–Clapeyron equation, the vapour enters first a

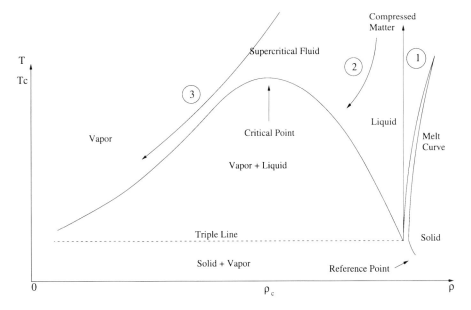

Fig. 2 Schematic T, ρ vapour–liquid–solid diagram. T_c and ρ_c indicate the critical point above which one cannot separate liquid from vapour. The reference point points to room-temperature conditions. The matter is shocked along path ①, and relaxed along the paths ② and ③

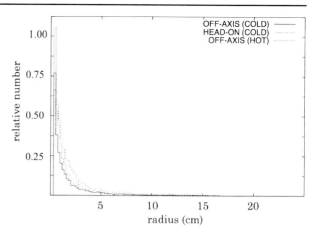

Fig. 3 Size distribution of particles (condensates and melt droplets) which result from runs 6, 11 and 17 using ANEOS. Particle sizes range from 1.7×10^{-6} cm to 22.3 cm for run 6, 1.6×10^{-6} cm to 21.4 cm for run 11, and from 1.6×10^{-2} cm to 20.7 cm for run 17

saturated and then a supersaturated (metastable) state. The vapour then undergoes a partial transformation into liquid droplets, where each droplet contains some average number of atoms. Their growth is only possible if their volume energy exceeds their surface energy. The number of liquid droplets formed is given by the nucleation rate which depends critically upon the surface tension σ. In our calculations we adopt $\sigma = 1,400$ erg/cm^2 for iron (Gail and Sedlmayr 1986) and $\sigma = 350$ erg/cm^2 for dunite (Elliot et al. 1963). It is beyond the scope of this paper to discuss this model in more detail, but the interested reader will find more in the original paper and in Anic et al. (2007).

On the other hand, if the hot vapour cools along path ②, we use the formalism provided by Grady (1982) to compute the decompression and fragmentation. We also check whether droplet formation is governed by dynamic fragmentation (Grady 1982) or the liquid to vapour transition (Melosh and Vickery 1991), or both. Here again we refer the reader to the original papers and to Anic et al. (2007) for more details.

The resulting distributions of droplet sizes obtained for runs 4 (head-on, "cold"), 11 (off-axis, "cold") and 17 (off-axis, "hot") are shown in Fig. 3.

For all three runs, the majority of the droplets are less than 5 cm in radius with a peak at or slightly below 1 cm. The differences between the three simulations, as far as the size distribution is concerned, are relatively small. Results obtained using a perfect gas EOS to compute the expansion lead to somewhat smaller condensates. In particular, the peak of the distribution is markedly below 1 cm. We conclude that giant impacts that lead to a suitable fractionation of a chondritic proto-Mercury produce essentially centimeter and subcentimeter sized particles in the ejecta.

4.2 Non-gravitational Forces

Several nongravitational mechanisms may perturb the orbit of dust particles. Because we assume that both our proto-Mercury and the impactor are quite large and have differentiated to form iron cores, we may presume that the impact occurs late enough that the nebular gas has dissipated. We may therefore neglect the effect of gas drag on the ejecta; drag due to the modern solar wind is significant only for particles of size less than one micron. Because the overwhelming majority of the condensates and (solidified) melt droplets are less than 1 cm in size, we may also neglect the Yarkovsky effect (see e.g. Bottke et al. 2000). We are left

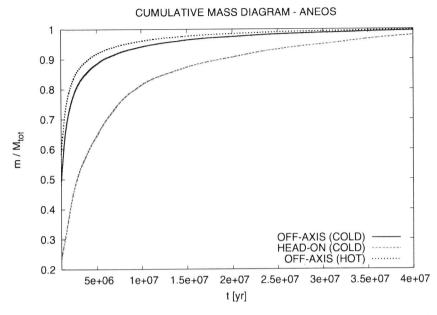

Fig. 4 The effect of Poynting–Robertson drag on particles with initial orbits and particle sizes determined from our SPH simulations and condensation calculations. Results for three different impact scenarios are shown, and the ANEOS equation of state was used in all cases. The figure shows the mass fraction of ejecta particles collected by the Sun as a function of time after the impact. In the slowest case, for a head-on impact with a cold proto-Mercury, the half-life of the particles is about 2.5 Myr

with direct photon pressure and the Poynting–Robertson effect. For particles of the densities considered here, the Poynting–Robertson effect is dominant for particle sizes greater than 1 micron.

Poynting–Robertson drag arises from the relativistic interaction of dust particles with solar photons. Robertson (1937) investigated the fate of small particles in circular orbits and set up the corresponding equations of motion. Wyatt and Whipple (1950) extended the method to the general case of elliptic orbits. Using the equations published in those papers we can calculate the time scales on which the condensates (and melt droplets) resulting from the simulations presented earlier disappear into the Sun.

Knowing the size distribution of the ejecta and knowing the corresponding material density, we need only the initial orbital elements of the condensates in order to actually compute their removal timescale. We obtain these orbital elements for each ejected SPH pseudo-particle by picking an arbitrary impact site somewhere along proto-Mercury's orbit which gives us the centre of mass velocity to which we add the ejection velocity as computed by the hydrodynamics code. We further assume that all particles are spherical and of uniform density and that they intercept radiation from the Sun over a cross-section πr^2 and isotropically reemit it at the same rate (thermal equilibrium). The relevant decay equations for this case can be found in Wyatt and Whipple (1950) and Fig. 4 shows the results of applying these equations to the simulated ejecta particles. For simplicity we have neglected the effects of finite size and rotation of the Sun (Mediavilla and Buitrago 1989). The time-scale for removal of particles with our calculated size distribution can be seen to be less than a few million years.

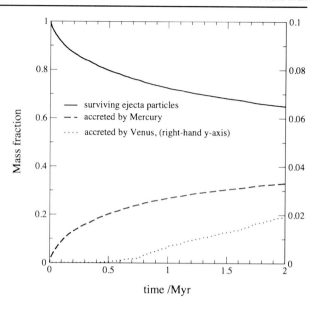

Fig. 5 Decay of a population of particles ejected from Mercury in a head-on collision (simulation run 6) at the perihelion of the planet's orbit. N.B. the fraction of particles colliding with Venus has been scaled up for better visibility

4.3 Collision Time and Gravitational Scattering

The ejecta from the collision initially have heliocentric orbits which still cross the orbit of Mercury. Unless they are removed from such orbits, e.g. by the Poynting–Robertson effect (see Sect. 4.2) or by gravitational scattering, most of the matter will be reaccreted by the planet over time and the resulting collisional fractionation will be too small to explain the planet's anomalous composition.

In order to study the dispersal and reaccretion of the ejected matter under the effects of gravity, a number of simulations were carried using the hybrid integrator MERCURY (Chambers 1999). We simulated the behaviour of a large population of ejected particles under the gravitational influence of Mercury (taken as being the mass of the currently observed planet), Venus, Earth, Mars and Jupiter, for a period of two million years. For lack of a better choice, these planets were placed on their current orbits. The ejected particles were treated as being massless, and were followed until they were either ejected from the inner Solar system (passing beyond the orbit of Jupiter), or collided with one of the planetary bodies. Their initial positions and velocities were computed using hermeocentric velocities chosen randomly from amongst the ejected SPH pseudo-particles, the choice of a position of proto-Mercury on its orbit at the time of the impact and the necessary coordinate transformation from a hermeocentric to a heliocentric frame of reference.

Two different series of integrations were run. The first simulation, which was the most detailed, followed the behaviour of 10,000 ejected particles over the two million year period for the case of a head-on collision. The second series of simulations used a smaller dataset (1,000 particles), but examined the effect of the collision location, the collision geometry (head-on vs. glancing, as described earlier) and the effect of scaling the mass of the remnant planet. Here we show only the results from the simulation involving the large number of particles (Fig. 5). The other simulations yield very similar results except for when the collision occurs at aphelion; in this case the rate of reaccretion is significantly less.

It can clearly be seen from Fig. 5 that the number of surviving particles decays over time, with the bulk of the removed material being lost to reaccretion by Mercury. However, the rate at which material reaccretes is particularly low—after two million years, 6,496 of the 10,000 particles remain in the simulation, which corresponds to a decay half-life of about 3.2 Myr. Of the 3,504 particles which were removed from the simulation over the course of the 2 million year period, 3,306 were reaccreted by Mercury, while 191 hit Venus, with the remaining 7 particles hitting the Sun or being ejected beyond the orbit of Jupiter. In longer simulations, particles were observed to impact the Earth, and an ever-increasing fraction were ejected from the system rather than being accreted, so it is clear that the particles slowly diffuse throughout the inner Solar system as a result of repeated encounters with the inner planets. These results are consistent with those obtained by Gladman (2003) for slightly different initial conditions. Warrel et al. (2003) found that hermeocentric and 1:1 Mercury mean motion resonance orbits can be stable for long time periods, but that ejecta with velocities only slightly greater than the escape velocity are likely to be reaccreted due to the necessity of successive close encounters with Mercury to achieve significant gravitational scattering. They did not, however, provide a numerical result.

In Sect. 4.2 we showed that the half-life of the condensates is of order 2.5 Myr before they are removed by the Poynting–Robertson effect. Over the same period of time, roughly 40% of particles are found to collide with Mercury. This implies that successful collisions are really those for which the post-collision R/I ratio is somewhat less than Mercury's present-day ratio probably in the range $0.3 \leq R/I \leq 0.4$ in order to allow for this reaccretion. No attempt was made to simulate a collision that would, after reacculation, lead to the exact R/I ratio. However, since our ratios bracket the desired value there would be no problem to find an appropriate collision. We conclude that giant collisions as envisioned here can indeed lead to significant long-term chemical fractionation.

5 Summary and Conclusions

We have confirmed, using SPH models with a significantly higher resolution than previous efforts, that a giant impact is capable of removing a large fraction of the silicate mantle from a roughly chondritic proto-Mercury. The size and velocity of the impactor were chosen to be consistent with predictions of planetary formation and growth, and a plausible Mercury can be obtained for several assumptions about initial temperatures and impact parameter.

We extended the previous work on the subject by addressing the fate of the ejecta in order to assess the fraction that could be reaccumulated by Mercury thereby changing again the fractionation achieved immediately after the impact. In particular, using a simple condensation model, we derived the expected size distribution of the ejected material after it cools following adiabatic expansion, and the subsequent dynamical evolution of the resulting particles. The loss of ejected particles into the Sun due to Poynting–Robertson drag was shown to be at least as efficient as reaccretion onto Mercury, and so the bulk density and composition that result from the giant impact would have been largely retained. The giant impact hypothesis for the formation of Mercury is thus entirely plausible.

Our simulations provide estimates of particle size and temperature, and gas density in the ejecta plume. First-order estimates of chemical mixing and loss of volatile elements could perhaps be undertaken with this information. Future simulations will concentrate on chemical fractionation resulting from the impact, and may have sufficient resolution to consider the effect of large ion lithophiles having been preferentially incorporated into a "crust" on the proto-Mercury. These more sophisticated simulations would then be able to make predictions about the isotopic and elemental composition of the modern Mercury. These predictions could then be tested, at least in part, by the data expected from the two coming Mercury spacecraft missions.

Confirming the collisional origin of the anomalous density of Mercury would go a long way toward establishing the current model of planetary formation through collisions which predicts giant impacts to happen during the late stages of planetary accretion. Hence, small Mercury has the potential to become a Rosetta stone for the modern theory of planet formation!

Acknowledgements The authors gratefully acknowledge partial support from the Swiss National Science Foundation.

References

J.D. Anderson, G. Colombo, P.B. Esposito, E.L. Lau, G.B. Trager, Icarus **71**, 337 (1987)
A. Anic, PhD Thesis, University of Bern, 2006
A. Anic, W. Benz, J. Horner, J.A. Whitby, Icarus (2007, in preparation)
A. Balogh, R. Grard, S.C. Solomon, R. Schulz, Y. Langevin, Y. Kasaba, M. Fujimoto, Space Sci. Rev. (2007, this issue). doi:10.1007/s11214-007-9212-4
S.S. Barshay, J.S. Lewis, Annu. Rev. Astron. Astrophys. **14**, 81 (1976)
W. Benz, in *The Numerical Modelling of Nonlinear Stellar Pulsations Problems and Prospects*, ed. by J.R. Buchler (1990), p. 269
W. Benz, A.G.W. Cameron, W.L. Slattery, Icarus **74**, 516 (1988)
W. Benz, A.G.W. Cameron, H.J. Melosh, Icarus **81**, 113 (1989)
W.F. Bottke, D.P. Rubincam, J.A. Burns, Icarus **145**, 301 (2000)
A.G.W. Cameron, Icarus **64**, 285 (1985)
A.G.W. Cameron, B. Fegley, W. Benz, W.L. Slattery, in *Mercury*, ed. by F. Vilas, C. Chapman, M.S. Matthews (University of Arizona Press, Tucson, 1988), 692 pp
J.E. Chambers, Mon. Not. R. Astron. Soc. **304**, 793 (1999)
J.F. Elliot, M. Gleiser, V. Ramakrishna, in *Thermochemistry for Steelmaking*, vol. II (Addison-Wesley, 1963)
B. Fegley Jr., A.G.W. Cameron, Icarus **82**, 207 (1987)
B. Fegley Jr., J.S. Lewis, Icarus **41**, 439 (1980)
H.P. Gail, E. Sedlmayr, Astron. Astrophys. **166**, 225 (1986)
B. Gladman, LPSC **XXXIV**, #1933 (2003)
K.A. Goettel, S.S. Barshay, in *The Origin of the Solar System*, ed. by S. Dermott (Wiley, Chichester, 1978), 611 pp
D.E. Grady, J.A.P. **53**, 322–325 (1982)
H. Harder, G. Schubert, Icarus **151**, 118 (2001)
J.S. Lewis, Earth Planet. Sci. Lett. **15**, 286 (1972)
J.S. Lewis, Science **186**, 440 (1974)
J.S. Lewis, in *Mercury*, ed. by F. Vilas, C. Chapman, M.S. Matthews (University of Arizona Press, Tucson, 1988), p. 651
E. Mediavilla, J. Buitrago, Eur. J. Phys. **10**, 127 (1989)
H.J. Melosh, A.M. Vickery, Nature **350**, 494 (1991)
P. Michel, W. Benz, P. Tanga, D. Richardson, Icarus **160**, 448 (2002)
J.J. Monaghan, Annu. Rev. Astron. Astrophys. **30**, 543 (1992)
J.P. Morris, J.J. Monaghan, J. Comput. Phys. **136**, 41 (1997)
Y.P. Raizer, J. Exp. Theor. Phys. **37**(6), 1229 (1960)
H.P. Robertson, Mon. Not. R. Astron. Soc. **97**, 423 (1937)
E.M. Shoemaker, in *Physics and Astronomy of the Moon*, ed. by Z. Kopal (Academic, New York and London, 1962)
J.V. Smith, Mineral. Mag. **43**, 1 (1979)
T. Spohn, F. Sohl, K. Wieczerkowski, V. Conzelmann, Phys. Space Sci. **49**, 1561 (2001)
A. Sprague, J. Warell, G. Cremonese, Y. Langevin, J. Helbert, P. Wurz, I. Veselovsky, S. Orsini, A. Milillo, Space Sci. Rev. (2007, this issue). doi:10.1007/s11214-007-9221-3
H.S. Thompson, S.L. Lauson, Sandia Labarotories Report SC-RR-710714 (1984)
H.C. Urey, Geochim. Cosmochim. Acta **1**, 209 (1951)
H.C. Urey, *The Planets* (Yale University Press, New Haven, 1952)
J. Warrel, O. Karlsson, E. Skoglov, Astron. Astrophys. **411**, 291 (2003)
S.J. Weidenschilling, Icarus **35**, 99 (1978)
G.W. Wetherill, in *Origin of the Moon*, ed. by W.K. Hartmann, R.J. Phillips, G.J. Taylor (Lunar and Planetary Institute, Houston, 1986), 519 pp
S.P. Wyatt Jr., F.L. Whipple, Astrophys. J. **111**, 134 (1950)

Mercury's Interior Structure, Rotation, and Tides

Tim Van Hoolst · Frank Sohl · Igor Holin ·
Olivier Verhoeven · Véronique Dehant · Tilman Spohn

Originally published in the journal Space Science Reviews, Volume 132, Nos 2–4.
DOI: 10.1007/s11214-007-9202-6 © Springer Science+Business Media B.V. 2007

Abstract This review addresses the deep interior structure of Mercury. Mercury is thought to consist of similar chemical reservoirs (core, mantle, crust) as the other terrestrial planets, but with a relatively much larger core. Constraints on Mercury's composition and internal structure are reviewed, and possible interior models are described. Large advances in our knowledge of Mercury's interior are not only expected from imaging of characteristic surface features but particularly from geodetic observations of the gravity field, the rotation, and the tides of Mercury. The low-degree gravity field of Mercury gives information on the differences of the principal moments of inertia, which are a measure of the mass concentration toward the center of the planet. Mercury's unique rotation presents several clues to the deep interior. From observations of the mean obliquity of Mercury and the low-degree gravity data, the moments of inertia can be obtained, and deviations from the mean rotation speed (librations) offer an exciting possibility to determine the moment of inertia of the mantle. Due to its proximity to the Sun, Mercury has the largest tides of the Solar System planets. Since tides are sensitive to the existence and location of liquid layers, tidal observations are ideally suited to study the physical state and size of the core of Mercury.

Keywords Mercury · Interior · Composition · Rotation · Libration · Tides

1 Introduction

With a mass of $M = 3.302 \times 10^{23}$ kg and a radius of $R = 2439 \pm 1$ km (Anderson et al. 1987), Mercury is the smallest terrestrial planet and has the second-largest mean density,

T. Van Hoolst (✉) · O. Verhoeven · V. Dehant
Royal Observatory of Belgium, Ringlaan 3, 1180 Brussels, Belgium
e-mail: tim.vanhoolst@oma.be

F. Sohl · T. Spohn
German Aerospace Center (DLR), Institute of Planetary Research, Rutherfordstr. 2, 12489 Berlin, Germany

I. Holin
Space Research Institute, Moscow, Russia

which indicates a large core. If the core consists mainly of iron, the core radius will be about 3/4 of the radius of the planet, resulting in a core-to-mantle size ratio that is larger than for the three other terrestrial planets. If created by a dynamo, the magnetic field observed by Mariner 10 is evidence for a liquid outer core and a strong indication for a solid inner core, which is also predicted by thermal evolution models (Schubert et al. 1988; Hauck et al. 2004; Breuer et al. 2007; Wicht et al. 2007).

The large core implies large differences in bulk composition of Mercury with respect to the other terrestrial planets, and suggests a different formation history (e.g. Benz et al. 2007; Taylor and Scott 2005). A scenario in which Mercury suffered a collision with another large protoplanet is presently the most popular (Taylor and Scott 2005), but other scenarios such as evaporation and condensation models can not be ruled out (for a detailed account of Mercury's formation, see Benz et al. 2007).

Here, we review the present knowledge on Mercury's deep interior and discuss the geodetic measurements that can advance our understanding of Mercury to the level of that of Mars, and even beyond. We do not review the formation, evolution or magnetic field generation in Mercury, although they are obviously and intimately linked to the interior structure and bulk composition, since these topics are treated elsewhere in this issue.

This review is organized as follows. Present constraints on Mercury's composition and internal structure are reviewed in Sect. 2, and possible interior models are described. The large core is probably the property in which Mercury differs most from the other terrestrial planets. In the next sections, we therefore discuss methods that can be used to derive properties of Mercury's core, in particular geodetic methods. The low-degree gravity field of Mercury gives information on the differences of the principal moments of inertia, which are a measure of the mass concentration toward the center of the planet (Sect. 3). These coefficients were determined by radio tracking Mariner 10 during its Mercury flybys in 1974 and 1975, but are too inaccurate for use in the interior structure models. The MESSENGER and BepiColombo mission to Mercury will improve these values to subpercentage level, but even then other geodetic data are needed to obtain the moments of inertia themselves instead of their differences to constrain models of the interior. In Sect. 4, we review the rotational behaviour of Mercury and show that the moments of inertia of the core and the mantle can be determined from observations of the orientation and rotation rate variations (librations), if the degree-two gravity coefficients are known. The gravity information is necessary because the rotation of Mercury depends on the solar gravitational torque on Mercury, whose components are linearly proportional to the degree-two coefficients J_2 and C_{22} of the planet's gravity field. With a known gravitational forcing, observations of Mercury's rotation can be used to estimate Mercury's inertia to rotation.

Tides provide another means to obtain insight into the deep interior of Mercury. Tides were already used in the early 20th century to show that the Earth's core is liquid (Jeffreys 1926). Due to its close distance to the Sun, tides on Mercury are larger than those on Earth, and tidal observations are an excellent tool for constraining the constitution of the interior. In Sect. 5, we discuss how constraints on Mercury's interior can be inferred from MESSENGER and BepiColombo tidal measurements. Conclusions are presented in Sect. 6.

2 Models of the Interior Structure and Composition

2.1 Interior Structure

From the analysis of Mariner 10 Doppler data and radio occultation observations, Mercury's total mass of 3.302×10^{23} kg and mean radius of 2439 ± 1 km was inferred. The determi-

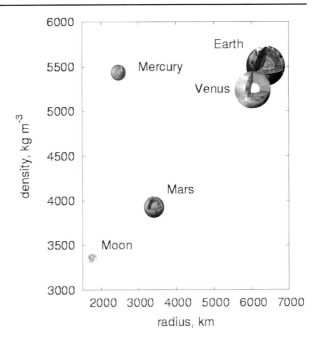

Fig. 1 Radius–density relation of the terrestrial planets and the Moon. Note the anomalous mean density of Mercury that implies an iron-rich interior and peculiar formation history of the planet

nations of the planet's mass and radius resulted in a mean density of 5430 ± 10 kg m^{-3} (Anderson et al. 1987), which is comparable to that of the Earth and Venus but much larger than those of the Moon and Mars (Fig. 1). Though substantially smaller in size, Mercury's surface gravity of 3.7 m s^{-2} almost equals that of the larger planet Mars.

If extrapolated to zero-pressure, Mercury's density is about 5300 kg m^{-3}, i.e., much higher than the uncompressed densities of Earth, Venus, and Mars which are about 4100, 4000 and 3800 kg m^{-3}, respectively. This suggests that Mercury contains a much larger proportion of heavier elements than any other terrestrial planet. The relative abundance of iron could for example be about twice that of the Earth (Wasson 1988). The existence of the weak intrinsic magnetic field and compressional surface features observed by Mariner 10 together with the large average density suggest that most of the iron is concentrated in a partially liquid iron-rich core with a radius of roughly 0.8 times the total radius of Mercury. The core occupies about half of the planet's volume corresponding to a core mass fraction of 2/3 relative to the planet's mass or about twice that of the Earth (Siegfried and Solomon 1974).

If Mercury were in hydrostatic equilibrium, its moment of inertia could be determined from the observed flattening as an additional constraint on the interior structure. The degree-two coefficients J_2 and C_{22} of the gravity field of Mercury were determined from Doppler and range data on the radio link between Mariner 10 and Earth during the first and third encounter of the spacecraft with Mercury in March 1974 and March 1975. A reanalysis by Anderson et al. (1987) resulted in $J_2 = (6 \pm 2) \times 10^{-5}$ and $C_{22} = (1.0 \pm 0.5) \times 10^{-5}$. The polar flattening caused by rotation would give a J_2-value two orders of magnitude smaller, indicating that J_2 is mainly due to nonhydrostatic effects. These effects are also larger than for the other slowly rotating planet Venus, which has $J_2 = (4.404 \pm 0.002) \times 10^{-6}$ and $C_{22} = (1.57 \pm 0.02) \times 10^{-6}$ (Konopliv et al. 1999). Moreover, ground-based radar ranging data suggest that the equatorial shape of Mercury is significantly elliptical (Anderson et al. 1996). These observations show that Mercury has not attained an equilibrium figure. Therefore, its shape and gravitational field cannot be used to infer the size of its metallic core. From a comparison between the equatorial shape and the gravitational equatorial ellipticity C_{22}, Anderson et al. (1996) have concluded that the crust could be 200 ± 100 km thick if Mercury's equatorial ellipticity were entirely compensated by Airy isostasy. The crust thickness has also been estimated from constraints on the heat flow into the base of the crust from observations of ancient faults (Nimmo 2002). Taking into account that the base of the crust does not melt, Nimmo and Watters (2004) derived an upper bound on the crustal thickness of 140 km. The large crust thicknesses deduced from Mercury's shape and surface tectonics are difficult to reconcile with the planet's magmatic evolution.

Models of the interior structure rely on the mass and mean radius of the planet since a value for the moment-of-inertia (MoI) factor is not available at present. A determination of the MoI factor, as envisioned by future missions to Mercury, would help distinguish an iron core from a more homogeneous distribution of iron in oxidized form within the planet (Schubert et al. 1988). The spectral characteristics and high albedo of the surface of Mercury are consistent with the existence of a metal-poor and possibly highly differentiated, feldspathic crust that contains less FeO and TiO_2 than the lunar highland crust. This is taken as evidence for the strong internal differentiation of the planet (Jeanloz et al. 1995; Sprague et al. 2007).

To construct depth-dependent models of the interior structure of Mercury, a spherically symmetric planet in perfect mechanical and thermal equilibrium is assumed. The following set of differential equations for mass m, iron mass m_{Fe}, mean moment of inertia θ, acceleration of gravity g, pressure p, and heat flux q can be derived from fundamental principles (Sohl and Spohn 1997):

$$\frac{dm}{dr} = 4\pi r^2 \rho, \tag{1}$$

$$\frac{dm_{Fe}}{dr} = x_{Fe}\frac{dm}{dr}, \tag{2}$$

$$\frac{d\theta}{dr} = \frac{8}{3}\pi r^4 \rho, \tag{3}$$

$$\frac{dg}{dr} = 4\pi G\rho - 2\frac{g}{r}, \tag{4}$$

$$\frac{dp}{dr} = -\rho g, \tag{5}$$

$$\frac{dq}{dr} = \rho\epsilon - 2\frac{q}{r}, \tag{6}$$

where r is the radial distance from the center of the planet, G the gravitational constant, ρ the density, x_{Fe} the concentration of iron per unit mass, and ϵ the specific heat production rate.

Heat is primarily carried by conduction across the stagnant outer portion of Mercury's silicate shell and the top and bottom thermal boundary layers of mantle convection (Breuer et al. 2007). The corresponding radial temperature gradient is given by

$$\frac{dT}{dr} = -\frac{q}{k}, \tag{7}$$

where k is the thermal conductivity. Within the convective portion of the silicate shell and the liquid outer core, the temperature gradient can be approximated by the adiabatic temperature gradient (Stacey 1977)

$$\frac{dT}{dr} = T \frac{\gamma}{K_S} \frac{dp}{dr}, \tag{8}$$

where $\gamma = \alpha K_S / \rho c_p$ is the thermodynamic Grüneisen parameter, c_p the specific heat, α the thermal expansion coefficient, and K_S the adiabatic bulk modulus.

The set of basic differential equations (1–8) can be separated into two subsets that are coupled through the density ρ. The mechanical properties of the interior are calculated from (1–5), while (6–8) give the thermal structure of the model. These equations have to be supplemented with equations of state to include pressure-induced compression and thermal expansion effects on the density. For example, a third-order isothermal Birch–Murnaghan can be used to correct for the pressure-induced compression, and temperature corrections can be done through the use of a thermal-pressure term, according to

$$p = \frac{3K_{0T}}{2} \left[\left(\frac{\rho}{\rho_0} \right)^{7/3} - \left(\frac{\rho}{\rho_0} \right)^{5/3} \right] \left\{ 1 + \frac{3}{4}(K_0' - 4) \left[\left(\frac{\rho}{\rho_0} \right)^{2/3} - 1 \right] \right\}$$
$$+ \alpha_0 K_{0T} (T - 298) \tag{9}$$

(Baumgardner and Anderson 1981). Here, K_T is the isothermal bulk modulus, K' its derivative with respect to pressure, and subscripts zero denote standard pressure and temperature conditions at $p = 0$, $T = 298$ K. Additional assumptions about the chemistry and densities of a basaltic crust, a more primitive mantle, and an iron-rich core are then required to construct models of the interior in accordance with the mass and mean density of the planet (Wood et al. 1981).

2.2 Composition

Mercury is the only terrestrial planet other than the Earth with a perceptible dipole magnetic field. The presence of an internally-generated magnetic field suggests that the iron core is at least partially liquid with an electrically conducting outer core of unknown thickness surrounding a solid inner core. Thermal evolution models indicate that the core would have solidified early in the history of Mercury unless a light alloying element such as sulfur were present (Breuer et al. 2007). A small amount of sulfur as suggested by Stevenson et al. (1983) is sufficient to depress the freezing point of the core alloy and is consistent with the refractory bulk composition Mercury should have acquired if it accreted in the hot innermost part of the solar nebula.

Most studies assume the core to be composed of iron (Fe) and sulfur (S), but other light elements (O, H, ...) and heavy (Ni) elements could also be present (although the lower pressures strongly reduce the solubility of, e.g., oxygen compared to the Earth). Sulfur has the important property that it lowers the melting temperature with respect to that of pure iron, contrary to oxygen, for example (Williams and Jeanloz 1990). The concentration of sulfur is unknown and depends critically on the origin of Mercury, in particular where the planetesimals from which the planet formed came from (Wetherill 1988). If Mercury formed close to its present position, its sulfur concentration would probably be very low (Lewis 1988). However, if Mercury formed in the same feeding zones as Earth, Venus, and Mars, its light element concentration could be higher and closer to that of those planets. The Earth's core has a light-element concentration of about 10 wt% (Poirier 1994), and for Mars, a sulfur concentration of 14 wt% in the core is often considered (Longhi et al. 1992), although smaller values even down to 0.4 wt% (Gaetani and Grove 1997) have been obtained.

As a consequence of the planet's cooling history, a solid inner core surrounded by a volatile-rich liquid outer core may have formed on Mercury. Inner-core growth by solid-iron precipitation occurs when the local core temperature drops below the local liquidus temperature. At the low core pressures in Mercury, sulfur strongly partitions in the liquid, and the inner core is expected to be composed of almost pure iron, if the initial sulfur concentration at formation was less than the eutectic concentration (Li et al. 2001). Therefore, the liquid outer core will gradually increase its light element concentration until it attains the eutectic composition, which is characterized by a sharp minimum in the liquidus curve. Upon further cooling, the outer core liquid gradually freezes, and solids of eutectic composition are deposited onto the solid inner core. In such an evolution stage, compositional buoyancy is absent in the outer fluid core, and it seems unlikely that a dynamo generating a global magnetic field can be effective. The magnetic field observation then suggests that Mercury's inner core radius should be smaller than that at which the core fluid becomes eutectic (see, e.g., Christensen 2006; for a detailed description in this issue, see Breuer et al. 2007 and Wicht et al. 2007).

Thermal history models taking into account parameterized convective heat transport through the mantle indicate sulfur concentrations of 1 to 5% to retain a liquid outer core at the present time (Stevenson et al. 1983; Schubert et al. 1988; Spohn 1991). More sophisticated models of mantle convection including pressure and temperature-dependent rheology demonstrate that the cooling history of a terrestrial planet is governed by the growth of its lithosphere while the deep interior remains relatively hot. These models compare well to the parameterized convection calculations but produce thicker outer core at identical sulfur concentrations. Depending on the stiffness of the mantle rheology, a liquid outer core is then sustained even for sulfur concentrations as small as 0.2% consistent with cosmochemical arguments in favor of a volatile-poor planet (Conzelmann and Spohn 1999; Spohn et al. 2001).

At the typical pressures and temperatures of Mercury's core, solid iron is in the fcc phase (γ iron, Anderson 2002), and FeS in the high-pressure phases FeS IV and FeS V (Fei et al. 1995). Moreover, at pressures between 14 GPa and 18 GPa, an intermediate iron-sulfide compound Fe_3S_2 forms (Fei et al. 1997), and, at a pressure of 21 GPa, two new additional compounds, Fe_3S and Fe_2S, were obtained by Fei et al. (2000). Density values for solid fcc iron and FeS IV are $\rho_{Fe} = 8094$ kg m^{-3} (Sohl and Spohn 1997) at standard conditions, i.e., atmospheric pressure and 25°C, and $\rho_{FeS} = 4940$ kg m^{-3} at zero pressure and 800 K (Fei et al. 1995). In the liquid state, the density values are somewhat smaller, but the density difference is not well known (e.g., Hixson et al. 1990; Sanloup et al. 2002). An estimate of the density difference of iron of 3.5% at core conditions, derived from data for density changes associated with phase transitions of pure iron at triple points, has been used in Mercury models by Van Hoolst and Jacobs (2003). Most often, the difference in densities between liquid and solid phases is neglected in planetary models (e.g., Harder and Schubert 2001), or relatively small values around 1% are taken, which are typical for ϵ-iron at the Earth's inner core boundary (Boehler 1996). However, in the Earth's core, iron is in the hcp phase (ϵ-iron) instead of the fcc phase (γ-iron), and the pressure is more than 10 times larger, about 330 GPa at the inner core boundary.

From measurements of lobate scarps on Mercury's surface, Strom et al. (1975) deduced that the radial contraction of Mercury after the heavy bombardment period is limited to about 2 km. This 2 km radial contraction of Mercury in the absence of large-scale magmatism about 4 Gyr ago may be linked to core shrinking due to solid inner core growth and mantle cooling governed by lithospheric thickening and sluggish mantle convection (Schubert et al.

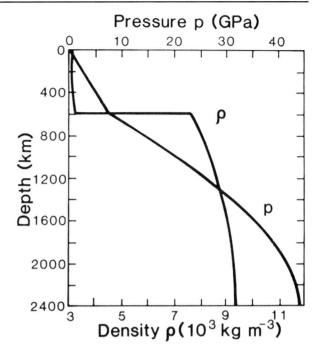

Fig. 2 Variation of pressure P and density ρ versus depth for a fully differentiated model of Mercury's interior. Note that a refractory bulk composition is assumed. (From Schubert et al. 1988, after Siegfried and Solomon 1974)

1988; Breuer et al. 2007). The density increase upon core solidification is a crucial parameter in the study of this effect, but is, unfortunately, not well known.

The mantle composition of Mercury is uncertain for lack of relevant observational data. Nevertheless, an estimate of the iron content of the mantle has been derived from Earth-based observations. Spectroscopic studies show that the surface of Mercury has a FeO content $\lesssim 3$ wt% (Sprague et al. 1994; Jeanloz et al. 1995). Since many smooth plains with low FeO content have morphological features consistent with a lava flow origin and the FeO content of lava is considered to be close to that of the mantle source region, Robinson and Taylor (2001) concluded that Mercury's mantle is equally low in FeO. Models for the bulk composition of the silicate shell have been proposed that satisfy this constraint and depend on the formation scenario of Mercury (see Taylor and Scott 2005). Mercury's mantle, like the mantle of Mars and the upper mantle of the Earth, essentially consists of olivine, pyroxene, and garnet, with relative proportions strongly dependent on the compositional model. Density discontinuities induced by major phase transitions should not be present in the mantle due to the small pressure increase with depth resulting in a pressure at the core-mantle boundary of at most about 8 GPa (Siegfried and Solomon 1974; Harder and Schubert 2001). It cannot be safely excluded, however, that compositional changes occur across the silicate mantle.

Model calculations show that the moment of inertia factor ranges from 0.325 for fully differentiated models with low sulfur content in the core and low mantle density to 0.394 for chemically homogeneous, undifferentiated models (see Fig. 2). The silicate shell comprising crust and mantle layers is at most 700 km thick, for a pure iron core and large mantle density (Siegfried and Solomon 1974; Harder and Schubert 2001; Spohn et al. 2001). The inner core radius of Mercury can vary between zero and the total core radius, but magnetic observations suggest an inner core to be present and smaller than that for an eutectic outer core. For low sulfur concentration in the core, this maximum core size is close to the total core size (Spohn

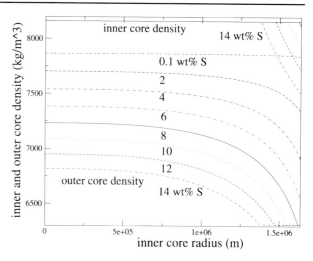

Fig. 3 Inner and outer core densities (in kg/m^3) as a function of inner core radius

et al. 2001; Van Hoolst and Jacobs 2003). Mean densities for the inner core and outer core as a function of inner core radius are shown in Fig. 3.

Future spacecraft missions together with Earth-based radar observations are expected to provide important new constraints on the internal structure of Mercury by determining its gravity field, large-scale topography, and tidal and rotational parameters with unprecedented accuracy (see Sects. 3, 4, and 5).

3 Gravity Field

The gravity field of Mercury is poorly known. Only the degree-two coefficients J_2 and C_{22} have been determined from Mariner 10 radio data with large error bars, and the higher-order coefficients are unknown. Only space missions can improve this situation. X-band radio tracking of MESSENGER in orbit around Mercury will be used to estimate the gravity field of Mercury up to degree 16 with an average resolution of about 500 km in the northern hemisphere and about 1500 km in the southern hemisphere. The expected accuracy on the degree-two coefficients J_2 and C_{22} is better than 1% (Solomon et al. 2001). The BepiColombo mission will use both X- and Ka-band links, so that effects on the radio links of the Earth's and possibly Mercury's ionosphere and solar plasma along the path can be corrected for. The Mercury Planetary Orbiter (MPO) of the mission is moreover equipped with an accurate accelerometer to determine the non-gravitational forces on it. BepiColombo's lower eccentricity orbit is also more adapted for estimating the gravity field than the MESSENGER orbit, which is highly eccentric. Simulations show that the gravity field will be well determined (maximum error of about 10%) up to degree 20 (corresponding to a spatial surface resolution of about 400 km, the pericenter altitude of BepiColombo), and that the J_2 and C_{22} coefficients will be known with a relative accuracy of about 0.01% (Milani et al. 2001).

The degree-two gravity coefficients are necessary for the interpretation of rotation variations in terms of Mercury's interior structure. The expected accuracies are more than sufficient for this purpose. Higher-order gravity coefficients can be used to study lateral heterogeneities in the mantle and crust. Detailed information, such as on the local crust and lithosphere thickness requires a joint interpretation of gravity and topography data (Wieczorek 2007). Topographic data will be obtained by the laser altimeters onboard the MESSENGER and BepiColombo spacecraft. Due to the large size of Mercury's core relative to the planet's size, long-wavelength undulations of the core mantle boundary driven by mantle convection should be detectable in the gravity field (Spohn et al. 2001; Breuer et al. 2007). The topography of the planet's core-mantle boundary may have important implications for the mechanical and electromagnetic coupling of the core and mantle of Mercury (Van Hoolst 2007) and may permit magnetic field generation by operating a thermoelectric dynamo (Stevenson 1987; Giampieri and Balogh 2002; Wicht et al. 2007).

4 Rotation

4.1 Observational Methods

4.1.1 Earth-Based Observations

An interesting new way to obtain information about the interior of Mercury is by estimating its spin dynamics through the Radar Speckle Displacement Interferometry (RSDI) technique (Holin 1988, 1992, 2004). RSDI is an optimized Earth-based technique to measure the instantaneous transverse (orthogonal to the Earth-planet line-of-sight) spin vector Ω of planetary mantles with minimum errors caused by thermal noise in radiotelescopes. The technique relies on the Maximum Likelihood approach.

Let us illuminate Mercury by an Earth-based powerful monochromatic transmitter at a wavelength of several centimeters. The received radar field (echo) on Earth will show irregularities (speckles) due to the scattering by Mercury's rough surface. Because of Mercury's rotation, the speckles move over the surface of the Earth in a 'frozen' state, and we can correlate the echoes received at different radiotelescopes to determine both the instanteneous rotation rate and the orientation of the spin vector of Mercury with high accuracy. Two parameters need to be measured: the time delay of the speckles and the time when the correlation is maximum, i.e., when the speckle trajectory is along the telescope baseline. For optimum radar estimation of the planetary spin (OREPS procedure) with minimum errors, a statistical procedure has been developed. In accordance with the Maximum Likelihood approach, one introduces a Likelihood functional L for the receiving interferometer. The optimum estimate for Ω is the vector that maximizes L. Due to the fact that with the monochromatic transmission the distribution of the echo from a strongly rough surface is Gaussian with zero mean, a Gaussian functional can be used for L (see Holin 1992 and references therein)

$$\ln L = -0.5 \iiiint W(\mathbf{r}_1, t_1; \mathbf{r}_2, t_2) y(\mathbf{r}_1, t_1) y(\mathbf{r}_2, t_2) d\mathbf{r}_1 d\mathbf{r}_2 dt_1 dt_2 + \text{const}, \qquad (10)$$

where integration is over the Earth surface through the observation time T and $y(\mathbf{r}, t) = x(\mathbf{r}, t) + n(\mathbf{r}, t)$ is the mixture of the radar echo field x received at space-time points $(\mathbf{r}, t)$ with input noise n. The function W satisfies the equation

$$\iint H(\mathbf{r}_1, t_1; \mathbf{r}_2, t_2) W(\mathbf{r}_2, t_2; \mathbf{r}_3, t_3) d\mathbf{r}_2 dt_2 = \delta(\mathbf{r}_1 - \mathbf{r}_3)\delta(t_1 - t_3), \qquad (11)$$

where δ is the Dirac function and H is the space-time correlation function (STCF) of the input mixture y:

$$H(\mathbf{r}_1, t_1; \mathbf{r}_2, t_2) = K(\mathbf{r}_1, t_1; \mathbf{r}_2, t_2) + N_0 \delta(\mathbf{r}_1 - \mathbf{r}_2)\delta(t_1 - t_2). \qquad (12)$$

Here, N_0 is the spectral density of "white" input noise n and K is the STCF of the radar echo field x defined by

$$K(\mathbf{r}_1, t_1; \mathbf{r}_2, t_2) = \langle x(\mathbf{r}_1, t_1), x^*(\mathbf{r}_2, t_2) \rangle, \qquad (13)$$

where the brackets $\langle , \rangle$ denote averaging over the surface roughness ensemble and $*$ denotes the complex conjugate.

The functional equations (10–12) present the optimum processing algorithm (OPA) that guarantees maximum accuracy in the presence of noise n. The limiting rms (root mean square) of joint estimates of the orientation η (absolute rms in radians) and magnitude Ω (relative rms) of Ω for nearly symmetrical objects like Mercury are determined respectively by

$$\sigma_\eta^{-2} = -\frac{\partial^2 \ln L}{\partial \eta^2} \quad \text{and} \quad \sigma_\Omega^{-2} = -\Omega^2 \frac{\partial^2 \ln L}{\partial \Omega^2}, \qquad (14)$$

where the second derivatives are calculated at the maximum of $\ln L$.

The asymptotic analytical solutions for σ_η and σ_Ω were derived from (10–12) by Holin (1992) for the two-element receiving interferometer and, due to the nearly spherical shape of Mercury, can be written in a single form $\sigma_\eta = \sigma_\Omega = \sigma$ as

$$\sigma = \frac{lv}{\pi q b \Omega d}, \qquad (15)$$

where l is the speckle diameter, v the velocity of speckle displacement (Holin 1988), q the output amplitude signal-to-noise ratio, b the interferometer baselength, and d the Earth–Mercury distance. Equation (15) describes the one-shot (measuring time of several tens of seconds) limiting accuracies and is consistent with known expressions for the limiting accuracy of time lag estimation of noisy random signals. For this reason, the above OREPS procedure can be treated as a speckle displacement technique (Holin 1992).

Another main feature of high precision RSDI is that it uses the space–time coherence of the echo on a global Earth scale. As follows from (10–12), OPA is determined in full by the STCF of the echo. Initial (Holin 1988) and recent (Holin 2004) investigations of the STCF showed that decorrelation ("boiling") of speckles is negligible (the loss in coherence is less than or comparable to 10^{-4}) during their displacement over the Earth's surface at global scales, and therefore the same Doppler variations can be observed, e.g., at Brussels and Tokyo or at Bern and Washington. In other words, the radar speckle pattern from Mercury is "frozen" all over the Earth, and baselengths as long as possible can be used to improve the accuracy in accordance with (15). Today's opportunities are related mostly to the fully steerable transmitting facility at Goldstone (USA). The one-shot accuracy (15) for the Goldstone/Green-Bank radar interferometer is about 10^{-5} (2 arcsec). Observations during 15–20 days within a single inferior conjunction of Mercury can improve the accuracy in the instantaneous obliquity, precession angle and 88-day libration amplitude by about 4 times, and the use of many conjunctions along with additional baselines, e.g., Goldstone (Japan), could lead to about 200 mas (milliarcseconds) accuracies in case of regular variations. For the about 2 arcmin obliquity and about 40 arcsec libration amplitude, the current radar facilities promise to give a relative accuracy of about 1% in both parameters. Further improvements in accuracy of obliquity and libration amplitude to about 0.1% (comparable with that for the low-degree gravitational coefficients to be determined from space missions to Mercury) can be obtained with a new radar transmitter to be constructed in, e.g., Euro-Asia or the north of Africa, that works with a variety of European radiotelescopes. Further

4.1.2 Space Mission Observations

In the MESSENGER mission, libration and obliquity will be determined from the topographic and gravitational shape, which will be obtained by the onboard laser altimeter and radio tracking of the spacecraft. The BepiColombo mission will in addition measure rotational displacements of selected spots on the surface of Mercury with a camera. Simulations show that the libration and obliquity of Mercury can be determined with an accuracy of a few arcsec from the topographic shape (Solomon et al. 2001). With the BepiColombo measurements, a similar accuracy can be reached by using a few camera observations of a single spot on Mercury's surface, and results with an accuracy below 1 arcsec may be expected (Milani et al. 2001; Pfyffer et al. 2006).

4.2 Spin–Orbit Resonance and Libration

Radar observations made at the Arecibo Observatory in Puerto Rico have shown that Mercury is in a 3:2 spin–orbit resonance, in which the mean rotation period (87.969 d) is exactly 2/3 of the orbital period (Pettengill and Dyce 1965; Colombo 1965). Mercury's rotation has most likely been slowed down as a result of tidal friction to rotation periods commensurate with the orbital period on a timescale much smaller than the age of the Solar System (e.g. Correia and Laskar 2004). Ultimately, tidal friction tends to drive Mercury's rotation speed to an equilibrium value that depends on the value of the eccentricity. Because of the large orbital eccentricity ($e = 0.206$), a final equilibrium rotation synchronous with the orbital motion is less likely than the 3:2 resonance (Colombo and Shapiro 1966; Goldreich and Peale 1966). The latter resonance is stable for non-negligible eccentricities because of the strong axial component of the solar torque on Mercury. Other stable resonances exist, but Mercury has the largest probability to be captured in the 3:2 resonance. The capture probability is strongly increased when the large variations in time of the eccentricity between zero and about 0.5 due to planetary influences are taken into account (Laskar 1994). Correia and Laskar (2004) numerically followed the evolution of Mercury for 1000 different, but close initial conditions and found that the 3:2 spin–orbit resonance is the most likely final state with a large probability of 55.4%, due to multiple passages through the 3:2 resonance. Especially the periods with large eccentricities are important, as they can lead to a faster orbital motion than the 3:2 resonant rotation speed near perihelion, causing the rotation rate of Mercury to speed up due to the torque on the tidal bulge and to pass again through the 3:2 resonance. As higher-order resonances have a larger critical eccentricity below which they loose stability than the 3:2 resonance, the excursions to low eccentricities during Mercury's history may also have led to an escape from a previous, higher-order resonance capture (Correia and Laskar 2004).

Because of the gravitational torque on the permanent figure of Mercury in its elliptical orbit, the rotation of Mercury is not constant and varies about a mean state. By expanding the solar torque in terms of orbital parameters in the conservation of angular momentum equation, variations in the libration angle $\gamma = \phi - \frac{3}{2}M$, where ϕ is the rotation angle and M the mean anomaly, can be expressed as

$$\frac{d^2\gamma}{dM^2} + \frac{3}{4}\frac{B-A}{C}e\left(7 - \frac{123}{8}e^2\right)\sin 2\gamma$$

$$= -\frac{3}{2}\frac{B-A}{C}\left[\left(1 - 11e^2 + \frac{959}{48}e^4 + \cdots\right)\sin M\right.$$
$$\left. - \frac{1}{2}e\left(1 + \frac{421}{24}e^2 + \cdots\right)\sin 2M + \cdots\right] + \cdots \quad (16)$$

(Peale 2005). We here assume zero obliquity and principal-axis rotation. At perihelion passage, γ is the angle between the direction of Mercury's long axis and the direction to the Sun, and it is considered to be very small. Equation (16) is an equation for a forced harmonic oscillator describing forced libration in longitude. The forcing is at 1 orbital period, 1/2 orbital period, and higher subharmonics of the orbital period, which have been neglected here.

When Mercury does not react as a single rigid body to external gravitational forcing, which is the case if Mercury has a liquid core, angular momentum equations for the separate solid and liquid layers are needed. If Mercury has one solid layer (mantle+crust, hereafter denoted by 'mantle') and one spherically symmetric liquid layer (the core), libration equation (16) remains valid with C replaced by C^m since only the mantle responds to the forcing. Pressure coupling between the core and the mantle due to an equatorially flattened core–mantle boundary does not change the libration of the mantle (Van Hoolst 2007), and other core–mantle couplings such as electromagnetic coupling, viscous coupling, and gravitational coupling between the mantle and a solid inner core have a negligible effect on the libration (Peale et al. 2002).

Since the ratio $(B - A)/C^m$ is small, the amplitude γ_1 of the 88-day libration can easily be obtained from the libration equation (16) as the coefficient of the $\sin M$ term in the right-hand member:

$$\gamma_1 = \frac{3}{2}\frac{B-A}{C^m}\left(1 - 11e^2 + \frac{959}{48}e^4 + \cdots\right) \quad (17)$$

(Peale 1972). The amplitude of the $\sin 2M$ term is about 11% of that (Jehn et al. 2004). The amplitude of the forced libration depends on the geophysically interesting ratio $(B - A)/C^m$. Since $B - A$ can be determined accurately from spacecraft orbiting Mercury, as will be done by MESSENGER and BepiColombo, libration gives access to the mantle moment of inertia, which is related to the size and density distribution of the mantle. Figure 4 shows the forced libration amplitude for several interior structure models and Fig. 5 represents the mantle moment of inertia relative to the total moment of inertia as a function of the outer core radius. The models assume a core with a composition ranging from almost pure iron to an iron-sulfur assemblage with 14 wt% sulfur, and an inner core radius anywhere between 0 km (entirely liquid core) and the radius of the core (entirely solid core) (Van Hoolst and Jacobs 2003). The forced libration amplitude is calculated by using (17) for the moment of inertia difference $B - A = 4 \times 10^{-5} MR^2$ (Anderson et al. 1987). The amplitude is about 21 arcsec for models with a solid core, and between 37 arcsec and 47 arcsec for models with a liquid core. Since the mantle moment of inertia is mainly determined by the core size or sulfur content of the core, the observation of the libration amplitude constrains the core composition, as illustrated in Fig. 5. A precision of 1 arcsec on the libration amplitude corresponds to a precision of about 2.5% on the mantle moment of inertia. We show below that the moment of inertia C can be determined even more accurately, and therefore the error on C^m/C is about the same as that on C^m. Error lines around the relative mantle moment of inertia for a chosen model (core bulk sulfur concentration of 4 wt% and inner core radius of 872 km were chosen) are included in Fig. 5 to quantify the improvement in the interior structure modeling of Mercury from libration observations. A strong reduction of possible

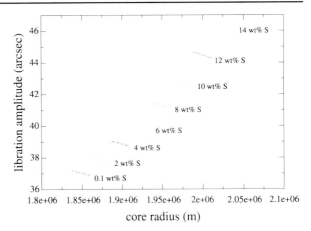

Fig. 4 Libration amplitude for models with a liquid core

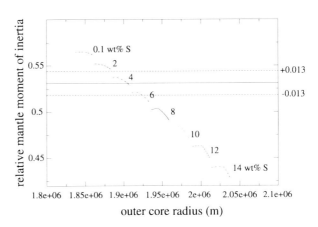

Fig. 5 Moment of inertia ratio C^m/C

models and a constraint on the sulfur content of the core with an error of a few wt% are expected.

Preliminary results of RSDI observations of Mercury with the radio telescopes at Goldstone, Green Bank, and Arecibo show a large forced libration amplitude of about 60 ± 6 arcsec (Margot et al. 2004). This amplitude is larger than the model values calculated above with $B - A = 4 \times 10^{-5} MR^2$ (see (17)), even for the largest liquid cores (see Fig. 4), and suggests that $C_{22} = (B - A)/4MR^2$ is at the high end of the Mariner 10 values (about 1.5×10^{-5}). By taking into account that the equatorial moment of inertia difference $B - A$ has a 50% uncertainty (Anderson et al. 1987), it can be concluded with very high probability that Mercury has a liquid core (Margot et al. 2004). The large uncertainty on $C_{22} = (1.0 \pm 0.5) \times 10^{-5}$ for the moment nevertheless prevents improving the moment of inertia estimate.

By averaging equation (16) over an orbital period, the periodic forcing terms in the righthand member vanish, and the equation reduces to the pendulum equation

$$\frac{d^2\gamma}{dM^2} + \frac{3}{4}\frac{B-A}{C^m}e\left(7 - \frac{123}{8}e^2\right)\sin 2\gamma = 0, \tag{18}$$

which has $\gamma = 0$ as a stable solution. This solution corresponds to the long axis of Mercury pointing toward the Sun at perihelion. When the long axis does not point toward the Sun at perihelion, the averaged gravitational torque on the permanent figure of Mercury tends to restore the alignment, and the long axis librates around the direction to the Sun at perihelion with a free libration period given by

$$P_{\text{free}} = \frac{2\pi}{n} \left[\frac{1}{3} \frac{C^m}{B-A} \frac{1}{e\left(\frac{7}{2} - \frac{123}{16}e^2\right)} \right]^{1/2}. \tag{19}$$

It depends through the same factor $(B - A)/C^m$ (although with a square root here) on the interior structure of Mercury and, if the free libration could be observed, offers another independent possibility to constrain the mantle moment of inertia. For an entirely solid Mercury, the period is equal to 15.830 years. As before, values $B - A = 4 \times 10^{-5} M R^2$ and $C = 0.34 M R^2$ were used. For the models with a liquid core used in Fig. 4, the free libration is essentially a mantle libration, and its period is shorter and between about 10.5 years and 12 years (Rambaux et al. 2007).

In the absence of excitation, free libration will be damped to zero by viscous and electromagnetic core–mantle coupling and to a lesser degree by tidal dissipation on a time scale of 10^5 years (Peale 2005). This damping time scale is short compared to the age of the Solar System, and, without recent excitation, the free libration is expected to be completely damped. Impact excitation is very unlikely since the average time span between impacts of sufficient size is about 10^9 years with current cometary fluxes. A recent impact would have left an impact crater of at least 20 km diameter for the excitation of an observable amplitude of 0.1 arcmin (Peale 2005). On the other hand, planetary perturbations provide a continuous source of excitation for libration since they change Mercury's orbit and cause the long axis of Mercury to be misaligned with respect to the direction to the Sun at pericenter. However, these librations are mainly at other periods and planetary perturbations can not excite the free libration to an observable level (Peale et al. 2007). The main librations excited by planetary perturbations have periods on the order of several years and a maximum total amplitude of about 30 arcsec (Peale et al. 2007). In future observations, the forced 88-day libration should be easily separable from them as they have different periods.

4.3 Orientation

Besides its unique equilibrium rotation rate, Mercury's orientation (orientation of the rotation vector) is also thought to occupy an equilibrium position. This so-called Cassini state 1 represents an extreme (equilibrium point) in the Hamiltonian of motion, made integrable by limiting the planetary perturbations to a single term corresponding to a regular motion of the ascending node (Cassini 1693). In Cassini state 1, the rotation axis and the orbit normal remain coplanar with the normal to the Laplace plane as both the rotation axis of Mercury and the orbit normal precess about the normal to the Laplace plane with a period of about 280,000 years (Colombo 1966; Peale 1969; Ward 1975). The Laplace plane is defined as the plane about which Mercury's orbit precesses with constant inclination between the two planes. Mercury is thus also in a 1:1 resonance between the secular precession rate of the node of Mercury's equatorial plane with respect to the ecliptic plane and the ascending node of the orbit (Beletskii 1972; Rambaux and Bois 2004). Tidal friction drives the spin of Mercury to Cassini state 1 from almost any initial condition on a time scale that is short compared to the age of the Solar System (Peale 1974).

The obliquity, or angle between the rotation axis and the orbit normal, of the final evolved Cassini state 1 can be expressed as

$$\epsilon_C = \sin I \left\{ \cos I + \frac{n}{\mu} \frac{MR^2}{C} \left[\frac{J_2}{(1-e^2)^{3/2}} + 2C_{22} \left(\frac{7}{2} e - \frac{123}{16} e^3 \right) \right] \right\}^{-1} \qquad (20)$$

(Peale 1969), where ϵ_C is the obliquity of Cassini state 1, I the inclination of the orbital plane to the Laplace plane, and μ the precession rate of Mercury's orbit about the normal to the Laplace plane. For the Mariner 10 gravitational coefficients and $C/MR^2 \approx 0.34$, the Cassini state obliquity is equal to about 1.6 arcmin (Peale 1988; Rambaux and Bois 2004). This value is below the measurement accuracy of the obliquity determinations from Mariner 10 observations (Klaasen 1976) and Earth-based radar measurements (Anderson et al. 1996), but well within reach of the RSDI technique. The small obliquity can also be measured by the MESSENGER mission (Solomon et al. 2001) and the BepiColombo mission, which is aiming for an arcsecond precision (Milani et al. 2001).

According to (20), the moment of inertia C is approximately inversely proportional to the obliquity ϵ_C of the Cassini state. Observational determination of the Cassini state obliquity therefore allows the determination of the polar moment of inertia, which is a strong constraint for interior structure models. Both C^m and C can be determined from the rotation observations because of the widely different time scales involved in the rotation variations. For the short-periodic librations, the core almost does not participate in the libration motion, and C^m can be obtained from the observations. On the other hand, the precession motion associated with the Cassini state is on a time scale of 10^5 yrs. Viscous core–mantle coupling is sufficiently large to couple mantle and core on such a long time scale (Peale et al. 2002), and therefore C is the relevant quantity for the Cassini state.

Preliminary results by Margot et al. (2006) of radar observations with the Goldstone and Green Bank telescopes yield a present obliquity of Mercury of 2.1 arcmin with a 5% error (about 6 arcsec, such as for the libration). By taking into account the uncertainties of the Mariner 10 values of the gravitational degree-two coefficients, and polar moment of inertia values (C/MR^2) between 0.32 and 0.37 (e.g. Harder and Schubert 2001), the theoretical Cassini state 1 obliquity values range between about 66 arcsec and 162 arcsec. Therefore, Mercury's orientation differs by at most about 1 arcmin from the Cassini state 1 obliquity, and it can be concluded that Mercury is very close to occupying the Cassini state 1. Since the observed Cassini state obliquity is at the high end of the theoretically possible values and increases with decreasing J_2 according to (20), J_2 is most likely at the low end of the Mariner 10 values. A set of values consistent with the observed forced libration amplitude and obliquity would be $J_2 \approx 4 \times 10^{-5}$ and $C_{22} \approx 1.5 \times 10^{-5}$ for $C/MR^2 \approx 0.34$ and $C^m/C \approx 0.5$.

Because of the large uncertainties in the gravitational coefficients of degree-two, on which the Cassini state obliquity also depends (see (20)), it is currently impossible to determine a precise value for the polar moment of inertia from the measurements of the obliquity. However, the space missions MESSENGER and BepiColombo are expected to measure the degree-two gravity coefficients with a precision better than 1% in the near future. The first MESSENGER flyby of Mercury, which can be used to improve the determination of the gravity coefficients, is scheduled for January 14, 2008. With very accurately known gravity coefficients, an expected 1% precision on the obliquity (1 arcsec precision for a signal of about 100 arcsec) would result in a 1% precision on the moment of inertia. This would be an order of magnitude improvement with respect to the present theoretical uncertainty on C and would strongly constrain interior structure models of Mercury (see Fig. 6).

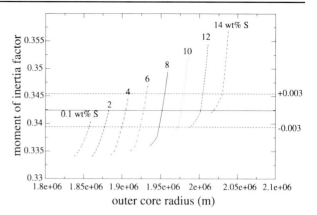

Fig. 6 Polar moment of inertia C/MR^2. Horizontal lines represent errors of 1% around a given value corresponding to the model with core bulk sulfur concentration of 4 wt% and inner core radius of 872 km

Since the measured obliquity is the actual obliquity of Mercury, and relation (20) is for the Cassini state obliquity, an accurate estimate of C can only be obtained when the difference between both obliquities is very small, preferably below the measurement accuracy of the obliquity. Deviations from the Cassini state can be caused by planetary perturbations, by short-periodic variations in the solar torque, and by excitation of free precession. The short-periodic variations in obliquity, similar to the nutations for the Earth and Mars are of the order of 0.1 arcsec or smaller (Carpentier and Roosbeek 2003; Rambaux and Bois 2004) and can be neglected. Free precession is the precession due to the solar torque of the rotation axis of Mercury about its position in the Cassini state if Mercury is close to the Cassini state (Peale 1974; Ward 1975; Rambaux and Bois 2004). If the much slower orbital precession is neglected, it can be compared to the precession of the Earth about the normal to the ecliptic. For a solid core, $C = 0.34MR^2$, and the Mariner 10 gravitational coefficients of Anderson et al. (1987), the precession period is 1062 years (Rambaux and Bois 2004; Peale 2005; D'Hoedt and Lemaitre 2005; Yseboodt and Margot 2006). If Mercury has a liquid core, the period is a factor C^m/C smaller. Tidal friction and dissipative core-mantle coupling will damp free precession, i.e., they will drive the rotation axis to the Cassini state on a time scale of about 10^5 yrs (Peale 1974; Ward 1975; Peale 2005). This short period suggests that free precession is either completely damped or recently excited (Peale 1974; Ward 1975). A possible excitation due to impacts is very unlikely (Peale 2005).

Orbital variations induced by planetary perturbations may cause deviations of the rotation axis of Mercury from the Cassini state because they change the Cassini state, and the rotation axis needs some time to adjust. Peale (2006) followed the spin position and the Cassini state position during short-time scale orbital variations over 20,000 years as well as during long-time scale variations over the past 3 Myrs, and showed that the spin axis remains within one arcsec of the Cassini state if it initially occupied the Cassini state. Therefore, it can be expected that accurate estimates of the polar moment of inertia on the order of 1% can be obtained from measurements of the instantaneous obliquity of Mercury.

5 Tides

5.1 Introduction

In a reference system with origin at the centre of mass of Mercury, the tide-generating potential V_T of the Sun at a point on the surface of Mercury with coordinate vector **r** can be written as

$$V_T(\mathbf{r}) = -\frac{GM_\odot}{d} \sum_{n=2}^{\infty} \left(\frac{r}{d}\right)^n P_n(\cos\psi), \tag{21}$$

where G is the universal gravity constant, $M_\odot$ the mass of the Sun, P_n the Legendre polynomial of degree n, and ψ the angle between the directions from the mass centre of Mercury to the mass centre of the Sun and the point on Mercury considered. Since the ratio r/d is very small, about 4×10^{-5}, the degree-two tides dominate. The ratio of the tidal potential V_T caused by the Sun on Mercury to that on the Earth can be estimated as $(R_M/R_E)^2 (d_E/d_M)^3 \approx 2.5$, where subscripts M and E stand for Mercury and the Earth, respectively. Because of its proximity to the Sun, Mercury has the largest body tides. The tidal potential on Mercury due the other planets can be neglected (Van Hoolst and Jacobs 2003). The largest effect is due to Venus and is about 4×10^{-6} smaller than the effect of the Sun.

An accurate tide-generating potential catalogue for Mercury has been derived from very precise ephemerides of the planets of our solar system (VSOP87, Bretagnon and Francou 1988) by Van Hoolst and Jacobs (2003). By expanding the Legendre polynomial in terms of spherical harmonics, the tidal potential (21) can be expressed as a sum of sectorial, tesseral, and zonal tidal waves. The sectorial waves (azimuthal order $m = 2$) are the largest since the relative motion of the Sun is approximately in the equatorial plane of Mercury (see Fig. 7). For a circular orbital motion and zero obliquity, the sectorial waves would be the only time-dependent tides. In such a configuration, the tidal pattern of maximum tidal displacement (expansion of Mercury) in the Mercury–Sun direction and minimum tidal displacement (compression) perpendicular to that can be expressed by a sectorial spherical harmonic of degree 2. The zonal waves ($m = 0$) are due to the varying distance between

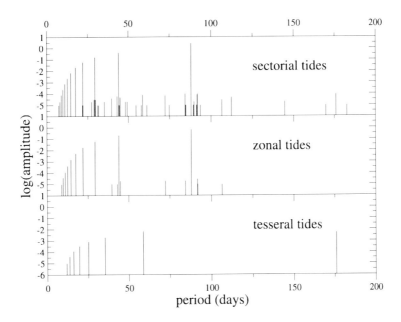

Fig. 7 Tidal spectrum for sectorial (*top panel*), zonal (*middle panel*), and tesseral (*bottom panel*) waves of degree 2. The main sectorial and zonal tidal waves have periods of 1/2 Solar Day on Mercury (SDM), 1/4 SDM, 1/6 SDM, 1/8 SDM, ... The main tesseral tidal waves have periods of 1 SDM, 1/3 SDM, 1/5 SDM, 1/7 SDM, ...

Mercury and the Sun resulting from the eccentricity of the orbit. The tesseral tides ($m = 1$) are at least a factor 100 smaller because of the very small obliquity of Mercury.

Because the sectorial tides have two maxima on Mercury, one in the direction to and the other in the direction away from the Sun, and the largest tidal wave of Mercury is sectorial, the main tidal period is half the mean solar day on Mercury (175.938 d). This period is exactly equal to twice the orbital period (87.969 d) and three times the sidereal rotation period (58.646 d) because of the 3:2 spin–orbit resonance of Mercury. A further consequence of the spin–orbit resonance is that the division of tidal waves in different period bands according to azimuthal number, as is well known for the Earth, can not be made for Mercury. The second-largest wave is zonal with a frequency equal to that of the largest sectorial tidal wave (87.969 d).

5.2 Tidal Response

The tidal potential on Mercury causes periodically varying surface displacements, gravity variations at the surface, and variations in the external gravitational potential field of Mercury. An estimate of the magnitudes of these effects can easily be made by considering only the direct tidal effects and neglecting the planet's response. Tidal displacements are most easily estimated for an equipotential surface and are then given by the ratio of the tidal potential to the gravity (Heiskanen and Moritz 1967), which is about 1.1 m for Mercury. The relative variations in the external potential field compared to Mercury's non-tidal gravity field V can be estimated by $V_T/V \approx (M_\odot/M)(R/a)^3 \approx 5 \times 10^{-7}$. By taking the gradient of the tide-generating potential, the tidal contribution to surface gravity can be estimated as $2V_T/R \approx 3 \times 10^{-6}$ m s^{-2}, or relative variations compared to Mercury's gravity ($g \approx 3.7$ m s^{-2}) of about 8×10^{-7}. The relative tidal variations are thus all on the order of several times 10^{-7}.

The reaction of the planet to the tidal forcing is described by Love numbers: Love number h for vertical surface displacements, Love number l for horizontal surface displacements, and Love number k for changes in the external gravitational potential. The surface displacement $\vec{u}$ can be described by

$$u_r = -h \frac{V_T}{g}, \tag{22}$$

$$u_\theta = -l \frac{1}{g} \frac{\partial V_T}{\partial \theta}, \tag{23}$$

$$u_\phi = -l \frac{1}{g \sin\theta} \frac{\partial V_T}{\partial \theta}, \tag{24}$$

where θ and ϕ are colatitude and longitude on Mercury, and only the degree-two part of the tidal potential is considered. An orbiter around Mercury at distance r from the mass center will sense a time-varying potential ΔV caused by the tide-generating potential itself and an additional potential due to the mass redistribution of the tidally-deformed planet. For the degree-two potential, it can be expressed as

$$\Delta V(r) = \left[1 + k \left(\frac{R}{r}\right)^3\right] V_T(r). \tag{25}$$

To be able to calculate the reaction of the planet to the tidal forcing, the adiabatic bulk modulus K_S and shear modulus μ profiles have to be known, besides the density profile.

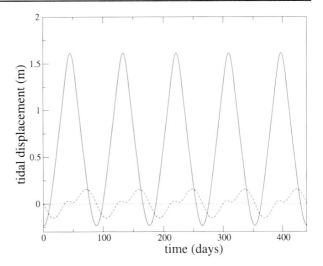

Fig. 8 Radial displacement (*solid line*), and horizontal displacement in North–South (*dotted line*) and East–West (*dashed line*) directions at the equator (zero longitude) starting on J2000.0

Alternatively, knowledge of seismic P and S velocities

$$V_P = \sqrt{\frac{K_S + \frac{4}{3}\mu}{\rho}}, \qquad (26)$$

$$V_S = \sqrt{\frac{\mu}{\rho}} \qquad (27)$$

inside the planet is required. For the mantle of the models used here, we assume that the seismic velocities have the same pressure dependence as the velocities in the Earth's upper mantle, which are taken from PREM (Dziewonski and Anderson 1981). In the core, bulk modulus K_S and shear modulus μ are calculated from the composition (concentration of sulfur), density, pressure, and temperature inside the core (see Sect. 2). For a given density profile, pressure is computed from the condition of hydrostatic equilibrium, and temperature is assumed to decrease linearly from 2000 K at the centre to 1600 K at the core mantle boundary (see Van Hoolst and Jacobs 2003 for further model details).

Figure 8 shows the displacement in radial and horizontal directions as a function of time for a three-layer model with an initial sulfur concentration of 4% and an inner core radius of 1000 km. The radial displacement is several times larger than the horizontal components and has a peak-to-peak amplitude of 1.85 m. With single-shot laser altimer measurements, these tidal displacements will be difficult to measure with the Messenger and BepiColombo missions to Mercury (Wu et al. 1995; Milani et al. 2001). However, repeated observations with the BepiColombo laser altimeter of the tides, which have a well-known phase and spatial pattern, are expected to lead to tidal amplitudes with a precision of a few (Christensen et al. 2006).

The tidal displacement differs according to the location on Mercury's surface and is largest on the equator, where the sectorial waves are largest and also the zonal waves contribute. Measurements of tidal surface displacements should therefore preferentially be made of spots in the equatorial region. At the poles, the tidal displacement is determined by the zonal tides only. The radial displacement there, as shown in Fig. 9, is always negative because it is calculated with respect to an equilibrium state without tides, and tides always

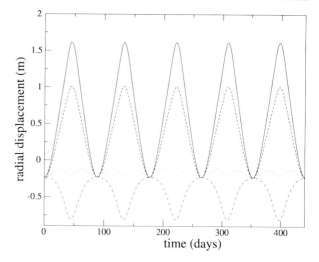

Fig. 9 Radial displacement as a function of time from J2000.0 for different latitudes: equator (*solid line*), 30° latitude (*dashed line*), 60° latitude (*dotted line*), and 90° latitude (*dashed-dotted line*)

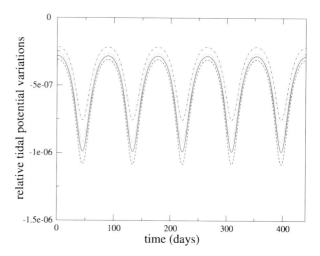

Fig. 10 Relative external potential variations at 400 km altitude and polar latitude as a function of time from J2000.0 for different models. *Solid line* ($x_s = 4$ wt% S, $R_{ic} = 1000$ km), *dashed line* ($x_s = 2$ wt% S, $R_{ic} = 1859$ km), and *dotted line* ($x_s = 14$ wt% S, $R_{ic} = 0$ km). The *dashed-dotted line* is for a typical model with a solid core (Love numbers vary only by a few percent for the solid models)

stretch Mercury towards the Sun, and compress the planet in the perpendicular (almost polar) direction.

Figure 10 shows the potential variations as a function of time relative to the main planetary gravitational potential $-GM/r$. The variations are much larger than the expected accuracy on the determination of the degree-two gravity potential of about 10^{-9} for the BepiColombo radio science experiment (Milani et al. 2001), but will be difficult to observe with MESSENGER (Solomon et al. 2001). A precision of 1% on the tidal potential variations is expected with BepiColombo (Milani et al. 2001).

5.3 Love Numbers

The dependence of tides on the interior of Mercury is fully and most adequately described by the Love numbers. The Love numbers h and k show similar though not equal dependence on interior structure parameters (Van Hoolst and Jacobs 2003), and we here only explain the Love number k. Typical values for models with a (partially) liquid core are about five times

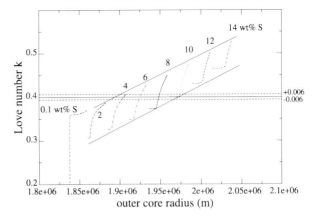

Fig. 11 Love number k as a function of core radius. *Lines* represent models with inner core radius varying from 0 km (*top*) to almost equal to the outer core radius. The values for models with a substantial liquid core fall in a band delineated by *two solid straight lines*

larger than those for models with an entirely solid core, and differences of up to 50% exist between the models with a substantial liquid core (see Fig. 11). These large differences show the excellent potential of tidal measurements for determining interior properties of Mercury.

Love numbers increase with increasing core radius and decreasing sulfur concentration in the outer core (Spohn et al. 2001; Van Hoolst and Jacobs 2003). An increasing core radius tends to increase k since the liquid layer, which is more easily deformed than the solid layers (smaller seismic velocities), is then closer to the surface. An increasing concentration of sulfur in the outer core leads to larger seismic velocities and hence decreases the Love number. The abrupt change in the dependence of the Love number k for a sequence of models with constant bulk sulfur composition and varying inner core radius occurs where the outer core changes from an eutectic composition to a composition with less sulfur. In such a sequence of models, the Love number increases strongly with increasing core radius for outer core concentration below eutectic because of the large changes in outer core composition, and decreases only slightly with decreasing core radius for models with a eutectic outer core composition. If the magnetic field on Mercury is due to a hydrodynamo, compositional core convection connected with inner core growth is required (see Breuer et al. 2007; Wicht et al. 2007). This can only happen if the outer liquid core does not have a eutectic composition, and only those models have to be considered then.

Milani et al. (2001) have simulated BepiColombo tidal potential measurements and concluded that the Love number k_2 can be determined with a precision of 6×10^{-3} or about 1%. This value is much smaller than the range of values for the Mercury models which is about 0.2, as is illustrated in Fig. 11, where a value of 0.4 for k with an error of 6×10^{-3} is indicated by horizontal lines. Only a selected sample of our models leads to a Love number k in that range, and the measurement of the tidally induced external potential variations of Mercury by BepiColombo would drastically reduce the number of possible models. For the models used here, a value of $k = 0.4$ would indicate that the core radius should be between about 1880 km and 1980 km and the bulk sulfur concentration between about 2 and 10 wt%. Determination of the Love number h from a joint analysis of BepiColombo laser altimeter data and radio science data with an equally good precision of 1% would further reduce this allowed set of models. To demonstrate this, we chose one model with $k = 0.4$ (core bulk sulfur concentration of 4 wt% and inner core radius of 872 km, the same model as used in Sect. 4) and calculated the associated displacement Love number $h = 0.739$. Adding a 1% error on h gives a set of allowed models for the h measurements. The overlapping set of models that satisfy the k and h constraints is much smaller than the individual model sets, as

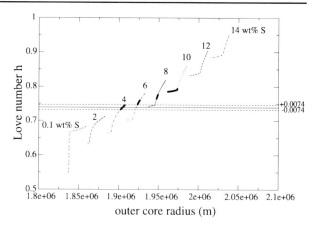

Fig. 12 Love number *h* as a function of core radius. *Lines* represent models with inner core radius varying from 0 km (*top*) to almost equal to the outer core radius. The *horizontal lines* delineate the set of models with Love number *h* in the interval 0.739(1 ± 0.01), and models that have a Love number *k* in the interval 0.4(1 ± 0.01) are shown in *large circles*

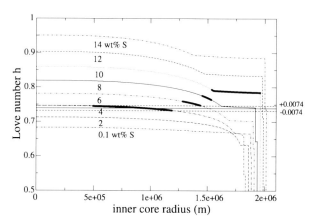

Fig. 13 Love number *h* as a function of inner core radius. The *horizontal lines* delineate the set of models with Love number *h* in the interval 0.739(1 ± 0.01), and models that have a Love number *k* in the interval 0.4(1 ± 0.01) are shown in *large circles*

is shown in Figs. 12 and 13, and inner core radius, outer core radius, and core composition can be determined much more accurately than with the results for one Love number only. Our results indicate that the concentration of light elements (S) in the core can be estimated with a precision of a few percent and the outer core radius with a precision of some tens of kilometers. The inner core size will be more difficult to estimate, and would require a large inner core (see Fig. 13). These quantities are of prime importance for the formation and evolution of Mercury. The core size and composition will provide a much better composition model of Mercury, and the inner core radius will constrain thermal evolution models.

6 Conclusion

Our knowledge of the interior structure, composition, and mineralogy of Mercury is limited, but major advances are expected from future ground-based observations and two space missions. Preliminary results for the libration and orientation of Mercury show with high probability that the core of Mercury is at least partially liquid and that Mercury is close to the Cassini state 1. Future Earth-based radar speckle displacement interferometry (RSDI) observations of Mercury and measurements by the MESSENGER and BepiColombo missions to Mercury will measure the libration and obliquity with an accuracy on the order of arcsec, and the space missions will provide an accurate map of the gravity field of Mercury and will measure the tides.

Four quantities that will be determined accurately in the near future from these observations and that are important for the deep interior of Mercury have been discussed here: the mantle and total planetary moment of inertia, and the Love numbers h (tidal radial displacement) and k (tidal potential variations). The expected precisions are about 1% for the total planetary moment of inertia and the two Love numbers, and about 2% for the mantle moment of inertia. The utility of these quantities in constraining Mercury's interior depends on their sensitivity to interior structure parameters. The tidal Love numbers depend most strongly on the deep interior (core) and have a present theoretical uncertainty of about 25% (see Fig. 11). An observational 1% error would enormously reduce the set of allowed interior structure models. The polar moment of inertia has a theoretical uncertainty of $\geq 5\%$ (see Fig. 6), and the relative mantle moment of inertia is presently uncertain by $\geq 15\%$ (see Fig. 5). With the expected observational accuracy, it will be possible to put tight constraints on the core size and composition of Mercury. Since the four quantities depend on the interior structure in different ways, the results of an analysis of all data together will be much more stringent for Mercury's interior than each individual value. It may be expected that the concentration of light elements (S) in the core can be estimated with a precision of a few percent and the outer core radius with a precision of some tens of kilometers. These results will have important implications for the formation and evolution of Mercury.

Note added in proof New values for the libration amplitude and obliquity have recently been obtained by Margot et al. (2007) from RSDI observations. The obliquity has about the value mentioned here (2.11 ± 0.1 arc minutes), but the libration amplitude is much smaller: 35.8 ± 2 arc seconds versus 60 ± 6 arc seconds. With these values, our suggestion that the C_{22} value is at the high end of the Mariner 10 values is no longer valid.

References

O.L. Anderson, in *Earth's Core, Dynamics, Structure, Rotation*, ed. by V. Dehant, K.C. Creager, S. Karato, S. Zatman. Geodynamics Series, vol. 31 (American Geophysical Union, Washington, 2002)
J.D. Anderson, G. Colombo, P.B. Esposito, E.L. Lau, G.B. Trager, Icarus **71**, 337–349 (1987)
J.D. Anderson, R.F. Jurgens, E.L. Lau, M.A. Slade, G. Schubert, Icarus **124**, 690–697 (1996)
J.R. Baumgardner, O.L. Anderson, Adv. Space Res. **1**, 159–176 (1981)
V.V. Beletskii, Celest. Mech. **6**, 356–378 (1972)
Benz et al., Space Sci. Rev. (2007, this issue)
R. Boehler, Annu. Rev. Earth Planet. Sci. **24**, 15–40 (1996)
P. Bretagnon, G. Francou, Astron. Astrophys. **202**, 309–315 (1988)
Breuer, et al., Space Sci. Rev. (2007, this issue). doi:10.1007/s11214-007-9228-9
G. Carpentier, F. Roosbeek, Celest. Mech. Dyn. Astron. **86**, 223–236 (2003)
G.D. Cassini, *De l'Origine et du Progrès de l'Astronomie, et de son usage dans la Géographie et dans la Navigation* (Imprimerie royale, Paris, 1693)
U.R. Christensen, Nature **444**, 1056–1058 (2006). doi:10.1038/nature05342
U.R. Christensen, C. Koch, M. Hilchenbach, Determining Mercury's tidal Love number h with laser altimetry. General Assembly of the European Geosciences Union, Vienna, 02-07/04/2006, abstract EGU06-A-05711, 2006
G. Colombo, Nature **208**, 575 (1965)
G. Colombo, Astron. J. **71**, 891–896 (1966)
G. Colombo, I.I. Shapiro, Astrophys. J. **145**, 296–307 (1966)
V. Conzelmann, T. Spohn, Bull. Am. Astron. Soc. **31**, 1102 (1999)
A.C.M. Correia, J. Laskar, Nature **429**, 848–850 (2004)
S. D'Hoedt, A. Lemaitre, in *Transits of Venus: New Views of the Solar System and Galaxy*, ed. by D.W. Kurtz. Proceedings of IAU Colloquium, vol. 196 (Cambridge University Press, Cambridge, 2005), pp. 263–270.

A.M. Dziewonski, D.L. Anderson, Phys. Earth Planet. Inter. **25**, 297–356 (1981)
Y. Fei, C.T. Prewitt, H. Mao, C.M. Bertka, Science **268**, 1892–1894 (1995)
Y. Fei, C.M. Bertka, L.W. Finger, Science **275**, 1621–1623 (1997)
Y. Fei, J. Li, C.M. Bertka, C.T. Prewitt, Am. Mineral. **85**, 1830–1833 (2000)
G.A. Gaetani, T.L. Grove, Geochim. Cosmochim. Acta **61**(9), 1829–1846 (1997)
G. Giampieri, A. Balogh, Planet. Space Sci. **50**, 757–762 (2002)
P. Goldreich, S. Peale, Astron. J. **71**, 425–438 (1966)
H. Harder, G. Schubert, Icarus **151**, 118–122 (2001)
S.A. Hauck II, A.J. Dombard, R.J. Phillips, S.C. Solomon, Earth Planet. Sci. Lett. **222**, 713–728 (2004)
W.A. Heiskanen, H. Moritz, *Physical Geodesy* (W.H. Freeman, San Francisco, 1967)
R.S. Hixson, M.A. Winkler, M.L. Hodgdon, Phys. Rev. B **42**(10), 6485–6491 (1990)
I.V. Holin, Radiofizika **31**, 515 (1988). English translation: I.V. Kholin, Radiophys. Quant. Elec. **31**, 371 (1988)
I.V. Holin, Radiofizika **35**, 433 (1992). English translation: I.V. Kholin, Radiophys. Quant. Elec. **35**, 284 (1992)
I.V. Holin, Sol. Syst. Res. **38**, 449 (2004)
R. Jeanloz, D.L. Mitchell, A.L. Sprague, I. de Pater, Science **268**, 1455–1457 (1995)
H. Jeffreys, Mon. Not. R. Astron. Soc. Geophys. Suppl. **1**, 371–383 (1926)
R. Jehn, C. Corral, G. Giampieri, Planet. Space Sci. **52**, 727–732 (2004)
K.P. Klaasen, Icarus **28**, 469–478 (1976)
A.S. Konopliv, W.B. Banerdt, W.L. Sjogren, Icarus **139**, 3–18 (1999)
J. Laskar, Astron. Astrophys. **287**, L9–L12 (1994)
J.S. Lewis, in *Mercury*, ed. by F. Vilas, C.R. Chapman, M.S. Matthews (The University of Arizona Press, Tucson, 1988)
J. Li, Y. Fei, H.K. Mao, K. Hirose, S.R. Shieh, Earth Planet. Sci. Lett. **193**, 509–514 (2001)
J. Longhi, E. Knittle, J.R. Holloway, H. Wänke, in *Mars*, ed. by H.H. Kieffer, B.M. Jakosky, C.W. Snyder, M.S. Matthews (The University of Arizona Press, Tucson, 1992), pp. 184–208
J. Margot, S.J. Peale, R.F. Jurgens, M.A. Slade, I.V. Holin, Earth-based measurements of planetary rotational states. American Geophysical Union, Fall Meeting 2004, abstract G33A-02, 2004
J. Margot, S.J. Peale, R.F. Jurgens, M.A. Slade, I.V. Holin, Observational proof that Mercury occupies a Cassini state. American Astronomical Society DPS meeting 38, 49.05, 2006
J.L. Margot, S.J. Peale, R.F. Jurgens, M.A. Slade, I.V. Holin, Science **316**, 710–714 (2007). doi:10.1126/science.1140514
A. Milani, A. Rossi, D. Vokrouhlicky, D. Villani, C. Bonnano, Planet. Space Sci. **49**, 1579–1596 (2001)
F. Nimmo, Geophys. Res. Lett. **29**(5) (2002). doi:10.1029/2001GL013883
F. Nimmo, T.R. Watters, Geophys. Res. Lett. **31**, L02701 (2004). doi:10.1029/2003GL018847
S.J. Peale, Astron. J. **74**, 483–489 (1969)
S.J. Peale, Icarus **17**, 168–173 (1972)
S.J. Peale, Astron. J. **79**, 722–744 (1974)
S.J. Peale, in *Mercury*, ed. by F. Vilas, C.R. Chapman, M.S. Matthews (Univ. of Arizona Press, Tucson, 1988), pp. 461–493
S.J. Peale, Icarus **178**, 4–18 (2005)
S.J. Peale, Icarus **181**, 338–347 (2006)
S.J. Peale, R.J. Phillips, S.C. Solomon, D.E. Smith, M.T. Zuber, Meteor. Planet. Sci. **37**, 1269–1283 (2002)
S.J. Peale, M. Yseboodt, J.L. Margot, Icarus **187**, 365–373 (2007)
G.H. Pettengill, R.B. Dyce, Nature **206**, 1240 (1965)
G. Pfyffer, N. Rambaux, A. Rivoldini, T. Van Hoolst, V. Dehant, Determination of libration amplitudes from orbit. Abstract P0031; EPSC2006-A-00568, European Planetary Science Congress 2006, 18–22/9/2006, Berlin, Germany
J.-P. Poirier, Phys. Earth Planet. Int. **85**, 319–337 (1994)
N. Rambaux, E. Bois, Astron. Astrophys. **413**, 381–393 (2004)
N. Rambaux, T. Van Hoolst, V. Dehant, E. Bois, Astron. Astrophys. **468**, 711–719 (2007)
M.S. Robinson, G.J. Taylor, Meteorit. Planet. Sci. **36**, 841–847 (2001)
C. Sanloup, F. Guyot, P. Gillet, Y. Fei, J. Geophys. Res. **107**(B11), 2272 (2002). doi:10.1029/2001JB000808
G. Schubert, M.N. Ross, D.J. Stevenson, T. Spohn, in *Mercury*, ed. by F. Vilas, C.R. Chapman, M.S. Matthews (Univ. of Arizona Press, Tucson, 1988), pp. 429–460
R.W. Siegfried, S.C. Solomon, Icarus **23**, 192–205 (1974)
F. Sohl, T. Spohn, J. Geophys. Res. **102**, 1613–1635 (1997)
S.C. Solomon, 20 co-authors, Planet. Space Sci. **49**, 1445–1465 (2001)
T. Spohn, Icarus **90**, 222–236 (1991)
T. Spohn, F. Sohl, K. Wieczerkowski, V. Conzelmann, Planet. Space Sci. **49**, 1561–1570 (2001)

A.L. Sprague, R.W.H. Kozlowski, F.C. Witteborn, D.P. Cruikshank, D. Wooden, Icarus **109**, 156–167 (1994)
A.L. Sprague et al., Space Sci. Rev. (2007, this issue). doi:10.1007/s11214-007-9221-3
F.D. Stacey, Geophys. Surv. **3**, 175–204 (1977)
D.J. Stevenson, Earth Planet. Sci. Lett. **82**, 114–120 (1987)
D.J. Stevenson, T. Spohn, G. Schubert, Icarus **54**, 466–489 (1983)
R.G. Strom, N.J. Trask, J.E. Guest, J. Geophys. Res. **80**, 2478 (1975)
G.I. Taylor, E.R.F. Scott, in *Treatise on Geochemistry*, vol. 1, ed. by A.M. Davis (Elsevier, Amsterdam, 2005), pp. 477–485
T. Van Hoolst, C. Jacobs, J. Geophys. Res. **108**(E11), 5121 (2003). doi:10.1029/2003JE002126
T. Van Hoolst, in *Treatise on Geophysics*, vol. 10 (Elsevier, 2007, in press)
W.R. Ward, Astron. J. **80**, 64–70 (1975)
J.T. Wasson, in *Mercury*, ed. by F. Vilas, C.R. Chapman, M.S. Matthews (Univ. of Arizona Press, Tucson, 1988), pp. 692–708
G.W. Wetherill, in *Mercury*, ed. by F. Vilas, C.R. Chapman, M.S. Matthews (University of Arizona Press, Tucson, 1988), pp. 670–691
Wicht et al., Space Sci. Rev. (2007, this issue). doi:10.1007/s11214-007-9280-5
M. Wieczorek, in *Treatise on Geophysics*, vol. 10 (2007, in press)
Q. Williams, R. Jeanloz, J. Geophys. Res. **95**, 19299–19310 (1990)
J.A. Wood, D.L. Anderson, W.R. Buck, W.M. Kaula, E. Anders, G.J. Consolmagno, J.W. Morgan, A.E. Ringwood, E. Stolper, H. Wänke, in *Basaltic Volcanism on the Terrestrial Planets*, ed. by B.V.S. Project (Pergamon, New York, 1981), pp. 633–699
X. Wu, P.L. Bender, G.W. Rosborough, J. Geophys. Res. **100**, 1515–1525 (1995)
M. Yseboodt, J.-L. Margot, Icarus **181**, 327–337 (2006)

Interior Evolution of Mercury

Doris Breuer · Steven A. Hauck II · Monika Buske ·
Martin Pauer · Tilman Spohn

Originally published in the journal Space Science Reviews, Volume 132, Nos 2–4.
DOI: 10.1007/s11214-007-9228-9 © Springer Science+Business Media B.V. 2007

Abstract The interior evolution of Mercury—the innermost planet in the solar system, with its exceptional high density—is poorly known. Our current knowledge of Mercury is based on observations from Mariner 10's three flybys. That knowledge includes the important discoveries of a weak, active magnetic field and a system of lobate scarps that suggests limited radial contraction of the planet during the last 4 billion years. We review existing models of Mercury's interior evolution and further present new 2D and 3D convection models that consider both a strongly temperature-dependent viscosity and core cooling. These studies provide a framework for understanding the basic characteristics of the planet's internal evolution as well as the role of the amount and distribution of radiogenic heat production, mantle viscosity, and sulfur content of the core have had on the history of Mercury's interior.

The existence of a dynamo-generated magnetic field suggests a growing inner core, as model calculations show that a thermally driven dynamo for Mercury is unlikely. Thermal evolution models suggest a range of possible upper limits for the sulfur content in the core. For large sulfur contents the model cores would be entirely fluid. The observation of limited planetary contraction (∼1–2 km)—if confirmed by future missions—may provide a lower limit for the core sulfur content. For smaller sulfur contents, the planetary contraction obtained after the end of the heavy bombardment due to inner core growth is larger than the observed value. Due to the present poor knowledge of various parameters, for example, the mantle rheology, the thermal conductivity of mantle and crust, and the amount and distribution of radiogenic heat production, it is not possible to constrain the core sulfur content nor

D. Breuer (✉) · M. Pauer · T. Spohn
DLR, Institut für Planetenforschung, Berlin, Germany
e-mail: doris.breuer@dlr.de

M. Pauer
Department of Geophysics, Charles University of Prague, Prague, Czech Republic

S.A. Hauck II
Department of Geological Sciences, Case Western Reserve University, Cleveland, OH 44106, USA

M. Buske
Max-Planck Institut für Sonnensystemforschung, Lindau, Germany

the present state of the mantle. Therefore, it is difficult to robustly predict whether or not the mantle is conductive or in the convective regime. For instance, in the case of very inefficient planetary cooling—for example, as a consequence of a strong thermal insulation by a low conductivity crust and a stiff Newtonian mantle rheology—the predicted sulfur content can be as low as 1 wt% to match current estimates of planetary contraction, making deep mantle convection likely. Efficient cooling—for example, caused by the growth of a crust strongly in enriched in radiogenic elements—requires more than 6.5 wt% S. These latter models also predict a transition from a convective to a conductive mantle during the planet's history. Data from future missions to Mercury will aid considerably our understanding of the evolution of its interior.

Keywords Mercury · Mantle convection · Thermal evolution · Volcanic activity

1 Introduction

Mercury, the innermost planet, is the least-explored planet in the solar system, yet one of the most interesting and puzzling ones, even in comparison with Pluto (regardless of Pluto's reclassification as a dwarf planet by the International Astronomical Union in 2006). Mercury was visited by the Mariner 10 spacecraft in 1974–1975 during three flybys. Data from this mission, though limited and variable in resolution, together with some earth-based observations (e.g., Sprague et al. 1994, 1997; Anderson et al. 1996; Harmon 1997) helped to improve our understanding of Mercury. Here we review the present state of knowledge of the interior evolution of the planet. We begin by summarizing relevant spacecraft and Earth-based observations of Mercury and their implications for the evolution of the interior.

Surface: The known hemisphere of Mercury is similar to the Moon; the morphologies of both surfaces are dominated by craters and are roughly 4 Ga old. Based on infrared spectral data, Sprague et al. (1994, 1997) proposed that Mercury may have an anorthosite-rich crust that formed as a flotation crust on a magma ocean just like the highlands crust on the Moon. The low iron content of the surface suggests that Mercury may be more differentiated than the other terrestrial planets (Vilas 1988). The intercrater plains, which are the most widespread surface type, are thought to have been formed by a global resurfacing event before the end of the late heavy bombardment, 4 to 4.2 Ga. However, no unambiguous evidence of volcanic activity, such as domes or rilles, has been observed, possibly due to the lack of high-resolution data. The wide distribution of basin ejecta has been proposed as an alternative mechanism for resurfacing, but no source is visible on the observed hemisphere. Lobate scarps resulting from thrust faults, and a lack of significant extensional faulting, indicates a contraction of the planetary radius by 1 to 2 km (e.g., Strom et al. 1975; Watters et al. 1998) since the end of the heavy bombardment, which was most likely induced by the cooling of Mercury's mantle and the partial freezing of its core.

Interior structure: The Mariner 10 flybys revealed that Mercury has an average density of 5,430 kg m^{-3} (Anderson et al. 1987), which suggests a relative core radius, R_c/R_P, of about 0.8 and thus a thin mantle shell of only about 600 km thickness (e.g., Spohn et al. 2001; Harder and Schubert 2001; Van Hoolst et al. 2007, this issue). The core may actually comprise the entire planet if a pure FeS core composition is assumed (Harder and Schubert 2001). By comparison, the Moon has a $R_c/R_P < 0.25$–0.3 and Mars and Earth have core radii of about half the surface radius. A value of $R_c/R_P \sim 0.5$ is also consistent with the bulk density of Venus despite not being uniquely constrained, as well as data for Io and Ganymede, after removal of the ice layer of the latter. The crustal thickness of Mercury crust is loosely constrained by a combination of results from Earth-based radar topography data and Mariner 10 gravity data that suggest a thickness of 100–300 km (Anderson et al. 1996). Studies employing stereo topographical information about the Mercurian surface (Nimmo and Watters 2004; Watters et al. 2004) suggest that the extensional faulting of observed lobate scarps is consistent with a mean crustal thickness of 90–140 km.

Magnetic field: Mercury features an internal magnetic field, which is rivaled in strength only by the Earth and Ganymede among solid bodies in the solar system. The internal field was measured during only two flybys of the Mariner 10 mission in 1974 and 1975 on its way to Mars and is therefore not very well known. The dipole moment calculated from the data of 4.9×10^{12} T m^3 is about three orders of magnitude smaller than that of the Earth field and the field strength at the equator is estimated to be about 350 nT or 1/100 of that of the Earth (Ness et al. 1974). The discovery of the internal field strongly suggests the presence of a fluid outer core (e.g., Stevenson et al. 1983) although a crustal source of the measured field cannot be excluded at present (Aharonson et al. 2004).

To understand the interrelation between various observations, in particular the existence of a weak magnetic field, the small amount of planetary contraction, and the thin mantle, a comprehensive study of the interior dynamics and of related processes is required. Fundamentally, the internal dynamics of terrestrial bodies are driven by the transport of heat from the interior to the surface. Heat generation and transport are the processes which connect the structure and chemistry of a planetary body with its dynamical behaviour, such as mantle convection, volcanism, tectonism, and magnetic field generation. The physical and chemical properties of planetary materials influence the mechanics of the planetary heat engine driving dynamics, while the dynamics in turn alter the physical and chemical properties of the planet's interior.

The interactions between various processes have been studied with thermal and thermo-chemical evolution models (e.g., Stevenson et al. 1983; Schubert et al. 1988; Hauck et al. 2004). In this paper, we review these models and also present some first results of 2D and 3D thermal convection models. The paper begins with a detailed description of the important observational constraints, such as estimates of the global contraction and heat flow, the crustal thickness and composition, and the magnetic field. In the next section we describe the main heat transport mechanisms relevant for Mercury, conductive and convective heat transport. For the latter heat transport mechanism, we introduce the main analytical tools, finite amplitude and parameterized convection models, as well as the relevant parameters and initial thermal conditions to calculate mantle dynamics and the thermal evolution. Then we discuss results for dynamics of Mercury's mantle and the thermo-chemical evolution of the mantle and core, including magnetic field generation. The chapter ends with a summary and a discussion of what we will learn from the upcoming missions with respect to the interior evolution of Mercury.

2 Observational Constraints on Thermal History Models

2.1 Planetwide Contraction and Heat Flow

One of the most significant outcomes of the Mariner 10 mission was the idea that Mercury's surface may hold a tectonic record directly related to the planet's cooling history (e.g., Strom et al. 1975; Solomon 1976; Cordell and Strom 1977; Solomon 1977). This concept is the result of the discovery of a prominent and curious set of tectonic features known as lobate scarps, which are asymmetric, arcuate, scarps that are 10–100's of kilometers long with relief of up to a kilometer or more (e.g., Strom et al. 1975; Watters et al. 1998) and are interpreted to be the surface expression of thrust faults. Due to the apparently random locations and orientations of lobate scarps (e.g., Strom et al. 1975; Cordell and Strom 1977) they have been interpreted to have been formed as a result of global contraction of the planet (Strom et al. 1975; Watters et al. 1998). Though recent work as reopened the idea that the scarps have some distinct orientation (Watters et al. 2004) the global contraction hypothesis remains viable because planetary lithospheres are rarely homogeneous and global contraction may be one of several sources of stress superimposed on Mercury's lithosphere. Lobate scarp formation is inferred to have begun after heavy bombardment and continued through smooth plains formation based on the observation that the scarps are not embayed by intercrater plains material (e.g., Cordell 1977; Strom 1997) and that they deform several large craters as well as smooth plains material, which makes up the youngest geologic unit on Mercury (Strom et al. 1975; Cordell 1977; Cordell and Strom 1977; Spudis and Guest 1988). Structural analysis of lobate scarps has deduced that they record $\sim$0.05–0.10% strain that in turn reflects approximately 1–2 km reduction in radius since the end of the heavy bombardment, with recent estimates favouring the low end of this range (Strom et al. 1975; Watters et al. 1998). Global contraction is likely the result of net cooling of the planet's interior, which leads to contraction through a decrease in internal temperatures and phases changes such as that associated with the solidification of an inner core (e.g., Solomon 1976, 1977; Schubert et al. 1988; Hauck et al. 2004). Cooling and complete solidification of a pure metallic iron core from an initially molten state would result in a $\sim$17 km decrease in radius (Solomon 1976), in contrast with observations. Mercury's modest amount of inferred contraction may either require a mechanism to strongly limit planetary cooling or, conversely, additional manifestations of global contraction not included in current estimates (e.g., Hauck et al. 2004).

The largest lobate scarp on the hemisphere imaged by Mariner 10, Discovery Rupes, is >500 km in length has also been used to estimate the paleo heat flow on Mercury (Watters et al. 2002). This thrust fault is assumed to have cut the entire elastic and seismogenic lithosphere when it formed ($\sim$4.0 Gyr ago). On Earth, the maximum depth of faulting is thermally controlled. Assuming the limiting isotherm for Mercury's crust is $\sim$600 to 900 K and it occurred at a depth of about 40 km, the corresponding heat flux at the time of faulting was $\sim$10 to 43 mW m^2. This is less than the heat flow in the old terrestrial oceanic lithosphere but greater than the present heat flux on the Moon. The results by Watters et al. (2002) strongly depend on the accuracy of the measured topography, which was derived from digital stereoanalysis of Mariner 10 imagery (Watters et al. 1998; Cook and Robinson 2000), using updated camera orientations (Robinson et al. 1999). The errors in the data are large and better estimates on the paleo heat flow, including those from different locations, can be obtained when gravity and topographic data are available from MESSENGER (Solomon et al. 2001) and BepiColombo (Grard and Balogh 2001).

2.2 Crust Thickness and Composition

Mercury's crust holds potentially important clues to the planet's internal evolution. In particular, the history and magnitude of crustal production, as well as its composition, may provide information on thermal history of the interior. Ground radar measurements performed between 1967 and 1990 provided an estimate of the global shape of Mercury—among other parameters the equatorial flattening of Mercury was determined to be $(540 \pm 54) \times 10^{-6}$ (Anderson et al. 1996). This together with Mariner 10 flyby gravity measurement of $C_{22} = (1.55 \pm 0.77) \times 10^{-5}$ (Anderson et al. 1987) was used to constrain the mean crustal thickness to be 200 ±100 km under the assumption of full isostatic compensation (Anderson et al. 1996). However, the approach used to obtain the crustal thickness seems to be uncertain since it does not give realistic values of the mean crustal thickness for Mars (Anderson et al. 1996) or (with updated values of equatorial flattening from the Clementine laser altimetry experiment (Smith et al. 1997)) for the Moon.

Another approach to constraining the mean crustal thickness is based on analysis of tectonic structures and their topographic relief. The assumed depth of faulting of the lobate scarps (elongated thrust faults presumably originated from the planetary contraction) of 30–40 km (Watters et al. 2002) suggests that irrespective of the crustal heat generation the mean crustal thickness at the time of the formation of the lobate scarps was $\leq$140 km and the effective elastic thickness 25–30 km (Nimmo and Watters 2004). Similar limits on the crustal thickness have been estimated by the study of topographic profiles of extensional troughs in Caloris basin. These troughs have a shape consistent with lateral flow of a dry plagioclase lower crust with a total crustal thickness of 90–140 km producing 70–90% topographical relaxation within likely limits of radiogenic heat flux (Watters et al. 2005).

Each estimate represents a broad range and new, more accurate gravity, topography, and stereo imagery data from MESSENGER and BepiColombo may substantively revise these results. Regardless, the total volume of crust represents some combination of primary crust, the result of crystal–liquid separation in a primordial silicate magma ocean, and secondary crust, derived from partial melting of the mantle. The persistence of a primary crust depends principally on any primary crust not being removed either by external sources such as a giant impact (e.g., Cameron et al. 1988; Wetherill 1988) or vaporization of the surface (Fegley and Cameron 1987) or internally by crustal recycling. Existence of a primordial crust is supported by reflectance spectra of the surface of Mercury. Those spectra are similar to that of the lunar highlands (Vilas 1988; Sprague et al. 1997), which are predominately plagioclase and were most likely formed by freezing of a magma ocean. Furthermore, the radar characteristics of Mercury's surface are also reminiscent of the lunar highlands (Harmon 1997). As the time-integrated result of mantle partial melting, the secondary crust provides information on the thermal history and composition of the mantle and hence can be compared in a relatively straightforward manner with models of the planet's thermal history (e.g., Hauck et al. 2004). Though the relative fraction of primary to secondary crust on Mercury is unknown, the total crustal thickness can provide an absolute upper bound on the amount of secondary crust produced by the mantle. At present, there are too few data on the composition of Mercury's crust—beyond inferences of a low FeO content (e.g., Jeanloz et al. 1995; Robinson and Taylor 2001) and its similarity to anorthositic lunar crust—to provide any significant constraints on the history of Mercury's interior. However, future determinations of surface mineralogy and the concentrations of heat-producing elements could provide important information for modelling efforts.

2.3 Magnetic Field

Magnetic field generation in the iron core of a planet is strongly coupled to the thermal evolution of the planet. Thus, knowledge about the magnetic field evolution can be used to constrain the planet's thermal evolution. During its first and third close encounters in with Mercury 1974 and 1975, Mariner 10 passed briefly through, and measured a small, but Earth-like, magnetosphere. The analyses of these observations revealed the presence of an internal field, with a dipole moment that is a factor of about 10^4 smaller than that of the Earth (Ness et al. 1974; Russel et al. 1988; see also the review in Connerney and Ness 1988). Prior to the detection of the magnetic field, it had been assumed that Mercury's core froze early in the planet's history; however, this assumption is not consistent with the idea that the field is the result of dynamo action in the core. The discovery of the internal field strongly suggests the presence of a fluid outer core, although a crustal source of the measured field cannot be excluded at present (Aharonson et al. 2004), nor can the more exotic proposal of a thermoelectric dynamo (Stevenson 1987; Giampieri and Balogh 2002). To prevent the core from freezing, the addition of radioactive heat sources into the core (e.g., Toksöz et al. 1978) or late core formation (e.g., Sharpe and Strangway 1976; Solomon 1977) have been suggested. A late core formation would support a cool initial state for Mercury but is at variance with accretion models (e.g., Schubert et al. 1988). However, the most likely reason for Mercury not having a totally frozen core is the incorporation of a light alloying element into its core that reduces the core melting temperature. Here, sulfur is the most likely candidate (Ringwood 1977; McCammon et al. 1983). The amount of sulfur depends on the accretion and formation scenario of Mercury. In the most conservative equilibrium condensation models there is no sulfur at all in Mercury's core (Lewis 1972; Grossman 1972), an assumption often used in earlier thermal evolution models (e.g. Siegfried and Solomon 1974; Sharpe and Strangway 1976; Solomon 1976, 1977). However, a low concentration of sulfur has been suggested due to radial mixing of planetesimals and/or nonequilibrium condensation models (Basaltic Volcanism Study Project 1981; Wetherill 1985). That Mercury's inner structure is consistent with its high density can actually be explained by a wide range of the sulfur concentration in the core, ranging from a pure iron core to a core with a pure iron sulfide composition (Harder and Schubert 2001).

A necessary condition for generating a self-sustained, hydromagnetic dynamo is convection in a fluid or partly fluid core. Core convection can basically occur in two ways.

(1) **Thermal convection** in the core, like thermal convection in the mantle, is driven by a sufficiently large super-adiabatic temperature difference between the core and the mantle. It occurs if the core heat flow exceeds that conducted along the core adiabat.
(2) **Compositional convection** can occur due to the release of positively buoyant material during the process of solid inner core freezing from a fluid core with non-eutectic composition (Braginsky 1964). Chemical convection and the associated generation of a magnetic field in the core occur if the temperature in the fluid (outer) core ranges between the solidus and the liquidus of the core material. Inner core growth permits outer core convection even when the heat flow through the core–mantle boundary is less than the heat carried by conduction along the adiabat. Furthermore, compositional convection is energetically more efficient at driving a dynamo because of the vagaries of a heat engine.

In either case, the core must be cooling and the cooling is controlled by the heat transport through the outer layers of the planet—mantle and crust. Thus, understanding the magnetic field history of Mercury also requires an understanding of the thermal evolution of the planet.

Current thermal evolution models for Mercury try to explain the present-day magnetic field in a fluid core—assuming that this field has a dynamo origin. Future missions will show whether this assumption is valid or whether the present magnetic field is due to remnant magnetization of the crust. Measurements of possible remnant magnetization of the crust are in any case of important for understanding the evolution of Mercury. It is assumed that the remnant magnetization can be established at the time of crust formation while an internal dynamo was active. An age determination of the magnetized and non-magnetized crust would thus allow us to reconstruct the magnetic field history as has been done for Mars (Acuña et al. 1999) and the Moon (Runcorn 1975). However, should Mercury have a present-day dynamo-generated field like the Earth, the separation of a possible additional remnant field might be difficult to establish from orbit, as would the reconstruction of the magnetic field history. Such a reconstruction is more likely if Mercury's present field is caused by crust magnetization.

3 Modes of Heat Transfer

3.1 Conductive and Convective Heat Transport

In a terrestrial planet like Mercury there are basically two main mechanisms for the transfer of heat—conduction and convection. The conductive heat transfer is a diffusive process wherein molecules transmit their kinetic energy to other molecules by colliding with them. In a medium with a spatial variation of temperature, heat is conducted from the warmer region to the colder. The basic relation for linear conductive heat transport states that the heat flux is directly proportional to the temperature gradient and the thermal conductivity. The latter parameter is a bulk property of the material that indicates its ability to conduct heat. It depends on temperature, pressure, composition and structure of the material. In a convecting, one-plate planet, the upper mantle boundary layer, also known as the stagnant lid, transports heat by conduction. Beneath the stagnant lid heat is transported by convection.

Convective heat transport is associated with the motion of the material and is more efficient in transporting heat than conduction. In a convecting medium, hot fluid flows into a cold region and heats it. Similarly, a cold fluid flows into a hot region and cools it. The motion is caused by the temperature-induced density variations. As a hot fluid is less dense than a cold fluid of the same material, the flow pattern in a terrestrial planet shows, in general, hot uprising material and sinking cold material. Convection starts when the buoyancy forces due the thermal density variations are stronger than the resisting forces like the viscous drag and the thermal conductivity of the material. The ability and strength of the convection is usually described by the Rayleigh number, which is proportional to the cube of the layer thickness and inverse proportional to the mantle viscosity. Convection sets in when the Rayleigh number of a terrestrial mantle is larger than the critical Rayleigh number. The higher the Rayleigh number the stronger the convection and the more efficient the heat transport. In contrast, for a terrestrial mantle with a Rayleigh number lower than a critical Rayleigh number, heat is transported by conduction alone. Among others, the critical Rayleigh number depends on the aspect ratio of the mantle shell (core radius / planetary radius). The thinner a shell becomes, the smaller the critical Rayleigh number (Zebib et al. 1983). Thus, Mercury's mantle with an aspect ratio of about 0.8 has the lowest critical Rayleigh number of all the terrestrial planets; nonetheless, Mercury's small size leads to a low intrinsic Rayleigh number for the mantle. Whether and how long convection took place in Mercury will be discussed in the following. For the modelling of the heat transfer in a terrestrial planet in general two different approaches are used, parameterized convection and finite amplitude models.

3.2 Parameterized Models

Thermal evolution calculations with 2D or 3D mantle convection codes are still very time-consuming on present-day computers. Because of the inherent complexity in these models it is often desirable to take an empirical approach and parameterize the convective heat transfer rate as a function of known quantities. Such parameterizations can be derived using simple theories, which result in scaling laws that describe the heat transport in the interior. As an output, global parameters such as the mean mantle and core temperature, mean heat flow of the mantle and core, and average mantle flow velocity can be obtained as a function of time (e.g., Stevenson et al. 1983; Schubert et al. 2001).

Our improved understanding of the heat transport mechanisms on terrestrial planets over the last two decades has led to repeated changes in the preferred scaling law used to model the thermal evolution of one-plate planets. Initially, the scaling law for a fluid with constant viscosity was used for one-plate planets (e.g., Stevenson et al. 1983). Since that time, it has been recognized that this scaling law models the heat transport in a planet where convection comprises the whole mantle, including the outer layers. In fact, such a model reasonably describes the heat transport in a planet with plate tectonics better than that in a one-plate planet (e.g., Schubert et al. 2001). The plate tectonic regime is expected to cool the planet very efficiently because the comparatively cold outer layers become recycled into the interior of the mantle by convection.

Subsequent attempts to model the heat transport in one-plate planets included the effects of a growing lithosphere in the models with parameterizations based on constant viscosity scaling laws (e.g., Schubert et al. 1990). The base of the non-convecting lithosphere is represented by a characteristic isotherm for the transition from viscous deformation to a rigid response to loads applied over geologic time scales (e.g., Schubert et al. 1992). Such a model represents the heat transport in a planet with a single plate on top of a convecting mantle. However, this model assumes that the lid coincides with the rheological lithosphere. The part of the upper mantle that is weaker than the rheological lithosphere is assumed to be constantly recycled within the mantle. In comparison to the recently derived stagnant lid parameterization, these models, therefore, represent a mechanism of lithospheric delamination. The efficiency of the heat transport by lithospheric delamination is in between that of plate tectonics and that of stagnant lid convection.

Recently, new scaling laws have been derived from convection models in layers of fluids with strongly temperature-dependent viscosities (e.g., Richter et al. 1983; Davaille and Jaupart 1993; Solomatov 1995; Moresi and Solomatov 1995; Grasset and Parmentier 1998; Reese et al. 1999; Solomatov and Moresi 2000). These scaling laws have been suggested to represent the heat transport in a planet with a single plate on top of a convecting mantle as it is assumed for Mercury. To model the thermal evolution of a one-plate planet with a stagnant lid parameterization, one solves the energy equation of the mantle and the core and uses the scaling law which relates the heat loss to the strength of the convection in the mantle. Detailed descriptions of the equations and the methods can be found in the literature, including the growth of a crust associated with a redistribution of the radioactive elements and the magnetic field evolution (Hauck and Phillips 2002; Hauck et al. 2004; Breuer and Spohn 2003, 2006).

3.3 Finite Amplitude Models

Finite amplitude models in 2D or 3D geometry provide detailed information on mantle dynamics and the associated heat transport via calculation of local parameters such as the temperature and velocity fields. In convection models mantle material is considered a highly viscous and incompressible fluid, which can be described by the equations of fluid dynamics. In the framework of the Boussinesq approximation, which is commonly used in thermal convection models, the mantle density is constant except for the buoyancy term (see Schubert et al. 2001). The conservation of mass, momentum and energy are given in the following non-dimensional equations:

$$\nabla \cdot \vec{u} = 0, \tag{1}$$

$$\nabla p = \nabla \cdot \left(\eta' \left(\nabla \vec{u} + \{\nabla \vec{u}\}^{\mathrm{T}} \right) \right) + Ra\, T \vec{e}_r, \tag{2}$$

$$\frac{\partial T}{\partial t} + \vec{u} \nabla T = \nabla^2 T + H, \tag{3}$$

where ∇ denotes the nabla operator, η' the viscosity, $\vec{u}$ the velocity vector, p the dynamic pressure, $\{\}^{\mathrm{T}}$ the tensor transpose, T the temperature and $\vec{e}_r$ the unit vector in radial direction. The Rayleigh number comparing the convection supporting and the impeding effects is defined as follows: $Ra = \alpha \rho_m^2 C_m g \Delta T d^3 / \eta_{\mathrm{ref}} \lambda$ with the thermal expansivity α, the thermal conductivity k, the density ρ_m, the heat capacity C_m of the mantle, the mantle or shell thickness d, the initial temperature contrast between the surface and the core mantle boundary ΔT, and g the acceleration of gravity. The dimensionless internal heating rate is denoted as $H = Q d^2 / \lambda \Delta T$ where Q is the volumetric heating rate (heat produced by the decay of radioactive elements), which decreases exponentially with time ($Q = Q_0 \exp(-\sigma t)$) with the decay rate σ and the initial value Q_0.

Although models of finite amplitude convection have improved in recent years, particularly with the consideration of complex rheologies and parallelization of the codes for efficiency in calculation times, to our knowledge only one 2D convection model with strongly temperature-dependent viscosity for Mercury has been published so far (Conzelmann 1999). We present 2D and preliminary 3D convection models in a spherical shell with temperature-dependent viscosity that also consider the cooling of the core and potential inner core growth. A detailed description of the model is given in the Appendix and the discussion of our results and how they compare with earlier studies is given in Sect. 5.

4 Early Conditions and Relevant Parameters for Thermo-Chemical Evolution

A challenge for modelling and understanding Mercury's internal evolution is a lack of knowledge of Mercury's early thermal conditions and of its chemical composition. The latter influences relevant parameters of the thermal evolution such as the amount of radioactive elements, the mantle rheology, and the thermal properties of the mantle and crust. Some, if not all, of these parameters, however, depend on the formation scenario which seems to be very special among the terrestrial planet, as indicated by the Mercury's large bulk density of ~ 5430 kg/m^{-3}. There are three general hypotheses for the formation of Mercury with its anomalously large density. The first scenario for forming Mercury's large bulk density involves aerodynamic sorting of iron and silicate particles in the solar nebula (Weidenschilling 1978). The other two models for Mercury's formation involve a later stage loss of silicate, and retention of iron, either by stripping of the outer, silicate layers of a larger, differentiated, proto-Mercury by an impact (e.g., Cameron et al. 1988; Wetherill 1988) or by a more exotic scenario that involves a late-state vaporization of silicates (Fegley and Cameron 1987).

Based on different formation scenarios, Taylor and Scott (2005) proposed several composition models that satisfy both the high density of Mercury and the low FeO content of its mantle (see also Van Hoolst et al. in this book). They differ strongly in relative content in olivine, pyroxenes and garnet and in the amount of radiogenic elements; leading to the possibility that measurements performed by the future missions MESSENGER and BepiColombo may strongly constrain the bulk composition of Mercury and potentially its mode of formation.

4.1 Early Thermal State of Mercury

The earliest thermal state of Mercury depends on two main processes: the accretion of the planet by extraterrestrial impacts and core formation and the associated release of gravitational potential energy during metal-silicate differentiation. Both processes may have occurred during the first few ten's of million years of the planet's evolution, as is suggested for Mars and Earth (e.g., Kleine et al. 2002). Accretion of material during the formation of planets is likely one of the largest sources of heat. The accretional heat is the energy accumulated during the burial of heat by impacts as the planet grows through the accretion of planetesimals. During accretion a temperature profile is generated where the temperature increases from the center towards the surface. This accretional energy is sufficient to melt the entire planet. However, the fraction of the available accretional energy that is retained inside the planet instead of radiating away is not known, but only 20% would suffice to ensure accretional melting of nearly all of Mercury (Schubert et al. 1988).

Core formation, the result of separation of metal from silicate materials due to differences in their densities, most likely occurred contemporaneously with or shortly after a planet's accretion. Recent studies on short-lived radio-nuclides (e.g., ^{182}Hf) suggest that the separation between silicates and iron occurred on sampled terrestrial bodies on the order of $1-5 \times 10^7$ years after accretion (e.g., Kleine et al. 2002). Planets are assumed to have initially accreted as a mixture of silicate and metal particles, based largely on the composition of chondrites. However, it is likely that at the late stage of accretion the planetesimals were already differentiated (e.g., Kleine et al. 2004; Baker et al. 2005). Estimates of the time scale of the core formation process suggest that for rapid separation both the silicate and the iron need to be fluid at least in the upper part of the planet (e.g., Stevenson 1990). The gravitational energy released by core formation is converted into thermal energy, which strongly heats the interior. During core formation, when iron sinks to the center, the temperature profile is inverted to decrease from the center towards the surface. The total energy released by the differentiation of a homogeneous planet into an iron-rich core and a silicate mantle can be estimated from the difference between the potential energy stored in a homogeneous planet after accretion and the potential energy of the differentiated, two-layer planet. If this energy is homogeneously distributed in the planetary interior—that is, if there is thermal equilibrium between the mantle and the core—estimates for the mean temperature increase are about 700 K for Mercury (Solomon 1979). Formation models that assume a large impact that stripped of the outer, silicate layers of a larger, differentiated, proto-Mercury or a late-state vaporization of silicates would suggest a much stronger temperature increase. In these models, the planetary radius at the time of core formation was larger and thus, the associated potential energy higher. On the other hand, the potential energy released by core formation may be smaller for a heterogeneous accretion in which planetesimals are already differentiated.

It is also likely that due to the assumed rapid core formation process the core was superheated with respect to the mantle. The excess temperature can be the consequence of adiabatic heating of the sinking iron since the core alloy has a thermal expansion coefficient which is 2–3 times larger than that of mantle silicates. Furthermore, the order of magnitude higher thermal conductivity of iron in comparison to silicate allows more efficient heating of the sinking iron blobs by viscous dissipation than of the silicate mantle through which the iron sinks.

In addition to the two main processes that provide early internal heating and differentiation, other energy sources may have contributed to the early heat budget such as the decay of short-lived radioactive isotopes like Al^{26} (Lee et al. 1976), electromagnetic heating (Sonett

Table 1 Adopted models for the abundances of heat-producing elements in the silicate fraction of Mercury from the various formation scenarios

Model	U (ppb)	Th (ppb)	K (ppm)
Condensation[1]	30	120	0
Late impact[2]	8	30	550
Vaporization[3]	0	400	0

[1] (Weidenschilling 1978) [2] (Cameron et al. 1988; Wetherill 1988) and [3] (Fegley and Cameron 1987). For further description of the models see Sect. 4.2

et al. 1975) and tidal dissipation. In conclusion, a hot initial state of Mercury with early core formation is the most likely scenario for Mercury. The exact temperatures are unknown, although it is often assumed that the temperature after core formation is close to the mantle solidus and to the liquidus adiabat in the core (Stevenson et al. 1983; Schubert et al. 1988; Spohn 1991). However, a heterogeneous accretion for Mercury may also imply cooler core temperatures that are below the core liquidus (Hauck et al. 2004).

4.2 Radioactive Heat Source Density

The three general hypotheses for the formation of Mercury also yield predictions for the abundances of the important heat-producing elements uranium (U), thorium (Th), and potassium (K) (Table 1). Therefore, though the elemental composition of the planet is unknown, the predictions of each of these scenarios for Mercury's formation can inform understanding of the amount of heat potentially produced within the planet over time. The aerodynamic sorting of iron and silicate particles in the solar nebula (Weidenschilling 1978) might have given the planet abundances of U and Th similar to the upper mantle of the Earth, though lacking much K (Basaltic Volcanism Study Project 1981). A large impact that strips off the outer, silicate layers of a larger, differentiated, proto-Mercury (e.g., Cameron et al. 1988; Wetherill 1988) could have left a silicate layer with near CI chondritic abundances of heat-producing elements (e.g., Lodders and Fegley 1998). A late-state vaporization of silicates would result in a Th-rich silicate layer lacking appreciable U and K (Fegley and Cameron 1987).

4.3 Mantle Viscosity

The rheology of the mantle is a primary factor affecting convection in terrestrial planets and a prerequisite for convection is that the material behaves like a fluid (i.e., flows under small differential stresses). The timescale over which the mantle is fluid-like is of utmost importance. Over short timescales the mantle is approximately elastic, consistent with the transmission of seismic waves, but over periods longer than ~ 1000 years the mantle behaves like a fluid with a viscosity that depends on the magnitude of the high temperatures and pressures. Unfortunately, such conditions, particularly the low strain rates, cannot be reproduced in the laboratory and thus the assumed rheology of planetary mantles is primarily based on extrapolation of experimental studies and analysis of terrestrial geodetic data. What is known is that the viscosity of mantle rocks under the wide variety of conditions in planetary mantles is dependent on temperature, pressure, stress, grain size, and composition.

The exponential dependence of the viscosity on the inverse absolute temperature is the most important parameter for understanding the role of mantle convection in transporting heat. The temperature dependence of the viscosity acts as a thermostat to regulate the mantle temperature. Any tendency of the mean temperature to increase is offset by an associated reduction in mantle viscosity, an increase in convective vigour, and a more efficient outward transport of heat. Similarly, a decrease in mantle temperature tends to increase mantle viscosity, reduce convective flow velocities, and decrease the rate of heat transfer. As a result of the sensitive feedback between mantle temperature and rheology, relatively small changes in temperature can produce large changes in heat flux, and the temperature is consequently buffered at nearly constant temperature (Tozer 1965).

In addition to the temperature dependence, the rheology in a planetary mantle can be described by two main creep mechanisms—diffusion creep and dislocation creep. For the case of diffusion creep the solid behaves as a Newtonian fluid where the viscosity is independent of the applied shear stresses. In contrast, for dislocation creep, the solid behaves as a non-Newtonian fluid where viscosity depends on shear stress. Indeed, viscosity tends to decrease with increasing shear stress, often nonlinearly. Which creep mechanism is valid in Mercury's mantle is not certain. Most laboratory studies of mantle deformation have concluded that dislocation creep is the applicable deformation mechanism in the upper Earth mantle and diffusion creep in the lower mantle (see Schubert et al. 2001). However, this assumption is not consistent with post-glacial rebound studies that favour diffusion creep for the upper mantle, too. Furthermore, laboratory measurements have shown that the pressure-dependence on viscosity cannot be neglected in a terrestrial mantle (Karato and Rubie 1997); even for the thin Mercury mantle. Thus, the viscosity of a terrestrial mantle can be described with the following relationship:

$$\eta = \frac{C}{\tau^{n-1}} \exp\left(\frac{A + pV}{RT_m}\right). \qquad (4)$$

Here C is a constant, τ is the stress tensor, n is the stress exponent ($n = 1$ for Newtonian rheology and a typical value of n is 3.5 for a non-Newtonian rheology), T_m is the mean temperature of the convecting fluid, R is the universal gas constant, A is the activation energy for creep (e.g., Weertman and Weertman 1975), p is the pressure and V is the activation volume. The values of the activation energy and volume depend on the mineralogical composition, which is basically unknown for Mercury and as mentioned earlier depends also on the formation scenario. However, should Mercury's mantle contain olivine then it would likely dominate the rheology of the silicate mantle (e.g., Karato and Wu 1993). It has been generally assumed that the Mercurian mantle is volatile-poor because of its refractory nature (Schubert et al. 1988). A dry mantle is also favored from thermal evolution models to explain the small planetary contraction (Hauck et al. 2004). For dry olivine, an activation energy of about 540 kJ/mole at pressures of 12 GPa (Karato and Wu 1993) and an activation volume of about 15 cm^3/mol (Karato and Rubie 1997) can be assumed. To obtain C for Newtonian rheology, a reference viscosity of about 10^{21} Pa s at a temperature of 1,600 K can be used. This value is typical for the upper Earth mantle and often used also for other terrestrial mantles (e.g., Schubert et al. 1988; Nimmo and Stevenson 2000; Breuer and Spohn 2003). It has been speculated, however, that the mantle of Mercury might be even stiffer than the Earth's mantle at the same temperature and pressure because of its refractory and volatile poor composition (Schubert et al. 1988), thus even a higher reference viscosity might be possible.

4.4 Thermal Properties of Mantle and Crust

The thermal conductivity depends on various factors like temperature, pressure, composition, and texture of the material. In the mantle, the phonon contribution, k_{lat}, and the radiative contribution, k_{rad}, contribute to the thermal conductivity.

$$k_m(T, P) = k_{lat}(T, P) + k_{rad}(T). \qquad (5)$$

The phonon contribution k_{lat} decreases with increasing temperature and increases with increasing pressure, whereas the radiative contribution k_{rad} increases with increasing temperature independent of pressure; for relevant temperatures in a terrestrial mantle, the phonon conductivity is much higher than the radiative contribution. As a consequence, a decrease of the thermal conductivity through the upper mantle and an increase through the lower mantle is expected. Hofmeister (1999) developed a model describing the temperature and pressure dependence of the mantle thermal conductivity under terrestrial conditions. Using this model for Mercury suggests that the thermal conductivity in the convecting mantle ranges between 2.7 and 3.5 W m^{-1} K^{-1}, a value below the usually assumed value of 4 W m^{-1} K^{-1}.

The crust of the terrestrial planets can act as a thermal insulator for the planet. First, it is enriched in radioactive elements and second its thermal conductivity is lower than that of mantle material. The thermal conductivity of the majority of compact volcanic materials ranges between 1.5 and 3.5 W m^{-1} K^{-1} at ambient temperatures (Clifford and Fanale 1985; Clauser and Huenges 1995) but decreases with temperature, similar to mantle material (Seipold 1992). For example, the thermal conductivity of typical compact basalt decreases from about 2 W m^{-1} K^{-1} at 270 K to about 1.5 W m^{-1} K^{-1} at 800 K. In addition to the temperature effects on the thermal conductivity, the structure of the material can change the thermal conductivity significantly. Fractured and porous materials have a reduced thermal conductivity in comparison to compact material. The upper crust of Mercury been fractured due to impact processes, in particular in the early period of heavy bombardment. These impacts resulted in the production of a porous megaregolith that extends to considerable depths. At the transition between fractured and coherent basement, the lithostatic pressure is sufficient to close all fractures and the intergranular pore space. The megaregolith extends to about 10 km for Mars and as the lithostatic pressure for Mars is similar to Mercury also a similar depth of the megaregolith can be assumed. It is expected that the porosity decreases exponentially from the dusty surface to this self-compaction depth (Binder and Lange 1980). As a consequence, the thermal conductivity decreases from the bottom of the megaregolith toward the surface. At low atmospheric pressures like Mercury's, the thermal conductivity at the surface can exhibit extremely low values. Remote thermal measurements for the Martian surface indicate soil thermal conductivities in the range of 0.075 to 0.11 W m^{-1} K^{-1} (Kieffer 1976).

4.5 Melting Temperature of the Mantle and the Core

The melting temperature of the mantle and core depends on its chemical composition. Although the mineralogy of Mercury's mantle can be very different from the composition of the terrestrial upper mantle, dominated by peridotite rocks, the solidus of the KLB-1 peridotite (Zhang and Herzberg 1994) is usually assumed in thermo-chemical evolution models of Mercury (Conzelmann 1999; Hauck et al. 2004). Several recent experiments show indeed that the solidus temperature is quite insensitive to both mineralogical and iron contents (Bertka and Holloway 1993; Hirschmann 2000; Herzberg et al. 2000; Schwab and Johnston 2001).

Mercury's core, by analogy with Earth, is thought to be made up of an alloy of Fe–Ni and another light element. Based on cosmochemical models for Mercury's formation (e.g.,

Basaltic Volcanism Study Project 1981; Wetherill 1985; Lewis 1988) sulfur is a likely candidate for that light element and it has received considerable attention in modelling efforts (e.g., Stevenson et al. 1983; Schubert et al. 1988: Harder and Schubert 2001; Hauck et al. 2004). Increasing amounts of sulfur in the core alloy, up to the eutectic point, result in a corresponding depression in the melting point (e.g., Usselman 1975; Boehler 1992; Fei et al. 1997). Sulfur would be essentially insoluble in a solid iron inner core, hence inner core growth leads to an increase in the concentration of sulfur in the outer core, via mass balance, and therefore a moderation of further growth due to increased melting point depression. Indeed, even a minimal amount of sulfur could stave off complete solidification of the core because that would require core mantle boundary temperatures falling below the Fe–FeS eutectic point of ~1,260 K. Consequently, models of Mercury's internal evolution that account for the presence of sulfur as an alloying element in the core predict that total amounts of inner core growth decrease with increasing total bulk core sulfur content (Stevenson et al. 1983; Schubert et al. 1988; Hauck et al. 2004).

To calculate the growth of an inner core in thermal evolution models, the decrease in the melting temperature due to the increase of sulfur in the outer core needs to be considered. At present, only the melting temperatures of pure iron (e.g., Boehler 1996) and the eutectic composition (e.g., Boehler 1996; Fei et al. 1997, 2000) have been measured. Thus, a parameterization for the melting depression with increasing sulfur at constant pressure is required Note that the melting temperatures for the eutectic composition differ between the work of Boehler (1996) and of Fei et al. (1997, 2000); therefore, different parameterizations exist for the two data sets (see Appendix and Table 3). Furthermore, the parameterizations all use melting curves that have a linear-dependence on sulfur content, which is probably not the case. Thus, the melting points used are probably underestimates of the true melting temperature except for pure Fe and eutectic Fe–FeS.

5 Mantle Dynamics and the Thermo-Chemical Evolution of the Mantle and the Core

5.1 Conduction versus Convection in the Mercury Mantle

The question of whether mantle heat transport in Mercury occurs by conduction or convection is controversial, and depends mainly on the assumption about the initial thermal state, the amount and distribution of radioactive elements, and the efficiency of heat transport. Early models that assume an undifferentiated, and thus cold Mercury, show that the planet cools conductively throughout its evolution (e.g., Siegfried and Solomon 1974; Fricker et al. 1976; Sharpe and Strangway 1976; Solomon 1976, 1977, 1979). Models that assume an initially hot and fully differentiated Mercury are in favour of convection (Solomon 1977; Stevenson et al. 1983; Schubert et al. 1988). This assumption has been confirmed by axial symmetric convection models with a temperature- and pressure-dependent and Newtonian viscosity and a radioactive heat source density consistent to the formation model of silicate vaporization (Conzelmann 1999). The results show that thermal convection is likely during the entire evolution of the planet although it can be very sluggish at present. Recent studies (Solomatov and Reese 2001; Hauck et al. 2004) that have considered a non-Newtonian rheology and crustal formation suggest an initially convecting mantle only during the early stages of evolution accompanied by extensive melting and differentiation. At some later time convection and melting cease and the planet cools in a conductive regime. The more crust produced and the more the mantle is depleted in heat sources the more likely (and earlier) the transition from a convective to a

Table 2 Parameter values used for the 2D axis-symmetric and 3D spherical convection models

Planet radius	r_p	$2{,}440 \times 10^3$ m
Core radius	r_{CMB}	$1{,}900 \times 10^3$ m
Shell thickness	d	540×10^3 m
Gravity	g	3.8 m/s^2
Mantle density	ρ_m	3,060 kg/m^3
Mantle heat capacity	C_m	1,297 J/kg K
Thermal expansivity	α	2.0×10^{-5} 1/K
Thermal conductivity	λ	4 W/K
Viscosity	η_{ref}	8.7×10^{22} Pa s
Activation energy	E_{act}	466.07 kJ/mol
Activation volume	V_{act}	7.43 cm^3/mol
Initial temperature contrast	ΔT	1,660 K
Surface temperature	T_{surf}	440 K
Initial temperature at mid-depth	T_{int}	1,700 K
Initial core temperature	T_{CMB}	2,100 K
Initial internal heating rate	Q_0	5.2373×10^{-8} W/m^3
Radioactive decay rate	σ	0.04951 Ga
Rayleigh number	Ra	698
Core density	ρ_c	8,380 kg/m^3
Core heat capacity	C_c	750 J/kg K
Iron density at core pressure	ρ_{ic}	8,412.27 kg/m^3
Sulfur density at core pressure	ρ_s	6,077.78 kg/m^3

conductive regime. This transition can be retarded or does not take place when the temperature dependence of the mantle conductivity is considered together with the isolating effect of a low-conductivity crust (Breuer 2006). The importance of the temperature dependence of the mantle conductivity and the isolating effect of a low-conductivity crust has already been demonstrated for the thermal evolution of Mars (Schumacher and Breuer 2006).

To study how a present-day convective structure would look, assuming a Newtonian rheology and sufficient heat sources are left in the present mantle of Mercury or a strong thermal isolation of the crust, we have performed 2D axial and 3D spherical convection models with a strongly temperature-dependent and Newtonian viscosity and using the parameters in Table 2 (Figs. 1, 3, 4, 5, 6 and 7). The results show the rapid growth of a thick stagnant lid (Fig. 3). The upper thermal boundary layer comprises nearly half of the mantle, thus mantle convection is restricted to a thin layer of about 250 to 300 km (Fig. 4). The convection pattern in the remaining convecting layer is dominated by small scales of degree 16 to 20 (Fig. 1). However, the convection pattern might be more large scale in the case of strongly pressure-dependent viscosity (Conzelmann 1999). Due to the small temperature gradient at the core–mantle boundary (Fig. 4), cold downwellings (instead of warm plumes) determine the flow pattern. One possibility to learn whether the Mercurian mantle is convecting, in addition to numerical modelling of the mantle flow, is the determination of the mantle structure and the core–mantle boundary shape with gravity and topography data.

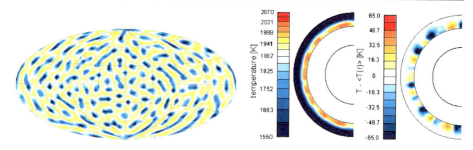

Fig. 1 Present-day convection pattern for an illustrative thermal evolution scenario calculated form an axis-symmetric model with temperature- and pressure-dependent Newtonian viscosity and a growing inner core (detailed description of the model is given in Sect. 3.2 and Appendix) with 1 wt% S in the core (melting curve parameterization based on data by Boehler (1996); see Table 3); silicate heat production given by Th abundances for a vaporization-dominated composition (see Table 1); and an initial mantle temperature of 1,700 K. *Left figure*: the radial velocity at $r = 1,996$ km. *Bluish colours* indicate downward motion, *reddish colours* upward motion. The contour step is 0.47 mm/a. *Middle*: meridional cut through the temperature field ($\phi = 40°$) after 4.5 Ga. *Right*: meridional cut ($\phi = 40°$) of the temperature anomaly ($T - \langle T(r) \rangle$)

Table 3 Parameter values to calculate the melting depression of a Fe–FeS core with increasing sulfur content at different pressure (10)

	a_S	T_{m0}	T_{m1}	T_{m2}
Conzelmann (1999)[1]	2	1835 K	1.215×10^{-11} Pa^{-1}	-8.0×10^{-23} Pa^{-2}
Hauck et al. (2004)[2]	2.4	1809 K	1.54×10^{-11} Pa^{-1}	-11.7×10^{-23} Pa^{-2}

[1] Based on data by Boehler (1996) and [2] based on data by Fei et al. (1997, 2000)

5.2 Mantle Cooling and the Temperature of the Mantle

Similar to the question of whether Mercury's heat transport occurs by conduction or convection, mantle cooling and the temperature distribution depend mainly on the amount of radioactive heat sources and their distribution as well as the efficiency of the heat transport. We discuss the temperature evolution of two illustrative models. First, we examine a parameterized convection model with early mantle convection that later transfers its heat only by conduction (Fig. 2) followed by a finite-amplitude convection model with convection throughout the entire history (Fig. 3).

For the former model, the chosen heat source density is consistent with the composition of the condensation model to explain the high density of Mercury (Table 1). The evolution of mantle and CMB temperatures (Fig. 2a) displays a rapid decrease in temperature from the initial state followed by more moderate cooling paralleling the decay in the concentration of heat-producing elements. The large initial drop in mantle temperature is due to the extraction of a substantial amount of partial melt from the mantle. The associated crust formation depletes the mantle in radioactive heat sources. Mantle convection ceases at about 3.3 Ga, after which CMB temperatures indicate a readjustment in the rate of heat loss due to the change in dominant mechanism of heat transport. An inflection at about 0.8 Ga is due to the energy released on first appearance of a solid inner core (Fig. 2c).

Figure 3 shows the evolution of a temperature profile from a 2D axisymmetric convection model assuming a heat source density consistent with the composition of the vaporization model (Table 1). An initial mantle temperature of 1,700 K has been assumed. Very rapidly,

Interior Evolution of Mercury

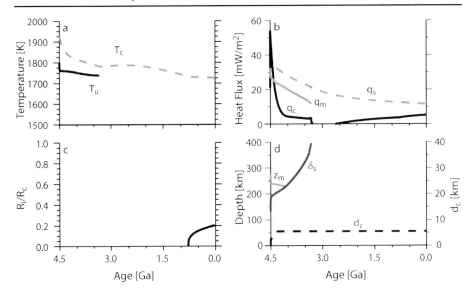

Fig. 2 Illustrative thermal evolution scenario with 8.5 wt% S in the core (melting curve parameterization based on data by Fei et al. (1997, 2000); see Table 2); silicate heat production given by U and Th abundances for a condensation-sequence-dominated composition (see Table 1); a non-Newtonian, pressure-dependent mantle rheology appropriate for dry olivine; partial melting of the mantle and melt transport to the crust; an initial mantle temperature of 1,800 K. (**a**) Temperatures T_c and T_u at the core–mantle boundary and the base of the thermal lithosphere, respectively. (**b**) Heat flux at the surface, q_s, base of the lithopshere, q_m, and core–mantle boundary, q_c. (**c**) Ratio of inner core to outer core radius. (**d**) Greatest depth of pervasive partial melting, z_m, lithospheric thickness δ_s, and crustal thickness, d_c (after Hauck et al. 2004)

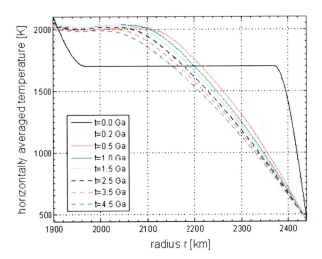

Fig. 3 Temperature profiles at different times in the Mercurian evolution for an axis-symmetric model. For further description see Fig. 1 caption

the lower mantle heats up (temperatures up to 2,026 K are reached) while the upper part cools down. After about 500 Myr, a "final" shape of the temperature profile is reached. In the subsequent evolution, the mantle cools down by thickening the lithosphere which includes the majority of the mantle; convection is restricted to a very thin layer of about 250 km. The temperature of the lower mantle decreases insignificantly and after 4.5 Gyr

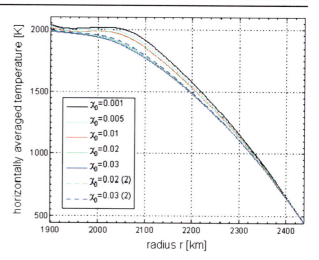

Fig. 4 Temperature profiles after 4.5 Ga for axis symmetric convection simulations with different initial sulfur content and different parameterizations of the melting curve. The initial sulfur content is given in the legend. The *solid lines* refer to the melting curve parameterization based on data by Boehler (1996) and the *dashed lines* refer to the fit of the melting curve using the data by Fei et al. (1997, 2000). For further description see Fig. 1 caption

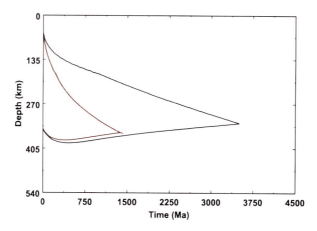

Fig. 5 Extent of partial melt zone in the Mercurian mantle as a function of time for an illustrative thermal evolution scenario with 2 wt% S in the core (melting curve parameterization based on data by Boehler 1996; see Table 2); silicate heat production given by U and Th abundances for a condensation-sequence-dominated composition (see Table 1); a Newtonian, pressure-dependent mantle rheology appropriate for dry olivine; partial melting of the mantle and melt transport to the crust; an initial mantle temperature of 1,800 K. (*Red lines*) constant thermal conductivity in crust and mantle of 4 W/m^{-1} K^{-1} and (*black lines*) a low-conductivity crust with $k_c = 2$ W/m^{-1}K^{-1} and a temperature and pressure dependent thermal conductivity in the mantle according to Hofmeister (1999)

the temperature in the isothermal part is still about 1985 K. The temperature gradient at the CMB is very small consistent with the dominance of cool downwellings in the convection pattern (Figs. 3 and 4).

Variations in bulk core sulfur content also play an important role in influencing the temperature evolution (Fig. 4) of the interior. A variation in the sulfur content leads to different sizes of the solid inner core and, therefore, to a different heat output from the core to the mantle (Fig. 7). The larger the inner core the more gravitational and latent heat due to inner core growth is released during the evolution. At a radius of 2,000 km the temperature is about 80 K higher for the model with a low sulfur content (0.1 wt.%), which shows a big

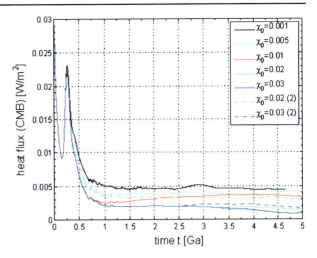

Fig. 6 Heat flux at the core–mantle boundary as a function of time for axis symmetric thermal evolution simulations assuming different initial sulfur contents χ_0. The *solid lines* refer to the melting curve parameterization based on data by Boehler (1996) and the *dashed lines* refer to the fit of the melting curve using the data by Fei et al. (1997, 2000). For further description see Fig. 1 caption

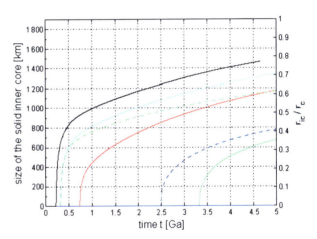

Fig. 7 Evolution of the inner core depending on the initial sulfur content and the parameterization of the melting curve (same legend as in Figs. 4 and 6). For further description see Fig. 1. The *dashed lines* refer to the melting curve parameterization based on the data by Fei et al. (1997, 2000)

inner core (Fig. 7), in comparison to the temperature for the model assuming 3 wt% sulfur in the core. Here, no solid inner core has been formed. The shape of the temperature profiles in the cases with high sulfur content indicates that convection is already quite weak.

5.3 Mantle Partial Melting, Crust Formation and Growth

Formation of crustal material is one of the more important and outwardly visible consequences of terrestrial planet evolution. Primary crustal material, formed from a magma ocean during the latter stages of planet assembly and differentiation, may sequester incompatible heat-producing elements in the crust if no recycling has occurred. On the Earth, most of the present-day crust is the result of extraction of partially melted mantle created during adiabatic decompression of upwelling material. Although Mercury may hold a primary crust, mantle convection there could also result in partial melting if the temperature of upwelling mantle rock exceeds the solidus, leading to formation of secondary crustal material. Effects of crust generated in this manner include a loss of thermal energy in the mantle through the latent heat of fusion and advection of heat as well as enrichment of the incompatible heat-producing elements in the surface layer, both of which lead to enhanced cooling of the mantle.

One consequence of inefficient heat transfer, particularly on a low-gravity planet like Mercury, is that average, sub-lithospheric temperatures may be pervasively above the solidus (e.g., Reese et al. 1998; Solomatov and Moresi 2000; Hauck and Phillips 2002; Hauck et al. 2004). This infers that any upwelling through such a region would result in generation of partial melt, as confirmed by the 2D and 3D convection models (e.g., Conzelmann 1999 and Fig. 3). The evolution of these models shows that partial melt is present in the deep mantle below the stagnant lid for a significant time. However, if mantle cooling due the extraction of partial melt and the depletion of the mantle in radiogenic heat sources is very strong, it is likely that the generation of partial melt is limited to the early evolution (Solomatov and Reese 2001; Hauck et al. 2004; Fig. 2d). This cooling effect may be compensated by the thermal insulation of a low-conductivity crust and possibly by tidal dissipation in a partially molten mantle. In the case where the cooling of Mercury's interior is significantly reduced— for example, due to the thermal insulation of the crust—the existence of a partial melt zone deep in the interior may be expanded. Also the temperature dependence of the mantle thermal conductivity (k_m) tends to reduce the efficiency of mantle cooling; thus, models show a much longer existence of a partial melt zone in comparison to evolution models with constant k_m (Fig. 5). Whether melt can still rise from the deep interior to the surface is not certain. For instance, a sluggish convection in the deep mantle may not introduce strong enough stresses for fractures to occur which are necessary for the melt to segregate.

The derived crustal thicknesses of the secondary crust varies between 3–70 km depending on the initial temperature distribution, the assumed amount of radioactive elements and the thermal conductivity in mantle and crust (Hauck et al. 2004; Breuer 2006). These thicknesses are significantly thinner than the estimated crustal thickness for Mercury, varying between 100–300 km (Anderson et al. 1996; Nimmo and Watters 2004; Watters et al. 2005). If one neglects the uncertainties in such models, either a peridotite-dominant mantle is inappropriate or much of the crust is primordial. Recent experiments show that the solidus temperature is quite insensitive to both mineral and iron contents (Bertka and Holloway 1993; Hirschmann 2000; Herzberg et al. 2000; Schwab and Johnston 2001). Thus, it is more likely that a large part of the crust is primordial. Though we lack a comprehensive study of the relative importance of primary versus secondary crust formation on the cooling history of Mercury, any remaining primary crust would have acted to facilitate early and long-term cooling of the planet via sequestration of heat-producing elements near the surface of the planet. This may be inconsistent with the limited amount of radial contraction Mercury may have experienced (Strom et al. 1975; Watters et al. 1998).

5.4 Core Cooling, Solidification and Magnetic Field Generation

Cooling of the metallic core of any terrestrial planet is a major component in its internal evolution; for Mercury it is of tantamount importance due to its large mass fraction of the planet and as a favoured source of the planet's magnetic field. A major aspect of core cooling is the precipitation of solid metallic iron to form a solid inner core from the molten, overlying outer core. Growth of the inner core depends primarily on the rate of cooling of the core by the mantle and the melting relationships of the core materials (i.e., how the melting temperature varies as a function of pressure and composition) (e.g., Stevenson et al. 1983; Labrosse 2003).

Mercury's magnetic field suggests (Ness et al. 1976), though not conclusively (cf. Aharonson et al. 2004), that the planet's core is at least partially fluid. A totally fluid core, however, is difficult to reconcile with the observed magnetic field because cooling by the mantle alone is likely insufficient to drive convective motions capable of generating a magnetic field. The heat flow is below the critical heat flow along the adiabat of about 11 mW m^{-1} K^{-1} (Schubert et al. 1988) that is necessary to sustain thermal convection in the core (Stevenson et al. 1983; Schubert et al. 1988; Hauck et al. 2004 and Figs. 2b and 6). Thus, compositional buoyancy generated during inner core growth may be necessary to create the field, which may place an upper bound on the amount of sulfur that the core may host. Depending on the assumed heat source density and its distribution as well as the mantle rheology, the temperature dependence of the mantle thermal conductivity, the thickness of the low-conductivity crust, and tidal dissipation in the mantle, the threshold for the bulk sulfur content (for larger values inner core growth does not begin until after 4.5 Gyr) varies strongly. The more efficient the cooling of the planet's interior, the higher the value of the maximal sulfur content. For instance, Hauck et al. (2004) calculated an upper value of ∼9–10 wt%. Their models assume a pressure-dependent, non-Newtonian rheology and consider crust formation with a depletion of the mantle in radioactive elements; thermal insulation of the crust is not considered. A lower threshold of less than 3 wt% is suggested with the 2D and 3D thermal convection models with a strongly temperature-dependent and Newtonian rheology (Fig. 7) and a heat source density consistent to the silicate vaporization model. These models, however, neglect crust formation and the associated redistribution of radioactive elements. A systematic study is needed to better constrain the upper value.

The evolution of the core size for the latter models is given in Fig. 7. The higher the initial sulfur content, the later the start of the inner core growth, and the smaller is the present size of the inner core. The shown results demonstrate the need for a better understanding of the melting depression between pure iron and the eutectic composition. The parameterization by Hauck et al. (2004) based on data by Fei et al. (1997, 299) leads to higher melting temperatures in comparison to the parameterization by Stevenson et al. (1983) and Conzelmann (1999) based on data by Boehler (1996). In the latter case the inner core formation starts 3 Ga later and the final size of the inner core is 500 km smaller assuming the same bulk sulfur content in the core.

Thermal evolution models for Mercury (Stevenson et al. 1983; Schubert et al. 1988; Conzelmann 1999; Hauck et al. 2004) infer a possible evolution of a dynamo-generated magnetic field. In the early evolution, a dynamo might be generated by thermal convection before the initiation of inner core growth and an associated chemical dynamo. Such an early thermal dynamo, however, could only be active when the core was superheated with respect to the mantle. It is unlikely that a core, which is in thermal equilibrium with the mantle, can start thermal convection and thus dynamo action (Breuer and Spohn 2003). Whether the dynamo action ceases for a time until inner core growth provides sufficient energy depends on the onset time of inner core growth. As shown earlier, the higher the sulfur content the later the onset of inner core growth (Fig. 7). A detection of remnant magnetized crust and the determination of its age with future missions could therefore help to constrain the amount of sulfur in the core and thus the formation scenario.

Like tidal heating in a partially molten mantle, a solid Fe inner core might be capable of generating an amount of tidal dissipation comparable to the heat loss along the adiabat in the outer core. The results by Schubert et al. (1988) show that only a large inner core generates significant heating. However, only 4% of the heat currently leaving the core is the result of tidal heating. Thus, the inner core growth rates and the temperatures are not significantly influenced by this small contribution.

A hydromagnetic dynamo—as assumed in many thermal evolution models—poses a problem in terms of the strength of the calculated field. Although there is presently no satisfying parameterization to calculate the magnetic field strength, estimates suggest a much larger field than observed (Stevenson et al. 1983; Schubert et al. 1988). Recent thin-shell dynamo models, however, have shown that planets with a large solid inner core (relative inner core size larger than about 0.8) can produce magnetic fields with Mercury-like field intensities (Heimpel et al. 2005; Stanley et al. 2005). Alternatively, a deep dynamo model that suggests a rather small inner core (smaller than 1,000 km in radius) was proposed by Christensen (2006). In this model, the dynamo operates only at depth and the associated dynamo field is strongly attenuated by the skin effect through a stable, conducting region of the upper core. An alternative way of generating a weak magnetic field in Mercury is by a thermo-electric dynamo (Stevenson 1987; Giampieri and Balogh 2002). This dynamo makes use of a thermally derived electromotive force set up at a distorted core–mantle boundary. Such a dynamo requires topography variations of the core–mantle boundary of the order of one kilometer due to mantle convection. Whether this kind of dynamo is active in Mercury cannot be concluded from the current magnetic data. It is, however, possible to detect core–mantle undulations by inversion of the global Mercurian gravity and topography field (Spohn et al. 2001 and Sect. 6). New insights into the magnetic field evolution and the dynamo mechanism of Mercury are expected from the future Mercury missions (see also the chapter by Wicht et al. 2007 in this volume).

5.5 Implications for Planetary Contraction

A major consequence of cooling terrestrial planet forming materials is a reduction in volume. Therefore, estimates of 1–2 km planetary contraction based on surface tectonics (e.g., Strom et al. 1975; Watters et al. 1998) represent an opportunity to constrain the internal history of Mercury via understanding the amounts and rates of cooling consistent with these estimates. Since the Mariner 10 era several workers have constrained a variety of thermal models from purely conductive (e.g., Solomon 1976) to mantle convection (e.g., Schubert et al. 1988) and ones that account for time periods when either mechanism may have dominated (Hauck et al. 2004). A cross-cutting conclusion that can be distilled from these studies is that while the mode of mantle cooling plays an important role in generating planetary contraction either it is not the sole governor of contraction or the estimates of contraction are less than has actually occurred on Mercury, or possibly both.

A significant part of planetary contraction—in addition to the contraction due to the mantle cooling—is the result of cooling the core and the solidification of an inner core. Cooling and complete solidification of an initially fluid core would result in a $\sim$17 km decrease in radius (Solomon 1976). Therefore, Mercury's small amount of inferred contraction is consistent only with a small inner core or a large inner core if most of the inner core was formed before the end of heavy bombardment. The onset time of inner core growth and the inner core size depends on the initial sulfur content (Fig. 7). In general, one can state that the more sulfur in the core the smaller the present inner core size. The observation of a small planetary contraction may provide a lower limit of the sulfur content. For smaller sulfur contents, the planetary contraction obtained after the end of the heavy bombardment and caused by the inner core growth is larger than the observed value of 1 to 2 km. However, the lowest amount of sulfur that is in accord with the small contraction depends also on the efficiency of mantle cooling. Due to mantle regulation of core cooling, slow cooling of the mantle also implies slow core cooling and a smaller solid inner core.

Parameters that control mainly the cooling behaviour of the mantle are the amount and the distribution of radioactive elements, the mantle rheology and conductivity, the thickness of the low-conductivity crust and the amount of tidal dissipation. As the natural radioactive decay of long-lived radioisotopes is the primary source of heat generation in the mantle,

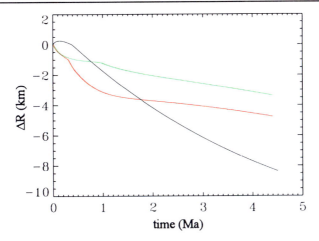

Fig. 8 Contraction of the surface as a function of time for a parameterized convection model based on constant viscosity law (*black line*), for a 2D axisymmetric convection model with strongly temperature dependent viscosity (*red line*) and a 2D axisymmeteric convection model with strongly temperature- and pressure-dependent viscosity (*green line*). The viscosity increases by a factor of 10 in the convecting mantle. Models further assume a concentration of radioactive heat sources consistent to the silicate vaporization model (Fegley and Cameron 1987) and 1% of sulfur in the core (melting curve parameterization based on data by Boehler 1996) (After Conzelmann 1999)

the abundances of these elements (e.g., ^{40}K, ^{232}Th, and ^{235}U, ^{238}U) in the mantle will also tend to mediate total amounts of cooling and cooling rates. An important example is the silicate vaporization scenario for Mercury's high-bulk density that predicts ^{232}Th as the sole source of radiogenic heat (Fegley and Cameron 1987); which, with its 14 Gyr half-life, has not had a significant decline in heat output and hence predicts less cooling and contraction of the planet (Hauck et al. 2004). For models with a significant fraction of the mantle's history being dominated by convection, the effective viscosity of the mantle is a critical parameter controlling the rate of cooling. Higher viscosities tend to result in lower cooling rates, and hence lower amounts of contraction. Under the assumption that olivine is the dominant mantle mineral it is clear that a thoroughly dry mantle is considerably stronger than a wet one (e.g., Karato and Wu 1993). In turn this results in models with a dry olivine mantle viscosity having less, ~50% or less, contraction than wet olivine models, and the dry models tend to more closely approach the current estimates of total contraction, and therefore recent work has rejected the idea of a wet mantle for Mercury (Hauck et al. 2004).

In fact, it has been suggested that only models assuming a dry Mercurian mantle and the silicate vaporization scenario can explain the small observed contraction with minimal 1–3 wt% sulfur in the core for models without crust formation (Conzelmann 1999; Fig. 8) and at least 6.5 wt-% sulfur if crust formation and a non-Newtonian rheology is considered (Hauck et al. 2004). Enhanced cooling of the mantle due to crust formation can be compensated by the temperature dependence of the mantle thermal conductivity and the inefficient heat transport of a low-conductivity crust (Fig. 5). Depending on the thermal conductivity of the mantle and crustal material, the thickness of the crust, the thickness of the regolith layer, in which the thermal conductivity is significantly reduced in contrast to compact material, mantle cooling can be very inefficient and totally compensate (or even overcompensate) for the enhanced cooling due to crust formation and the depletion of radioactive elements in the mantle. Thus, as long as we do not know the relevant parameter values the sulfur content in the core cannot be restricted.

It is, however, difficult to make concrete conclusions regarding the cooling history of Mercury based merely on current estimates of Mercury's total contraction. Those strain estimates are based upon imagery of only 45% of the surface and do not account for other, potentially subtle, recorders of strain, such as long-wavelength folding. Indeed, Hauck et al. (2004) found that even a factor of two more strain than the current estimates of 0.05–0.10% makes it difficult to discriminate among models even for the concentration of heat-producing elements.

6 Summary and Outlook for Future Missions

We have reviewed our present knowledge of the interior evolution of Mercury. While the known density of the planet gives us at least some knowledge of the gross interior structure (see also chapter by van Holst et al. in this volume), prior to the arrival of MESSENGER and BepiColombo the evolution is largely unconstrained because of a lack of data capable of discriminating between possible evolution scenarios. It is generally agreed from accretion models that Mercury started hot, which also implies early mantle convection (Schubert et al. 1988). The observation of the magnetic field provides probably the best constraint on the interior evolution, assuming that it is generated by a core dynamo and not a remnant field, in that it may indicate a currently convecting outer core.

Another constraint for the interior evolution, although uncertain, is the inferred small planetary contraction from the lobate scarps. The small contraction provides the opportunity to constrain the concentration of radioactive elements, the mantle rheology and the minimal amount of sulfur in the core. Current models suggest that only a dry mantle and Th-rich mantle composition (U and K are strongly depleted) as it is assumed from the silicate vaporization model is consistent with the small planetary contraction of 1 to 2 km. The minimal amount of sulfur, however, remains unconstrained.

The sulfur content in the core depends on the competing effects between efficient cooling of the planet due to crust formation together with the associated depletion of the mantle in radioactive heat sources and inefficient cooling such as the temperature dependence of the mantle thermal conductivity (the thermal conductivity decreases with increasing temperature) and the thermal insulation of a low-conductivity crust. A sulfur content of more than 6.5 wt% is required if one considers cooling by crustal formation (but not its thermal insulation, i.e., the thermal conductivity of the crust is equal to the mantle's conductivity) and a constant mantle thermal conductivity of 4 W m^{-1} K^{-1}. The threshold of sulfur in that case can be even higher if an enriched primordial crust exists and further enhances the depletion of the mantle in heat sources. These models also suggest a change in the heat transport mechanism. In the early evolution the mantle transport heat by convection but changes later in to a conductive regime. The sulfur content can be as low as 1 wt% in the case of inefficient mantle cooling if the temperature dependence of the mantle thermal conductivity is considered and the crust acts as a thermal insulator. The lower the mantle cooling efficiency the smaller the sulfur content in the core must be in order to explain the observations; furthermore, this low efficiency would also delay any transition to a conductive regime. The efficiency of thermal insulation increases with an increase of the crust thickness and the thickness of the regolith layer. In particular, the regolith layer can exhibit low thermal conductivities that are up to two orders of magnitudes lower than that of the same compact material. In the case of inefficient mantle cooling, mantle convection exists during the entire evolution and a partial melt zone is likely to exist during a substantial period of time in the evolution if not until present.

Mercury's dynamical state is dominated by tidal interaction with the Sun. Thus, it is reasonable to assume that tidal heating might be a heat source in addition to the radioactive heat sources. Present-day tidal heating in Mercury is a consequence of rotation with respect to the Sun with a period of 87.96 days and the eccentricity of the orbit with a mean eccentricity of 0.206. It is commonly assumed that tidal heating in the mantle is unimportant at the present time (e.g., Schubert et al. 1988). However, this assumption is based on the current understanding that the present mantle is solid and behaves inelastically for tidal forces; the possibility of partial melt has not been adequately considered. Assuming that partial melt in Mercury's mantle was or is still present (Fig. 5), tidal dissipation may contribute to the heat budget of the planet.

Thermal evolution models indicate that the existence of a present-day magnetic field for Mercury is associated with the growth of an inner core—assuming that this field has internal origin. A present-day, purely thermal convection generated dynamo is unlikely since model results show a slow cooling core with less heat flow than can be conducted along the core adiabat (Stevenson et al. 1983; Schubert et al. 1988; Hauck et al. 2004; Figs. 2b and 6). In fact, a thermal dynamo is in general difficult to generate and likely only if the core is superheated with respect to the mantle due to the core formation process. If a thermal dynamo ever existed, it shut off very early in the evolution as the heat flow out of the core decreases rapidly during the first few hundred million years. Whether in the subsequent evolution an inner core can grow and thus a compositional dynamo can be initiated depends strongly on the temperature evolution and the core composition.

At the current stage, only future missions may help to better understand the interior evolution of Mercury. In the following, we present a short overview of the necessary data and what we may learn from them.

6.1 Magnetic Field Data

One of the primary goals of the two missions to Mercury (MESSENGER and BepiColombo) is to clarify the nature of the planet's magnetic field. The data collected by two of Mariner 10's flybys are not conclusive. The BepiColombo mission with its two orbiters, one of them (the Mercury Magnetospheric Orbiter MMO) committed to exploring the magnetosphere and the other (the Mercury Planetary Orbiter) still equipped with a magnetometer is particularly suited to answer that question (Wicht et al., this issue). However, as we discussed earlier, a magnetic field does not necessarily imply a convecting mantle, though a convecting core is likely if the magnetic field originates there, due to the possibility that the inner core growth may power a dynamo while the mantle is conductively cooling. However, convection in the mantle can be indirectly detected by looking for the gravity signal of core mantle boundary undulations and the associated density variations due to mantle convection (see the following).

The magnetometer on board of the MMO will further be able to detect a possible remnant magnetized crust—if one exists at all. Together with a determination of the relative ages of the crust with crater counting, the magnetic field history of Mercury can be obtained. This in turn will help to constrain the sulfur content in the core and the planetary formation scenario.

6.2 Gravity and Topography Data

Radio science and laser altimeters on the orbiters are important instruments for measuring the gravity field and topography. Mercury with its presumably large iron-rich core and thin silicate mantle provides a good candidate to reconstruct the long-wavelength structure of the core–mantle boundary (CMB) from gravity/topography inversion. It is expected that the convective mantle flow induces stresses and thus undulations at the interface (i.e., an inward deflection below downwellings and an upward below upwellings). Therefore, any information on the CMB shape could provide useful constraints on the mantle dynamics and the proposed thermoelectric dynamo (Stevenson 1987; Giampieri and Balogh 2002). A quantitative study that compared signals generated by the surface topography and CMB topography show that the CMB gravitational signal can be dominant over the surface topography signal at the longest wavelengths ($\ell = 2$–8) (Spohn et al. 2001). For smaller wavelengths as suggested by the convection models described here, the CMB signal is smaller than the crustal gravity signal. Thus, any information about small-scale undulations of the CMB requires a reduction of the crustal field from a measured gravity field. In addition to the difficulty of reducing the crustal field, the model by Spohn et al. (2001) neglects the gravity signal induced by lateral density variations in the convecting mantle, that is, the density is smaller in hot upwellings and larger in cold downwellings as compared to the average value.

Assuming that one is able to remove the crustal field, an inversion method of the reduced measured gravity field, which is caused by density variations in the mantle and by core–mantle undulations, is needed to distinguish between these two contributions. The amplitudes of deformation at the core mantle boundary depend on the strength of convection, which on the other hand depends on mantle viscosity and thermal buoyancy. The latter is caused by density variations due to lateral temperature variations. For Earth, the complete system of a mantle flow and dynamic topography at the boundaries (surface, phase transitions, CMB) has been extensively studied using the framework of internal loading theory (e.g. Hager and Clayton 1989). This method uses surface gravity and topography measurements, density distribution derived from seismic tomography data, and plate velocities to reconstruct the viscosity structure of the mantle. In the case of Mercury, like for other terrestrial planets, the lack of these data does not allow for such an inversion and the ambiguity of the mantle density structure derived from the gravity/topography inversion has to be reduced. One simple approach is to assume a radially averaged (instead of full 3D) structure of the mantle (Pauer et al. 2006).

If such a simplification is employed (Pauer et al. 2007), then properties of such an inversion could be checked through forward/inverse modeling of Mercury's mantle density structure (together with the induced CMB shape) derived from 3D spherical convection models. Such a study shows first of all the dependency of the gravity anomalies amplitudes on the dominant mantle convection pattern wavelength (since the high-degree signal is strongly attenuated as shown in Fig. 9) and on the thickness of an active mantle convection zone (since that is the gravity signal inducing zone). If we are not able to distinguish in the measurements the dynamic surface topography, another unconstrained parameter is then the mantle viscosity profile, which must be chosen a priori. That preliminary study (Pauer et al. 2007) demonstrates that there are two main free parameters, the viscosity profile and the thickness (or depth) of the convecting mantle. These parameters can be constrained together with models of the interior structure and dynamics. In any case future measurements of the gravity field up to degree 20 or even higher will allow us to constrain better the CMB structure.

The forthcoming Mercury missions will provide the J2 gravity term, the obliquity of the rotation axis and the amplitude of libration, data necessary to determine the core radius as outlined by Peale (1988). Though this determination will be non-unique it will provide an additional constraint on the interior structure, the ratio Cm/C. Furthermore, significant progress has recently been made on this front by Margot et al. (2007) who measured the obliquity and amplitude of physical libration using ground-based radar and determined that Mercury's outer core is indeed molten. Future measurements of the degree-two gravitational harmonics by MESSENGER and BepiColombo will significantly improve these results.

Interior Evolution of Mercury

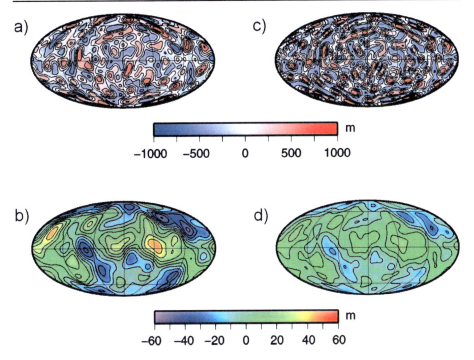

Fig. 9 (**a**) CMB undulations generated by synthetic mantle density distribution dominated by $\ell = 8$ structures (max. anomalies ± 1, 2%) in the model with 300 km thick lithosphere, mantle density 3,350 kg m^{-3}, core radius 1,890 km, core density 8,000 kg m^{-3} and constant mantle viscosity. (**b**) Corresponding surface geoid anomalies. (**c**) Same as for (**a**) but with density distribution dominated by $\ell = 13$ and (**d**) corresponding surface geoid anomalies

BepiColombo will attempt to measure the solar tides of Mercury. With an amplitude of about 1 m the tidal deformation is indeed within the reach of a laser altimeter. This is also true for the gravity signal of the tides. These data will allow a calculation of the tidal Love numbers k and h. The Love numbers are useful as constraints on the interior structure. Their imaginary parts will constrain the rheology of the mantle and possible tidal heating in the mantle. However, it is still doubtful that the time variation of the tidal signal can be accurately resolved. The tides of Mercury also offer the possibility to estimate the sulfur concentration of the core with the precision of a few percent (Van Hoolst et al. 2007, this issue).

The measured gravity field and the topography can also be used to estimate the crustal thickness and the elastic lithosphere (Belleguic et al. 2005). The elastic lithopshere is the upper layer of a terrestrial planet that supports stresses over geologically long periods ($\sim 10^8$ years). Since the base of the elastic (mechanical) lithosphere can be defined by an isotherm which depends on the rheology (composition) and the strain rate (McNutt 1984), its thickness allows estimates of the surface heat flow. It is important to note that the elastic lithosphere obtained by gravity and topography data, does not represent the current state of the thermal lithosphere. It rather shows the thermal state at the time of formation of the considered geological structure. Thus, the elastic lithosphere thickness provides constraints on models of thermal evolution.

6.3 Camera and Spectrometer Data

Cameras together with laser altimeters on both missions will measure tectonic deformations of the crust such as the lobate scarps with an accuracy of a few meters or better on a global scale. As we showed earlier, these measurements are necessary to better constrain the contractional tectonic, and hence the cooling history of the planet. Spectrometers on the Mercury missions will measure the composition and the mineralogy of the crustal rock. These data will be decisive in elucidating the formation scenario and the very early history of the planet. They will also constrain crust growth models through the dependence of the latter on crust physical parameters.

It should be stressed that 55% of the planet is unexplored and that the known 45% have been imaged only at a poor resolution, comparable with lunar images before Apollo. It is entirely possible that hidden in the surface is evidence of the volcanic history of Mercury that will be very important in the light of the theoretical evidence for partial melting in the mantle.

6.4 In Situ Measurements

While MESSENGER and BepiColombo will provide important constraints for models of Mercury's interior it must also be said that many important questions will not likely be solved; these questions will require landers or even networks of landers. The BepiColombo mission for some time planned to include a small lander which was later dropped for technical and financial reasons. Mission studies performed by the European Space Agency have shown that landing on Mercury is extremely difficult and expensive. Nevertheless, the detailed interior structure including the possible asthenosphere can best be studied with the help of seismometers. And the interior energy balance can best be assessed with planetary heat flow measurements.

Appendix

In the 2D axial and 3D spherical convection models presented here the viscosity is strongly temperature dependent, and the viscosity η varies with depth as a function of the horizontally averaged temperature and the lithostatic pressure following an Arrhenius law:

$$\eta = \eta_0 \exp\left(\frac{E_{act} + V_{act} g \rho_m (r_p - r)}{R \langle T(r) \rangle}\right), \quad (6)$$

where R is the gas constant, E_{act} the activation energy and V_{act} the activation volume which describes the pressure dependence of the viscosity. η_0 normalizes the profile so that a reference value η_{ref} is reached at mid-depth at a reference temperature of 1,573 K. This viscosity law provides the ability to model the evolution of a stagnant lid, the region of the mantle which does not flow. In that upper layer, heat is transported only by conduction.

A free-slip mechanical boundary condition is used at the CMB and a no-slip condition is implemented at the surface that assures the stagnant lid regime of a one-plate planet. For the temperature boundary conditions, we use a fixed temperature at the surface while the temperature at the CMB decreases according to how much heat is transported out of the core by the mantle. To model the core–mantle temperature, we also solve the energy equation for the core. The cooling of the core is equal to the heat flux out of core (transported by the mantle) and the energy released if an inner core is growing (mass m), that is, latent heat L and the gravitational energy E_G.

$$V_c \rho_c C_c \frac{dT_{CMB}}{dt} = A_c k \cdot \frac{dT}{dr}\bigg|_{CMB} - (L + E_G)\frac{dm}{dt}, \quad (7)$$

where V_c, A_c, ρ_c and C_c are the volume, the surface, the density and the heat capacity of the core and T_{CMB} is the temperature at the CMB.

The latent heat release per mass is $L = 250$ kJ kg^{-1} (Schubert et al. 1988). Gravitational energy contributes to the energy balance because the core contains some light elements like sulfur (see Sect. 2.3). On the iron-rich side of the eutectic composition a pure iron core freezes out and the light component gets enriched and buoyant in the liquid part. We employ the following equation for the gravitational energy given by (Schubert et al. 1988):

$$E_G = \frac{2\pi G r_{CMB}^2 \chi_0 \Delta \rho}{(1-\xi^3)^2} \left(\frac{\rho_{ic}}{\rho_S}\right) \left[\frac{1}{5}(1-\xi^5) - \frac{\xi^2}{3}(1-\xi^3)\right], \quad (8)$$

where G is the gravitational constant, $\Delta\rho$ the difference in density between the light component (ρ_S) and the iron (ρ_{ic}). $\xi = r_{ic}/r_{CMB}$ denotes the aspect ratio between the inner core radius (r_{ic}) and the core radius (r_{CMB}) and χ_0 is the initial value of the light element mass concentration in the outer core.

The temperature in the core is assumed to follow an adiabat, so that the radius of the inner core is then determined as the intersection point of the core adiabat and the melting curve for the iron–sulfur alloy. The adiabatic temperature in the core is parameterized as a function of the pressure p (Stevenson et al. 1983):

$$T_{ad}(p) = T_{CMB} \frac{1 + 8 \cdot 10^{-12} p - 3.9 \cdot 10^{-23} p^2}{1 + 8 \cdot 10^{-12} p_{CMB} - 3.9 \cdot 10^{-23} p_{CMB}^2}, \quad (9)$$

where p_{CMB} is the pressure at the CMB.

Only the melting temperature of pure iron (Boehler 1996) and the eutectic composition (Boehler 1996; Fei et al. 1997, 2000) has been measured, thus, a parameterization for the melting depression with increasing sulfur at constant pressure is needed. The melting depression of iron due to the light element concentration χ is described in a parameterized melting curve as follows:

$$T_m(p) = T_{m0}(1 - a_S \chi)(1 + T_{m1} p + T_{m2} p^2), \quad (10)$$

where a_S, T_{m0}, T_{m1}, T_{m2} are constant fitting parameters. With a growing inner core the sulfur content in the liquid shell gets enriched by $\chi = \chi_0 r_{CMB}^3/(r_{CMB}^3 - r_{ic}^3)$. Note that the melting temperatures for the eutectic composition differ between the work of Boehler (1996) and of Fei et al. (1997, 2000) and therefore different parameter values of a_S, T_{m0}, T_{m1}, T_{m2} have to be chosen (Table 3).

To convert the pressure at the intersection point of melting curve and core adiabat into the radius of the inner core a pressure model is needed. Using that the gravity in the core follows rg_{CMB}/r_{CMB} the inner core radius can be derived by the following equation:

$$r_{ic} = \sqrt{2[p(r=0) - p(r=r_{ic})]r_{CMB}/\rho_c g_{CMB}}. \quad (11)$$

In this approach which is used in the here presented models the pressure at the CMB is 6.7 GPa and 42.1 GPa in the center of the planet. The gravity at the CMB g_{CMB} is 4.45 m/s^2.

The different pressure models change the adiabatic temperature structure in the core (using the described parameterization) which gives another source for uncertainty of the size of the inner core. With the lower pressure at the CMB the adiabat gives higher temperatures in the interior, so that the onset of inner core growth is delayed and smaller cores are expected.

The change of the mass of the inner core can be derived from the change of the inner core radius with time:

$$\frac{dm}{dt} = 4\pi r_{ic}^2 \rho_c \frac{dr_{ic}}{dt} = 4\pi r_{ic}^2 \rho_c \frac{dr_{ic}}{dT_{CMB}} \frac{dT_{CMB}}{dt}. \quad (12)$$

The derivative of the inner core radius with respect to the temperature at the CMB can be calculated by equating the melting curve and the adiabat at the transition pressure to the inner core. This expression enters in the energy balance of the core which can be rearranged to a differential equation for the evolution of the temperature of the core in the thermal evolution models.

The initial temperature distribution follows a boundary layer profile with high temperatures in the interior and thin boundary layers. Random small scale disturbances are used to initialize the convection process.

References

M.H. Acuña, J.E.P. Connerney, N.F. Ness, R.P. Lin, D. Mitchell, C.W. Carlson, J. McFadden, K.A. Anderson, H. Rème, C. Mazelle, D. Vignes, P. Wasilewski, P. Cloutier, Science **284**, 790–793 (1999)
O. Aharonson, M.T. Zuber, S.C. Solomon, Earth Planet. Sci. Lett. **218**, 261–268 (2004)
J.D. Anderson, R.F. Jurgens, E.L. Lau, M.A. Slade, III, G. Schubert, Icarus **124**, 690–697 (1996)
J.D. Anderson, G. Colombo, P.B. Esposito, E.L. Lau, G.B. Trager, Icarus **71**, 337–349 (1987)
J. Baker, M. Bizzarro, N. Wittig, J. Connelly, H. Haack, Nature **436**, 1127 (2005)
Basaltic Volcanism Study Project, *Basaltic Volcanism on the Terrestrial Planets* (Pergamon, New York, 1981), 1286 pp
V. Belleguic, P. Lognonné, M.A. Wieczorek, J. Geophys. Res. **110**, E11005 (2005), doi:10.1029/2005JE002437
C.M. Bertka, J.R. Holloway, J. Geophys. Res. **98**, 19,755–19,766 (1993)
A.B. Binder, M.A. Lange, J. Geophys. Res. **85**, 3194–3208 (1980)
R. Boehler, Earth Planet. Sci. Lett. **111**, 217–227 (1992)
R. Boehler, Phys. Earth Planet. Int. **96**, 181–186 (1996)
S.I. Braginsky, Geomag. Aeron. **4**, 698–712 (1964)
D. Breuer, Thermo-chemical evolution of Mercury, EPSC 2006, Berlin, Germany, Sept. 18th–22th, Talk EPSC2006-A-00755, 2006
D. Breuer, T. Spohn, J. Geophys. Res.-Planets **108**(E7), 5072 (2003). doi:10.1029/20002JE001999
D. Breuer, T. Spohn, Planet. Space Sci. **54**, 153–169 (2006)
A.G.W. Cameron, Jr. B. Fegley, W. Benz, W.L. Slattery, in *Mercury*, ed. by F. Vilas et al. (University of Arizona Press, Tucson, 1988), pp. 692–708
C. Christensen, Nature **444**, 1056–1058 (2006)
C. Clauser, E. Huenges, *Thermal Conductivity of Rocks and Minerals, Rock Physics and Phase Relations*, A Handbook of Physical Constants. AGU Reference Shelf 3, 1995
S.M. Clifford, F.P. Fanale, Lunar Planet. Sci. **XVI** 144–145 (1985)
J.E.P. Connerney, N.F. Ness, in *Mercury* (Univ. of Arizona Press, Tucson, 1988), pp. 494–513
V. Conzelmann, Thermische Evolution des Planeten Merkur berechnet unter Anwendung verschiedener Viskositätsgesetze, Ph.D. Thesis, University Münster, 1999
A.C. Cook, M.S. Robinson, J. Geophys. Res. **105** 9429–9443 (2000)
B.M. Cordell, Tectonism and the interior of Mercury, Ph.D. thesis, University of Arizona, Tucson, 1977, 124 pp
B.M. Cordell, R.G. Strom, Phys. Earth Planet. Inter. **15** 146–155 (1977)
A. Davaille, C. Jaupart, J. Fluid Mech. **253** 141–166 (1993)
B. Fegley Jr., A.G.W. Cameron, Earth Planet. Sci. Lett. **82** 207–222 (1987)
Y. Fei, C.M. Bertka, L.W. Finger, Science **275** 1621–1623 (1997)

Y. Fei, J. Li, C.M. Bertka, C.T. Prewitt, Am. Mineral. **85** 1830–1833 (2000)
P.E. Fricker, R.T. Reynolds, A.L. Summers, P.M. Cassen, Nature **259** 293–294 (1976)
G. Giampieri, A. Balogh, Planet. Space Sci. **50** 757–762 (2002)
R. Grard, A. Balogh, Planet. Space Sci. **49** 1395–1407 (2001)
O. Grasset, E.M. Parmentier, J. Geophys. Res. **103** 18,171–18,181 (1998)
L. Grossman, Geochim. Cosmochim. Acta **36**, 597–619 (1972)
B.H. Hager, R.W. Clayton, in *Mantle Convection: Plate Tectonics and Global Dynamics*, ed. by W.R. Peltier (Gordon and Breach, New York, 1989), pp. 675–763
H. Harder, G. Schubert, Icarus **151**, 118–122 (2001)
J.K. Harmon, Adv. Space Res., **19**, 1487–1496 (1997)
S.A. Hauck II, A.J. Dombard, R.J. Phillips, S.C. Solomon, Earth Planet. Sci. Lett. **222**, 713–728 (2004)
S.A. Hauck III, R.J. Phillips, J. Geophys. Res. **107**, 5052 (2002). doi:5010.1029/2001JE001801
M.H. Heimpel, J.M. Aurnou, F.M. Al-Shamali, N. Gomez Perez, Earth Planet. Sci. Lett. **236**, 542–557 (2005)
C.T. Herzberg, P. Raterron, J. Zhang, Geophys. Geochem. Geosyst. **1** (2000). doi:10.129/2000GC000089
M.M. Hirschmann, Geophys. Geochem. Geosyst. **1** (2000). doi:10.129/2000GC000070
A.M. Hofmeister, Science **283**(5408), 1699 (1999)
R. Jeanloz, D.L. Mitchell, A.L. Sprague, I. de Pater, Science **268**, 1455–1457 (1995)
S. Karato, D.C. Rubie, J. Geophys. Res. **102**, 20111–20122 (1997)
S.-I. Karato, P. Wu, Science **260**, 771–778 (1993)
H.H. Kieffer, Science **194**, 1344–1346 (1976)
T. Kleine, C. Münker, K. Mezger, H. Palme, Nature **418**, 952–955 (2002)
T. Kleine, K. Mezger, H. Palme, E. Scherer, C. Munker, AGU, Fall Meeting 2004, Abstract P31C-04, 2004
S. Labrosse, Phys. Earth Planet. Interiors **140**, 127–143 (2003)
T. Lee, D.A. Papanastassiou, G.J. Wasserburg, Geophys. Res. Lett. **3**, 109–112 (1976)
J.S. Lewis, Science **186**, 440–443 (1972)
J.S. Lewis, in *Mercury*, ed. by F. Vilas et al. (University of Arizona Press, Tucson, 1988), pp. 651–666
K. Lodders, B. Fegley Jr., *The Planetary Scientist's Companion* (Oxford University Press, New York, 1998), 371 pp
J.L. Margot, S.J. Peale, R.F. Jurgens, M.A. Slade, I.V. Holin, Science **316**, 710–714 (2007)
C.A. McCammon, A.E. Ringwood, I. Jackson, Geophys. J. Roy. Astron. Soc. **72**, 577–595 (1983)
M.K. McNutt, J. Geophys. Res. **89**, 11180–11194 (1984)
L.-N. Moresi, V.S. Solomatov, Phys. Fluids **7**, 2154–2162 (1995)
N.F. Ness, K.W. Behannon, R.P. Lepping, Y.C. Whang, K.H. Schatten, Science **185**, 151–160 (1974)
N.F. Ness, K.W. Behannon, R.P. Lepping, Y.C. Whang, Icarus **28**, 479–488 (1976)
F. Nimmo, D. Stevenson, J. Geophys. Res. **105**, 11969–11979 (2000)
F. Nimmo, T.R. Watters, Geophys. Res. Lett. **31**, L02701 (2004)
M. Pauer, O. Fleming, K. Čadek, J. Geophys. Res. **111**(E11), E1100 (2006). doi:10.1029/2005JE002511
M. Pauer, D. Breuer, T. Spohn, Subsurface structure of Mercury—Expected results from gravity/topography analyses (2007, submitted)
S.J. Peale, in *Mercury*, ed. by F. Vilas et al. (University of Arizona Press, Tucson, 1988), pp. 494–513
C.C. Reese, V.S. Solomatov, L.N. Moresi, J. Geophys. Res. **103**, 13643–13658 (1998)
C.C. Reese, V.S. Solomatov, L.-N. Moresi, Icarus **139**, 67–80 (1999)
F.M. Richter, H.C. Nataf, S.F. Daly, J. Fluid Mech. **129**, 183 (1983)
A.E. Ringwood, Geochem. J. **11**, 111–135 (1977)
M.S. Robinson, M.E. Davies, T.R. Colvin, K.E. Edwards, J. Geophys. Res. **104**, 30 (1999)
M.S. Robinson, G.J. Taylor, Meteorit. Planet. Sci. **36**, 841–847 (2001)
S.K. Runcorn, Nature **253**, 701–703 (1975)
C.T. Russel, D.N. Baker, J.A. Slavin, in *Mercury*, ed. by F. Vilas, C.R. Chapman, M.S. Matthews (Univ. Press of Arizona, Tucson, 1988), pp. 514–561
G. Schubert, M.N. Ross, D.J. Stevenson, T. Spohn, in *Mercury*, ed. by F. Vilas et al. (Univ. Press of Arizona, Tucson, 1988), pp. 429–460
G. Schubert, D. Bercovici, G.A. Glatzmeier, J. Geophys. Res. **95**, 14105–14129 (1990)
G. Schubert, S.C. Solomon, D.L. Turcotte, M.J. Drake, N.H. Sleep, in *Mars*, ed. by H.H. Kieffer, B.M. Jakobsky, C.W. Snyder, M.S. Matthews (University of Arizona Press, Tucson, 1992), pp. 147–183
G. Schubert, D.L. Turcotte, P. Olson, *Mantle Convection in the Earth and Planets* (Cambridge University Press, Cambridge, 2001), 956 pp
S. Schumacher, D. Breuer, J. Geophys. Res. **111**, E02006 (2006). doi:10.1029/2005JE002429
B.E. Schwab, A.D. Johnston, J. Petrol. **42**, 1789–1811 (2001)
U. Seipold, Phys. Earth Planet. Int. **69**(3–4), 299-303 (1992)
H.N. Sharpe, D.W. Strangway, Geophys. Res. Lett. **3**, 285–288 (1976)
R.W. Siegfried, S.C. Solomon, Icarus **23**, 192–205 (1974)

D.E. Smith, M.T. Zuber, G.A. Neumann, F.G. Lemoine, J. Geophys. Res. **102**, 1591–1611 (1997)
V.S. Solomatov, Phys. Fluids **7**, 266–274 (1995)
V.S. Solomatov, L.-N. Moresi, J. Geophys. Res. **105**, 21795–21817 (2000)
V.S. Solomatov, C.C. Reese, Mantle convection and thermal evolution of Mercury reviseted, in *LPI Conference Mercury: Space Environment, Surface, and Interior*, Chicago, 2001
S.C. Solomon, Icarus **28**, 509–521 (1976)
S.C. Solomon, Phys. Earth Planet. Inter. **15**, 135–145 (1977)
S.C. Solomon, Earth Planet. Sci. Lett. **19**, 168–182 (1979)
S.C. Solomon, R.L. McNutt Jr., R.E. Gold, M.H. Acuña, D.N. Baker, W.V. Boynton, C.R. Chapman, A.F. Cheng, G. Gloeckler, J.W. Head III, S.M. Krimigis, W.E. McClintock, S.L. Murchie, S.J. Peale, R.J. Phillips, M.S. Robinson, J.A. Slavin, D.E. Smith, R.G. Strom, J.I. Trombka, M.T. Zuber, Planet. Space Sci. **49**, 1445–1465 (2001)
C.P. Sonett, D.S. Colburn, K. Schwartz, Icarus **24**, 231–255 (1975)
T. Spohn, F. Sohl, K. Wieczerkowski, V. Conzelmann, Planet. Space Sci. **49**, 1561–1570 (2001)
T. Spohn, Icarus **90**, 222–236 (1991)
A.L. Sprague, R.W.H. Kozlowski, F.C. Witteborn, D.P. Cruikshank, D.H. Wooden, Icarus **109**, 156–167 (1994)
A.L. Sprague, D.B. Nash, F.C. Witteborn, D.P. Cruikshank, Adv. Space Res. **19**, 1507–1510 (1997)
P.D. Spudis, J.E. Guest, in *Mercury*, ed. by F. Vilas et al. (University of Arizona Press, Tucson, 1988), pp. 118–164
S. Stanley, J. Bloxham, W.E. Hutchinson, M.T. Zuber, Earth Planet. Sci. Lett. **234**, 27–38 (2005)
D.J. Stevenson, Earth Planet. Sci. Lett. **82**, 114–120 (1987)
D.J. Stevenson, in *Origin of the Earth*, ed. by H.E. Newsom, J.H. Jones (Oxford University Press, New York, 1990), pp. 231–249
D.J. Stevenson, T. Spohn, G. Schubert, Icarus **54**, 466–489 (1983)
R.G. Strom, Adv. Space Res. **19**, 1471–1485 (1997)
R.G. Strom, N.J. Trask, J.E. Guest, J. Geophys. Res. **80**, 2478–2507 (1975)
G.J. Taylor, E.R.D. Scott, in *Treatise on Geochemistry, vol. 1, Meteorites, Comets and Planets*, ed. by M.A. Davis (Elsevier, Amsterdam, 2005), pp. 477–485
M.N. Toksöz, A.T. Hsui, D.H. Johnston, Thermal evolution of the Moon and the terrestrial planets, in *The Soviet–American Conference on Cosmochemistry of the Moon and Planets*, NASA SP-370, 1978, pp. 245–328
D.C. Tozer, Phil. Trans. Roy. Soc. **258**, 252–271 (1965)
T.M. Usselman, Am. J. Sci. **275**, 278–290 (1975)
T. Van Hoolst, F. Sohl, I. Holin, O. Verhoeven, V. Dehant, T. Spohn (2007), this issue
F. Vilas, in *Mercury*, ed. by F. Vilas, C.R. Chapman, M.S. Matthews (University of Arizona Press, Tucson, 1988), pp. 622–650
T.R. Watters, M.S. Robinson, A.C. Cook, Geology **26**, 991–994 (1998)
T.R. Watters, R.A. Schultz, M.S. Robinson, A.C. Cook, Geophys. Res. Lett. **29**(11), 1542 (2002). doi:10.1029/2001GL014308
T.R. Watters, M.S. Robinson, C.R. Bina, P.D. Spudis, Geophys. Res. Lett. **31**, 04701 (2004)
T.R. Watters, F. Nimmo, M.S. Robinson, Geology **33**(8), 669–672 (2005). doi:10.1130/G21678.1
J. Weertman, J.R. Weertman, Annu. Rev. Earth Planet. Sci. **3**, 293–315 (1975)
S.J. Weidenschilling, Icarus **35**, 99–111 (1978)
G.W. Wetherill, Science **228**, 877–879 (1985)
G.W. Wetherill, in *Mercury*, ed. by F. Vilas et al. (University of Arizona Press, Tucson, 1988), pp. 670–691
J. Wicht, M. Mandea, F. Takahashi, U.R. Christensen, M. Matsushima, B. Langlais (2007), this issue
A. Zebib, G. Schubert, J.L. Dein, R.C. Paliwal, Geophys. Astrophys. Fluid Dyn. **23**, 1–42 (1983)
J. Zhang, C. Herzberg, J. Geophys. Res. **99**, 17,729–17,742 (1994)

The Origin of Mercury's Internal Magnetic Field

J. Wicht · M. Mandea · F. Takahashi · U.R. Christensen ·
M. Matsushima · B. Langlais

Originally published in the journal Space Science Reviews, Volume 132, Nos 2–4.
DOI: 10.1007/s11214-007-9280-5 © Springer Science+Business Media B.V. 2007

Abstract Mariner 10 measurements proved the existence of a large-scale internal magnetic field on Mercury. The observed field amplitude, however, is too weak to be compatible with typical convective planetary dynamos. The Lorentz force based on an extrapolation of Mariner 10 data to the dynamo region is 10^{-4} times smaller than the Coriolis force. This is at odds with the idea that planetary dynamos are thought to work in the so-called magnetostrophic regime, where Coriolis force and Lorentz force should be of comparable magnitude. Recent convective dynamo simulations reviewed here seem to resolve this caveat. We show that the available convective power indeed suffices to drive a magnetostrophic dynamo even when the heat flow though Mercury's core–mantle boundary is subadiabatic, as suggested by thermal evolution models. Two possible causes are analyzed that could explain why the observations do not reflect a stronger internal field. First, toroidal magnetic fields can be strong but are confined to the conductive core, and second, the observations do not resolve potentially strong small-scale contributions. We review different dynamo simulations that promote either or both effects by (1) strongly driving convection, (2) assuming a particularly small inner core, or (3) assuming a very large inner core. These models still fall somewhat short of explaining the low amplitude of Mariner 10 observations, but the incorporation of an additional effect helps to reach this goal: The subadiabatic heat flow through Mercury's core–mantle boundary may cause the outer part of the core to be stably stratified, which would largely exclude convective motions in this region. The magnetic field, which is

J. Wicht (✉) · U.R. Christensen
Max-Planck Institute for Solar-System Research, 37191 Kaltenburg-Lindau, Germany
e-mail: wicht@mps.mpg.de

M. Mandea
GeoForschungsZentrum Potsdam, Telegrafenberg, 14473 Potsdam, Germany

F. Takahashi · M. Matsushima
Department of Earth and Planetary Sciences, Tokyo Institute of Technology, 2-12-1 Ookayama,
Meguro-ku, Tokyo 152-8551, Japan

B. Langlais
UMR CNRS 6112, Laboratoire de Planétologie et Géodynamique, Nantes Atlantique Universités, 2 Rue de la Houssinière, 44000 Nantes, France

small scale, strong, and very time dependent in the lower convective part of the core, must diffuse through the stagnant layer. Here, the electromagnetic skin effect filters out the more rapidly varying high-order contributions and mainly leaves behind the weaker and slower varying dipole and quadrupole components (Christensen in Nature 444:1056–1058, 2006). Messenger and BepiColombo data will allow us to discriminate between the various models in terms of the magnetic fields spatial structure, its degree of axisymmetry, and its secular variation.

Keywords Mercury: magnetic field · Mercury: dynamo

1 Introduction

Our knowledge of Mercury's internal magnetic field is scarce. Data from two close flybys by Mariner 10 in 1974 and 1975, half of the first flyby being rendered useless due to strong magnetospheric perturbations, is all we have. These data give only limited information about the internal field structure (Connerney and Ness 1988), but the existence of an internal field was a surprise in itself.

Prior to the Mariner 10 encounters, it was thought that Mercury had no active dynamo. If the planet ever had a dynamo it would have ceased to work when the rapid cooling of the comparatively small planet led to a completely frozen core. This idea was disproved by the Mariner 10 measurements, suggesting that at least part of the metal core must still be liquid for a convective dynamo to operate. Recent ground-based observations of Mercury's forced libration are also in favor of a partly liquid core, since their large amplitude requires a mechanical decoupling of core and mantle (Margot et al. 2007). The low observed field strength, however, seems incommensurate with typical convective dynamos. The Earth's magnetic field, for example, is two orders of magnitude stronger.

This motivated scientists to look for alternative models. Stevenson (1987) and later Giampieri and Balogh (2002) suggested a thermoelectric mechanism: Temperature differences at the core–mantle boundary caused by topographic undulations would give rise to a thermoelectric electro-motive force that drives electric currents in core and mantle. These currents are the source for a toroidal magnetic field that in turn is converted into the observed poloidal field by convective flows too sluggish to support a classical self-sustained dynamo. The parameterized models by Giampieri and Balogh (2002) show that the resulting field strength is compatible with the observations. The field structure is determined by the undulation of the core–mantle boundary and the structure of the convective flow. A significant correlation between gravity and magnetic field pattern detected by future Mercury missions would be in favor of the thermoelectric dynamo model (Giampieri and Balogh 2002), assuming that the gravity signal reflects the core–mantle boundary shape.

Remanent crustal magnetization has also been suggested as a possible source for the global magnetic field observed by Mariner 10. Since Mars Global Surveyor discovered a surprisingly strong magnetization of the Martian crust (Acuña et al. 1999), the Hermian field amplitude seems within reach of remanent magnetization models (Aharonson et al. 2004). However, the magnetization of the Martian and Earth's crust is rather complex and small scale (see Sects. 2.1 and 2.2). The pattern found on Earth is a result of long-term tectonic and magnetic processes conserving the varying direction of a reversing magnetic field. Variations in the magnetic susceptibility of lithospheric rocks are responsible for additional smaller modulations. The reason for the complexity of Martian magnetization is less clear, but similar mechanisms are typically evoked.

A simple magnetization pattern compliant with Mariner 10 observations is possible only if Earth-like tectonic processes are absent and the dynamo is not reversing during the acquisition of magnetization. Mercury's surface is predominantly shaped by impact processes, so that the first condition is fulfilled. The latter condition also seems little restrictive, given the fact that many published dynamo simulations never reverse. However, we will see in the following that Mercury's dynamo may operate in a regime where magnetic field reversals are expected. A theorem by Runcorn (1975) states that a uniform spherical shell magnetized by an internal source that is subsequently removed has no external field. Having ruled out tectonics, one therefore has to invoke a different mechanism that breaks the uniformity on a larger scale. Aharonson et al. (2004), for example, suggested that the strong latitudinal and longitudinal variations in Mercury's mean surface temperature lead to a varying thickness of the magnetized crustal layer because the Curie temperature is reached at different depths. A key feature that will allow us to identify crustal magnetization—once MESSENGER and BepiColombo provide new data— is the correlation of magnetic field features with topographic entities like craters (Aharonson et al. 2004).

The weakness of Mercury's internal magnetic field has the consequence that external and internal contributions are harder to separate than in the geomagnetic field, where they differ by orders of magnitude. The external field can reach amplitudes comparable to the dynamo field strength above the planet's surface (Fujimoto et al. 2007). So far, external field models were simply scaled down versions of Earth's magnetospheric dynamics. This, however, may fall short in the Mercury case. Field contributions due to electric currents induced in the core by external field variations and due to convective alteration of the external field also have to be considered (Glassmeier et al. 2007). Exploring the tight link of internal and external sources may require us to develop a dynamo model that incorporates the interaction with the external field.

After giving an overview of different magnetic fields of terrestrial planets in Sect. 2, we concentrate on convective dynamo simulations for the remainder of the article. Numerical simulations geared to model the geodynamo successfully reproduce the geomagnetic field strength and are thus not directly suited to explain the weaker Hermian field (Christensen and Aubert 2006). Typical estimates for the field strength of convectively driven planetary dynamos likewise predict much stronger fields than observed by Mariner 10 (Stevenson 1987). We review this issue in Sect. 4 where we also discuss the power available to drive Mercury's dynamo. Although some researchers challenged the applicability of a convective dynamo model (Stevenson 1987; Schubert et al. 1988), more recent numerical simulations have resolved the caveat by exploring the influence of possible differences in Earth's and Mercury's interior on the dynamo process. Smaller as well as larger inner cores are assumed, and both can help to reduce the global field strength (at satellite altitude) compared to an Earth-like relative inner-core size. We present both types of models in Sect. 5 after having introduced the numerical dynamo problem in Sect. 3. Simulations of Mercury's thermal evolution suggest that the inner core still grows while the core–mantle boundary temperature gradient is subadiabatic. The convection, which is chemically driven in the deeper regions of the fluid shell, could therefore be stably stratified due to thermal effects in the outer part of the core. We present dynamo models that assume this mixed type of convection in Sect. 6. A discussion in Sect. 7 concludes this article and assesses how future Messenger and BepiColombo measurements could distinguish between the different models.

2 Earth-Like Planetary Magnetic Fields

Nowadays, measurements of planetary magnetic fields are mainly devoted to understanding the planet's interior structure and dynamics as well as the near-planetary environment. The existence of an interior field is a strong indicator that at least part of the iron core must still be liquid and convects vigorously enough to drive a dynamo. Variations in the internal magnetic field bear valuable, and often the only available, information on the core dynamics.

The planetary dynamos in our solar system seem to be quite diverse. Jupiter's and Earth's magnetic fields are both dominated by the axial dipole contribution but differ significantly in strength. The dipole plays no special role in the fields of Uranus and Neptune and is also significantly tilted. Saturn's magnetic field, on the other hand, seems to be extremely axisymmetric and somewhat weak. Venus and Mars have no active dynamos at present, and Mercury's field seems to be large scale but is exceptionally weak. Explaining these differences is an interesting challenge for planetary research in general and dynamo modeling in particular.

To get an understanding of the variability of terrestrial dynamos, we outline the key features of the magnetic fields of Earth, Mars, Venus, and Mercury, and also add Earth's Moon for good measure. Planetary fields may contain contributions from different external and internal sources. The interaction between the internal magnetic field and the solar wind results in a current system that is responsible for external magnetic fields (Fujimoto et al. 2007). Magnetized crust is an additional potential magnetic field source. Separating and understanding the different contributions is a formidable task that requires as many data in space and time as possible.

Naturally, the data coverage is by far the best for our home planet. More than 200 years of continuous observations of the vector magnetic field and measurements from several spacecrafts providing global geographical coverage have established a good, but still incomplete, knowledge of the field morphology and its time evolution. The case is different for the other planets, where spacecraft are the only means of measuring the magnetic field. Sometimes, as in the case of Mars Global Surveyor, we are lucky and the spacecraft orbits a planet for several years. In other cases our knowledge is based on only one flyby.

2.1 The Earth

The Earth's magnetic field is the sum of several internal and external contributions: Crustal and dynamo fields are of internal origin, ionospheric and magnetospheric fields form the external part. Various data sets from different sources are available to describe and separate these contributions and range from historic magnetic measurements to recent high-quality data provided by magnetic observatories (worldwide but unevenly distributed) and the magnetic satellites MAGSAT, SAC-C, Ørsted, and CHAMP.

Figure 1 shows the total intensity of the dynamo field at Earth's surface, which is the by far dominant field contribution. It reaches magnitudes of approximately 60,000 nT (nano tesla) near the poles, and decreases to values less than 25,000 nT in the South-Atlantic Anomaly region. Time variations of the core field, the secular variation, mainly occur on decades and longer time scales, but geomagnetic data have also revealed more abrupt changes in the trend of the secular variation, that last about one year and have been called geomagnetic jerks or secular variation impulses (Mandea et al. 2000).

The lithospheric field, shown in Fig. 2, is considered to be constant in time in comparison and originates from the crust and upper mantle. Crustal magnetization, both remanent and induced, may reach amplitudes of some 1,000 nT and cover large areas. The Kursk anomaly

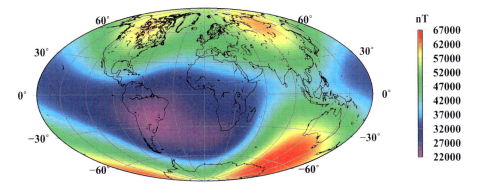

Fig. 1 The geomagnetic field intensity at the Earth's surface for epoch 2005.0 as given by the CHAOS model (Olsen et al. 2006)

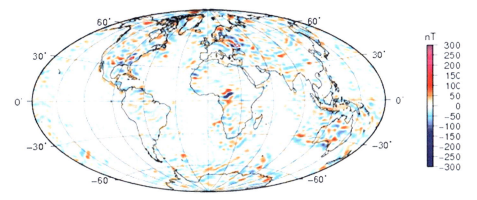

Fig. 2 Total intensity field based on the lithospheric field model MFX4 (Lesur and Maus 2006)

in the Ukraine is a prominent example. Banded Iron Formations and other rock types peculiar to the Proterozoic era are common in surface exposures of these regions. In the ocean area, striped anomalies are associated with sea-floor spreading; this pattern is characterized by a nearly constant magnetization direction over many hundreds of km and can be correlated with the dipole reversal sequence over millions of years. More details on magnetic field contributions of internal origin can be found in Mandea and Purucker (2005).

At distances of more than two Earth radii, the geomagnetic field can be approximated by a centered dipole that is inclined by 11° with respect to the planet's rotation axis. Closer to the planet, at altitudes between about 50 and 600 km, lies the ionosphere. Ultraviolet light from the Sun ionizes atoms in the upper atmosphere, making the day side of the ionosphere significantly more electrically conducting than the night side. Strong electric currents circulating in the sunlit hemisphere generate magnetic fields that contribute with intensities of up to 80 nT to the surface field. The respective daily cycle gives rise to the so called solar-quiet variations. Solar storms and other more energetic nonquiet events linked to the activity of the Sun may create magnetic fields up to a few thousand nT.

The Earth's magnetosphere, which extends for about 10 Earth radii in the direction towards the Sun and much further on the night side, hosts a complex system of electric currents (Fujimoto et al. 2007). It is dominated by the equatorial ring current (Glassmeier et al.

2007) whose contribution closer to the planet can be described by a uniform field that continuously varies with periods of six months and one year. These variations induce electric currents in Earth's interior which in turn produce the induced magnetic field that depends on the electrical conductivity structure of the planetary body. Secondary induced fields are, for example, caused by the motion of electrically conducting seawater through the Earth's main magnetic field (Tyler et al. 1997). Though the respective fields amount to only a few nT in magnitude, the effects of the regular lunar semidiurnal tide could nevertheless be discerned in geomagnetic satellite observations (Tyler et al. 2003). Large-scale ocean currents, such as the Antarctic Circumpolar Current, are also expected to produce signatures of up to a few nT at satellite altitude.

2.2 Mars

Though Mars has no active dynamo, a highly irregular field of lithospheric origin was detected. The most likely scenario is that the crust acquired an intense remanent magnetization while cooling in the presence of a still-active strong dynamo early in the planet's history (Arkani-Hamed 2007). Recent magnetic field maps (Langlais et al. 2004; Connerney et al. 2005) are based on compiled measurement by the Mars Global Surveyor (see Fig. 3). Amplitudes of up to 250 nT are reached at 400 km altitude. External contributions are weaker and result from the interaction between the solar wind and the thin Martian atmosphere. Most of these interactions take place on the day side and cause a highly variable dynamic magnetosphere both in intensity and in altitude.

Mars' magnetic field is about one order of magnitude more intense than the lithospheric component of the terrestrial field. No surface measurements are currently available, but model predictions suggest that it could locally reach up to 10,000 nT or more (Langlais et al. 2004), which is equivalent to the core field at Earth's surface.

Figure 3 shows a map of the vertical component of the magnetic field. The Martian magnetization is highly inhomogeneous, the southern highlands being apparently more magnetized than the northern lowlands. The largest volcanic edifices (Tharsis, Elysium, Olympus) as well as the largest impact craters (Hellas, Argyre, Isidis) are devoid of significant magnetic field at satellite altitude, which is generally attributed to local demagnetization of an otherwise magnetized crust. Thermal demagnetization due to volcanoes

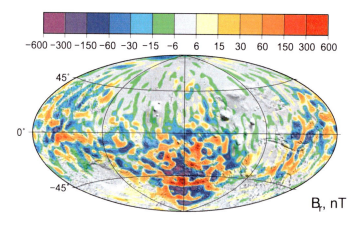

Fig. 3 Vertical component of the Martian field at 200 km altitude (Langlais et al. 2004)

and impacts as well as impact excavations are likely explanations (Artemieva et al. 2005; Carporzen et al. 2005).

The magnetization identified in a Martian meteorite dated at 4 Ga supports the idea of an early dynamo (Weiss et al. 2004) which was most likely driven by convection. Cooling of the mantle early in the planet's history and the associated increase in mantle viscosity severely lowered the vigor of mantle convection. This, in turn, reduced the heat flow though the core–mantle boundary and eventually caused the dynamo to shut off (Breuer and Spohn 2003; Buske and Christensen 2007). Thermal evolution simulations suggest that the Martian core is still completely liquid, so that a chemically driven dynamo can be ruled out (Stevenson et al. 1983; Spohn et al. 2001; Williams and Nimmo 2004; Buske and Christensen 2007; see also Sect. 4).

2.3 The Moon

Early measurements showed that the Moon has no global magnetic field but possesses a magnetic field of lithospheric origin. The lunar magnetic field has been studied indirectly via the natural remanent magnetization of the returned lunar samples, and directly with magnetometers placed on the surface or in low-altitude orbits by the Apollo 15 and 16 missions. These measurements reveal widespread lunar magnetism on scales ranging up to many tens of kilometers. Paleointensity measurements on the returned samples suggest that the lunar surface field was comparable in intensity to the present-day terrestrial surface field about 3.6 to 3.8 Gy ago. Recent measurements by Lunar Prospector refined our understanding of the lunar magnetic field (Hood et al. 2001). Relatively weak fields (up to some tens of nT at spacecraft altitude) were mostly recorded antipodal to the largest lunar basins (Imbrium, Crisium, Serenitatis) (Anderson and Wilhelms 1979). This suggests that these magnetic anomalies are caused by the ejecta of impact processes, that were deposited at antipodal locations in the presence of an impact-amplified magnetic field (Hood et al. 2001).

2.4 Venus

Despite its relative similarity to the Earth in terms of size and density, Venus is devoid of any intrinsic magnetic field. Mariner 5 approached the planet to 1.4 Venus radii in 1967 and detected only signatures of solar wind deflection. The Pioneer Venus Orbiter mission measured the magnetic field during its first years of operation between 1979 and 1981. Low-altitude observations ($\simeq$150 km) on both the night and day sides proved that the observed field can be explained by the solar wind interaction with the planet alone; any internal field is insignificant at satellite altitude (Luhmann 1986).

2.5 Mercury

Mariner 10 recorded Mercury's magnetic field during the two close flybys Mariner I and Mariner III. Figure 4 displays the measurements in a body-centered Cartesian coordinate system; the Z-direction points northward and the X-direction points towards the Sun. Encounter I had its closest approach of 1.29 R_M at a latitude of 2° south and recorded a maximum field strength of 98 nT. Encounter III had its closest approach at 1.13 R_M and a latitude of 68° north and recorded a maximum field strength of 400 nT. The increased fluxes of energetic particles detected during the second half of Mariner I were attributed to a magnetospheric substorm (Siscoe et al. 1975), which also disturbed the magnetic recordings (see Fig. 4). Mariner I probably also led the spacecraft through the tail current sheet which

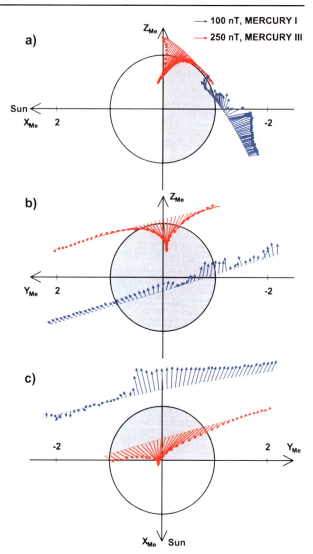

Fig. 4 The Mercury magnetic field as observed during the first (*blue*) and third (*red*) flybys of Mariner 10

complicates the extraction of internal field contributions (Fujimoto et al. 2007). A higher latitude was therefore chosen for Mariner III to reduce these problems.

The data scarcity and uncertain external field contributions render Mercury's internal magnetic field models as highly ambiguous (Connerney and Ness 1988). External field contributions can reach the same order of magnitude as the internal field. Moreover, internal magnetic fields from induction due to external field variations and/or convective modification of the external field may not be neglected (Glassmeier et al. 2007). Commonly, the planetary surface field configuration is expressed in Schmidt normalized spherical harmonic coefficients g_{lm} and h_{lm} that denote the cosine and sine contributions of degree l and order m, respectively. Most authors fitted axial dipole g_{10} and axial quadrupole g_{20} components to the data from the encounter tracks that lie inside a scaled-down Earth-like magnetosphere (Fujimoto et al. 2007).

Estimates of the axial dipole component range from $g_{10} = -260$ nT to $g_{10} = -350$ nT, those for the axial quadrupole components range from $g_{20} = -120$ nT to zero (see Connerney and Ness 1988, for an overview of Hermian field models). Connerney and Ness (1988) explored the ambiguities within a model family that represents the internal field by degree $l = 1$ harmonics plus an axial quadrupole term and assumes a degree $l = 1$ external field. They found that the model vector of least sensitivity to the specific flyby track III has the relative internal components:

$$g_{10} = 0.76, \quad g_{11} = 0.23, \quad h_{11} = 0.03, \quad g_{20} = -0.50. \tag{1}$$

This result can be interpreted such that a change in the dipole contribution of any internal field model by Δg_{10} would lead to an equivalently valid model as long as the axial quadrupole is also changed by $\Delta g_{20} = -0.50/0.76 \Delta g_{10} = -0.66 \Delta g_{10}$. When taking into account that size and shape of the magnetosphere put additional constraints on the model coefficients, this ambiguity explains the range of internal field models found in the literature. The data do also allow for a significant harmonic order $m = 1$ component, which could give rise to a dipole tilt that Ness (1979) estimated to $14 \pm 5°$. Significant higher order contributions seem unlikely at satellite altitude given the fact that the magnetic field vector changed rather smoothly during encounter III. Because of their stronger decay with radius, these contributions could nevertheless contribute significantly in Mercury's dynamo region.

3 Numerical Dynamo Models

Self-consistent numerical dynamo models solve for convection and magnetic field generation in a rotating spherical shell filled with an electrically conducting fluid. Evolution equations for flow **u**, magnetic field **B**, and the codensity variable b are formulated in a frame of reference corotating with the planetary mantle. The codensity combines dynamically relevant density differences due to thermal and compositional origin. The equations are made dimensionless by assuming scales that seem appropriate to the problem. We mostly follow the scaling chosen by Christensen et al. (2001), taking the shell thickness $D = r_o - r_i$ as length scale, the viscous diffusion time D^2/ν as time scale, $(\rho_o \mu \lambda \Omega)^{1/2}$ as the magnetic scale, and ρ_o as the codensity scale.

The physical properties of the problem are r_i and r_o, the inner and outer boundary radii respectively, mean outer-core density ρ_o, magnetic permeability μ, magnetic conductivity σ, magnetic diffusivity $\lambda = 1/(\mu \sigma)$, planetary rotation rate Ω, and kinematic viscosity ν. These properties are condensed into five dimensionless parameters that govern the problem: Ekman number $E = \nu/(D^2 \Omega)$, Prandtl number $Pr = \nu/\kappa$, magnetic Prandtl number $Pm = \nu/\lambda$, Rayleigh number $Ra = g_o \Delta b D/(\nu \Omega)$, and aspect ratio $\eta = r_i/r_o$. Here, κ stands for a diffusivity that combines thermal and compositional effects (see the following). Gravity g_o at the outer boundary is taken as the reference value, and we assume that gravity varies linearly with r. The choice of scales and dimensionless parameters is always somewhat arbitrary, and different approaches can be found in the literature.

The evolution of flow **u** are described by the Navier–Stokes equation that formulates the momentum balance:

$$E \left(\frac{\partial \mathbf{u}}{\partial t} + \mathbf{u} \cdot \nabla \mathbf{u} \right) + 2\hat{\mathbf{z}} \times \mathbf{u} + \nabla \Pi = E \nabla^2 \mathbf{u} + Ra \frac{\mathbf{r}}{r_o} b + \frac{1}{Pm} (\nabla \times \mathbf{B}) \times \mathbf{B}. \tag{2}$$

The modified pressure Π combines the nonhydrostatic pressure and centrifugal forces. Maxwell's laws and Ohm's law can be condensed into the induction equation:

$$\frac{\partial \mathbf{B}}{\partial t} - \nabla \times (\mathbf{u} \times \mathbf{B}) = \frac{1}{\mathrm{Pm}} \nabla^2 \mathbf{B}. \tag{3}$$

A simplified induction equation is solved in the conducting inner core, replacing flow $\mathbf{u}$ by the inner core rotation. Appropriate matching conditions guarantee the continuity of the magnetic field and the horizontal electric field at the inner core boundary (Wicht 2002).

Magnetic field and flow field are divergence free,

$$\nabla \cdot \mathbf{B} = 0, \quad \nabla \cdot \mathbf{u} = 0. \tag{4}$$

The latter condition is the simplified form of the continuity equation in the Boussinesq approximation, where density changes are retained only in the essential buoyancy term that drives convection in (2). This approximation is justified by the smallness of density variations in planetary iron cores.

Convection is driven by buoyancy effects resulting from two types of density variations. Freezing of the solid inner core releases light elements whose solubility is lower in solid than in liquid iron, sulfur is the most likely light constituent in the Hermian core (Breuer et al. 2007). Variations $\chi' = \chi - \chi_o$ in the mean relative weight concentration of these light elements drive the so-called chemical or compositional convection. Here, $\chi_o = m_S/(m_S + m_{Fe})$ is the homogeneous reference concentration in the outer core; m_S and m_{Fe} are the total mass of sulfur, assumed to be the light constituent, and of iron, assumed to be the heavy constituent, respectively. Temperature variations, the driving source for thermal convection, have three principal origins: latent heat from inner core freezing, secular cooling, and heat production due to radiogenic elements. The simulations are concerned only with the superadiabatic temperature contribution T' that actually drives convective motions. Codensity b combines thermal and chemical density variations in one variable:

$$b = \alpha T' + \beta \chi'. \tag{5}$$

Here, α is the thermal expansivity and β is the equivalent chemical expansivity $\beta = -\rho_o(\rho_{Fe} - \rho_S)/\rho_{Fe}\rho_S$, where ρ_{Fe} and ρ_S are the densities of iron and sulfur in the outer core respectively (see also Sect. 4).

In principle, T' and χ' obey two individual transport equations, which can be combined under the assumption that the effective (turbulent) thermal diffusivity κ_T and compositional diffusivity κ_χ are identical (Braginsky and Roberts 1995; Kutzner and Christensen 2004):

$$\frac{\partial b}{\partial t} + \mathbf{u} \cdot \nabla b = \frac{1}{\mathrm{Pr}} \nabla^2 b + \epsilon. \tag{6}$$

The source/sink term ϵ represents various effects: possible radiogenic heat production, secular cooling, the destruction of compositional differences by mixing of the core fluid, and the loss of potential thermal buoyancy by the adiabatic gradient. On time average, ϵ must balance the codensity flux through the boundaries.

Different convective driving types can be modeled by choosing appropriate combinations of ϵ and outer and inner boundary conditions for the codensity variable b (Kutzner and Christensen 2004). A simple and therefore common choice is to impose the codensity contrast Δb over the simulated shell in combination with $\epsilon = 0$. The inner boundary will then serve as a source, while the outer boundary forms the complementary sink. This approach was chosen for the thick-shell dynamos by Heimpel et al. (2005) as well as the thin-shell dynamos by Takahashi and Matsushima (2006) presented in Sect. 5.

Recent thermal evolution models suggest that the temperature gradient at the core–mantle boundary may actually be subadiabatic because the Hermian mantle developed a thick lithosphere early in the planets history, preventing the planet from cooling more efficiently. The core is unlikely to be completely solid for the same reason but has most likely cooled sufficiently to develop an inner core (Breuer et al. 2007). This leaves chemical convection as the main driving force for the dynamo. The subadiabatic thermal core mantle boundary gradient translates into a negative codensity flux within the dynamo model outlined earlier ($\partial b/\partial r > 0$), i.e. both outer and inner boundaries serve as a source of b. We explore this interesting case in more detail in Sects. 4 and 6.

Appropriate flow and magnetic field boundary conditions close the system of differential equations ((2), (3), (4), (6)) that constitute the dynamo problem. So-called rigid flow boundary conditions are commonly employed, which implies that the radial flow component vanishes and that the fluid is corotating with the respective boundary. Stanley et al. (2005), however, assumed that the viscous shear stresses vanish at the boundaries, which defines an alternative condition for the horizontal flow components. This approach is motivated by the fact that viscous effects are much larger in the simulations than in planetary cores. Some authors (Zhang and Busse 1988; Kuang and Bloxham 1997) therefore have argued that neglecting the viscous boundary layers is dynamically more realistic than overestimating their effects.

Numerical dynamo simulations have the problem that they cannot run at realistic values of the control parameters. In particular, the kinematic viscosity has to be chosen several orders of magnitude too large in order to damp smaller scale turbulence that cannot be resolved numerically. As a result, Ekman and magnetic Prandtl numbers are significantly larger in the models than would be appropriate. A simple way to parameterize small-scale turbulent mixing is to assume larger effective diffusivities that are of comparable magnitude for all diffusive effects (Braginsky and Roberts 1995). This argument is commonly cited to justify the combination of thermal and compositional effects into one codensity variable and motivates the choice of $Pr = 1$ in numerical dynamo simulations. It does, however, not suffice to justify the significantly larger Ekman numbers required to compensate the inadequate numerical resolution. Table 1 lists parameter values for Mercury along with typical values assumed by the numerical simulations presented in the following. The value of the magnetic Reynolds number $Rm = UD/\lambda$ is not known for Mercury since we cannot access the (RMS) core flow amplitude U. It is generally believed, however, that convective dynamos can only operate when Rm is large enough, say when Rm exceeds a critical value of 50.

We refer to Takahashi et al. (2003) and Christensen and Wicht (2007) for details on the numerical methods.

Table 1 Parameter values for Mercury and for the dynamo simulations presented here. Only Christensen (2006) provides a Ro_ℓ value for his dynamo model which is listed here

Parameter	Mercury	Simulations
E	10^{-12}	10^{-3}–10^{-5}
Pr (thermal)	0.1	1
Pr (compositional)	100	1
Pm	10^{-6}	$1-5$
Rm	≥ 50	100–1,000
Ro_ℓ	10	0.2

4 Estimating Planetary Magnetic Field Strengths

Estimates of planetary magnetic field strengths are either based on the power available to drive the dynamo or are derived from force balances in the Navier–Stokes equation (2). We start with outlining the latter method.

4.1 Mercury's Elsasser Number

Dynamo theory distinguishes weak-field and strong-field cases. These adjectives relate to the role of the Lorentz force in the Navier–Stokes equation (2). Strong-field dynamos are characterized by a leading order force balance between Coriolis force, pressure gradient, and the Lorentz force. The Elsasser number Λ, which measures the ratio of Lorentz to Coriolis force, is therefore of order one in this so-called magnetostrophic regime:

$$\Lambda = \frac{B^2}{\mu \lambda \rho_o \Omega} \approx 1. \tag{7}$$

With B we denote the typical (RMS) magnetic field strength in the fluid shell.

Planetary dynamos are thought to operate in the strong-field regime and two arguments are commonly cited to justify this assumption. The first argument simply assumes that the Lorentz force has to enter the leading order balance to accomplish magnetic field saturation. Suppose that Lorentz forces were so weak that their back reaction on the fluid flow could be neglected. The dynamo equation can then be regarded as an eigenvalue problem for a given velocity field **u**. The system is said to work as a kinematic dynamo when the set of eigen-solutions contains an exponentially growing mode. The amplitude of the respective eigen-vector will increase until the Lorentz force modifies the flow sufficiently to saturate the growth, which is supposedly the case when the Lorentz force enters the leading order force balance.

The second argument relies on the fact that a strong Lorentz force is needed to help convection by balancing the Coriolis force at least partly, thereby releasing the Taylor–Proudman constraint that severely restricts the flow vigor (Hollerbach 1996). The Lorentz force itself allows the flow to gain the strength necessary for the dynamo process and also increases the length scale of the flow, two effects that have been identified in dynamo simulations in Cartesian geometry (Rotvig and Jones 2002; Stellmach and Hansen 2004). However, these effects have not been found in the spherical-shell simulations that so successfully explain the geomagnetic field (Christensen and Aubert 2006), even though the Lorentz force indeed balances the Coriolis force to a good degree (Aubert 2005).

In order to access the Elsasser number Λ of a planetary core we have to learn how to estimate the field strength in the dynamo region based on magnetic field measurements. Three different Elsasser numbers are introduced to discuss this issue: The classical Elsasser number Λ is based on the RMS field strength in the dynamo regions, Λ_{CMB} is the value calculated with the RMS field strength at the core–mantle boundary (CMB), and the dipole Elsasser number Λ_D uses only the dipole contribution to the CMB field. The latter is assumed to be representative of the larger scale field, which is all we know in the case of Mercury. Including the relatively well-determined geomagnetic field as well as dynamo simulations into our discussion allows us to explore the different aspects of this problem.

Using recent geomagnetic field models for the year 2000 (Maus et al. 2006) we arrive at $\Lambda_{CMB} \approx 0.15$. Several issues prevent us from generalizing this value to represent core conditions. Earth's crustal magnetization contaminates any field contribution beyond spherical harmonic degree $l = 14$. Consequently, these components are discarded in the internal field models, and their contribution to the Elsasser number is not known a priori. We also face the additional problem that there is no clear-cut way to extrapolate the poloidal magnetic field into the conducting dynamo region and, in addition, we lack any information on the toroidal field contribution.

Dynamo simulations provide the full magnetic field information and can therefore help to get an idea how Λ_{CMB} and Λ_D should be extrapolated to obtain the core value Λ. We first examine models that are geared to explain the geomagnetic field. These models are characterized by an inner core that occupies 35% of the total core radius and by convection that fills the whole outer core. A very Earth-like field geometry and amplitude can, for example, be found for $Ra = 18 \times Ra_c$ (Ra_c is the critical Rayleigh number for onset of convection), $E = 3 \times 10^{-4}$, $Pr = 1$, and $Pm = 3$. This dynamo is representative of several nonreversing dipole-dominated cases examined by Christensen et al. (1999), Kutzner and Christensen (2004), and recently by Christensen and Aubert (2006). The time averaged CMB Elsasser number based on spherical harmonic degrees $l \leq 14$ is $\Lambda_{CMB} \approx 0.16$. The value increases to $\Lambda_{CMB} \approx 0.25$ when all spherical harmonic contribution are taken into account (the model is truncated at $l = 85$). The RMS Elsasser number in the fluid shell is significantly larger, $\Lambda = 5.16$, which suggests that the dynamo operates in the strong-field regime. About two thirds of the magnetic energy is contributed by the toroidal field. Examination of similar Earth-like dipole-dominated cases leads to the following rules of thumb: (a) including degrees beyond $l = 14$ roughly doubles Λ_{CMB}; (b) the value based on the internal field strength is about an order of magnitude larger than the CMB value; and (c) toroidal and poloidal fields are of comparable amplitude. Translated to the geomagnetic field, this would suggest that Earth's Elsasser number is about $\Lambda = 3$.

The dipole Elsasser number Λ_D, assumed to be representative of the coarse field information in the case of Mercury, amounts to about $\Lambda_D = 0.05$ for recent geomagnetic field models and the dipole-dominated dynamo model mentioned earlier. When increasing the Rayleigh number in the simulations, the dynamo shows Earth-like reversal behavior with longer stable polarity epochs being interspersed by shorter excursions and reversals (Kutzner and Christensen 2002). We analyze a case at $Ra = 26 \times Ra_c$. During stable polarity epochs, the dipole Elsasser number is somewhat lower than for the nonreversing case at $\Lambda_D \approx 10^{-2}$. It decreases significantly to $\Lambda_D \approx 3 \times 10^{-3}$ during excursion or reversals where the dipole loses its dominance (Kutzner and Christensen 2002; Christensen and Aubert 2006). We have averaged over several 10,000 years in both cases; Λ_D can be somewhat lower or larger at times. The unstable-dipole period is also representative of the regime found at even higher Rayleigh numbers where a weak dipole constantly reverses and clear dipole-dominated stable epochs are missing altogether (Kutzner and Christensen 2002). The magnitude of roughly $\Lambda \approx 6$ in both the stable and the low-dipole epochs shows that the dynamo still operates in the strong field regime and demonstrates that the magnetostrophic balance can also be established by small scale magnetic field contributions.

Whether a dynamo operates in the dipole-dominated or in the constantly reversing regime where the dipole contribution is only marginal seems to be controlled by the local Rossby number, $Ro_\ell = U/(\Omega \ell)$, where ℓ is a characteristic length scale of the flow (Christensen and Aubert 2006; Olson and Christensen 2006). The transitional value depends on various conditions, but is probably always less than 0.12. Olson and Christensen (2006) suggested a scaling law that relates Ro_ℓ to the fundamental control parameters and estimated a value $Ro_\ell \approx 10$ for Mercury. This puts Mercury into the nondipolar regime, in apparent contradiction to the observed dominance of the axial dipole and quadrupole contributions. We come back to this point in Sect. 6.

Table 2 Physical parameters assumed for Mercury. Values have been taken from Spohn et al. (2001), Conzelmann (1999), and Braginsky and Roberts (1995)

Quantity	Symbol	Value
Mean planetary radius	R_M	2,240 km
Core radius	r_o	1,850 km
Rotation rate	Ω	1.24×10^{-6} sec^{-1}
Mean mantle density	ρ_m	3,350 kg m^{-3}
Surface acceleration	g_M	3.8 m sec^{-2}
Sulfur density	ρ_S	6,077 kg m^{-3}
Iron density	ρ_{Fe}	8,412 kg m^{-3}
Thermal expansivity	α	3×10^{-5} K^{-1}
Heat capacity	c	675 J (kg K)$^{-1}$
Latent heat	L	250×10^3 W kg^{-1}
Thermal diffusivity	κ_T	7×10^{-6} m^2 sec^{-1}
Compositional diffusivity	κ_χ	10^{-8} m^2 sec^{-1}
Kinematic viscosity	ν	10^{-6} m^2 sec^{-1}
Electrical conductivity	σ	6×10^5 S m^{-1}
Magnetic permeability	μ	$4 \times 10^{-7} \pi$ N A^{-2}

Representing the Hermian field by a dipole contribution of $g_{10} = -325$ nT and assuming the properties listed in Table 2 we arrive at an Elsasser number of $\Lambda_D \approx 5 \times 10^{-5}$. This is more than two orders of magnitude lower than the lowest value ($\Lambda_D \approx 10^{-2}$) in the Earth-geared dynamo simulations that still produce fields where the dipole contribution is stronger than the higher order components. To pose it differently, when using the extrapolations derived from geodynamo models we would limit the Elsasser number in Mercury's core to values less than $\Lambda \approx 10^{-2}$. This result could be interpreted in two ways: It could mean that Mercury's dynamo is not operating in the strong-field regime thought to be typical for planetary dynamos and found to be typical for geodynamo simulations. This conclusion motivated the development of the alternative dynamo models mentioned in the introduction. A second possibility is that the extrapolations fail because they are based on simulations that neglect important differences between Earth's and Mercury's interior. Potential candidates are the different relative size of Mercury's inner core and the particular type of convection owed to the fact that the core–mantle boundary heat flow is probably subadiabatic. These two issues are explored in Sects. 5 and 6, respectively.

4.2 Dynamo Power Budget

We proceed with exploring the dynamos power budget and related field strength estimates. Since the magnetic diffusivity generally exceeds the kinematic viscosity by several orders of magnitude, the energy dissipation is likely dominated by Ohmic effects in planetary dynamo regions. Balancing the energy input or power P with Ohmic dissipation may then serve to estimate the field strength:

$$P \approx \frac{\lambda V_o B^2}{\mu l_\lambda^2}. \tag{8}$$

Here, we have introduced the length scale l_λ at which the magnetic energy is supposedly dissipated most effectively (Christensen and Tilgner 2004), and we furthermore assume that Ohmic decay is mainly concentrated in the outer core volume V_o. Unfortunately, both power P and diffusion length scale l_λ are hard to constrain for planetary cores.

Christensen and Tilgner (2004) explored Ohmic dissipation in a suit of dynamo models and suggested that magnetic energy is predominantly dissipated at a length scale

$$l_\lambda \approx (1.74/Rm)^{1/2} D. \tag{9}$$

This result is supported by the fact that the Karlsruhe dynamo experiment seems to obey the same scaling (Christensen and Tilgner 2004). Equation (9) connects l_λ with the convective vigor, a relationship that was investigated more closely by Christensen and Aubert (2006) based on an even larger suite of dynamo simulations than the one explored by Christensen and Tilgner (2004). Christensen and Aubert (2006) suggested that the flow amplitude is determined by the codensity flux $Q_{b,i}$ at the inner core boundary, and they proceeded to derive an empirical scaling that we also adopt here:

$$Rm \approx 0.83 \frac{Pm}{E} Ra_{b,i}^{0.41}. \tag{10}$$

The buoyancy or codensity flux Rayleigh number $Ra_{b,i}$ is defined as:

$$Ra_{b,i} = \frac{1}{4\pi r_o r_i} \frac{g_o}{\rho_o \Omega^3 D^2} Q_{b,i}. \tag{11}$$

This scaling has been derived for dipole-dominated dynamos driven by vigorous convection that involves the whole fluid shell. Mercury's dynamo may actually fall into a different category, but the dependence of length scale l_λ on the flow vigor and the dependence of the flow vigor on the codensity flux is certainly suggestive. The actual scaling factors and also, perhaps, the exponents may differ. By combining (7) with (8), (9), and (10) we arrive at a scaling that establishes the dependence of Elsasser number Λ on codensity flux and power:

$$\Lambda \approx 2Pm\, E^2 P' Ra_{b,i}^{-2/5}. \tag{12}$$

The dimensionless power $P' = PD^4/(V_o \rho_o \nu^3)$ has been rescaled using $\rho_o V_o$ as the mass scale. Christensen and Aubert (2006) pointed out that the buoyancy flux Rayleigh number is proportional to the nondimensional power, and they determined the factor of proportionality for the case when convection is driven by an imposed temperature contrast. The situation in Mercury's core is more complex because the combined action of thermal and compositional buoyancy may stabilize the outer part of the fluid core. In the following we derive the expression that links the buoyancy flux $Q_{b,i}$ to the heat flow at the core–mantle boundary and the relevant core properties. Furthermore, we determine the form factor F that relates P' to $Ra_{b,i}$, which will finally allow us to calculate the Elsasser number for a given buoyancy flux.

When estimating the total power available to drive the dynamo, we are interested in the long-term evolution of a mean reference state rather than the short-term convective fluctuations. Formally, this reference state represents a time-average over periods long enough to average out convective fluctuation but shorter than the relevant thermal and chemical evolution time scales. The laterally homogeneous reference state is characterized by a hydrostatic pressure and density profile $p(r)$ and $\rho(r)$, respectively. Furthermore, it is assumed that convection is strong enough to establish a chemically homogeneous and adiabatic reference state $T_a(r)$ in the lower convective part of the outer core (discussed later).

Compositional as well as thermal sources are available for driving the dynamo. Inner core growth and the associated release of light elements provide the compositional gravitational power P_C, which is subsequently converted to heat $Q_C = P_C$ by viscous effects. Additional (pseudo) heat sources are the change of internal energy due to secular cooling Q_S and the

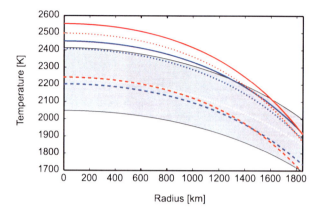

Fig. 5 Iron–sulfur melting curves assumed by Stevenson et al. (1983) (*blue*) and Hauk et al. (2004) (*red*), respectively. Values for sulfur weight fractions $\chi = 0.001$ (*solid*), $\chi = 0.01$ (*dotted*), and $\chi = 0.05$ (*dashed*) are shown. The area between the core adiabats with core–mantle boundary temperatures of $T_o = 1,700$ K and $T_o = 2,000$ K, respectively is marked in grey, white dots indicate the adiabat with $T_o = 1,850$ K (Stevenson et al. 1983; Schubert et al. 1988; Hauk et al. 2004)

release of latent heat Q_L arising from inner core freezing. Ohmic heating and adiabatic heating have been neglected since they cancel when integrated over the core (Lister and Buffett 1995).

The total power available to the dynamo is given by:

$$P = P_C + c_i(Q_i - Q_{A,i}) + c_o(Q_o - Q_{A,o}). \qquad (13)$$

This expression assumes that the heat flux $Q_i = Q_L + Q_{S,i}$ coming in through the inner boundary r_i is redistributed homogeneously throughout the outer core, so that the respective Carnot efficiency is given by $c_i = 1 - \bar{T}/T_i$. $Q_{S,i}$ represents effects of inner core secular cooling, $Q_{A,i}$ is the adiabatic heat flux through the inner boundary, T_i is the temperature of the inner core surface, and $\bar{T}$ is the mass-averaged temperature (Lister and Buffett 1995). The superadiabatic heat flux $(Q_o - Q_{A,o})$ at the outer boundary contributes with Carnot efficiency $c_o = \bar{T}/T_o - 1$ to the available power; $Q_o = Q_L + Q_S + Q_C$ and $Q_{A,o}$ are the total and adiabatic heat flux at $r = r_o$, respectively.

The inner core radius is given by the crossing point of the core adiabat and the melting curve of the iron–sulfur mixture thought to be present in Mercury's core. However, neither the core adiabat nor the iron–sulfur melting curve are well determined. Figure 5 compares melting curves adopted by Stevenson et al. (1983) and Hauk et al. (2004) for three different core sulfur weight fractions: $\chi = 0.001$ (solid), $\chi = 0.01$ (dotted), and $\chi = 0.05$ (dashed). The core sulfur content is largely unknown and depends on the distance at which the planetesimal formed that eventually built Mercury. Closer to the Sun less volatiles, including sulfur, would have been incorporated. Values from $\chi = 0$ to $\chi = 0.1$ have been considered by, for example, Hauk et al. (2004), Breuer et al. (2007).

Thermal evolution simulations (Stevenson et al. 1983; Schubert et al. 1988; Hauk et al. 2004; Breuer et al. 2007) suggest that Mercury's present core–mantle boundary temperature may lie between roughly $T_o = 1,800$ K and $T_o = 2,000$ K. Some extreme models consider temperatures approaching $T_o = 1,700$ K (Hauk et al. 2004). Figure 5 shows the debated adiabatic range along with an example adiabat for $T_o = 1,850$ K and demonstrates how crucially r_i depends on the melting-curve model, in particular for larger CMB temperatures.

Sulfur contents beyond $\chi = 0.05$ could mean that Mercury's core is still completely liquid, which is hard to reconcile with the convectively driven dynamo models considered here. We adopt the melting curve and adiabat parameterizations proposed by Stevenson et al. (1983) and, rather than depicting specific inner core radii, span the range $0 < \eta < 0.8$. Furthermore, we explore the three initial sulfur weight fractions $\chi_0 = M_S/(M_{Fe} + M_S) = 0.001, 0.01$, and 0.05, where M_S and M_{Fe} denote the total mass of sulfur and iron in the whole core, respectively. The sulfur weight fraction χ increases with growing inner core size: $\chi = \chi_0/(1 - \eta^3)$.

The inner core radius changes in time due to the drop of the adiabat, caused by secular core cooling, and due to the drop of the melting curve with increasing χ. These two effects establish the dependence of the rate of CMB temperature change $\dot{T}_o$ on the inner core growth rate $\dot{r}_i$:

$$\dot{T}_o = F_T \dot{r}_i. \tag{14}$$

The proportionality factor F_T depends on r_i and the adopted core model. Equation (14) allows us to write the heat change due to the secular-cooling Q_S in terms of the inner core growth rate:

$$Q_S = V_c \rho_c c \dot{T}_o = F_S \dot{r}_i. \tag{15}$$

For simplicity, we neglect the density difference between inner and outer core and use the mean core density ρ_c; V_c is the total core volume. Moreover, we assume that the inner core cools homogeneously.

The chemical power P_C can be derived in two alternative ways: from calculating the change of the gravitational energy due to inner core growth (Schubert et al. 1988) or from considering the change of the reference state as described by the Navier–Stokes equation (2) and codensity equation (6) (Lister and Buffett 1995). Both methods lead to the same expression:

$$P_C = 6\pi g_o \delta\rho \frac{\rho_o}{\rho_{Fe}} r_o^4 \left(\frac{1}{5}(1 - \eta^5) - \frac{1}{3}\eta^2(1 - \eta^3) \right) \frac{\dot{r}_i}{r_i} = F_C \dot{r}_i, \tag{16}$$

where $\delta\rho = \rho_i - \rho_o$ is the density difference between inner and outer core. We assume that the inner core contains no sulfur and that iron has the same density in the inner and outer core, so that the inner core density is $\rho_i = \rho_{Fe}$. We furthermore neglect the small effects due to outer core shrinking (Schubert et al. 1988).

Using the dependence of inner core mass change $\dot{m}_i$ on the inner core growth rate,

$$\dot{m}_i = 4\pi \rho_i r_i^2 \dot{r}_i, \tag{17}$$

we can formulate the latent heat release $Q_L = L\dot{m}_i$ in terms of $\dot{r}_i$:

$$Q_L = 3LV_i \rho_i \frac{\dot{r}_i}{r_i} = F_L \dot{r}_i, \tag{18}$$

where V_i is the inner core volume. Having established the dependence of all three heating/cooling contributions on $\dot{r}_i$ we compare the relative contributions to the heat budget, $Q_C/Q_S = F_C/F_S$ and $Q_L/Q_S = F_L/F_S$, in Fig. 6. The heat gain due to the conversion of chemical power is at least an order of magnitude smaller than the secular cooling effect, while latent heat and secular cooling effects are of comparable size. The difference between contributions due to chemical convection and due to latent heat release grows with increasing initial sulfur content χ_0.

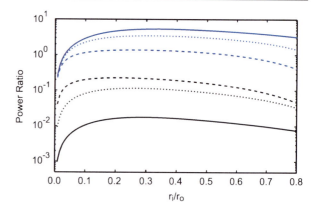

Fig. 6 Contributions of compositional gravitational energy (*black*) and of latent heat (*blue*) to the heat flow budget relative to that of secular cooling. Sulfur weight fractions $\chi_0 = 0.001$ (*solid*), $\chi_0 = 0.01$ (*dotted*), and $\chi_0 = 0.05$ (*dashed*) have been considered

We can now derive the absolute inner core growth rate and the individual contributions to the CMB heat flux using (15), (16), and (18):

$$\dot{r}_i = Q_o/(F_C + F_L + F_S) \approx Q_o/(F_L + F_S). \tag{19}$$

Above equations allow to derive the power avaible to drive Earth's dynamo where reasonable estimates of the CMB heat flux Q_o and the inner core radius r_i are available. For Mercury, where both quantities are unknown, we have to rely on thermal evolution models that couple mantle and core evolution via Q_o and T_o to form a combined model. Stevenson et al. (1983) and Hauk et al. (2004) parameterized the thermal evolution of both core and mantle; more recent simulations (Breuer et al. 2007) employ fully 3D mantle convection simulations. These models suggest that Mercury's CMB heat flow Q_o may only have been super-adiabatic for a relatively brief period in the planet's history. The early formation of a thick lithosphere severely limits the heat flow, whose present day value may not exceed 5 mW/m². This is more than a factor two lower than the adiabatic value estimated to be about 12 mW/m² (Stevenson et al. 1983; Hauk et al. 2004), which has the unfortunate consequence that we cannot directly apply the power estimate (13).

A subadiabatic CMB heat flow suggests that the upper part of the core could be stably stratified, provided the stabilizing positive thermal buoyancy gradient exceeds the destabilizing negative compositional buoyancy gradient. The thermal gradient may change its sign to also become destabilizing deeper in the outer core. We have outlined in Sect. 3 how thermal and compositional effects can be combined into a codensity variable b. Formulating a power balance based on the codensity flux allows us to incorporate the effects of a subadiabatic outer boundary. We assume that convection is restricted to a region $r_i \leq r \leq r_n$, where r_n is the neutral buoyancy radius $\partial \bar{b}(r_n)/\partial r = 0$. The over-bar indicates an average over a spherical surface and over time. The relative smallness of the gravitational power allows us to neglect its contribution to the total CMB heat flux so that the inner core growth rate can be approximated by $\dot{r}_i \approx Q_o/(F_L + F_S)$. Given $\dot{r}_i$ we know $\dot{m}_i$ and can thus determine the inner core codensity or buoyancy flux:

$$Q_{b,i} = \left(\frac{\rho_{Fe} - \rho_o}{\rho_{Fe}} + \frac{\alpha L}{c} \right) \dot{m}_i + \frac{\alpha}{c}(Q_{S,i} - Q_{A,i}). \tag{20}$$

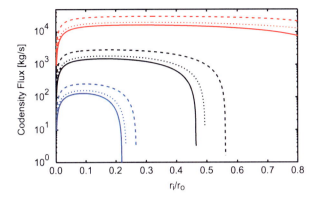

Fig. 7 Codensity or buoyancy flux at the inner boundary for the nine different models explored here. Cases for a CMB heat flux of 0.1 mW/m^2, 1 mW/m^2, and 10 mW/m^2 are shown in *blue*, *black*, and *red*, respectively. The corresponding CMB buoyancy fluxes are $-3,822$, $-21,021$, and $-22,741$ kg s^{-1} respectively. *Solid*, *dotted*, and *dashed lines* symbolize the inner codensity flux for initial sulfur weight fractions of $\chi_0 = 0.001$, $\chi_0 = 0.01$, and $\chi_0 = 0.05$ respectively

Fig. 8 Relative size of the convective zone $r_i \leq r \leq r_n$ for a core–mantle boundary heat flux of 0.1 mW/m^2 (*blue*), 1 mW/m^2 (*black*), and 10 mW/m^2 (*red*) and initial sulfur weight fractions of $\chi_0 = 0.001$ (*solid*), $\chi_0 = 0.01$ (*dotted*), and $\chi_0 = 0.05$ (*dashed*)

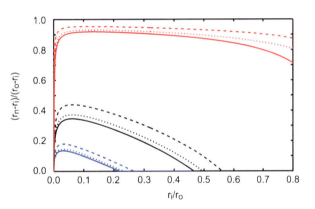

The codensity flux through the outer boundary is determined by the difference between the prescribed heat flow and the adiabatic heat flow:

$$Q_{b,o} = \alpha/c(Q_o - Q_{A,o}). \tag{21}$$

Solving the time-averaged and horizontally averaged codensity diffusion equation, that describes the evolution of the reference state subject to the flux boundary conditions $Q_{b,o}$ and $Q_{b,i}$ then allows us to determine r_n. Figure 7 shows inner and outer codensity fluxes for CMB heat flows that cover three orders of magnitude, from mildly subadiabatic at $q_o = 10$ mW/m^2 to heavily subadiabatic at $q_o = 0.1$ mW/m^2. The release of latent heat increases quadratically with inner core size r_i (see (18)), while fixing the CMB heat flux limits the rate of latent heat release. Increasing the size of the inner core therefore leads to a decrease of the inner core growth rate, which in turn diminishes the codensity flux $Q_{b,i}$ that may ultimately become negative. A negative value of $Q_{b,i}$ means that the complete outer core is stably stratified leaving no convective energy to drive a dynamo.

Figure 8 shows the relative size of the convective zone, $(r_n - r_i)/(r_o - r_i)$, for the cases explored here. Assuming that Mercury's dynamo is indeed driven convectively we can ex-

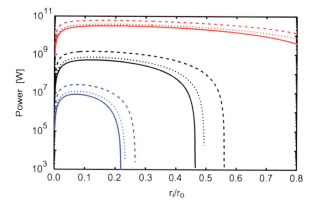

Fig. 9 Power available to drive the dynamo for the nine different cases explained in the caption of Fig. 7

clude, for example, inner core radii beyond 55% of the total core radius when the CMB heat flux is only 1 mW/m² and beyond 25% when the flux is as low as 0.1 mW/m². The convective zone amounts to at most roughly 45% and 20% of the shell thickness, respectively, for these two cases.

The total power available to drive the dynamo can finally be calculated by modifying the chemical power expression (16) to also include the thermal buoyancy contribution,

$$P = \frac{3}{2} \frac{g_o}{r_o} \frac{r_n^2}{1-\xi^3} \left(\frac{1}{5}(1-\xi^5) - \frac{1}{3}\xi^2(1-\xi^3) \right) Q_{b,i}, \tag{22}$$

with $\xi = r_i/r_n$.

Here we integrate over gravitational energy changes in the convective region $r_i \leq r \leq r_n$ but assume that ρ_o is equivalent to a homogeneous distribution of light elements over the whole outer core. Figure 9 shows the dependence of the available power P on aspect ratio $\eta = r_i/r_o$. For $q_o = 1$ mW/m², the power amounts to about 1 GW for smaller inner cores and drops off rapidly for inner cores that fill more than 50% of the total core radius. For $q_o = 0.1$ mW/m², the power is roughly two orders of magnitude lower and drops off for relative inner core radii beyond 25%. Ample power seems available for CMB fluxes of $q_o = 10$ mW/m² or larger.

The combination of (12) and (22) establishes the scaling of the Elsasser number Λ with the inner boundary buoyancy-flux Rayleigh number (11):

$$\Lambda \approx F Pm\, E^{-1} Ra_{b,i}^{3/5}, \tag{23}$$

where F is the form function:

$$F = \frac{9}{5} \frac{\eta^3}{1-\eta^3} \frac{1 - 5/3\xi^2 + 2/3\xi^5}{\xi^2(1-\xi^3)}. \tag{24}$$

To derive the Elsasser number in the convective region we replace V_o in (12) with the appropriately modified volume V_n of the convective region; Fig. 10 shows the respective results.

It is generally believed that convective dynamos cannot work at magnetic Reynolds numbers below, say, $Rm = 50$. This additional condition would limit the inner core radius to roughly 10% and 40% of the total core for $q_o = 0.1$ mW/m² and $q_o = 1$ mW/m², respectively. We have used (10) to calculate Rm here.

The Elsasser numbers are typically of order one or larger in all cases. Small values are only possible for very tiny inner cores, $r_i/r_o < 0.01$, which is an unlikely scenario. Small

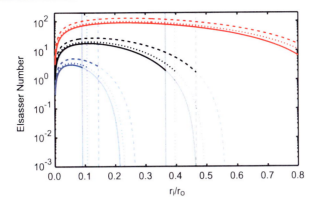

Fig. 10 Elsasser numbers for the power estimates in Fig. 9. *Vertical lines* mark aspect ratios beyond which the magnetic Reynolds number would fall below $Rm = 50$ for CMB heat flows of 0.1 mW/m^2 (*blue*) and 1 mW/m^2 (*black*), respectively. We switch to pale colors if $Rm < 50$. CMB heat flows of 10 mW/m^2 (*red*) are uncritical in this respect since Rm always exceeds the value $Rm = 50$ estimated to be a minimum requirement for convective dynamos

Elsasser numbers are also formally predicted for larger inner cores, close to the point where the codensity flux becomes negative. However, the magnetic Reynolds number is already too low for a self-sustained dynamo when the predicted Λ drops below one.

In conclusion, the available power most likely suffices to drive a classical magnetostrophic planetary dynamo with an Elsasser number of order one or larger. This result is reassuring but not compliant with the measured magnetic field strength. Two effects can explain the differences: First, the Elsasser numbers given in Fig. 10 concern the RMS field strength in the convective region, which is more remote from the planetary surface than the CMB. Assuming that the magnetic field is dominated by the dipole contribution, an upward continuation to the CMB reduces the Elsasser number by about four orders of magnitude when the CMB heat flux is as low as $q_o = 0.1$ mW/m^2. In this particular case, the geometric effect is large enough to explain Mercury's low field strength because the dynamo is confined to a thin convective layer that surrounds a small inner core and is remote from the CMB. Thermal evolution simulations, however, suggest that the CMB heat flux is at least $q_o = 1$ mW/m^2 (Breuer et al. 2007). In this case, the geometric effect amounts to only two orders of magnitude, too small to bridge the gap between our Elsasser number estimates and the observed field strength. In Sect. 6 we will demonstrate that this goal will nevertheless be reached when we take into account that (a) Mercury's dynamo probably works in the nondipolar regime (discussed earlier) and that (b) the time-dependent magnetic field is attenuated by the electromagnetic skin-effect in the stratified conducting outer layer.

5 Thin Shell and Thick Shell Dynamos

We pointed out earlier that the size of Mercury's inner core is basically unknown. Several authors therefore explored how different inner core sizes would affect the dynamo process in models that were originally developed with the geodynamo in mind. A reduction of the externally observed field strength was found both for dynamos operating in a thin shell and dynamos operating in a thick shell, i.e. dynamos assuming rather small or rather larger inner cores.

Thick shell dynamos were explored by Heimpel et al. (2005). They realized that the ratio Λ_{CMB}/Λ decreases for smaller inner cores in their model. Although all their dynamos operate in the strong field regime, i.e $\Lambda \geq 1$, Λ_{CMB} reaches values as low as 10^{-2} for $\eta = 0.15$ which is the smallest aspect ratio they explored. This decrease can be traced to two causes: (1) the growing fraction of toroidal magnetic field produced by an ω-effect, and (2) the local concentration of poloidal field generation. At $\eta = 0.15$ radial convective motion is mainly concentrated in only one convective column that is also the center of poloidal magnetic field production. Consequently, the degree $m = 1$ field component is somewhat pronounced. The toroidal field, on the other hand, is associated to a global zonal flow around the inner core. Roughly 75% of the magnetic energy at the CMB is carried by the dipole contribution, the field is thus still predominantly large scale and Λ_D is not significantly smaller than Λ_{CMB}. The dipole tilt, judging from their Fig. 5a, seems compatible with the value $14 \pm 5°$ inferred for Messenger data (Ness 1979). It should be mentioned that their simulations ran at only marginally supercritical Rayleigh numbers with respect to the onset of convection. A more supercritical convective motion will start to fill the whole shell and possibly lead to a different solution.

Simulations of dynamos that operate in thin shells have been carried out by Stanley et al. (2005) and Takahashi and Matsushima (2006). Stanley et al. (2005) explored aspect ratios between $\eta = 0.7$ and $\eta = 0.9$ and reported that thin-shell dynamos promote the production of a small-scale poloidal field as well as a strong axisymmetric toroidal magnetic field. While their dynamos operate in the strong-field regime, the observable dipole Elsasser number goes down to $\Lambda_D = 10^{-3}$. The reason why the ratio Λ_D/Λ lies below the value found by Heimpel et al. (2005) and the value discussed for the larger Ra cases presented in Sect. 4 is the dominant toroidal magnetic field in their simulations, which is up to an order of magnitude stronger than the poloidal field and establishes the magnetostrophic balance. As in the work of Heimpel et al. (2005), the low Λ_D simulations presented by Stanley et al. (2005) operate at Rayleigh numbers close to the onset of dynamo action. At somewhat larger Rayleigh numbers the ratio of toroidal to poloidal field strength decreases and the Elsasser number Λ_D increases. It should be noted that the neglect of viscous boundary layers (3) or the use of hyper-diffusivities seem to promote the stronger toroidal field (Stanley et al. 2005) not found in other dynamo simulations.

Takahashi and Matsushima (2006) also explored dynamos operating in a thin shell at $\eta = 0.7$ and obtained dipolar as well as nondipolar dynamos. We present two examples here that ran at different Rayleigh numbers, $Ra = 3.50 \times Ra_c$ (lower Ra case) and $Ra = 5.83 \times Ra_c$ (larger Ra case), but otherwise identical parameters: $E = 10^{-3}$, $Pm = 5$, and $Pr = 1$. The increase in Rayleigh number causes a change in the convective motion that goes along with a change in magnetic field geometries. A dipole dominated dynamo is found at the lower Rayleigh number, while the dipole has lost its dominance in the larger Ra case.

Figure 11(a) shows a snapshot of the radial magnetic field at the core surface in the lower Rayleigh number case. The field geometry is characterized by a longitudinal array of strong magnetic flux patches at mid-latitudes that corresponds to the outer rim of the tangent cylinder (TC) in both the northern and southern hemispheres. The tangent cylinder is an imaginary cylinder that is aligned with the rotation axis and touches the inner core equator. The field patches are created by flows converging where cyclonic convective columns connect with the CMB. These flows concentrate magnetic field lines and proceed down the columnar axis towards the equator (Olson et al. 1999). Because the convective columns are attached to the inner core, its size can possibly be inferred from the magnetic field morphology at the CMB. However, we note that, since the small-scale magnetic fields contributions decay more rapidly with distance from the source region, high-resolution measurements at satellite

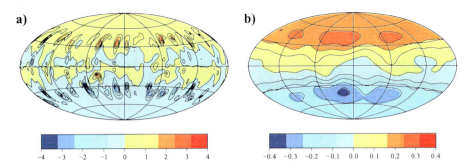

Fig. 11 The radial component of the magnetic field at (**a**) core surface and at (**b**) Mercury's surface level for $Ra = 3.50 \times Ra_c$ in the thin-shell model by Takahashi and Matsushima (2006)

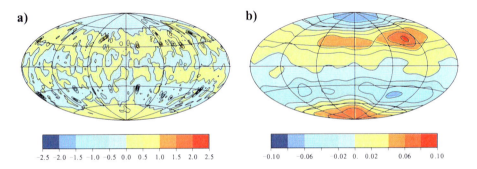

Fig. 12 Same as Fig. 11 but for the higher Rayleigh number case at $Ra = 5.83 \times Ra_c$

altitude are required in order to discern these features. This is demonstrated by Fig. 11(b) where the now elongated flux patches are already much harder to identify at the planetary surface level (Takahashi and Matsushima 2006).

Figure 12 displays the magnetic field in the nondipolar case at a higher Rayleigh number. The structure exhibits several smaller strong flux patches distributed irregularly over the core surface. The increased field complexity is much less apparent at planetary surface level shown in Fig. 12(b). Such a complex magnetic field generally goes back to the increased spatial and temporal complexity of convective motion.

Columnar convective motion, which plays a key role in generating the magnetic field (Olson et al. 1999), occurs only outside the TC in thicker to intermediate shells. The region inside the TC is basically stagnant until plume-like convection starts at higher Rayleigh numbers. These localized upwellings typically create inverse magnetic field opposing the mean field direction outside the TC and tend to be highly time-dependent (Wicht and Olson 2004; Wicht 2005).

In a thin spherical shell, however, the region inside the TC occupies a much larger fraction, so that the localized and inverse field created by plumes rising at higher Ra has a bigger impact on the overall field geometry. Figure 12 clearly shows the large inverse field region inside the TC in both hemispheres, which are reflected in strong field components of octupolar and higher (odd) degrees. This is documented by the magnetic power spectra shown in Fig. 13. The magnetic power density W_l contained in spherical harmonic degree l

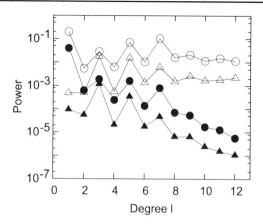

Fig. 13 Magnetic power spectra with respect to spherical harmonic degree *l* at the core surface (*open symbols*) and at planetary surface altitude (*filled symbols*) for dipolar (*circles*) and nondipolar (*triangles*) dynamos in the model by Takahashi and Matsushima (2006)

is defined as

$$W_l(r) = (l+1)\left(\frac{r_M}{r}\right)^{2l+4} \sum_{m=0}^{l}\left[(g_{lm})^2 + (h_{lm})^2\right]. \quad (25)$$

The spectrum of the nondipolar dynamo is characterized by nondipolar components at $l = 3$ and $l = 5$ at the core surface and by $l = 3$ at the planetary surface. The axial octupole component actually dominates the field. Figure 13 also demonstrates that the field is generally smaller in the higher Ra case than for the dipole-dominated field (Takahashi and Matsushima 2005). Once again, the dynamo Elsasser number Λ is of order one in the high Ra case, and the dipole Elsasser number is about $\Lambda_D = 10^{-2}$.

While the models of Takahashi and Matsushima (2006) cannot ultimately explain the smallness of the observed field, they suggest that a precise measurement of Mercury's magnetic field may help to estimate the planet's inner core size, and thereby highlight the usefulness of global high-quality measurements.

6 Dynamo in a Partly Stable Core

Christensen (2006) addressed the question of how the presence of a stably stratified zone in the outer part of Mercury's core, as suggested by the subadiabatic CMB heat flow, would affect the dynamo process. Two dynamo models with different aspect ratios $\eta = 0.35$ (case 1) and $\eta = 0.50$ (case 2) have been explored. In both cases a nondimensional buoyancy flux of $\partial b/\partial r = -1$ is specified at the inner core boundary, but instead of prescribing the buoyancy flux at the CMB the volumetric sinks are set to $\epsilon = -1.235$ in case 1 and to $\epsilon = -1.5$ in case 2. Boundary condition $b = 0$ is enforced at the CMB. This approach allows the CMB buoyancy flux to vary in time, its mean values amount to -4 and -2.5 times the inner core boundary flux, respectively, and therefore fall within the range suggested by Fig. 7 for CMB heat flows of a few mW/m². The neutral buoyancy radius r_n is such that only the lower 40% of the outer core are convectively unstable and the upper 60% are stable in both cases.

Since it is not possible to run dynamo models at the actual planetary values of the fundamental control parameters, care has been taken to ensure that the arguably most important nondimensional number, the magnetic Reynolds number, is in a realistic range for Mercury. The analysis by Christensen and Aubert (2006) suggests that Rm should obey scaling

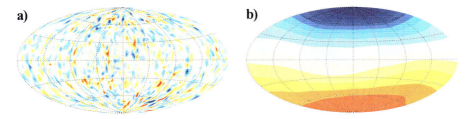

Fig. 14 Snapshot of the radial magnetic field in case 2. Panel (**a**): at radius $r = r_i + 0.37D$, i.e. near the top of the dynamo region with contour interval 40,000 nT; panel (**b**): at the planetary surface ($r = 1.34r_o$) with contour interval 200 nT. Physical units of the magnetic field have been calculated assuming $\sigma = 8 \times 10^5$ S m^{-1} and $\rho = 8{,}200$ kg m^{-3}

law 10. The combination ($Pm\, E^{-1} Ra_{bi}^{0.41}$) is of order 1,000 for the Hermian parameter values assumed here, and the same value is obtained for the chosen model parameters: (Ra_b, E, Pm, Pr) = $(1.08 \times 10^{-5},\, 3 \times 10^{-5},\, 3,\, 1)$ in case 1 and $(2 \times 10^{-4},\, 10^{-4},\, 3,\, 1)$ in case 2. The model value of the magnetic Reynolds number should thus match the planetary value. Another important consideration is that the models should be in a regime with a sufficiently large value of the local Rossby number Ro_ℓ to make sure that the dynamo operates in the nondipole dominated regime. Although the value appropriate for Mercury, $Ro_\ell \approx 10$, cannot be reached in the numerical simulations, Ro_ℓ is large enough to ensure that the dynamo operates in the correct regime (Olson and Christensen 2006).

Though active convection is restricted to the unstable lower part of the outer core, the circulation nevertheless penetrates into the upper region in the form of Taylor columns aligned with the rotation axis. The respective flow is almost entirely horizontal (toroidal) and weaker than in the convection region (Christensen 2006). Figure 14 shows a snapshot of the radial magnetic field for model case 2. As expected, the field is strong and small-scaled in the dynamo region without a clearly discernable dipole component (upper panel). However, the field is dominated by an axial dipole component at the planetary surface where the field intensity is considerably weaker. While the core Elsasser number is approximately $\Lambda = 2.5$, the CMB value is significantly smaller at $\Lambda_{CMB} = 2 \times 10^{-4}$. This is remarkably close to the observed value, which is about twice as small.

The strong dipole dominance of the field outside the core is also obvious in the magnetic power spectrum (see Fig. 15(b), circles), which decays rapidly with increasing harmonic degree l. In contrast, the spectrum inside the core (crosses) peaks at degree 4, where the energy is four times larger than at $l = 1$. The spectrum is rather flat out to $l = 30$. In case 1, the power spectrum inside the core is similar to that of case 2 (Fig. 15(a)). The surface spectrum again shows a rapid decay for $l > 3$, but the dipole contribution is weaker than in case 2. On time average, dipole and quadrupole components are of comparable strength, but one or the other may dominate at some instance in time (Christensen 2006). The axisymmetric components are usually more prominent than nonaxisymmetric components in the outside field. They clearly dominate in case 2 and, on average, carry about 2/3 of the energy in case 1. In contrast, they contribute only 6% to the poloidal magnetic energy inside the dynamo region. The average surface field is about a factor five weaker than the observed field in case 1.

One reason for the differences in spectral power between the field in the dynamo region and the surface field is the geometric decrease with radius, according to $r^{-(l+2)}$ for a potential field. However, the reduction in amplitude is larger than that of simple geometric decay for all field components in case 1 and for all components except the axial dipole in case 2. The dynamo-generated magnetic field must essentially diffuse through the stable conducting

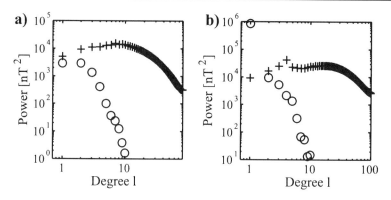

Fig. 15 Time-average magnetic power spectra versus harmonic degree n for (**a**) case 1 and (**b**) case 2. Circles refer to the planetary surface and *crosses* to the mean energy inside the fluid core, scaled down by a factor 10^{-4}

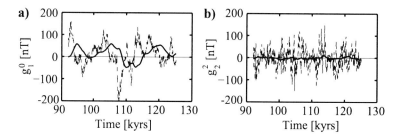

Fig. 16 Axial dipole coefficient (**a**) and coefficient for a nonaxial quadrupole component (**b**) versus time for case 1. *Full lines*: Gauss coefficient at Mercury's surface; *broken lines*: equivalent coefficients for the poloidal field at $r = r_i + D/2$, scaled down by a factor 0.1

layer, ignoring for simplicity induction effects by the horizontal flow in this layer. The field is time-dependent and attenuated by the skin effect.

Figure 16(a) compares the time variation of the axial dipole field slightly above the dynamo region and at the planetary surface for case 1. Rapid fluctuations are completely eliminated, whereas slow variations on a 10,000 yrs time scale penetrate to a limited degree. At the surface they are phase-shifted by about 2,500 yrs, as expected for the skin-effect. Nonaxisymmetric components of the dynamo field and, in general, higher multipole components $l > 2$ fluctuate rapidly on short time-scales (Fig. 16(b)). Though these contributions dominate in the dynamo region, they are strongly suppressed in the surface field. The characteristic time scale of magnetic field changes in dynamos has been found to vary with the inverse magnetic Reynolds number (Christensen and Tilgner 2004). Since Rm is in the correct range in the models the time-scales should also be about correct.

In case 2, the dipole field is much more stable and, in contrast to case 1, does not reverse during the 86,000 yrs worth of simulation. Since the DC part of the dipole field is not affected by the skin effect, case 2 retains a comparatively strong dipole outside the core. The enhanced dipole stability could be a consequence of the increased magnetic inertia represented by the larger solid inner core (Hollerbach and Jones 1993; Takahashi and Matsushima 2006), which might prevent short-term fluctuations of the dipole in the dynamo region from growing into complete polarity reversals.

Because of the filtering effect of the stably stratified layer secular-variation, time scales of the externally observable field are much longer than the intrinsic time scales of the dynamo process. Significant variations of Mercury's field can thus only be expected on time scales of centuries or longer. Both the dipolar field geometry of case 2 and the strong axial quadrupole contribution that exists at some times in case 1 are compatible with the limited knowledge about Mercury's field structure (Connerney and Ness 1988). Although the surface field is too weak in case 1 and too strong in case 2, moderate changes such as decreasing the thickness of the stable layer in case 1 or driving convection more strongly to force the dipole to reverse in case 2 could bring the field strength into line with the observed intensity.

7 Conclusion

Dynamo theory suggests that convective planetary dynamos operate in the strong-field regime where Lorentz forces and Coriolis forces are of comparable magnitude and establish the so-called magnetostrophic balance. In other words, the Elsasser number Λ, the ratio of Lorentz to Coriolis forces, should be of order one. While this seems to be true for Earth, Jupiter, Saturn, and possibly also the ice giants (Stevenson et al. 1983), Mercury violates this rule on first sight. An extrapolation of the weak field measured by Mariner 10 to the dynamo region suggests that Λ is as low as 10^{-4}.

Alternative explanations therefore promote, for example, crustal magnetization (Aharonson et al. 2004) or a thermoelectric mechanism (Stevenson 1987; Giampieri and Balogh 2002). We have reviewed several convective dynamo simulations, which suggest that the discrepancy may come about because the observed surface field does not reflect the true field strength within the core. Two possible reasons for this are: (1) the magnetostrophic balance is established to a substantial degree by small-scale magnetic field that has not been resolved by the observations, and (2) the balance is mainly maintained by the toroidal field, which cannot be detected outside the core. Significant small-scale contributions and/or strong toroidal fields are promoted by either particularly small inner cores (thick shell dynamo) (Heimpel et al. 2005) but also by large inner cores (thin shell dynamo) (Stanley et al. 2005; Takahashi and Matsushima 2006). The discrepancy between the Elsasser number Λ_D, based on the observable large scale dipole field, and Λ, based on the RMS field in the dynamo region, can reach three orders of magnitude (Stanley et al. 2005).

Analyses by Christensen and Aubert (2006) and Olson and Christensen (2006) suggest that a parameter called the local Rossby number Ro_ℓ decides whether a dynamo falls into one of two distinct families. Dipole-dominated fields are produced for small local Rossby numbers but the dipole loses its dominance when Ro_ℓ is increased beyond about 0.12. The value $Ro_\ell \approx 10$ estimated for Mercury puts the planet into the second category. Analyses of a dynamo operating in this second regime, but with an Earth-like relative inner-core size, shows that the dynamo Elsasser number is still of order one but that the large scale dipole value Λ_D is nearly three orders of magnitude lower.

Thin-shell, thick-shell, and also high-Ro_ℓ dynamos produce magnetic fields with significant small-scale contributions. These dynamos can possibly be ruled out should Messenger and BepiColombo measurements confirm the impression that Mercury's field is dominated by large-scale components, i.e. dipole and quadrupole. Should the measurements, on the other hand, resolve smaller scale field patches, they may help to constrain the size of Mercury's inner core, since these features tend to coalesce at the tangent cylinder that touches the inner-core equator.

Christensen (2006) explored a somewhat different dynamo model that is motivated by estimates of the core power budget. Thermal evolution simulations suggest that the heat flow at the planet's core–mantle boundary may be subadiabatic. The available power nevertheless suffices to drive a magnetostrophic dynamo ($\Lambda = 1$) due to chemical differentiation, latent heat release, and secular cooling. However, a subadiabatic CMB heat flow has the consequence that the outer part of the core may be stably stratified. Christensen (2006) simulated cases where only the lower 40% of the core convect actively; he also made sure that Ro_ℓ is large enough for the dynamo to operate in the nondipole-dominated regime. Key points of this model are: (1) the field in the dynamo region is strong but is dominated by small scales with rather weak dipole and quadrupole contributions, (2) the field is time-dependent with small-scale components varying more rapidly than low-order components, and (3) the dynamo field must diffuse through the stable conducting region. Here the rapidly varying high-order components are filtered out by the skin effect, whereas the slowly varying dipole and quadrupole components pass with some attenuation.

While the resulting field obeys $\Lambda \approx 1$ in the dynamo region, its surface value is close to the measured amplitude, so that these models successfully explain the planet's measured low field strength (Christensen 2006). Two other magnetic field features seem to distinguish the models by Christensen (2006) and those that do not incorporate a stagnant outer-core region (Stanley et al. 2005; Heimpel et al. 2005; Takahashi and Matsushima 2006). First, the smaller scale field contributions are significantly less pronounced, and second, secular variation takes place on time scales of some centuries rather than decades. Thus the model with a partly stagnant fluid core would be supported should Messenger and BepiColombo find little magnetic field change compared to the Mariner 10 epoch and also confirm that the field is indeed predominantly large scale.

References

M.H. Acuña, J.E.P. Connerney, N.F. Ness, R.P. Lin, D. Mitchell, C.W. Carlson, J. McFadden, K.A. Anderson, H. Réme, C. Mazelle, D. Vignes, P. Wasilewski, P. Cloutier, Global distribution of crustal magnetization discovered by the Mars Global Surveyor MAG/ER experiment. Science **284**, 790–793 (1999)

O. Aharonson, M.T. Zuber, S. Solomon, Crustal remanence in an internally magnetized non-uniform shell: A possible source for Mercury's magnetic field? Earth Planet. Sci. Lett. **218**, 261–268 (2004)

K.A. Anderson, D.E. Wilhelms, Correlation of lunar farside magnetized regions with ringed impact basins. Earth Planet. Sci. Lett. **46**, 107–112 (1979)

J. Arkani-Hamed, Magnetization of Martian lower crust: Revisited. J. Geophys. Res. **112**(E11), 5008 (2007)

N. Artemieva, L. Hood, B. Ivanov, Impact demagnetization of the Martian crust: Primaries versus secondaries. Geophys. Res. Lett. **32**, L22204 (2005)

J. Aubert, Steady zonal flows in spherical shell dynamos. J. Fluid Mech. **542**, 53–67 (2005)

S.I. Braginsky, P.H. Roberts, Equations governing convection in Earth's core and the geodynamo. Geophys. Astrophys. Fluid Dyn. **79**, 1–97 (1995)

D. Breuer et al., Interior evolution of Mercury. Space Sci. Rev. (2007). doi:10.1007/s11214-007-9228-9

D. Breuer, T. Spohn, Early plate tectonics versus single-plate tectonics on Mars: Evidence from magnetic field history and crust evolution. J. Geophys. Res. (Planets) **108**, 8–1 (2003)

M. Buske, U.R. Christensen, (2007). Three-dimensional convection models for the thermal evolution of the martian interior (2007, in prep.)

L. Carporzen, S. Gilder, R. Hart, Palaeomagnetism of the Vredefort meteorite crater and implications for craters on Mars. Nature **435**, 198–201 (2005)

U. Christensen, J. Wicht, Numerical dynamo simulations, in *Core Dynamics*, Treatise on Geophysics (Elsevier, 2007)

U. Christensen, P. Olson, G.A. Glatzmaier, Numerical modeling of the geodynamo: A systematic parameter study. Geophys. J. Int. **138**, 393–409 (1999)

U.R. Christensen, A deep rooted dynamo for Mercury. Nature **444**, 1056–1058 (2006)

U.R. Christensen, J. Aubert, Scaling properties of convection-driven dynamos in rotating spherical shells and applications to planetary magnetic fields. Geophys. J. Int. **166**, 97–114 (2006)

U.R. Christensen, A. Tilgner, Power requirement of the geodynamo from ohmic losses in numerical and laboratory dynamos. Nature **429**, 169–171 (2004)

U.R. Christensen, J. Aubert, F.H. Busse, P. Cardin, E. Dormy, S. Gibbons, G.A. Glatzmaier, Y. Honkura, C.A. Jones, M. Kono, M. Matsushima, A. Sakuraba, F. Takahashi, A. Tilgner, J. Wicht, K. Zhang, A numerical dynamo benchmark. Phys. Earth Planet. Interiors **128**, 25–34 (2001)

J.E.P. Connerney, N.F. Ness, Magnetic field and interior, in *Mercury*, ed. by F. Vilas, C.R. Chapman, M.S. Matthews (The University of Arizona Press, Tucson, 1988), pp. 494–513

J.E.P. Connerney, M.H. Acuña, N.F. Ness, G. Kletetschka, D.L. Mitchell, R.P. Lin, H. Rme, Tectonic implications of Mars crustal magnetism. Proc. Nat. Acad. Sci. **102**, 42 (2005)

V. Conzelmann, *Thermische Evolutionsmodelle des Planeten Merkur berechnet unter der Anwendung verschiedener Viskositätsgesetzte*. Ph.D. thesis, Westfälische Wilhelms Universität Münster (1999)

M. Fujimoto, W. Baumjohann, K. Kabin, R. Nakamura, J.A. Slavin, N. Terada, L. Zelenyi, Space Sci. Rev. (2007, this issue). doi:10.1007/s11214-007-9245-8

G. Giampieri, A. Balogh, Mercury's thermoelectric dynamo model resvisisted. Planet. Space Sci. **50**, 757–762 (2002)

Glassmeier, K.-H., Grosser, J., Auster, H.-U., Constantinescu, D., Narita, Y., Stellmach, S., Electromagnetic induction effects and dynamo action in the Hermean system. Space Sci. Rev. (2007). doi:10.1007/s11214-007-9244-9

S.A. Hauk, A.J. Dombard, R.J. Phillips, S.C. Solomon, Internal and tectonic evolution of Mercury. Earth Planet. Sci. Lett. **222**, 713–728 (2004)

M.H. Heimpel, J.M. Aurnou, F.M. Al-Shamali, N. Gomez Perez, A numerical study of dynamo action as a function of sperical shell geometry. Phys. Earth Planet. Interiors **236**, 542–557 (2005)

R. Hollerbach, On the theory of the geodynamo. Phys. Earth Planet. Interiors **98**, 163–185 (1996)

R. Hollerbach, C.A. Jones, Influence of the Earth's inner core on geomagnetic fluctuations and reversals. Nature **365**, 541–543 (1993)

L.L. Hood, A. Zakharian, J. Halekas, D.L. Mitchell, R.P. Lin, M. Acuña, A.B. Binder, Initial mapping and interpretation of lunar crustal magnetic anomalies using Lunar Prospector magnetometer data. J. Geophys. Res. **106**, 27,825–27,839 (2001)

W. Kuang, J. Bloxham, An Earth-like numerical dynamo model. Nature **389**, 371–374 (1997)

C. Kutzner, U.R. Christensen, From stable dipolar to reversing numerical dynamos. Phys. Earth Planet. Interiors **131**, 29–45 (2002)

C. Kutzner, U.R. Christensen, Simulated geomagnetic reversals and preferred virtual geomagnetic pole paths. Geophys. J. Int. **157**, 1105–1118 (2004)

B. Langlais, M.E. Purucker, M. Mandea, The crustal magnetic field of Mars. J. Geophys. Res. **109** (2004).

V. Lesur, S. Maus, A global lithospheric magnetic field model with reduced noise level in the polar regions. Geophys. Res. Lett. **33** (2006)

J.R. Lister, B.A. Buffett, The strength and efficiency of thermal and compositional convection n the geodynamo. Phys. Earth Planet. Interiors **91**, 17–30 (1995)

J.G. Luhmann, The solar wind interaction with Venus. Space Sci. Rev. **44**, 241 (1986)

M. Mandea, M. Purucker, Observing, modeling, and interpreting magnetic fields of the solid Earth. Surv. Geophys. **26**, 415–459 (2005)

M. Mandea, E. Bellanger, J.-L. Le Mouël, A geomagnetic jerk for the end of 20th century? Earth Planet. Sci. Lett. **183**, 369–373 (2000)

J.L. Margot, S.J. Peale, R.F. Jurgens, M.A. Slade, I.V. Holin, Large longitude libration of Mercury reveals a molten core. Science **316**, 710–00 (2007)

S. Maus, M. Rother, C. Stolle, W. Mai, S. Choi, H. Lühr, D. Cooke, C. Roth, Third generation of the potsdam magnetic model of the earth (POMME). Geochem. Geophys. Geosyst. **7**, 7008 (2006)

N.F. Ness, The magnetic field of Mercury. Phys. Earth Planet. Interiors **20**, 209–217 (1979)

N. Olsen, H. Lühr, T.J. Sabaka, M. Mandea, M. Rother, L. Tofner-Clausen, S. Choi, Geophys. J. Int. **166**, 67–75 (2006)

P. Olson, U.R. Christensen, Dipole moment scaling for convection-driven planetary dynamos. Earth Planet. Sci. Lett. **250**, 561–571 (2006)

P. Olson, U. Christensen, G.A. Glatzmaier, Numerical modeling of the geodynamo: Mechanism of field generation and equilibration. J. Geophys. Res. **104**, 10,383–10,404 (1999)

J. Rotvig, C.A. Jones, Rotating convection-driven dynamos at low Ekman number. Phys. Rev. E **66**(5), 056308 (2002)

S.K. Runcorn, An acient lunar magnetic dipole field. Nature **253**, 701–703 (1975)

G. Schubert, M.N. Ross, D.J. Stevenson, T. Spohn, Mercury's thermal history and the generation of its magnetic field, in *Mercury*, ed. by F. Vilas, C.R. Chapman, M.S. Matthews (The University of Arizona Press, Tucson, 1988), pp. 651–666

G. Siscoe, N.F. Ness, C.M. Yeates, Substorms on Mercury? J. Geophys. Res. **80**, 4359 (1975)

T. Spohn, M.H. Acuña, D. Breuer, M. Golombek, R. Greeley, A. Halliday, E. Hauber, R. Jaumann, F. Sohl, Geophysical constraints on the evolution of Mars. Space Sci. Rev. **96**, 231–262 (2001)

T. Spohn, F. Sohn, K. Wieczerkowski, V. Conzelmann, The interior structure of Mercury: What we know, what we expect from BepiColombo. Planet. Space Sci. **49**, 1561–1570 (2001)

S. Stanley, J. Bloxham, W.E. Hutchison, M.T. Zuber, Thin shell dynamo models consistent with Mercury's weak observed magnetic field. Earth Planet. Sci. Lett. **234**, 341–353 (2005)

S. Stellmach, U. Hansen, Cartesian convection driven dynamos at low Ekman number. Phys. Rev. E **70**(5), 056312 (2004)

D.J. Stevenson, Mercury's magnetic field: A thermoelectric dynamo? Earth Planet. Sci. Lett. **82**, 114–120 (1987)

D.J. Stevenson, T. Spohn, G. Schubert, Magnetism and thermal evolution of terrestrial planets. Icarus **54**, 466–489 (1983)

F. Takahashi, M. Matsushima, Dynamo action in a rotating spherical shell at high Rayleigh numbers. Phys. Fluids **17**, 076601 (2005)

F. Takahashi, M. Matsushima, Dipolar and non-dipolar dynamos in a thin shell geometry with implications for the magnetic field of Mercury. Geophys. Res. Lett. **33**, L10202 (2006)

F. Takahashi, M. Matsushima, Y. Honkura, Dynamo action and its temporal variation inside the tangent cylinder in MHD dynamo simulations. Phys. Earth Planet. Interiors **140**, 53–71 (2003)

R. Tyler, L. Mysak, J. Oberhuber, Electromagnetic fields generated by a three dimensional global ocean circulation. J. Geophys. Res. **102**, 5531–5551 (1997)

R. Tyler, A. Maus, H. Lühr, Satellite observations of magnetic fields due to ocean tidal flow. Science **299**, 239–241 (2003)

B.P. Weiss, S.S. Kim, J.L. Kirschvink, R.E. Kopp, M. Sankaran, A. Kobayashi, A. Komeili, Ferromagnetic resonance and low-temperature magnetic tests for biogenic magnetite. Earth Planet. Sci. Lett. **224**, 73–89 (2004)

J. Wicht, Inner-core conductivity in numerical dynamo simulations. Phys. Earth Planet. Interiors **132**, 281–302 (2002)

J. Wicht, Palaeomagnetic interpretation of dynamo simulations. Geophys. J. Int. **162**, 371–380 (2005)

J. Wicht, P. Olson, A detailed study of the polarity reversal mechanism in a numerical dynamo model. Geochem. Geophys. Geosyst. **5**, 3 (2004)

J.-P. Williams, F. Nimmo, Thermal evolution of the Martian core: Implications for an early dynamo. Geology **32**, 97 (2004)

K.-K. Zhang, F.H. Busse, Finite amplitude convection and magnetic field generation in in a rotating spherical shell. Geophys. Astrophys. Fluid Dyn. **44**, 33–53 (1988)

The Surface of Mercury as Seen by Mariner 10

G. Cremonese · A. Sprague · J. Warell · N. Thomas ·
L. Ksamfomality

Originally published in the journal Space Science Reviews, Volume 132, Nos 2–4.
DOI: 10.1007/s11214-007-9231-1 © Springer Science+Business Media B.V. 2007

Abstract The Mariner 10 spacecraft made three flyby passes of Mercury in 1974 and 1975. It imaged a little less than half of the surface and discovered Mercury had an intrinsic magnetic field. This paper briefly describes the surface of Mercury as seen by Mariner 10 as a backdrop to the discoveries made since then by ground-based observations and the optimistic anticipation of new discoveries by MESSENGER and BepiColombo spacecraft that are scheduled for encounter in the next decade.

Keywords Planets: Mercury · Space vehicule · Instrumentation: imager · Mercury surface · Mariner 10

1 Introduction

This paper is devoted to the description of Mercury's surface as seen by Mariner 10. Mariner 10 imaged just less than half of the surface. From those images we learned that Mercury is a heavily cratered planet with an ancient surface that dates back to a period before accretionary heat was fully dissipated as evidenced by a scarp system that indicates global contraction.

G. Cremonese (✉)
INAF-Osservatorio Astronomico, vic. Osservatorio 5, 35122 Padova, Italy
e-mail: gabriele.cremonese@oapd.inaf.it

A. Sprague
Lunar and Planetary Laboratory, University of Arizona, Tucson, AZ 85721, USA

J. Warell
Dept. Astronomy & Space Phys., Uppsala University, Uppsala, Sweden

N. Thomas
Physikalisches Institut, University of Bern, Bern, Switzerland

L. Ksamfomality
IKI, Moscow, Russia

Mercury poses severe thermal and dynamical challenges to observation by spacecraft, and to date the planet has been visited by only one spacecraft, Mariner 10. The Mariner 10 mission was not intended to orbit Mercury because it would be travelling so fast past the planet (50 km/s) that it would require a huge amount of fuel to slow down the spacecraft enough to put into orbit. The size of the retrorocket would have to be equivalent to a medium-sized launch vehicle of that era. For almost 30 years we have waited for new technology to overcome the obstacles of cruise and orbital insertion to Mercury.

At last, two more missions are scheduled to encounter and orbit the planet in the next decade. MESSENGER, a NASA mission, launched on August 2004, will undergo orbital insertion in March 2011. The European Space Agency (ESA) approved the new mission, Bepi-Colombo, to Mercury pointing to a cornerstone, the n.5. in the year 2000. BepiColombo, scheduled for launch in August 2013, will orbit the planet beginning in September 2019. In the meantime, ground-based observations of Mercury are systematically increasing our knowledge of this enigmatic planet. In order to better appreciate the new ground-based discoveries and the challenges and scientific goals of the MESSENGER and BepiColombo missions, we benefit from becoming familiar with the Mercury seen by Mariner 10.

2 The Mariner 10 Mission

The mission plan for Mariner 10 was the most complex for any planetary mission up to that time. A gravity-assist trajectory technique was needed to obtain an economically acceptable mission. This technique allows a spacecraft to change both its direction and speed without using a valuable fuel, thereby saving time and leaving more weight for the scientific payload. A single gravity-assist can provide more delta V than a full rocket stage.

In February 1970 the mechanical engineer Giuseppe Colombo, of the University of Padova, Italy, noted that after Mariner 10 flew by Mercury, its orbital period around the Sun would be quite close to twice Mercury's orbital period. Therefore, he suggested that a second planet encounter could therefore be accomplished. After having confirmed this suggestion, the JPL carefully selected the Mercury flyby points in order to get a gravity correction able to return to Mercury six months later. The number of flybys depended upon the fuel available for midcourse corrections and attitude control. Mariner 10 achieved three encounters with Mercury before running out of fuel. This strategy limited the view of Mercury's surface to the same half of the planet (longitudes 10–190°).

In the very narrow launch window, NASA chose November 3, 1973, so that the spacecraft encountered Mercury at a time when it could view the planet about half lit (quadrature). Viewing Mercury at this phase made it easier to distinguish surface features by their shadows. The trajectory relied on Venus's gravitational field to alter the spacecraft's flight path and causing it to fall closer to the Sun and cross Mercury's orbit at the precise time needed to encounter the planet.

The flight plan called for the upper-stage Centaur rocket to be turned off for 25 minutes shortly after launch from Kennedy Space Center. Then a second ignition thrust the Mariner spacecraft in a direction opposite to the Earth's orbital motion, providing the spacecraft with a lower velocity relative to the Sun than the Earth's orbital velocity. This allowed it to be drawn inward by the Sun's gravitational field and achieve an encounter with Venus. After a few months, Mariner 10 approached Venus from its night side, passing over the sunlit side and, slowed by Venus's gravitational field, falling inward toward the Sun to rendezvous with Mercury.

The Mariner 10 spacecraft evolved from more than a decade of Mariner technology, beginning with the Venus mission in 1962 and culminating with the Mars orbiter in 1971.

Mariner 10 would be the last of the Mariner spacecraft to fly. Like the other Mariners, it consisted of an octagonal main structure, solar cell panels, a battery for electrical power, nitrogen gas jets for three-axis attitude stabilization and control, star and Sun sensors for celestial reference, S-band radio (12.6 cm wavelength) for command and telemetry, a high- and low-gain antenna, and a hydrazine rocket propulsion system for trajectory corrections.

The weight of the spacecraft was 534 kg, including 29 kg of hydrazine and 30 kg for the adapter to the launch vehicle. The scientific payload had a mass of 78 kg. The Mariner project modified the spacecraft for a specific mission toward Mercury because it had to approach the Sun at the closest distance ever achieved by a spacecraft. It was subjected to insolation up to 4.5 times greater than at Earth, requiring thermal control to maintain temperatures at a level that would not damage the spacecraft systems. As the spacecraft approached the Sun, the panels were rotated to change the angle at which light fell on them to maintain a suitable temperature of about 115°C.

Mariner 10 was able to handle up to 118 kb/s of imaging data and 2,450 bit/s for any other data, using a X-band antenna and the capability of transmitting telemetry on both X and S bands.

2.1 The Imaging System

The Mariner 10 imaging system consisted of two vidicon cameras, each with an eight-position filter wheel. The vidicons were attached to long focal length Cassegrain telescopes, which were mounted on a scan platform for accurate pointing. These telescopes provided narrow-angle and high resolution images.

The design team came up with an attractive, but risky, solution: treble the focal length of the Mariner 6/7 design and use twin camera systems on alternate 42-second readout cycles. This way, many high-resolution images under good low lighting could be captured well before and well after passing the darkened surface at closest approach. In addition, the TV team was able to persuade the telemetry engineers at JPL that television images can be quite interpretable even with considerable "salt-and-pepper" telemetry noise. Hence, the communication bandwidth for Mariner 10 could be increased greatly within existing power and antenna capabilities by accepting a much higher noise level. This meant that in addition to acquiring a full tape recorder, as had the earlier Mars Mariners, Mariner 10 could also transmit the 115,000-bps output video signal directly to Earth during many intervals. As a consequence, Mariner 10 would return many thousands of extraordinarily sharp images of Mercury, rather than the few hundred lower-resolution frames that otherwise would have been the case. Trebling the focal length of an existing optical design is never easy, but in this case the team was constrained as well for volume and configuration reasons to keep the rather short overall length (550 mm) of the Mariner 6/7 telescope nearly the same for Mariner 10. Thus, the increase in focal length from 500 to 1500 mm had to be accomplished entirely by secondary mirror magnification at the front of the telescope, where there would maximum vulnerability to thermal gradients.

The primary objective of the imaging experiment was to study the physiography and geology of Mercury's surface; determine accurately its size, shape, and rotation period; evaluate its photometric properties; and search for possible satellites and color differences on its surface.

The vidicon imaging system was spatially nonuniform in bias and dark current, as well as being nonlinear at the extremes of the light transfer curve. Prelaunch flat-field images acquired at varying exposure times allowed for the derivation of a nonlinearity and sensitivity nonuniformity correction, while an average of inflight images of deep space corrected for system offset.

Table 1 Main characteristics of the vidicon cameras

Focal length	1500 mm (62 mm)[a]
F ratio	f/8.4
Shutter velocity range	33.3 ms to 11.7 s
Field of view	$0.38° \times 0.47°$ ($9° \times 11°$)[a]
Vidicon target area	9.6×12.35 mm
Line scans per picture	700
Number of pixel per line	832
Number of bits per pixel	8
Filters and central wavelength	Clear (487 nm), UV (355 nm), blue (475 nm), minus UV (511 nm), orange (575 nm)

[a]Wide angle imaging

Fig. 1 Sketch of the vidicon cameras onboard Mariner 10

Table 1 shows the main characteristics of the imaging system, and Fig. 1 presents a sketch of the cameras.

During the three close encounters with Mercury in 1974–1975, the imaging system has acquired images of 40–45% of the total surface at spatial resolutions between 1 and 4 km/px, with highest resolutions obtained in selected areas down to about 100 m/px and due to the resonances of orbital encounter geometry the viewed longitude range was 10–190°.

3 Cratering

The global view of Mercury's surface, as revealed by the Mariner 10 camera during its first encounter (Mercury I) with the planet, at first glimpse appears lunar-like, covered with impact craters (Murray et al. 1974). Like the Moon, Mercury shows several large multi-ring structures, such as the Caloris basin whose eastern half was captured by Mariner 10 cameras. In both ways comparable to, but also different from the lunar surface, Mercury features vast smooth plains, in many places with a lower density of impact craters. Crater forms on Mercury are similar to their lunar counterparts (Spudis and Guest 1988).

On planetary surfaces we can see simple craters that are the smallest hypervelocity impact structures, they are bowl-shaped in form and have sharp rims and over-turned stratigraphy in the ejecta blanket. The morphology of these structures is controlled mainly by the strength of the substrate.

As impact energy increases, the target loses strength and there is a collapse of the walls and an uplift of the crater interior creating the complex craters. These are consequently formed in a regime where gravity is the dominant factor. Complex craters are characterized by terraced walls that are the surficial manifestation of subsurface faults, and they also show central peaks that contain material brought up from deep beneath the crater. They have smaller depth-to-diameter ratios because the increased importance of gravity collapse results in more uplift of floor. In these craters the average depth is about one-tenth of the diameter.

The next morphological step up in the energy scale corresponds to basins, which transition from structures that have a central peak and rings, peak-ring basins, to no central peak and multiple rings, multi-ring basins. In the transition to a peak ring basin, the central peak collapses to form a small ring that increases with increasing impact kinetic energy. In a multi-ring basin the number of rings scales with the impact energy and the mechanical properties of the near-surface layer into which the impact occurred. Basins have even smaller depth-to-diameter ratios than complex craters due to more central uplift.

As for simple-complex craters, the diameter of the complex crater-to-basin transition also depends on gravity, but the morphology of large multi-ring basins cannot be attributed to gravity alone.

The transition diameter from simple to complex crater forms on Mercury is 10.3 km (Pike 1988). In contrast to lunar craters, however, continuous ejecta and secondary crater fields are rarely found on Mercury, due to the higher surface gravity (Gault et al. 1975).

However, impact debris re-impacts the surface with higher velocities and hence has a stronger effect on eroding pre-existing landforms (Gault et al. 1975; Spudis and Guest 1988).

In detail, Mercury's craters have morphological differences from those on the Moon and Mars, partly due to differences in gravity and impactor environment (e.g., higher velocity impacts on Mercury), but most of the differences are probably due to the different geological processes that erode and degrade craters after they have formed on the various planets.

Potential sources for the impactors that formed Mercury's craters are numerous. In principle, the size distributions and the impact rates could have varied with time and in ways not necessarily correlated with the cratering histories of other bodies. Sources include: the Near-Earth asteroids and their cousins (of which only three have yet been found), which orbit entirely interior to Earth's orbit (termed Apoheles); short- and long-period comets, including sun-grazers; vulcanoids, an as-yet-undiscovered hypothetical population of remnant planetesimals from accretionary epochs, orbiting mainly inside Mercury's orbit; and secondary cratering by ejecta from basins and large primary craters. Endogenic crater-forming processes (e.g., volcanism) are also possible.

Interesting crater studies of Mercury can be based on topographic information derived from the shadows of craters in the Mariner 10 images in order to get the depth-to-diameter ratio. This ratio can be used for investigating if terrain types may be related to different values.

Figure 2 shows two examples of measurements of depth and diameter of craters according to Pike (1988), who collected morphologic characteristics of 316 impact craters, and Andre' and Watters (2006) who analyzed 173 craters.

Craters can be used as relative age markers by counting their numbers and size distributions on a planetary surface. To use craters in the dating of surfaces one must consider the rate of crater production and obliteration. To assess the crater population, the principal piece of information is the measured number of craters as a function of diameter over planetary surfaces over all ages. The principal piece of information in crater counting is the number of craters per unit area as a function of diameter.

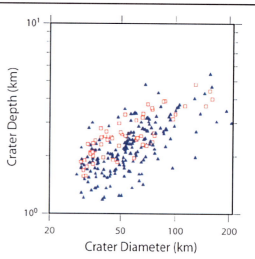

Fig. 2 Log–log plot of the depths and diameters of mature complex hermean craters. *Blue triangles* represent data from Andre' and Watters (2006) and *red squares* represent data from Pike (1988)

Neukum (1983) and Neukum and Ivanov (1994) proposed an analytical function to describe the cumulative number of craters with diameters larger than a given diameter D per unit area. This function was constructed from pieces of impact crater-size-frequency distribution (SFD) data measured in different areas of various ages on the Moon. Hence, the Neukum production function (NPF) implicitly assumes a constant shape of the production SFD during all lunar history (Neukum et al. 2001).

The well-investigated size-frequency distributions for lunar craters may be used to estimate the SFD for projectiles which formed craters on terrestrial planets and on asteroids (Neukum et al. 2001). The result shows the relative stability of these distributions during the past 4 Gyr. The derived projectile size-frequency distribution is found to be very close to the size-frequency distribution of Main-Belt asteroids as compared with the recent Spacewatch asteroid data and astronomical observations as well as data from close-up imagery by space missions (Ström et al. 2005). It means that asteroids (or, more generally, collisionally evolved bodies) are the main component of the impactor family. A cratering chronology model is established that can be used as a safe basis for modelling the impact chronology of other terrestrial planets, especially Mercury (Ivanov 2001; Hartmann and Neukum 2001).

The general conclusion from some models is that the impact cratering rate on Mercury does not seem to differ more than ±50% from the lunar cratering rate in the same diameter bins. However, the shift of crater diameters and diameters of strength/gravity and simple/complex crater transition change the shape of the Mercurian production function in comparison with the lunar PF (Strom et al. 1975).

Figure 3 reports an example of few impact cratering chronology models, as discussed by Neukum et al. (2001).

Wagner et al. (2001) carried out new crater size–frequency measurements on the Mariner 10 images for various geologic units, using a recently updated crater production function polynomial and impact cratering chronology model derived for Mercury (e.g. Ivanov et al. 2001), in order to reassess the time-stratigraphic system established (McCauley et al. 1981; Spudis and Guest 1988).

The geologic units identified are: the densely cratered terrain (highlands), craters and basins, as the Caloris that is the youngest and largest one (about 1300 km in diameter) known so far. The results are reported in Table 2.

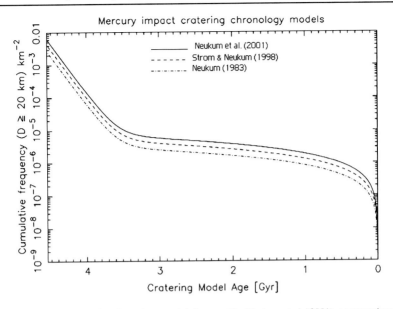

Fig. 3 Mercury impact cratering chronology model discussed by Neukum et al. (2001), compared to models published earlier (Neukum 1983; Ström and Neukum 1988). Cumulative frequency for $D \geq 20$ km plotted versus cratering model age in Giga-years (Gyr)

Table 2 Cumulative frequencies (for $D > 10$ km) and associated cratering model ages for major geologic units, craters and basins. Older values from (Ström and Neukum 1988) compared to updated values (Wagner et al. 2001). Uncertainties for cumulative frequencies (Ström and Neukum 1988; Wagner et al. 2001) are on the order of 20–30%, translating into model age uncertainties of 0.03–0.06 Gyr (Ström and Neukum 1988; Wagner et al. 2001)

Geologic unit	Cum. frequency ($D > 10$ km)		Crat. model age (Gyr)	
	Ström and Neukum (1988)	Wagner et al. (2001)	Ström and Neukum (1988)	Wagner et al. (2001)
Kuiper	–	(4.04e-6)	–	(1.0)
Mansur	–	(2.31e-5)	–	(3.5)
Caloris	6.85e-5	7.51e-5	3.85	3.77 ± 0.06
Beethoven	1.53e-4	1.22e-4	3.98	3.86 ± 0.05
Tolstoj	2.65e-4	2.51e-4	4.06	3.97 ± 0.05
Pushkin	3.45e-4	2.72e-4	4.10	3.98 ± 0.06
Haydn	3.65e-4	2.76e-4	4.11	3.99 ± 0.06
Dostojewskij	5.49e-4	2.75e-4	4.17	3.99 ± 0.06
Chekhov	4.04e-4	4.15e-4	4.12	4.05 ± 0.08
Highlands	5.99e-4	4.81e-4	4.18	4.07 ± 0.03

4 Global Contraction and Tectonics

One of the most important results of Mariner 10's imaging was the discovery that Mercury exhibits tectonic features. These features indicate Mercury's global contraction and provide enough geologic evidence to place the event at the end of the late heavy bombardment of

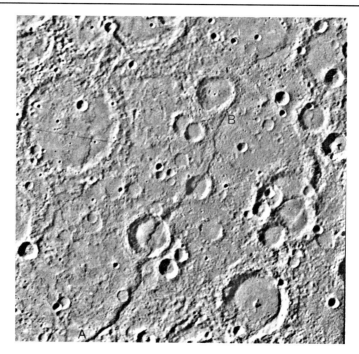

Fig. 4 Discovery fault discovered in Mariner 10 images during the third encounter in 1975. It traverses Mercury's surface for about 400 kilometers (measured from A to B) and has scarp faces of up to 1.5 km high. Figure from Strom and Sprague (2003)

the inner solar system and continuing until after the smooth intercrater plains formed (about 3.8 billion years ago) (Strom et al. 1975). The most obvious evidence discovered in Mariner 10 imagery was a long, sinuous scarp system extending for more than hundreds of km over Mercury's surface. Cliffs along the scarp appear to be 1.5 to 3 km high at some locations. The scarps cut crater rims (Fig. 4) and other landforms in such a way that permits us to be certain that they are a result of thrust faulting on a huge scale. The cliffs are rounded and in some cases have deformed the land forms they transect. Almost every region imaged by Mariner 10 shows examples of lobate scarps that were emplaced after the heavy bombardment and formation of the intercrater plains. The extensive system of thrust faults indicates that Mercury underwent a period of contraction that resulted in a decrease in surface area estimated to be about 31,000 to 63,000 km^2 following a shrinking diameter of 1–2 km (Strom et al. 1975).

Scarps younger than the emplacement of the intercrater plains have been identified, in particular, in pre-Tolstojan intercrater plains and Tolstojan and Calorian smooth plains units (Watters et al. 2004). This indicates that contraction associated with core formation and cooling continued throughout this period.

Another important tectonic discovery resulting from study of Mariner 10 images was evidence of a change in the shape of Mercury's lithosphere as a result of tidal despinning (cf. Melosh and McKinnon 1988). The evidence is a grid of lineaments (valleys, ridges, scarps, linear portions of central peaks, etc.) that roughly trends from the Caloris Basin antipode around to the Caloris Basin (Thomas et al. 1988). The formation of the grid is proposed to follow in two steps. First, the change in the surface area during global contraction caused a system of thrust faults while tidal despinning of the planet created various lineament struc-

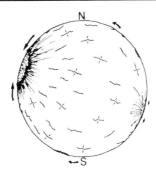

Fig. 5 Lateral flow of Mercury's lithosphere radially toward the center of Caloris Basin as basin melt cooled and subsided may have imprinted the lineament trend seen at Mercury. Figure adapted from Thomas et al. (1988)

Fig. 6 The EUV albedo (*circles with error bars*) of Mercury measured in eight bands by the UV instrument on Mariner 10. The BVRI measurements at right are from Harris (1961) with boxes from incoming and outgoing Mariner 10 images

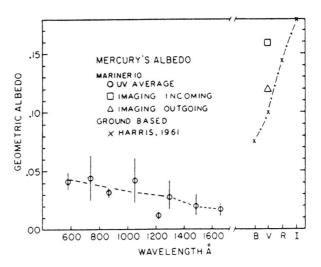

tures because of changes in shape of the lithosphere (Melosh and McKinnon 1988). Then, as illustrated by Fig. 5, the stretching of the lithosphere toward the center of the Caloris Basin following its formation, imprinted the grid trend (Thomas et al. 1988).

The part of the global system of thrust faults seen by Mariner 10 and the high relief ridges associated with them have been analyzed with respect to azimuthal and spatial distribution to constrain thermal models and to determine the origin of tectonic stresses (Melosh and McKinnon 1988). Some trends in the faults, lineaments and other features do not seem to be associated with despinning but rather with long-lived tectonic uplift bulge resulting in horst and graben extentional features in the Tolstoj-Zeami region (Thomas 1997). The implication of these features is that they are associated with a deep and long-lived, large-scale internal activity (Thomas 1997).

Other tectonic structures are also present. Following the excavation of Caloris Basin (and the lithospheric flow radially toward the basin center), basin concentric and basin radial ridges formed in response to basin interior lavas undergoing subsidence and compression (Watters et al. 2004). Wrinkle ridges formed as surface area decreased. Transecting the wrinkle ridges is a network of polygonal troughs exhibiting characteristics of graben formed from extensional stresses introduced by basin floor uplift (cf. Melosh and McKinnon 1988).

Recent modeling indicates that during the time of floor uplift, Mercury's crustal thickness must have already been thick enough to prevent the graben network destruction (Watters et al. 2004).

Regional thrust faults have been used to estimate the elastic lithospheric thickness given model assumptions regarding compositional and thermal physical parameters. Remarkably, models indicate the elastic lithosphere to be roughly 40 to 125 km (cf. Nimmo and Watters 2004 and references therein). Mercury's crust is estimated to be from 125 to 140 km thick, a range corresponding to a greater fraction of the total mantle volume than the Moon, Mars or Venus (Nimmo and Watters 2004).

The hilly and hummocky terrain antipodal to Caloris Basin is another region of tectonic, geologic, and physical interest. Because it exists antipodal to the Caloris Basin and shows no preferential directional trends in the lineations transecting the region, it does not appear to be a region of either regional compression or extension. Rather, a plausible formation scenario is that the random blockiness is a result of the convergence of seismic waves following the impact forming Caloris Basin (Shultz and Gault 1974).

It is thought that Mercury is no longer tectonically active. The geologic evidence from Mariner 10 indicates that the planet ceased tectonism sometime after core formation and the emplacement of the intercrater plains, probably between 3.8 and 3.2 by ago.

5 Surface Scattering Properties

The surface scattering and material properties of Mercury were studied from ultraviolet-optical images and thermal infrared radiance data obtained with the Television Photography (Murray et al. 1974; Soha et al. 1975) and the Two-Channel Infrared Radiometer experiments (Chase et al. 1976).

In addition to a primary objective of morphologic mapping and geologic studies of the Mercurian surface, the television camera also provided flux calibrated image data from which scattering and compositional parameters were inferred. For these studies, the ultraviolet (355 nm) and orange (575 nm) filters, as well as a UV polarizing filter, were primarily employed.

5.1 Light-Scattering Properties

From the orange image data, Hapke et al. (1975) determined the average normal albedo of Mercury's surface to be 0.14 at 554 nm. Pointing angle constraints for Mariner 10 restricted the phase angle coverage to between 80 and 110 degrees, and photometric results were thus derived from disk images of the planet rather than phase curve data or a combination of both. Brightness scans across the luminance equator extracted from first encounter images of the incoming hemisphere, which was found to be primarily covered by intercrater plains, were very similar to brightness profiles of lunar highlands at corresponding phase angles which had previously been obtained with the same camera. Images of the outgoing hemisphere, covered with smooth plains at the terminator and intercrater plains near the limb, indicated that the photometric properties of smooth plains more closely resemble the lunar maria than heavily cratered highland units.

Normal albedos at 5 degree phase angle and 554 nm wavelength were calculated by Hapke et al. (1975) assuming that the lunar photometric function of Hapke (1966) was valid for all types of geologic units. The intercrater plains were found to have albedos of 0.14 which was similar to lunar highlands, while the somewhat darker smooth plains were brighter than most lunar maria. Two types of smooth plains were found—bright plains with albedos of about 0.20, and dark plains in and around the Caloris basin with albedos of around 0.13. Immature ray craters were found to have albedos systematically higher than lunar examples, with values of 0.35 not uncommon, and with the floor of Kuiper being the brightest with a normal albedo of 0.45. These values are slightly higher, particularly for bright ray craters, than the normal albedos previously cited in the preliminary report by Murray et al. (1974).

A search for regions of anomalous relative linear polarization signal was performed by comparing the UV imaging and the UV polarizing filter images, having the polarization direction oriented in the principal scattering plane. Hapke et al. (1975) reported that no strongly polarizing regions could be found to a size limit of 20 km. Polarization differences across the surface were determined to be generally less than 15%, with the smallest polarizations at brighter craters, suggestive of a less mature surface with a somewhat higher crystalline rock content.

The bland appearance of the polarization ratio images were in stark contrast to color ratio images formed from ultraviolet- and orange-filter data. It was found that bright ray craters and their ejecta were more blue than the surrounding areas by about 12%, while other bright regions were generally more red than darker areas. No color differences could be detected across large scarps, supporting the view that these features were not due to lava flow fronts.

From the albedo and color properties of the surface, Hapke et al. (1975) made a number of conclusions which likely hold true even today: the high brightness and blue color of bright ray craters was inferred to be due to a material which is only weakly absorbing (like feldspar or quartz) and low in Ti, Fe^{3+} and metallic iron. As already pointed out by Murray et al. (1974), the albedo contrast between Mercury's smooth plains and the intercrater and highly cratered units is much smaller than between maria and highlands on the Moon, which implies that the smooth plains may be less abundant in Ti and Fe than the average lunar maria. It was further concluded that no high-Fe, high-Ti maria similar to the lunar counterparts seem to exist on the Mariner 10 hemisphere. These interpretations of the Mariner 10 data signify a surface considerably less similar to the lunar than reported by Murray et al. (1974), who found "surprising similarity to the Moon in regional color variations as well as albedo variations".

Hapke (1977) used brightness profiles along the luminance equator of the Moon and Mercury to derive the average maximum slope angle of the surface. The value for Mercury, about 25 degrees, was shown to be about half that for the Moon, attributed to the effect of the stronger surface gravity of Mercury on the angle of repose of a cohesive soil. Hapke (1977) also showed that the photometric darkening towards the poles detected on both Mercury and lunar far-side images obtained by Mariner 10 could be adequately explained by the effect of crater shadowing, and that the effect is not due to a systematic dependence of normal albedo with latitude. This result implied that the major physical cause of maturation darkening of the surfaces of both bodies is vapor-deposition reduction and formation of submicroscopic metallic iron particles in grain rims due to micrometeoritic impacts, rather than solar wind sputtering. If efficient, the latter effect would likely have produced darkening with a latitude dependency due to the magnetic field strength anisotropy, with large solar wind fluxes being primarily directed towards the polar regions by the inferred dipolar field.

In continuation of the photometric modelling work and its application, Hapke and Wells (1981) showed that a theoretical expression for the bidirectional reflectance of a particulate surface (Hapke 1981) was adequate to explain most of the relative brightness profiles of Mercury as derived from Mariner 10 image data. The model, assuming isotropically single-scattering particles of the same absorption coefficient, still predicted the occurrence of a strong limb surge which could not be observed for atmosphereless bodies with dark and rough porous regoliths (Hapke 1977). Hapke (1984) presented a rigorous mathematical formalism to take into account the effects on the bidirectional reflectance of macroscopic shadowing on arbitrarily tilted surface sections, and showed that surface roughness angles of 20 degrees could explain the observed absence of a limb surge in Mariner 10 photometric profiles of Mercury.

Further to this work, Bowell et al. (1989) compared and contrasted the photometric models of Hapke (1986 and references therein) and that of Lumme and Bowell (1981 and references therein) and their applications to the Mariner 10 brightness profiles of Mercury and ground-based integral phase curves of Mercury and the Moon. The fits provided by these models were shown to be good and nearly identical, although it was stressed that, due to the number of model parameters and their complex contribution to the brightness of a resolved or integral planetary surface, the solutions are generally not unique.

The highest albedo craters identified by Hapke et al. (1975) were studied by Dzurisin (1977). He proposed that the anomalously bright and structurally well-confined patches on floors of craters 50 km in diameter and larger were caused by an endogenic process as deduced from their morphologic and photometric attributes. This conclusion may suggest that local material, possibly originally subsurface, distinct in the chemical and/or textural properties from the surrounding surface material, may have been extruded through subsurface cracks generated at the impacts.

The Mariner 10 image data from the first inbound trajectory was fully recalibrated by Robinson and Lucey (1997) to remove image artifacts present in the originally reduced data set, in order to study subtle color variations across the surface. Using boundary conditions from ground-based integral phase curve photometry and optical spectroscopy, they removed the photometric function of the resolved disk with the model of Hapke (1986), based on the photometric parameters for the Moon (Helfenstein and Veverka 1987). The derived photometric solution had a mean surface roughness value of 15 degrees, with the other parameters being very lunar-like. Using this recalibration, they found the brightest ray craters to have normal albedos of 0.29, which is twice the global average for the observed hemisphere and considerably lower than previously found by Hapke et al. (1975). The difference could not be explained but was suggested to be due to the improved calibration techniques.

Brightness profiles of Mariner 10 images were reanalyzed by Mallama et al. (2002) and Warell (2004) in a comparison with ground-based photometry of Mercury and the Moon. In modelling the Mariner 10 image data with Hapke's (1993, 2002) bidirectional reflectance function, both authors revealed that the brightness scans were calibrated about 9% too bright compared to ground-based data, which had already been indicated by Hapke et al. (1975). The average surface roughness slope angle was determined to 16 degrees by Mallama from his new extended V-band phase curve observations, and to 8–15 degrees by Warell (2004) from a combination of Mallama's phase curve data and Warell's ground-based disk-resolved images. The smaller range of roughness values found by Warell (2004) was explained as due to light primarily scattered from smooth plains near Mercury's north pole at the very highest phase angles. Values near 15 degrees are consistent with both the Mariner 10 brightness scans and the integral phase curve in general and representative of terrains mainly consisting of cratered highlands and intercrater plains. This difference in surface roughness between geologically different surface units may indicate unique modes of origin, and is consistent with the suggestion that smooth plains are extrusive lava deposits.

Scanned and intensity-corrected photographic copies of Mariner 10 original images were used by Shevchenko (2004, 2006) to study the relative brightnesses and photometric functions of different geologic units. It was found that in a diagram of the value of the photometric function versus the relative image brightness, three types of units are seen as different trends in the plot. This was interpreted as a result of varying surface roughness caused primarily by particles with different size distributions.

Blewett et al. (2007) used the recalibrated Mariner 10 ultraviolet and orange image data of Robinson and Lucey (1997) to create two spectral parameter images sensitive to the abundance of spectrally neutral opaque phases and the degree of maturity and/or FeO content in the crust. Robinson and Lucey (1997) originally used these images to study plains units with unique properties and argued that one of them, showing low opaque abundance and embayment relations compared to the surroundings, was consistent with the presence of effusive volcanic material. Blewett et al. (2007) performed further work on these data to examine impact-related features across the first encounter incoming hemisphere, and found extended areas of low-opaque material to be present at depth and excavated by some impact craters, as well as small immature craters that had not excavated such low-opaque material. These relations suggested a two-layer crustal model, in which the low-opaque layer is located at a depth of 3–4 km, analogous to the stratification of the lunar highland crust. Furthermore, a location of geologically young but opaque-rich material was found, possibly related to lunar dark-halo impact craters or a pyroclastic deposit origin.

The work of Blewett et al. (2007) must be put in relation to that of Warell and Valegård (2006), who studied similarly constructed spectral parameter images for the poorly known hemisphere obtained with data from 1-m Swedish Solar Telescope (see Ksanfomality, this volume). Though the spatial resolution was two orders of magnitude lower than that of Mariner 10 images, one advantage was the extended near-infrared spectral coverage allowing separation of the maturity/FeO ambiguity. It appears that geologic features similar to those identified on the Mariner 10 hemisphere are also present on the less-known hemisphere particularly with respect to opaque-rich and mature areas. Such findings are consistent with the presence of more iron- and titanium-rich material that may have been excavated from depth. If so, volcanic plains may have formed at locations scattered globally across the surface of Mercury.

5.1.1 The Extreme Ultraviolet Albedo of Mercury's Surface

The UV albedos of Mercury's surface were measured by the Mariner 10 UV spectrometer (Broadfoot et al. 1976; Wu and Broadfoot 1977) and, at the first encounter, were found to be similar to the UV albedo of the Moon (Broadfoot et al. 1974). Further detailed analysis after the second and third encounters showed Mercury's surface to be of considerably lower albedo as shown in Fig. 6, for Mercury and in Fig. 7 for the ratio of Mercury's albedo to that of the Moon.

Although there are uncertainties in the absolute geometric albedo as shown by error bars in the EUV albedo, the values of the ratio of Mercury to the Moon are thought to be quite good because instrumental systematic errors are removed by the ratio technique.

5.2 Material Properties

Measurements carried out with the Mariner 10 infrared radiometer provided information on the temperature and thermal properties of the crust (Chase et al. 1976). Observations were made on the first pass and primarily the unilluminated hemisphere was observed. It was found that Mercury is essentially indistinguishable from the Moon, with the derived thermal skin depth, electric skin depth and dielectric loss tangent all within the ranges found for the Moon. Observed variations in the derived surface temperature along the scan direction was attributed to differences in thermal inertia and rock coverage. One of the local enhancements

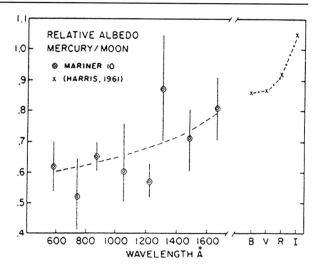

Fig. 7 The relative geometric EUV albedos of Mercury and the Moon as measured by the UV spectrograph on Mariner 10 and from Harris (1961)

of thermal inertia, found to display strong radar backscattering properties by Zohar and Goldstein (1974), was interpreted as due to a young ray crater. With respect to rock coverage, it was estimated that most of the surface was covered with about 2% exposed bare rock, while for some local hot regions the rock coverage was estimated to be up to a few times greater. Based on the relation between the loss tangent and the measured ilmenite content in lunar soil samples found by Olheoft and Strangway (1975), the estimated loss tangent of Mercury implied about 20% FeO + TiO$_2$ present in the crust. The results of ground-based cm wavelength observations of Mercury's surface and subsequent modelling of thermal and dielectric properties of the regolith by Mitchell and de Pater (1994) present quite a different view of Mercury's surface chemistry. The entire suite of ground-based observations puts an upper limit of 3% for FeO and TiO2 in Mercury's regolith. This subject is discussed in detail by Sprague et al. (2007, this volume).

6 Topography

The space exploration of terrestrial planets has recently demonstrated the importance of 3D rendering of planetary surfaces (e.g. Barnett and Nimmo 2002; Watters et al. 2002; Watters 2003; Neukum et al. 2004; Plescia 2004). Morphology is of paramount importance for better defining the main geological units of the planets, structural features linked to global and local tectonics, impact crater populations and, if present, volcanic edifices.

Then a digital elevation model (DEM) is needed when correcting images for illumination effects because it allows accurate calculation of incidence, emission, and phase angles on a pixel-by-pixel basis.

Up to now there are very few studies on the Mercury topography due to the small data set and the quite poor quality of the images to perform stereo reconstruction, in terms of spatial resolution, solar illumination, and view angles.

During all three flybys the same hemisphere and illumination conditions were presented to Mariner 10. Some of the images overlap and were taken from widely separated viewpoints, thus providing stereo coverage. Example stereo images (Davies et al. 1978) and a catalog of stereo pairs (Jet Propulsion Laboratory (JPL) 1976) have been published. However, at the time that these stereo pairs were published, the camera position and orientation data were still preliminary, and some of these pairs have extremely weak stereo.

Further interesting work was done by Cook and Robinson (2000) that used newly refined camera position and orientation data (Davies et al. 1996; Robinson et al. 1997, 1999a), finding additional useful stereo pairs never before identified. They compiled a new catalog of stereo pairs and produced a map of stereo coverage (Robinson et al. 1999b). They were able to provide approximately 24% of the planet mapped topographically to better than ±1 km height accuracy (for a single matched point), and 6% of the surface mapped to better than ±400 m height accuracy.

Recently, Andre' et al. (2005) used image pairs from the Mariner 10 data set to make stereo reconstructions of the Beethoven and Tolstoj basins from which they concluded that their topography is similar to that of lunar-mare filled basins. These maps are, however, coarse. Laser altimetry and dedicated stereo imaging from the MESSENGER and BepiColombo space missions will provide high spatial resolution products at better than 10 m vertical accuracy. A Digital Terrain Model—with a grid size lower than 300 m—of the entire planet surface; will also be implemented using new technologies and stereo reconstruction techniques. At present, there is no adequate analog. Clementine data from the Moon was too poor to be used for this purpose.

References

S.L. Andre', T.R. Watters, Proc. Lunar Planet. Sci. Conf. [CDROM], 37th, abstract 2054 (2006)
S.L. Andre', T.S. Watters, M.S. Robinson, Geophys. Res. Lett. **32**, L21202 (2005)
D.N. Barnett, F. Nimmo, Icarus **157**, 42 (2002)
D.T. Blewett, B.R. Hawke, P.G. Lucey, M.S. Robinson, J. Geophys. Res. **112**, 2005 (2007)
E. Bowell, B. Hapke, D. Domingue, K. Lumme, J. Peltoniemi, A.W. Harris, *Asteroids II, Proc. of the Conference* (University of Arizona Press, Tucson, 1989), p. 524
A.L. Broadfoot, S. Kumar, M.J.S. Belton, M.B. McElroy, Science **185**, 166 (1974)
A.L. Broadfoot, D.E. Shemansky, S. Kumar, Geophys. Res. Lett. **3**, 580 (1976)
S.C. Chase Jr., E.D. Miner, D. Morrison, G. Muench, G. Neugebauer, Icarus **28**, 565 (1976)
A.C. Cook, M.S. Robinson, J. Geophys. Res. **105**, 9429 (2000)
M.E. Davies, S.E. Dwornik, D.E. Gault, R.G. Strom, *Atlas of Mercury*. NASA Spec. Publ., vol. SP-423 (1978), p. 127
M.E. Davies, T.R. Colvin, M.S. Robinson, K. Edwards, Bull. Am. Astron. Soc. **28**, 1115 (1996)
D. Dzurisin, Geophys. Res. Lett. **4**, 383 (1977)
D.E. Gault, J.E. Guest, J.B. Murray, D. Dzurisin, M.C. Malin, J. Geophys. Res. **80**, 2444 (1975)
B. Hapke, Astron. J. **71**, 386 (1966)
B. Hapke, Conference on Comparisons of Mercury and the Moon, Houston, TX, Nov. 15–17 (1977)
B. Hapke, J. Geophys. Res. **86**, 3039 (1981)
B. Hapke, Icarus **59**, 41 (1984)
B. Hapke, Icarus **67**, 264 (1986)
B. Hapke, *Theory of Reflectance and Emittance Spectroscopy* (Cambridge University Press, Cambridge, 1993)
B. Hapke, Icarus **157**, 523 (2002)
B. Hapke, E. Wells, J. Geophys. Res. **86**, 3055 (1981)
B. Hapke, G.E. Danielson Jr., K. Klaasen, L. Wilson, J. Geophys. Res. **80**, 2431 (1975)
D. Harris, in *Planets and Satellites*, ed. by G. Kuiper, B. Middlehurst (University of Chicago Press, Chicago, 1961), pp. 273–342
W.K. Hartmann, G. Neukum, Space Sci. Rev. **96**, 165 (2001)
P. Helfenstein, J. Veverka, Icarus **72**, 342 (1987)
B.A. Ivanov, Space Sci. Rev. **96**, 87 (2001)
B.A. Ivanov, G. Neukum, R. Wagner, Workshop on Mercury: Space Environment, Surface, and Interior, Chicago, IL, 2001, p. 65
Jet Propulsion Laboratory (JPL), MVM '73 stereo data package user guide. Pasadena, CA, 1976
K. Lumme, E. Bowell, Astron. J. **86**, 1694 (1981)

A. Mallama, D. Wang, R.A. Howard, Icarus **155**, 253 (2002)
D. Mitchell, I. de Pater, Icarus **110**, 2–32 (1994)
B.C. Murray, M.J.S. Belton, G.E. Danielson, M.E. Davies, D.E. Gault, B. Hapke, B. O'Leary, R.G. Ström, V. Suomi, N. Trask, Science **185**, 169 (1974)
J.F. McCauley, G.G. Guest, J.E. Schaber, N.J. Trask, R. Greeley, Icarus **47**, 184 (1981)
H.J. Melosh, W.B. McKinnon, in *Mercury*, ed. by F. Vilas, C.R. Chapman, M.S. Matthews (University of Arizona Press, Tucson, 1988), p. 374
G. Neukum, Meteoritenbombardement and Datierungplanetarer Oberachen. Habilitation Dissertation for faculty membership. Univ. of Munich, 1983, p. 186
G. Neukum, B.A. Ivanov, in *Hazards Due to Comets and Asteroids*, ed. by T. Gehrels (Univ. of Arizona Press, Tucson, 1994), p. 359
G. Neukum, B.A. Ivanov, W.K. Hartmann, Space Sci. Rev. **95**, 55 (2001)
G. Neukum, R. Jaumann, H. Hoffmann, E. Hauber, J.W. Head, A.T. Basilevsky, B.A. Ivanov, S.C. Werner, S. van Gasselt, J.B. Murray, T. McCord, The HRSC Co-Investigator Team, Nature **432**, 979 (2004)
F. Nimmo, T.R. Watters, Gephys. Res. Lett. **31**, L02701 (2004). doi:10.1029/2003GL018847
G.R. Olheoft, D.W. Strangway, Earth Planet. Sci. Lett. **24**, 394 (1975)
R.J. Pike, in *Mercury*, ed. by F. Vilas, C.R. Chapman, M.S. Matthews (The University of Arizona Press, Tucson, 1988), p. 165
J.B. Plescia, J. Geophys. Res. **109**, E03003 (2004). doi:10.1029/2002JE002031
M. Robinson, P.G. Lucey, Science **275**, 197 (1997)
M.S. Robinson, M.E. Davies, T.R. Colvin, K.E. Edwards, *Proc. Lunar Planet. Sci. Conf., 28th*, 1997, pp. 1187–1188
M.S. Robinson, M.E. Davies, T.R. Colvin, K.E. Edwards, J. Geophys. Res. **104**, 30852 (1999a)
M.S. Robinson, A.S. McEwen, E. Eliason, E.M. Lee, E. Malaret, P.G. Lucey, *Proc. Lunar Planet. Sci. Conf.* [CDROM], 30th, abstract 1931, 1999b
V.V. Shevchenko, Sol. Syst. Res. **38**, 441 (2004)
V.V. Shevchenko, Adv. Space Res. **38**, 589 (2006)
P.H. Shultz, D.E. Gault, NASA Tech. Memo., X-62 (388) (1974)
J.M. Soha, D.J. Lynn, J.J. Lorre, J.A. Mosher, N.N. Thayer, D.A. Elliott, W.D. Benton, R.E. Dewar, J. Geophys. Res. **80**, 2394 (1975)
P.D. Spudis, J.E. Guest, in *Mercury*, ed. by F. Vilas, C.R. Chapman, M.S. Matthews (The University of Arizona Press, Tucson, 1988), p. 118
R.G. Strom, N.J. Trask, J.E. Guest, J. Geophys. Res. **80**(7), 2478 (1975)
R.G. Ström, G. Neukum, in *Mercury*, ed. by F. Vilas, C.R. Chapman, M.S. Matthews (The University of Arizona Press, Tucson, 1988), p. 336
R.G. Ström, R. Malhotra, T. Ito, F. Yoshida, D.A. Kring, Science **309**, 1847 (2005)
R.G. Strom, A.L. Sprague, *Exploring Mercury: The Iron Planet* (Springer, Praxis, Chichester, 2003), p. 216
P.G. Thomas, P. Masson, L. Fleitout, in *Mercury*, ed. by F. Vilas, C.R. Chapman, M.S. Matthews (University of Arizona Press, Tucson, 1988), p. 401
P.G. Thomas, Planet. Space Sci. **45**, 3 (1997)
R.J. Wagner, U. Wolf, B.A. Ivanov, G. Neukum, Workshop on Mercury: Space Environment, Surface, and Interior. Chicago, IL, 2001, p. 106
J. Warell, Icarus **167**, 271 (2004)
J. Warell, P.-G. Valegård, Astron. Astrophys. **460**, 625 (2006)
T.R. Watters, R.S. Schultz, M.S. Robinson, A.C. Cook, Geophys. Res. Lett. **29**(11), 1542 (2002). doi:10.1029/2001GL014308
T.R. Watters, J. Geophys. Res. **108**, 5054 (2003). doi:10.1029/2002JE001934
T.R. Watters, M.S. Robinson, C.R. Bina, P.D. Spudis, Geophys. Res. Lett. **31**, L04701 (2004). doi:10.1029/2003GL019171
H.H. Wu, A.L. Broadfoot, Geophys. Res. Lett. **82**, 759–760 (1977)
S. Zohar, R.M. Goldstein, Astron. J. **79**, 85 (1974)

Radar Imaging of Mercury

John K. Harmon

Originally published in the journal Space Science Reviews, Volume 132, Nos 2–4.
DOI: 10.1007/s11214-007-9234-y © Springer Science+Business Media B.V. 2007

Abstract Earth-based radar has been one of the few, and one of the most important, sources of new information about Mercury during the three decades since the Mariner 10 encounters. The emphasis during the past 15 years has been on full-disk, dual-polarization imaging of the planet, an effort that has been facilitated by the development of novel radar techniques and by improvements in radar systems. Probably the most important result of the imaging work has been the discovery and mapping of radar-bright features at the poles. The radar scattering properties of these features, and their confinement to permanently shaded crater floors, is consistent with volume backscatter from a low-loss volatile such as clean water ice. Questions remain, however, regarding the source and long-term stability of the putative ice, which underscores the need for independent confirmation by other observational methods. Radar images of the non-polar regions have also revealed a plethora of bright features, most of which are associated with fresh craters and their ejecta. Several very large impact features, with rays and other bright ejecta spreading over distances of 1,000 km or more, have been traced to source craters with diameters of 80–125 km. Among these large rayed features are some whose relative faintness suggests that they are being observed in an intermediate stage of degradation. Less extended ray/ejecta features have been found for some of the freshest medium-size craters such as Kuiper and Degas. Much more common are smaller (<40 km diameter) fresh craters showing bright rim-rings but little or no ray structure. These smaller radar-bright craters are particularly common over the H-7 quadrangle. Diffuse areas of enhanced depolarized brightness have been found in the smooth plains, including the circum-Caloris planitiae and Tolstoj Basin. This is an interesting finding, as it is the reverse of the albedo contrast seen between the radar-dark maria and the radar-bright cratered highlands on the Moon.

Keywords Mercury · Radar · Mercury surface · Delay-Doppler mapping · Polar ice

---

J.K. Harmon (✉)
National Astronomy and Ionosphere Center, Arecibo Observatory, Arecibo, PR 00612, USA
e-mail: harmon@naic.edu

1 Introduction

Our greatest single source of information on Mercury has, of course, been the Mariner 10 reconnaissance of 1974–1975. Much of what we have learned since then has come from Earth-based radar observations. Radar has not played the same crucial role for Mercury as it has for the exploration of cloud-shrouded Venus. Nevertheless, it has proved a valuable complement to the Mariner data, providing reflectivity images of the entire surface, altimetric profiles around the equator, and spin libration measurements bearing on the state of the core. Radar has provided most of what we know about the hemisphere left unimaged by Mariner 10. It has also revealed evidence of ice deposits in the permanently shaded floors of polar craters. The need for confirmation of polar ice has provided a major rationale for the MESSENGER and BepiColombo missions to Mercury and for the inclusion of neutron spectrometers in their instrument packages. It has also spurred on Earth-based and spacecraft searches for polar ice on the Moon.

The purpose of this paper is to review the current state of Mercury radar imaging and to discuss the implications for our knowledge of the planet's surface. The emphasis will be on the more recent imaging results from the Arecibo radar, because of their high resolution. However, imaging done with the Goldstone radar and Very Large Array (VLA) will also be discussed, as they have made important discoveries and also complement the Arecibo images in several respects. The best Arecibo imaging has been done since the 1998 completion of a major upgrade to the telescope and S-band radar system. The pre-upgrade Arecibo results have been presented elsewhere and will be mentioned only in an historical context. Preliminary or short reports on the post-upgrade Arecibo imaging have been presented in various venues (Harmon 2001; Harmon and Campbell 2002; Harmon 2004; Rice and Harmon 2004). The most thorough compilation of the post-upgrade Arecibo imaging of Mercury's equatorial and midlatitude regions is given in a recent paper by Harmon et al. (2007). Many of the results from that paper are also presented here, although in less detail and organized by topic rather than by region. A detailed presentation of the post-upgrade polar imaging results from 1998–1999 is given in Harmon et al. (2001). Here we will review those results as well as report on some of the polar imaging done in the years since.

A review of Mercury radar imaging is especially timely in light of the upcoming spacecraft flybys and orbital mapping missions. We have identified many radar features that warrant a closer look with spacecraft imagers, and the most impressive of these happen to be located in the Mariner-unimaged hemisphere (or "MUH"). New spacecraft imaging will, of course, supersede radar for mapping terrae incognitae. At the same time, it should open up interesting new opportunities for comparing radar imaging with spacecraft imaging and spectrometry.

2 Historical and Technical Background

The earliest radar observations of Mercury were simple detections aimed at measuring the planet's radar cross section and distance. Early measurements were also made of the frequency or Doppler spectrum of the echo, more recent examples of which are shown in Fig. 1. Since the Doppler broadening of the echo is caused by the apparent rotation of the planet as viewed from Earth, a measurement of the total bandwidth can give an estimate of the planet's rotation rate. Pettengill and Dyce (1965) used this method to establish Mercury's non-synchronous rotation, which remains one of the most important discoveries of radar astronomy.

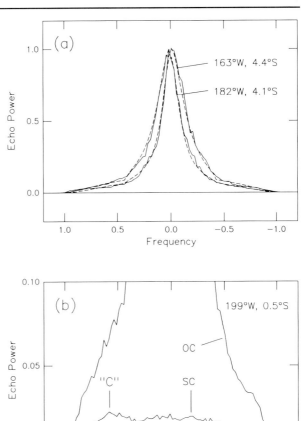

Fig. 1 Doppler spectra of the Mercury echo, from Arecibo radar observations. (**a**) OC spectra (*solid lines*) from two different dates, with corresponding scattering-law fits (*dashed lines*). (**b**) OC/SC spectrum pair from one date. The position of feature "C" is denoted on the SC spectrum. The OC spectrum is truncated at 10% of the maximum amplitude. The number pairs are the sub-Earth positions, and the frequency scale is in units of the limb Doppler. This figure is reproduced from Harmon (1997)

The Doppler spectrum can also be used to infer radio scattering properties of the planet surface. Normally one transmits a circularly polarized wave and receives the echo in both (orthogonal) circular polarizations. It is customary to use the terms OC (for Opposite Circular) and SC (for Same Circular) to denote the receive polarization senses that are the opposite of, or the same as, the transmitted sense, respectively. The OC (or "polarized") echo is dominated by a strong central peak associated with quasispecular reflections off large (superwavelength-scale) surface undulations (see Fig. 1a). This echo trails off into broad spectral wings associated with a diffuse component representing high-angle scattering off small (wavelength-scale) surface roughness elements distributed over the entire visible disk of the planet. The width of the OC central peak can be used to estimate the r.m.s. slope of the large-scale undulations (Fig. 1a), and the amplitude of the peak can give an estimate of the quasispecular reflectivity or effective surface dielectric constant. The SC (or "depolarized") echo is associated entirely with diffuse backscatter off wavelength-scale roughness. This component completely lacks the central specular peak and (since the diffuse component is only partially depolarized) remains weaker than the OC wings all the way out to the planet

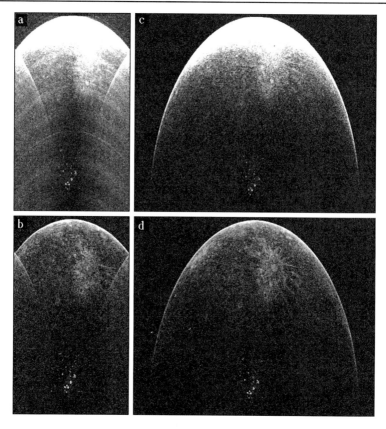

Fig. 2 Delay-Doppler arrays of the Mercury echo, from Arecibo radar observations, illustrating the two different observing modes: (a) standard-code, OC polarization; (b) standard-code, SC polarization; (c) long-code, OC polarization; (d) long-code, SC polarization. The standard-code data are from July 3, 1999 and the long-code data are from June 16, 2000. The horizontal axis is Doppler and the vertical axis is delay. The echo's leading edge is at top, the putative north polar ice spots are at bottom center, and Feature "A" is the prominent rayed feature. The gray scale is normalized to the peak polar echo brightness, so the OC specular echo at the leading edge is saturated. Note the echo folding from Doppler aliasing and OC streaking from code sidelobe leakage in the standard-code arrays. This figure originally appeared in Harmon (2002)

limbs (see Fig. 1b). Surface variations in diffuse backscatter tend to show up better in the SC spectrum, thanks largely to the absence of quasispecular glare. The first dual-polarization radar observations reported for Mercury were those of Goldstein (1970, 1971), made with the Goldstone S-band (λ13-cm) radar. Goldstein identified two main bumps in the SC spectrum, centered near planet longitudes 240°W and 345°W. The 240°W feature (dubbed "C") can be seen in a more recent spectrum obtained with the Arecibo radar (Fig. 1b). Although the "Goldstein features" are much weaker than some SC features seen in Mars spectra, they demonstrated that Mercury might be a worthy target of future full-disk radar imaging.

The first radar images of Mercury were obtained by Zohar and Goldstein (1974) with the Goldstone S-band radar. These images were made using the so-called "delay-Doppler mapping" method. A pulsed or coded transmission is used to resolve the echo in time delay, and then a spectral analysis is performed on each delay. This partitions the echo into a two-dimensional array in delay-Doppler space (Fig. 2). An image is then formed by mapping from delay-Doppler space into real-space (planetary latitude-longitude) coordinates. The

Zohar and Goldstein (1974) imaging, and later Goldstone imaging by Clark et al. (1988), mapped out the specular component of the OC echo within a few degrees of the sub-Earth point. This type of imaging, which responds to specular highlights, can delineate large-scale surface relief (such as crater rims) and provides some semblance of the view given by an optical image. Similar OC delay-Doppler observations were made at Arecibo in the late 1970s and early 1980s (Harmon et al. 1986; Harmon and Campbell 1988), although these were used primarily for obtaining altimetric profiles along the sub-Earth track and not for imaging.

Delay-Doppler mapping enables one to obtain planetary radar images with a resolution much finer than the dimensions of the telescope beam. The method suffers, however, from three major drawbacks. First, a $\csc(\theta)$ stretching of the latitude resolution causes severe image distortion within a few degrees of the "Doppler equator" (the roughly E–W line passing through the sub-Earth point and perpendicular to the apparent spin axis). Second, there is a north–south mapping ambiguity that causes the images to appear folded about the Doppler equator. Third, when employing the delay-Doppler technique in its standard implementation, one necessarily suffers delay folding and/or Doppler aliasing of the echo if the target is "overspread" (i.e., if the product of its delay depth and Doppler bandwidth exceeds unity). The "N-S ambiguity" can be avoided for the Moon, and mitigated for Venus, by offsetting the telescope beam to either side of the Doppler equator. This is not possible for Mercury, however, owing to its small apparent disk size and weaker echoes. Mercury is not a highly overspread planet, at least at S-band, and aliasing is not a problem when one is only imaging the sub-Earth region. However, overspreading poses serious problems for any full-disk imaging at S-band or higher frequencies and for practically all imaging projects with X-band ($\lambda 3.5$ cm) radars.

An alternative to delay-Doppler imaging is to form a radar image using an interferometer. This method avoids the aforementioned drawbacks of delay-Doppler, but is limited by the coarse spatial resolution provided by the synthesized beam. Muhleman et al. (1991) used a bistatic arrangement, involving VLA imaging of Goldstone X-band transmissions, to make the first radar images of highly overspread Mars in 1989. Two years later, the same team used this technique to make full-disk, dual-polarization images of Mercury with a best resolution of about 150 km (Slade et al. 1992; Butler et al. 1993; Butler 1994; Muhleman et al. 1995). The SC images from the August 1991 Goldstone-VLA observations are shown in Figs. 3 and 4. The left-hand image in Fig. 3, obtained at a sub-Earth longitude of 353.5°W, showed

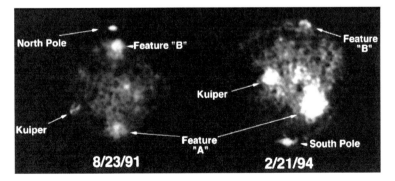

Fig. 3 Goldstone-VLA radar images (SC polarization) of Mercury from observations on (*left*) August 23, 1991 (sub-Earth point at 353.5°W, 11.0°N), and (*right*) February 21, 1994 (sub-Earth point at 15.7°W, 10.6°S). The 1991 image originally appeared in Slade et al. (1992). Images courtesy of B.J. Butler and M.A. Slade

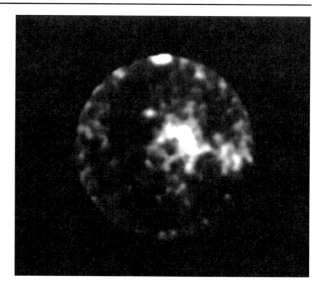

Fig. 4 Goldstone-VLA radar image of Mercury (SC polarization) from August 8, 1991 (sub-Earth point at 253°W, 11°N). Feature "C" is the large irregular bright structure just right of center, and the small bright spot just to its northwest is the crater in Fig. 18b. The bright spot at top is the north polar ice feature. This image originally appeared in Slade et al. (1992)

that one of the two Goldstein features is actually the combined effect of two bright spots with roughly the same longitude but widely separated latitudes. The southern and northern spots were later dubbed "A" and "B", respectively, by Harmon (1997). Figure 4, with a sub-Earth longitude of 253°W, showed that the other Goldstein feature (dubbed "C" by Harmon) is associated with a large irregular bright feature just north of the equator. The 1991 Goldstone-VLA imaging also produced the surprising discovery of a very bright spot at the north pole (see Fig. 3, left-hand image, and Fig. 4), whose location and peculiar radar properties immediately suggested the existence of polar ice deposits (Slade et al. 1992).

In the spring of that same year, a program of Arecibo S-band radar observations was begun with the aim of making dual-polarization, full-disk images of Mercury. To eliminate Doppler aliasing of the overspread echoes, a new "long-code" delay-Doppler method was used which had been tested out the previous year to make the first delay-Doppler images of Mars. This method uses a non-repeating transmission code and intra-code spectral analysis to circumvent the sampling-theorem limitations of "standard" delay-Doppler (which employs a repeating code and inter-code spectral analysis). The long-code method also eliminates all deterministic self-clutter from code-sidelobe leakage, but introduces a random self-clutter that, while negligible in the SC polarization, raises the OC noise level by up to 50%. For a comparison of Mercury delay-Doppler arrays obtained with the two methods, see Fig. 2. For more details on delay-Doppler and the long-code method, see Harmon et al. (1999) and Harmon (2002). The 1991 long-code observations were successful, and August observations made concurrently with the Goldstone-VLA observations immediately confirmed the north polar feature (Harmon and Slade 1992). Some of the same non-polar features were also seen, and an analysis of the earlier Arecibo data from that spring revealed a south polar bright spot from the floor of Chao Meng-Fu crater (Harmon and Slade 1992). The south polar feature was later confirmed by Goldstone-VLA images from 1994 (Fig. 3, right-hand image). Improved Arecibo images, made by summing 1991 and 1992 long-code data, mapped the poles at 15-km resolution and resolved out the putative ice deposits into more than a score of separate host craters (Harmon et al. 1994; Harmon 1997). The improved Arecibo images also revealed the non-polar feature "A" to be a Tycho-class impact crater with a prominent ray system, although the nature of non-polar features "B" and "C" remained ambiguous (Harmon 1997; see also Campbell 2002).

Immediately upon completion of a major upgrade to the Arecibo telescope in 1998 (which substantially improved the S-band radar sensitivity), we resumed Mercury imaging at yet higher resolution. Standard-code delay-Doppler observations in 1998–99 were used to image the north polar "ice" at 1.5–3 km resolution (Harmon et al. 2001). These images revealed many additional polar features, including some small unresolved spots and some spots at relatively low latitudes. Upon the acquisition of a long-code capability in 2000, we commenced a multi-year program of dual-polarization, full-disk radar imaging of the planet with the upgraded Arecibo telescope. Results of the non-polar imaging work done as part of this program were reported recently by Harmon et al. (2007) and make up a large part of the results presented in the next section (Sect. 3). These same observations have also produced some polar images, and these are reviewed along with the 1998–1999 polar imaging in Sect. 4.

3 Equatorial and Midlatitude Imaging

In this section we review imaging results for the non-polar regions, that is, for the region within ±65° latitude. For this we will draw heavily on the Arecibo long-code imaging from 2000–2005, because of its high resolution. However, we will also refer extensively to the Goldstone-VLA images, because of their lack of N–S ambiguity and the fact that they were obtained at a different radar wavelength. Also, the VLA images have conveniently been compiled into a single map (Butler 1994), which is reproduced in Fig. 5. From this map one can see that most of the prominent non-polar bright features are located in the MUH, which extends west from 190°W to 10°W longitude. However, we will also refer to, and make comparisons with, known features in the Mariner-imaged hemisphere (or "MIH") spanning 10°–190°W longitude. To assist in this, we reproduce in Fig. 6 a schematic map of the non-polar MIH region from the *Atlas of Mercury* (Davies et al. 1978), with additional annotation for some optically bright features; following Harmon et al. (2007), those rayed craters on

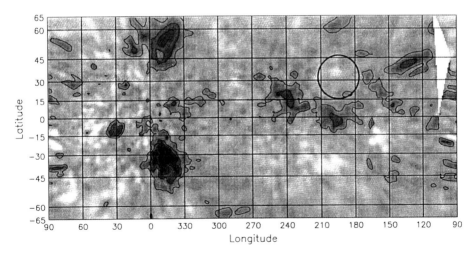

Fig. 5 Radar brightness contour map (SC polarization) of Mercury from Goldstone-VLA observations. *Darker shading* denotes higher radar brightness. The rim of Caloris Basin is shown (*circle*). This figure is from Butler (1994)

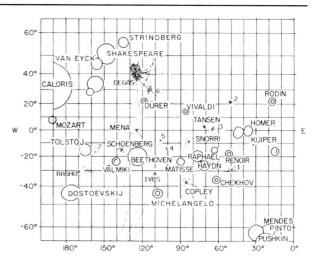

Fig. 6 Map schematic of Mercury's major basins and ray systems in the MIH. The map is adapted from Fig. 8 of the Atlas of Mercury (Davies et al. 1978) Some additional crater names have been added, along with the "MRC" numbers for the unnamed rayed craters. This figure originally appeared in Harmon et al. (2007)

this map that lack proper names have been assigned numbers on Fig. 6 and will be denoted by "MRC-#", where "MRC" stands for "Mariner Rayed Crater" and # is the crater number.

The Arecibo images shown include both OC and SC images, with surface resolutions in the range 2.5–7.5 km. The OC images are dominated by specular reflection within about 20° of the sub-Earth point and are best for rendering surface relief. A composite (quasispecular + diffuse) scattering law has been divided out of the OC images to reduce quasispecular glare and gray-scale dynamic range. Depolarized (SC) radar brightness is influenced primarily by decimeter-scale surface roughness, although dielectric effects may also contribute. Our working assumption is that radar-bright craters represent fresh impacts whose rough ejecta have yet to be broken down by impact gardening. Although the Arecibo images are N-S ambiguous, this ambiguity can be easily resolved for individual features by comparing images taken at different sub-Earth latitudes or by identifying smearing of foldover features in multi-day averages, and this case-by-case approach is the one adopted here. A formal algorithm for making unambiguous images exists, but this can be difficult to implement and inevitably results in image degradation.

3.1 Major Radar-Rayed Craters

Mercury's depolarized (SC) radar appearance is dominated largely by a number of (apparently) fresh impact features showing radar rays and other extended radar-bright ejecta. These include the Goldstein features ("A", "B", "C") as well as some smaller or fainter rayed structures. The locations of these craters are given in Table 1.

3.1.1 Crater "A"

"A" is the southerly of the two large bright patches just east of the Mariner 10 approach terminator (10°W Long.) in the VLA depolarized images in Fig. 3. A similar view to that in the right-hand image in Fig. 3 is given in the Arecibo SC image in Fig. 7. The Mariner approach terminator bisections this image north-south, with the MUH to the east. The image is dominated by "A" in the southeast quadrant. Closeup OC and SC images of "A" are shown in Fig. 8. The OC image shows a central 85-km-diameter crater (347.5°W, 34°S) with a central peak. The bright ejecta features show up best in the SC image. These include a

Table 1 Major radar-rayed craters

Name	Long.	Lat.	Diam. (km)
"A"	347.5	−34.0	85
"B"	343.5	+58.0	95
Kuiper	31.5	−11.0	60
	11.0	+62.0	140
"C"	246.0	+11.0	125
"Ghost"	242.0	−26.5	85
Bashō	170.5	−32.0	70
Degas	127.0	+37.5	45
MRC-6	117.2	+27.6	45
Bartok	135.0	−29.0	80
	327.0	+8.0	68

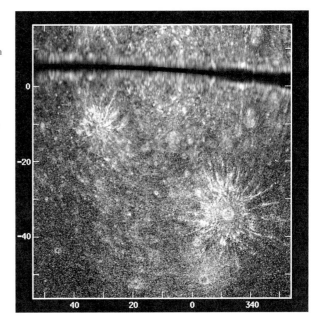

Fig. 7 Arecibo image (SC polarization) of the southern portion of the Mariner-approach region. The observations are from June 1–3, 2001. The Doppler equator corresponds to the dark swath at top. The Mariner 10 approach terminator approximately bisects the image vertically. The dominant features are Kuiper Crater in the *upper left* and "A" in the *lower right*. N/S-ambiguity foldover from feature "B" appears at bottom (south of "A"). This image originally appeared in Harmon et al. (2007)

bright rim collar, a fainter surrounding halo, and a complex ray system. The radial extent of the ejecta from this crater roughly agrees with the size of the feature in the VLA images. This is arguably the most spectacular radar crater feature in the solar system and probably the freshest large impact on Mercury.

3.1.2 Kuiper

Kuiper is the bright feature seen to the northwest of "A" in the VLA (Fig. 3) and Arecibo (Fig. 7) images. Optically, Kuiper was the dominant bright crater in the Mariner 10 approach images as well as the brightest mercurian feature seen by Mariner (Hapke et al. 1975). Presumably, this crater (31.5°W, 11°S) is one of the freshest impacts on the planet. This impact appears fresh to the radar as well, as evidenced by the prominent bright ejecta

Fig. 8 Arecibo image of Feature "A" in the (**a**) OC and (**b**) SC polarizations. The OC image is from June 1–3, 2001. The SC image is from June 1–2, 2001 and May 14, 2002. These images originally appeared in Harmon et al. (2007)

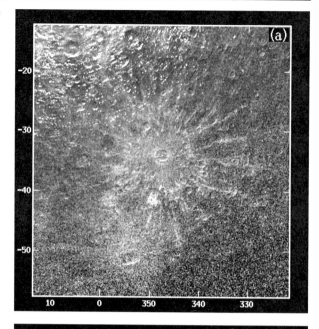

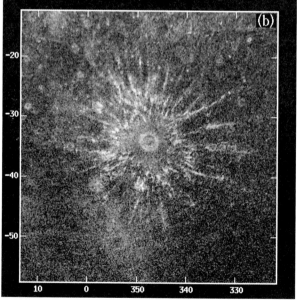

and rays seen in the SC image in Fig. 9. There is good correlation between the optical and radar rays west of the crater, and the radar ray running NNE of the crater coincides with a faint optical ray segment. Kuiper is designated as c_5 class (De Hon et al. 1981), the freshest crater class in the USGS geologic mapping sequence. It is interesting to note that other c_5-class craters to the east of Kuiper in Fig. 9 (including Mahler at 19°W, 19°S, Dvořák at 12.5°W, 9.5°S, and Donne at 14°W, 3°N) show little or no bright SC structure.

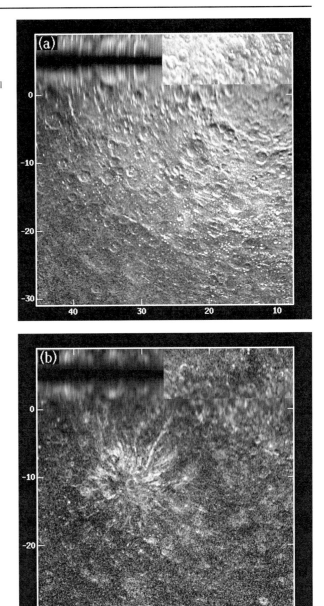

Fig. 9 Arecibo image of the southeast portion of the H-6 Quadrangle in the (**a**) OC and (**b**) SC polarizations. The data are from June 1–3, 2001, with a strip insert at top right from April 14–17, 2004. Kuiper Crater is at center left at 31.5°W, 11°S and Dvořák Crater is located at 12.5°W, 9.5°S. These images originally appeared in Harmon et al. (2007)

3.1.3 Crater "B"

Feature "B" bore little resemblance to "A" in the pre-upgrade Arecibo images (Harmon 1997), and the possibility of it being a shield volcano was even considered. However, the post-upgrade images now show "B" to be an impact feature. "B" can be seen in the northeast corner of Fig. 10, which is derived from the same observations as for Fig. 7 but which shows the mapping north of the Doppler equator. Closeup OC and SC images of "B" are shown

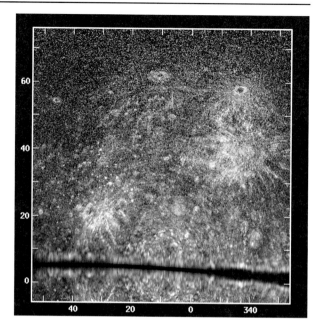

Fig. 10 Same as Fig. 7, except covering the region north of the Doppler equator. The dominant features are "B" (*upper right*), a large rayed crater west of "B" (*upper center*), and a cluster of small craters (*left center*). Note also the ambiguity foldover from Kuiper (below the cluster) and from "A" (below "B"). This image originally appeared in Harmon et al. (2007)

in Fig. 11. The central OC feature shows a circularity consistent with a 95-km-diameter impact crater (343.5°W, 58°N). The crater shows a bright rim ring and a surrounding halo that appears significantly brighter than that around "A". Beyond the halo are streaks and patches to the south and west that appear to be ejecta rays similar to, if less distinct than, those from "A". The crater shows no central peak, but rather a lobate or terrace structure on the west side of the crater floor. This may account for the rather non-circular appearance of the central crater in SC images, which contributed to the initial uncertainty as to whether this is an impact feature.

3.1.4 Crater "C"

Although "C" is one of the most prominent SC features in the Goldstone-VLA images (Figs. 4 and 5), it has been difficult to characterize from Arecibo images. This is because of its proximity to the Doppler equator and the relatively large planet distance at those times when this longitude was visible from Arecibo. The feature can be seen in the large-scale SC images in Fig. 12, where it can be seen near the Doppler equator and just left of center. From this image, "C" appeared more like a cluster of small, fresh impacts than a single dominant impact. However, more recent images now suggest that "C" arises from a large source crater at 246°W, 11°N. An SC image from 2004 (Fig. 13a) shows the crater's rim ring and surrounding dark halo on the Doppler equator. A weak OC image from 2002 (Fig. 13b) shows the crater itself, which is 125 km in diameter and shows specular glints from a central peak and eastern rim wall. The SC image in Fig. 13a shows ray features that appear to radiate from this crater. However, comparison of Figs. 12b and 13a indicates that "C" is an asymmetric feature, with most of the bright ejecta lying north of the source crater. This is consistent with the location and shape of the feature in the VLA images (Figs. 4, 5). Our proposed source crater coincides with a prominent ring feature in one of the old radar maps of Zohar and Goldstein (1974), and a possible association between their feature and feature "C" was, in fact, suggested by Butler et al. (1993) and Butler (1994).

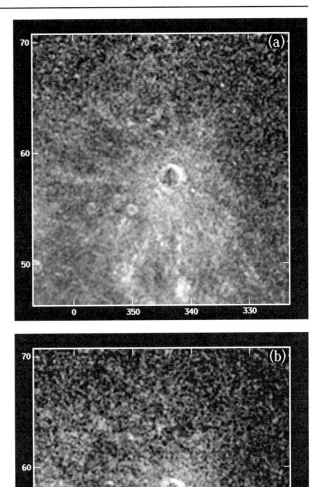

Fig. 11 Arecibo image of Feature "B" in the (**a**) OC and (**b**) SC polarizations. The data are from June 1–3, 2001. A 5×5-pixel smoothing was applied to reduce noise. These images originally appeared in Harmon et al. (2007)

3.1.5 "Ghost Features"

Several craters have been found that show extensive ray systems similar to "A" but which are much fainter. One of these can be seen due south of feature "C" in Fig. 12b. An OC image from 2002 (not shown) reveals an 85-km-diameter central crater at 242°W, 26.5°S (Harmon et al. 2007). The SC images (including Fig. 12b) show a dark crater floor, a bright rim collar surrounded by a dark halo, and a dense outer burst of faint rays.

Fig. 12 Arecibo radar image of the region west of Caloris Basin in (**a**) OC and (**b**) SC polarizations. The data are from July 7–8, 2001. These images originally appeared in Harmon et al. (2007)

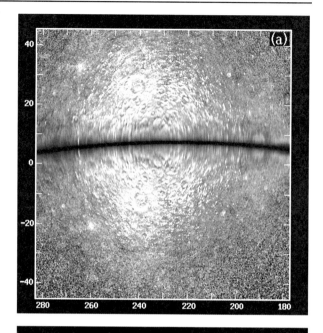

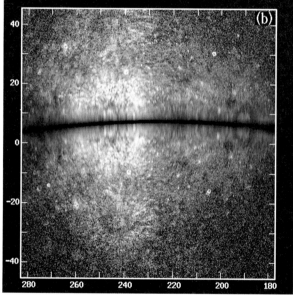

Another notable feature is a large structure centered at 11°W, 62°N, to the west of feature "B" in Fig. 10. This region was not imaged by Mariner 10. The 140-km-diameter central crater shows a bright floor, a central peak, and a surrounding dark halo. Outside the dark halo is a large and diffuse bright halo, along with some filamentary rays extending to the southwest. The high incidence angles probably contribute to the faintness of the feature. This object probably accounts for the bright contours extending west of "B" in the Goldstone-VLA map (Fig. 5).

Fig. 13 (a) Arecibo radar image (SC polarization) of the Feature-"C" region from observations on August 8–9, 2004. (b) Arecibo radar image (OC polarization) centered on the Feature-"C" source crater (246°W, 11°N) from data taken on April 22–30, 2002. These images originally appeared in Harmon et al. (2007)

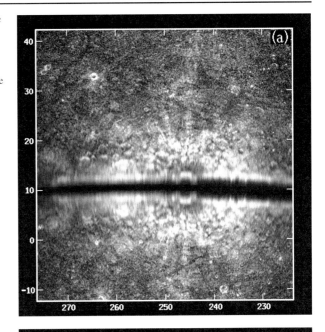

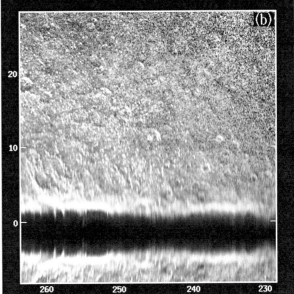

The best example of a radar ghost feature in the MIH is Bartok (135°W, 29°S), a c_5-class crater that shows a very bright rim and bright ejecta (including rays) in the Mariner images. Radar images showing this feature are given in Harmon et al. (2007). The feature appears very similar to the ghost feature south of "C", but is even fainter.

3.1.6 Bashō and Degas

Craters Bashō (170.5°W, 32°S) and Degas (127°W, 37.5°N) are among the most prominent of the rayed craters seen by Mariner 10, with Degas being particularly striking. Both of these

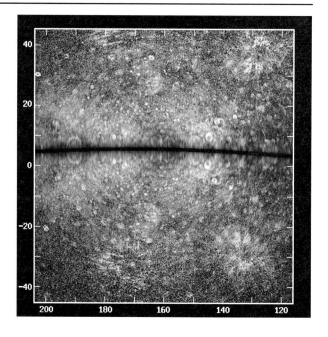

Fig. 14 Arecibo radar image of the circum-Caloris region (SC polarization) from observations on June 9–10, 2002. This image originally appeared in Harmon et al. (2007)

craters also show extended ejecta and rays in the Arecibo radar images. This can be seen in the large-scale SC image in Fig. 14, which shows the region extending east of the Mariner 10 departure terminator (190°W Long.). A closeup OC/SC image pair of the Degas region is also shown in Fig. 15. The radar appearances of these two medium-size (70 and 45 km diameter) craters are very similar. Both show irregular ray haloes that are offset from the crater by about five crater radii. They also show a bright crater floor rather than the dark floor and bright rim collar seen for many other fresh craters, both larger and smaller. All of Degas's radar rays coincide with optical rays, although the converse is not the case. For both craters, the optical rays are mostly longer than the radar rays.

3.1.7 Other Rayed Craters

Two rather more modest radar-rayed craters have been found in the Arecibo images. The first of these (117.2°W, 27.6°N) can be seen southeast of Degas and NNE of Dürer in Fig. 15. This crater (MRC-6 in Fig. 6) showed bright ejecta and rays in the Mariner 10 images. The crater is about the same size as Degas and shows a similar, but fainter, ray/ejecta pattern. Unlike Degas, it shows a bright rim ring. Curiously, it is only classified as c_4 freshness type in the USGS geologic maps (Guest and Greeley 1983).

The other rayed crater is located in the MUH at 327°W, 8°N. It appears just north of the Doppler equator in the large-scale SC image of the western MUH region in Fig. 16b. A better closeup view, taken when the Doppler equator was farther south in 2005, is given by OC/SC image pair in Fig. 17. The feature consists of a 68-km-diameter central crater and a surrounding halo of short rays and bright ejecta. The ray/ejecta halo is similar in size and form to that of MRC-6, although the central crater SC feature is fainter. This feature coincides with a prominent bright spot in optical telescopic images (Baumgardner et al. 2000; Warell and Limaye 2001), so it is likely to have a bright optical ray system.

Fig. 15 Arecibo radar images of Degas Crater and environs in (**a**) OC and (**b**) SC polarizations. The data are from June 2, 2002. Degas Crater is just north of center. Southeast of Degas are the rayed crater "MRC-6" and the two-ring basin Dürer. The gray scale of the SC image has been truncated at 30% of the maximum brightness. These images originally appeared in Harmon et al. (2007)

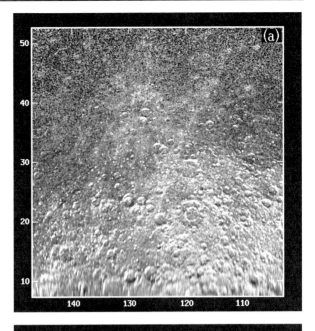

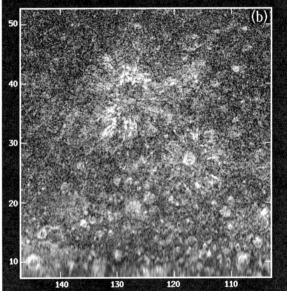

3.2 Radar-Bright Craters

Much more numerous than the radar-rayed craters are craters showing radar-bright rims and/or floors but no rays (or, at most, a single faint ray). Most of these are smaller than the radar-rayed craters. Two of the "radar-bright" craters in the Arecibo images correspond to prominent unresolved bright spots in the VLA images. One of these (339°W, 12°N) can be seen midway between features "A" and "B" in the August 1991 VLA image in Fig. 3, and

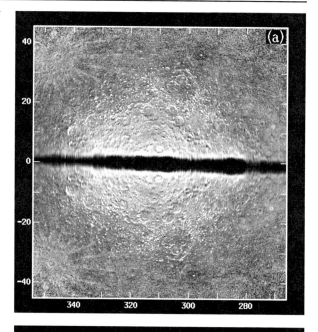
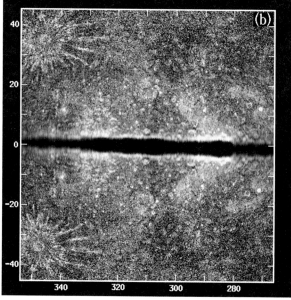

Fig. 16 Arecibo radar images of a central portion of the MUH region in (**a**) OC and (**b**) SC polarizations. The data are from May 6–7, 2002. These images originally appeared in Harmon et al. (2007)

just north of the Doppler equator on the west side of the Arecibo image in Fig. 16. The closeup OC/SC Arecibo image pair in Fig. 18a shows this to be a 12-km-diameter crater with a bright ring around the rim. The other VLA bright spot (264°W, 32°N) is located northwest of feature "C" in Fig. 4 and can be seen in the Arecibo images in Figs. 12 and 13. The closeup Arecibo images of this feature (Fig. 18b) show a 30-km-diameter crater with a bright rim ring, while Fig. 13 shows a single faint ray extending toward the northeast. A similar bright-rim crater (272°W, 14°S) can be seen in Figs. 12 and 16 and in the closeup

Fig. 17 Arecibo radar images of a rayed crater (327°W, 8°N) in (**a**) OC and (**b**) SC polarizations. The data are from March 24–25, 2005. These images originally appeared in Harmon et al. (2007)

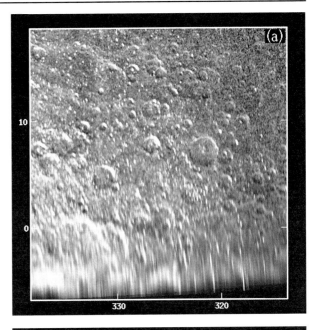

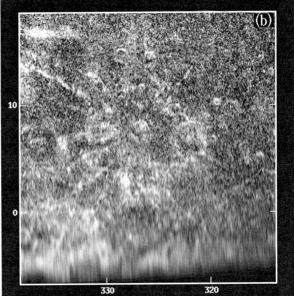

pair in Fig. 18c. Another, located in the western floor of Caloris Basin at 203°W, 30°N (Figs. 12 and 14), is shown in closeup in Fig. 18d. Farther east in Fig. 14 can be seen two prominent bright rim rings from craters Balzac (145°W, 10°N) and Theophanes (143°W, 4°S). A closeup of Theophanes is shown in Fig. 18e.

Still farther east, we find a particularly heavy concentration of radar-bright craters in the central region of the MIH. This region, much of which is located in the H-7 USGS map quadrangle, is shown in the SC image in Fig. 19. Inspection of this and other Arecibo

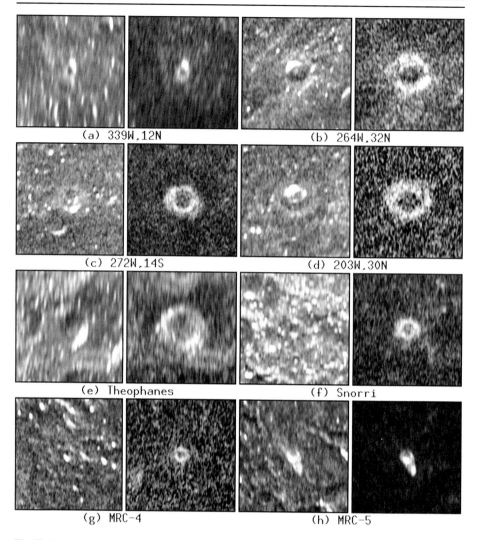

Fig. 18 Arecibo closeup image pairs (OC *on left*, SC *on right*) for small to medium-size radar-bright craters. Each image measures 4°×4°. The crater names or locations are given under each OC/SC pair. This montage is adapted from a figure that originally appeared in Harmon et al. (2007)

SC images shows that nearly all of the smallish "Mariner rayed craters" (Fig. 6) are radar-bright. Furthermore, most of these optically rayed craters show up as rayless, bright-ring features in the SC radar images. Three examples are shown in closeup in Fig. 18: (f) Snorri (83.5°W, 8.5°S); (g) MRC-4 (101°W, 13.2°S); (h) MRC-5 (105.7°W, 8.5°S). The last of these, MRC-5, is unique in showing a bright lobe to the southeast in addition to a rim ring; this is, in fact, the brightest (non-polar) depolarized radar feature on the planet and possibly the freshest small impact crater.

Mariner 10 image mosaics of the H-7 Quadrangle (Davies et al. 1978) showed a high concentration of optically bright and rayed craters of small to medium size. This region was also observed when the flyby geometry was at high sun angles (Davies et al. 1978; King and Scott 1990), so it might be concluded that the density of bright features might

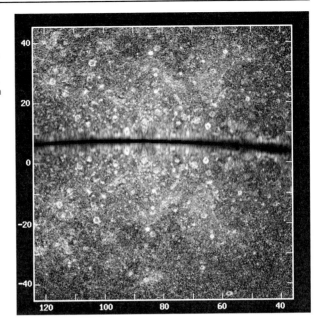

Fig. 19 Arecibo radar image (SC polarization) of a central portion of the MIH region. The data are from June 13, 2001. The gray scale has been truncated at 30% of the maximum brightness. This image originally appeared in Harmon et al. (2007)

simply be an artifact of the high solar illumination. However, the high concentration of radar-bright features over H-7 suggests that the high density of fresh impacts in this region is real. In addition to this large-scale concentration, we have also found some smaller clusterings of small, radar-bright craters. Two examples of these are a cluster centered at 168°W, 32°N (Fig. 14) and 25°W, 30°N (Fig. 10). These clusters are discussed in more detail in Harmon et al. (2007).

3.3 Circum-Caloris Smooth Plains

Besides radar-bright craters and ejecta, the region east of the Mariner 10 departure terminator (Fig. 14) shows some large, diffuse areas of enhanced brightness. The most prominent of these are the two butterfly-shaped features folded about the Doppler equator near 150°W and 180°W Longitude. Comparison with other images with different sub-Earth latitudes indicates that the western feature comes mainly from the south, i.e., from the circum-Caloris smooth plains in Tir Planitia, whereas the eastern feature comes mainly from the Budh Planitia smooth plains north of the Doppler equator. Another bright patch (surrounding Degas) in the NE corner of Fig. 14 is associated with the smooth plains of Sobkou Planitia. Also, a smaller bright patch (164°W, 16°S) was found in the floor of Tolstoj Basin. Although Tolstoj is clearly an impact feature, the interior floor has been assigned the same "smooth plains" ("ps") designation as for the circum-Caloris planitiae in the geologic map of Schaber and McCauley (1980). Furthermore, Spudis and Guest (1988) argued that "the interior of Tolstoj is flooded with smooth plains that clearly postdate the basin deposits".

Closeup images of the Tir-Tolstoj region are shown in Fig. 20. The smooth plains of Tir Planitia and Tolstoj Basin are clearly seen in the northwest and southeast quadrants, respectively, of Fig. 20a, while heavily cratered terrain runs diagonally across the image center. The SC image (Fig. 20b) shows the enhanced depolarized brightness of the Tir and Tolstoj plains as well as the relative darkness of the intervening cratered terrain. Portions of Tir Planitia are also bright in the VLA map (Fig. 5), as was noted and discussed by Butler

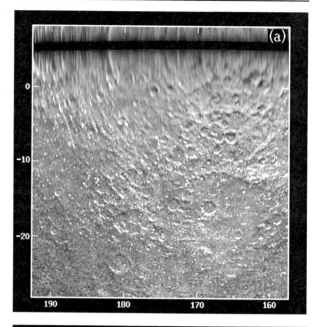

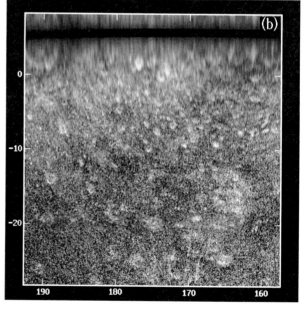

Fig. 20 Arecibo radar images of the Tir-Tolstoj region in (**a**) OC and (**b**) SC polarizations. The data are from June 9–10, 2002. Tir Planitia is in the northwest quadrant, Tolstoj Basin is in the southeast, and Mozart Crater is in the northwest corner. These images originally appeared in Harmon et al. (2007)

et al. (1993) and Butler (1994). The VLA map indicates that the enhanced brightness from Tir continues on west of 190° into the MUH. The Arecibo images in Figs. 12 and 14 also support this, as does a more recent SC image obtained in 2005 (Fig. 21). Extension of the southern circum-Caloris smooth plains into the MUH had been inferred earlier from Arecibo radar altimetry, which showed smooth downwarped topography along the equator and west of 210°W that was ascribed to subsidence under the load of a smooth plains fill (Harmon et al. 1986; Harmon and Campbell 1988; Harmon 1997).

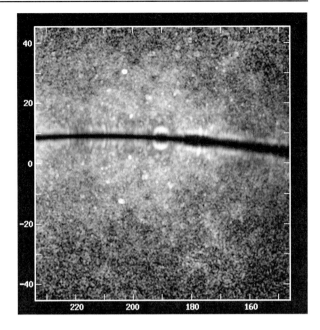

Fig. 21 Arecibo radar image of the circum-Caloris region (SC polarization) from observations on July 15–16, 2005. A 3×3-pixel smoothing was applied to reduce noise. This image originally appeared in Harmon et al. (2007)

3.4 Large Basins

3.4.1 Caloris

Caloris Basin, the most prominent impact structure seen by Mariner 10, shows no obvious large-scale SC radar features. This was already apparent from the Goldstone-VLA imaging (Slade et al. 1992; Butler et al. 1993; Butler 1994) as well as from pre-upgrade Arecibo imaging (Harmon 1997). The post-upgrade Arecibo images show Caloris to be generally radar-dark, as can be seen from the northeast corner of Fig. 12b and the northwest corner of Fig. 14. The only unambiguous bright SC features from Caloris are a few small to medium-size bright craters, the most prominent of which is the bright-ring crater at 203°, 30°N. Butler et al. (1993) found a bright OC feature centered on the southeast rim of Caloris in the X-band VLA images, while the Arecibo S-band images show little evidence of such a feature. However, Arecibo OC images (not shown) from 1998 and 2005 do reveal a 100-km-diameter crater at 193.5°W, 26°N that appears to be the largest impact crater inside Caloris. The Arecibo images (Fig. 7) show no obvious SC feature from the "hilly and lineated" terrain at the Caloris antipode (centered near 20°W, 30°S).

3.4.2 "Skinakas"

Ksanfomality (2004) presented evidence for a large circular basin centered at 280°W, 8°N, based on optical telescope imaging from Skinakas Observatory in Crete (see also Ksanfomality et al. 2005). Ksanfomality argues that this basin, which he has informally dubbed "Skinakas", is a two-ring structure with an inner ring diameter of 1,000 km. Recent Arecibo images provide strong supporting evidence for this basin. The OC image in Fig. 22a shows a smooth, roughly circular region around the Doppler equator between 275°W and 293°W longitudes. Some highlighting on the western side of this feature may

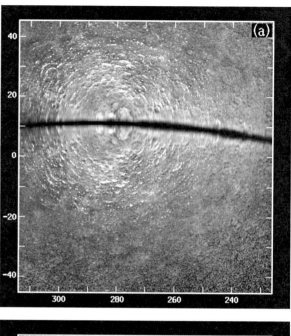
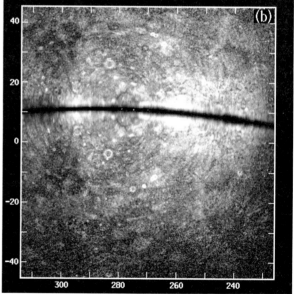

Fig. 22 Arecibo radar images of a central portion of the MUH region in (a) OC and (b) SC polarizations. The data are from August 14–15, 2004. The gray scale of the SC image has been truncated at 50% of the maximum brightness. These images originally appeared in Harmon et al. (2007)

even be specular glinting off the inner basin rim. A portion of the floor of this putative basin also showed smooth, flat topography in Arecibo altimetry (Harmon et al. 1986; Harmon and Campbell 1988). The SC image in Fig. 22b indicates the basin has enhanced depolarized brightness, although the SC image in Fig. 16b suggests this brightness is nonuniform and concentrated in the southwest and northeast portions of the basin floor. A large circular feature (250 km diameter) centered on the Doppler equator at 278°W in Fig. 22 is probably a large impact crater superimposed on the Skinakas basin floor. This feature coincides with a 2.5-km-deep crater feature in the old Arecibo altimetry (Harmon et al. 1986; Harmon and Campbell 1988); it is a likely candidate for a two-ring basin, and is listed as such in the next section (Sect. 3.5).

3.5 Two-Ring Basins

Two-ring basins are among the most distinctive impact features in the Mariner images. These can also be identified in the Arecibo OC images, as with Dürer in Fig. 15a. Harmon et al. (2007) also identified several other two-ring basins in the MUH. The most prominent of these is located at 303°W, 28°N in the Arecibo image of the MUH in Fig. 23. It also shows up in Figs. 16 and 22. With inner and outer ring diameters of 156 km and 310 km, this basin rivals Homer as the largest two-ring basin on the planet. This basin was actually discovered in 1980 (B.A. Burns and D.B. Campbell, unpublished data) in one of the few radar interferometric imaging observations of Mercury ever attempted at Arecibo. The SC images (Figs. 16b, 22b, 23b) show the basin interior to be radar-dark. This basin is also dark in optical telescopic images (Baumgardner et al. 2000; Mendillo et al. 2001). Other two-ring basins have been found in the MUH at 241°W, 27°N (Fig. 12a) and 273°W, 3°N (Fig. 22a). Other candidates in the MUH needing confirmation are located at: 315°W, 18°S; 278°W, 10°N (see Sect. 3.4.2); 304°W, 14°S; and 259°W, 21°N. Finally, there is the case of the large crater Mozart (190.5°W, 8°N), which, while prominent, is only just visible on the departure terminator in the Mariner images. Mozart is implied to be a two-ring basin by the NASA/USGS maps (Davies et al. 1978; Schaber and McCauley 1980), and is also listed as such by Pike (1988). The prominent ring feature seen in the SC radar image in Fig. 14 appears to be coming from the interior base of Mozart's outer ring. One can also see a fainter inner feature that may be from the inner ring.

3.6 Dark-Halo Craters

Several mercurian craters show a distinctive dark halo beyond the rim (or rim-ring). Some of these appear similar to dark-halo craters seen in radar images of the Moon (Ghent et al. 2005). Probably the most prominent of these is located at 340°W, 9°S in Fig. 23. A much smaller one can be seen in the same image at 325°W, 35°N. In Figs. 7 and 9 can be seen a dark halo around the moderately bright crater Hitomaro (16°W, 16°S) and around a crater at 358°W, 4°S. Three dark-halo craters can be seen superimposed on the bright Tir Planitia smooth plains in Figs. 14, 20, and 21. These include the large crater Mozart (190.5°W, 8°N) and two smaller craters at 180°W, 4.5°S and 195°W, 10°S. Dark-halo craters are lacking in the central MIH (Fig. 19), with only one example at 75°W, 1.5°N standing out. As one can see here, the mercurian dark-halo craters, like their lunar counterparts, are not especially common. The lunar dark haloes have been attributed to burial by a homogeneous ejecta blanket devoid of radar scatterers (Ghent et al. 2005), and a similar explanation may apply to Mercury.

3.7 Discussion: Equatorial and Midlatitude Imaging

Among the more interesting findings from the full-disk radar imaging of Mercury is the wide variety of radar-bright features showing up in the diffuse/depolarized echo. For the non-polar regions, the bright features are associated with either (1) ejecta from presumably fresh impact craters or (2) smooth plains. It is worthwhile considering both sources of radar brightness in somewhat more detail, citing radar results for the Moon for purposes of comparison.

Fig. 23 Arecibo radar images of a central portion of the MUH region in (**a**) OC and (**b**) SC polarizations. The data are from August 19, 2004. The gray scale of the SC image has been truncated at 50% of the maximum brightness. These images originally appeared in Harmon et al. (2007)

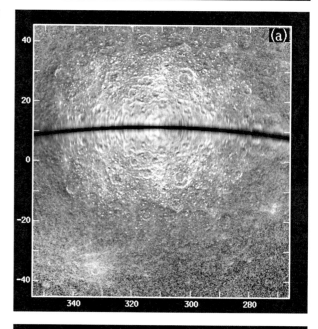

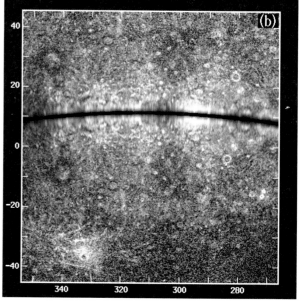

Mercury is similar to the Moon in the prominence and variety of radar-bright impact features. However, there are some important differences between the two bodies. For example, it now appears that some of the larger (diameter >80 km) mercurian impact craters are born with complex radar-bright ejecta fields consisting of dense systems of narrow rays. Such craters are not typical of the Moon. Tycho, the most prominent radar-rayed crater on the Moon and comparable in size to crater "A", shows only a few very long radar rays (Zisk et al. 1974). The Mariner images showed mercurian craters to have more compact ejecta fields and a higher density of secondary craters than is the case for lunar craters (Gault et al. 1975). Explanations offered for these differences include Mercury's stronger gravity (Gault et al. 1975) and higher impactor velocities resulting from the planet's deeper location in the Sun's potential well (Schultz 1988). Whatever the explanation, its seems likely that the relative prominence and complexity of the radar-bright ejecta from the larger mercurian craters is consistent with cratering characteristics noted earlier from the Mariner 10 optical imaging.

Smaller (<40 km diameter) fresh mercurian craters tend not to show radar rays or other extended radar-bright ejecta. Instead, they commonly exhibit compact, radar-bright rim rings surrounding a darker floor (although some bright-floored craters are also seen). Such bright-ring craters have also been seen in radar observations of the Moon (Zisk et al. 1974; Thompson et al., 1981, 1986) and Venus (Campbell and Burns 1980), although a higher percentage of fresh lunar craters show bright floors rather than bright rim rings. To explain bright-ring lunar craters, Thompson et al. (1986) argued that the flat floors of fresh craters may initially be rough and radar-bright, and then darken as impact gardening smooths the surface. The rims would remain rough and radar-bright for a longer time "… owing to preferential downslope movement of fines and continual exposure of fresh blocks." If, as Schultz (1988) argues, fresh-crater degradation from impact gardening proceeds more rapidly on Mercury than on the Moon, then one would expect mercurian craters to evolve to the bright-ring stage more rapidly and bright-floored craters would be less common. Mercury's stronger gravity may also be a factor. Gault et al. (1975) noted a greater degree of post-cratering rim collapse on Mercury than on the Moon, which they attribute to the stronger gravity. It is possible that this stronger post-impact rim collapse and mass wasting may contribute to the darkening of crater floors and to the brightening of the rims.

Fresh mercurian craters show a strong correlation between crater diameter and radar morphology. The radar features with the most extensive ejecta and most elaborate ray systems are associated with large craters in the 80–140 km diameter range; simpler and less extended ejecta/ray systems are associated with medium-size craters in the 45–70 km diameter range; still smaller (<40 km) fresh craters tend to be rayless features with bright rim rings. It is not surprising that large impactors produce the most extensive radar ray systems, since these high-energy impacts are the more likely to produce extended blocky ejecta as well as the secondary cratering shown to be associated with some of the radar-bright lunar rays (Pieters et al. 1985). It is also not surprising that smaller craters can show prominent optical rays but no radar rays. Optical rays are thought to represent material that is compositionally different from, or less mature than, the surface upon which it is emplaced (Pieters et al. 1985; Grier et al. 2001; Hawke et al. 2004). Radar rays are believed to be primarily structural in nature, associated with enhanced diffuse/depolarized backscatter from wavelength-scale surface (or near-surface) roughness (Thompson et al. 1981; Pieters et al. 1985). Composition is less likely to affect radar brightness because (1) compositional differences tend to produce less dielectric contrast in the radio than in the optical, and (2) thin surface coatings that could strongly affect optical albedo would appear invisible to radar (which can penetrate to depths of several meters or more before scattering). Hence, the smaller fresh impacts might produce sprays of fine material that show up as optical rays, but lack the energy to throw out boulder-size ejecta (or produce secondary craters) at comparable distances from the primary crater. Another possibility is that the smaller mercurian craters are born with radar rays, but that these rays are erased by impact gardening at a rate that is (1) faster than the optical ray fading rate and (2) faster than the fading rate for radar rays from larger impacts. The latter would be in line with the suggestion by Thompson et al. (1981) that radar-bright crater features on the Moon have a shorter lifetime for smaller craters than for larger ones, which seems plausible if the ejecta deposits from the smaller impacts are thinner or if smaller blocks are broken down faster. Grier et al. (2001) made a similar case for a crater-size dependence of the maturation time for the optical brightness of lunar craters. Support for a difference between radar and optical maturation rates comes from the fact that a wide variation in radar brightness has been observed amongst c_5-class craters (Harmon et al. 2007). In other words, a significant fraction of the radar maturation time scale is contained within the c_5-class age span. This effect is most noticeable for medium to large-size craters.

The enhanced depolarized brightness of the mercurian smooth plains is an interesting finding that begs an attempt at explanation. This is especially true in that just the reverse radar albedo contrast is seen for the Moon, where the lunar maria appear radar-dark relative to the highlands (Zisk et al. 1974; Thompson 1987; Campbell et al. 1997). The best current explanation for the radar darkness of the lunar maria is that the relatively high iron and titanium content of the mare lavas increases their electrical lossiness and reduces backscatter from subsurface roughness (Campbell et al. 1997). While it appears likely that the mercurian smooth plains are lava flows (Strom 1979), it is also quite likely that these flows are compositionally different from, and less electrically lossy than, lunar mare lavas. Supporting this is the fact that Mercury's surface in general exhibits a low microwave opacity (Mitchell and de Pater 1994) indicative of a highly differentiated, possibly non-basaltic crust that is relatively transparent to radio waves (Jeanloz et al. 1995). Of course, this does not explain why the mercurian smooth plains have higher depolarized radar brightness than the adjacent intercrater plains, especially as there is no evidence that the microwave opacity of the circum-Caloris region (which contains the highest concentration of smooth plains) is lower than the global average (Mitchell and de Pater 1994; Jeanloz et al. 1995). This suggests that subsurface transparency may not be the only determining factor in the SC brightness of the smooth plains. For example, the smooth plains may have rougher small-scale surface texture than the intercrater plains. If this texture is associated with rough lava flow surfaces, then the enhanced SC brightness could be a weaker version of the high depolarized brightness seen for the younger Martian lava flows (Harmon et al. 1999). Alternatively, the smooth plains may have retained more of their surface roughness because, being younger than the intercrater plains, they are less degraded by impact gardening. It is also possible that the rough scattering elements in the smooth plains have a higher dielectric constant than the intercrater plains because of compositional differences or higher densities.

4 Polar Imaging

The radar brightness of Mercury's poles represents an entirely different phenomenon from the non-polar radar backscatter discussed in the previous section, and therefore warrants separate treatment. Although the polar bright spots are found to be associated with impact craters, detailed mapping shows that the enhanced backscatter is confined to the permanently shaded portions of the crater floors and is not associated with conventional backscatter from rough ejecta. Since permanent shading implies a temperature effect, the most likely source of the radar brightness is enhanced backscatter off deposits of some cold-trapped volatile such as water ice. Supporting the ice hypothesis is the fact that the high radar albedos and circular polarization inversion of the mercurian bright spots is also characteristic of the radar backscatter from the three icy Galilean satellites and Mars's southern icecap. Therefore, we will provisionally use the term "ice" when referring to Mercury's polar radar features, while keeping in mind the fact that the ice hypothesis remains unproven and that alternative explanations that have been proposed may be viable.

In the next two subsections we will review observational results for the north and south poles, respectively. We then conclude with a discussion of the current state of theories of the polar radar features, with an emphasis on the ice hypothesis.

4.1 North Pole

The north pole has been the more intensively studied of the two poles, largely because of its better visibility from Earth at those times when the planet is observable with the Arecibo telescope. In fact, the north pole has been observable at near-optimal aspect and distance during every year since the 1998 completion of the Arecibo upgrade, whereas the south pole has only recently returned to view at Arecibo.

4.1.1 Host Craters and Pole Location

The best pre-upgrade Arecibo image of the north pole, obtained by averaging images from 14 observing dates in 1991–1992, was presented and discussed in Harmon et al. (1994). Although this image has been superseded by the post-upgrade images (and, hence, is not reproduced here), its 15-km resolution was sufficient to resolve the north polar bright feature into smaller crater-size spots. More than a dozen of these spots were located in the MIH, and all of these could be identified with specific "host craters" in the Mariner 10 images and NASA/USGS maps. However, the crater assignments could only be made after shifting the pole down by 1.6° of latitude (68 km) along the 0°W longitude meridian (and also making some other allowances for Mariner image scale distortion). This correction to the Mariner-based pole position did not seem unreasonable, given that the NASA/USGS maps (Davies et al. 1978; Grolier and Boyce 1984) quoted discrepancies of up to 40 km (1.0°) with respect to the Mercury Control Net, and given that the control net itself had a standard error of 25 km (0.6°) at the north pole (Davies and Batson 1975; Harmon et al. 1994). Note that when we refer to the pole position here we mean the location relative to mapped topographic features and not the direction of the spin axis on the sky. The assumed spin axis does define the radar grid, however, and to make our images we have assumed Mercury's obliquity to be zero. Not only is this a good assumption based on dynamical arguments, but also is found to give the most consistent polar feature locations when comparing radar images obtained from different orbital aspects.

The next advance in the radar imaging of the north pole came with the early post-upgrade Arecibo observations of 1998–1999 (Harmon et al. 2001). The higher sensitivity of the upgraded radar was exploited to make delay-Doppler images of the pole at improved spatial resolutions of 3 km (for the 1998 images) and 1.5 km (for the 1999 images). These new images confirmed all of the bright features seen in the pre-upgrade imaging and also revealed many new features. As before, all of the bright spots located in the MIH were found to coincide with known craters. Figure 24 shows an SC image from Harmon et al. (2001) that was formed by summing images from August 16–17, 1998 and July 25–26, 1999. Superimposed on this image are circles representing the rims of the respective host craters. Again, proper positioning of these crater circles required making adjustments to the pole location and polar coordinate grid. Harmon et al. (2001) did this by applying an affine transformation to the revised Mariner-based polar map grid of Robinson et al. (1999), using the radar-bright spots from the smallest craters as fiducial points. This relocated the north pole to the rim of crater E, approximately 65 km from the original Mariner-based pole and 15 km from the newer Mariner-based pole of Robinson et al. (1999). This radar-based pole position, which has an accuracy of about 2 km, is very close to the pole position deduced from

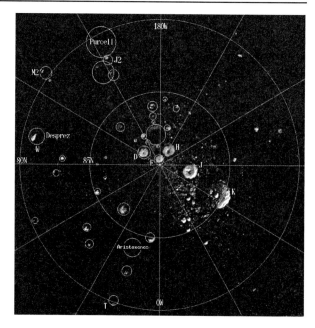

Fig. 24 Arecibo radar image of the north pole of Mercury, formed by summing SC images from August 16–17, 1998 and July 25–26, 1999. The radar-based coordinate grid has an accuracy of about 2 km. The original Mariner-based pole is located at 191°W, 88.46°N on this more accurate radar-based grid. The locations of host-crater rims (*circles*) are based on an adjustment to their respective positions on a revised Mariner-based grid (see text for explanation). Adapted from a figure that originally appeared in Harmon et al. (2001)

the pre-upgrade imaging (Harmon et al. 1994). The original Mariner-based pole is located at 191°W, 88.46°N on the new grid in Fig. 24; conversely, the new pole is located at 11°W, 88.46°N on the old Mariner-based maps of Davies et al. (1978) and Grolier and Boyce (1984).

The first X-band delay-Doppler images of the north pole were made with the Goldstone radar in 1999 and 2001, using the long-code method to mitigate the large overspreading at that shorter wavelength (Slade et al., 2000, 2001; Harcke et al. 2001; Harcke 2005). These 6-km-resolution images showed most of the major polar features seen in the Arecibo S-band images and also confirmed the Arecibo-based pole position.

4.1.2 Ice Distribution within Craters

One can get some idea of the ice distribution within individual craters simply from inspection of Fig. 24. The larger and more circular radar features are associated with the larger craters at latitudes poleward of 87–88°N, examples being features D, E, H, and J. Note, in particular, the MIH craters D and E, whose radar features occupy a large portion of their rim circles (with a slight offset to the south). The inference is that most of the floor area of these craters is permanently shaded and, hence, covered in ice. As one goes to lower latitudes in Fig. 24, the radar features occupy progressively smaller portions of the craters' southern interiors, which is consistent with confinement of ice deposits to the southern (north-facing) interior rim walls and any southern floor regions shaded by those rims. At 80–82° latitude the radar features are reduced to mere slivers under the south rims, as with W (Despréz), M2, and J2. Crater K is an example of a large (90 km diameter) crater at an intermediate latitude (85°N) whose ice deposits, while extensive, are confined to the south rim and southern floor region (Fig. 24).

A closer view of the ice distribution in the higher-latitude craters is provided by the smaller-scale images in Figs. 25 and 26. Figure 25, which is from Harmon et al. (2001), was obtained from observations on July 25–26, 1999, while Fig. 26 is a previously unpublished

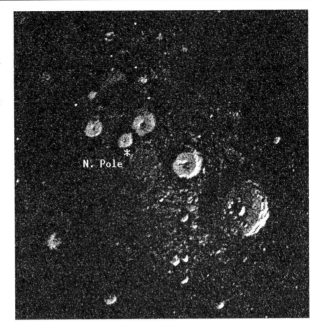

Fig. 25 Arecibo SC radar image (1.5-km resolution) of the north polar region of Mercury, from observations on July 25–26, 1999. The image shows the major near-pole features (D, E, H, J, K) and their environs. The radar illumination is from the upper left (138°W Long.). The north pole location is denoted (∗). Adapted from a figure that originally appeared in Harmon et al. (2001)

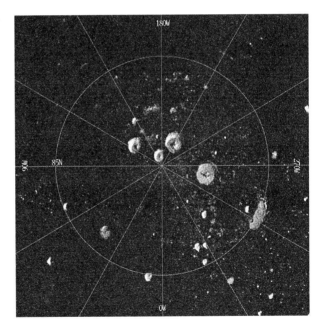

Fig. 26 Arecibo SC radar image (1.5-km resolution) of the north polar region of Mercury, from observations on August 14–15, 2004. The radar illumination is from the right (281°W Long.)

image from August 14–15, 2004. With mean sub-Earth longitudes of 138°W and 282°W, respectively, these images were obtained at near-opposite radar illumination directions. In Fig. 25 (where the radar illumination is from the upper left) note the echo highlights from the south (pole-facing) rim walls of craters J and K. One can even discern terraces on the south rim of crater K. This indicates that ice in these craters collects on the protected north-facing rim slopes as well as the floors. Craters D, E, and H show little or no such rim highlighting, because the north rim walls facing the radar are not permanently shaded and, hence, contain no ice, while the southern rim walls are in radar shadow. In Fig. 26, where the radar illumination is from the other direction (282°W), highlighting from ice on the north-facing southern rim walls of craters D, E, and H is now visible, while the south rim ice in craters J and K is now lost in radar shadow. Note the lack of highlighting from the north rim of crater J, again an indication that this rim receives some sunlight during the mercurian day. In the case of the larger and lower-latitude crater K, most of the crater floor is periodically sunlit and the permanently-shaded, radar-bright region is confined to the segment defined by the region between the south rim and a chord cutting across the crater floor at about 1/3 crater diameter from the south rim. This is precisely what one expects geometrically for permanent shading in a crater of this size and latitude, and this striking agreement had been noted earlier by Harmon et al. (2001) from their 1998 image (not shown). The ice boundary in the floor of crater K does not show up as well in the 1999 image (Fig. 25) because of the extremely high incidence angles at that radar illumination direction. Finally, note that craters D, E, H, and J all have central peaks that cast radar shadows or show radar highlights. The highlights indicate that ice must collect on the slopes of these central peaks. Crater K does not show a central peak (which is normal for a crater this large), although it does show a prominent bright crater in its central floor. This crater, labelled "Z" by Harmon et al. (2001), was the brightest polar feature seen in the 1998–1999 images.

Immediately surrounding the bright features in craters D, E, H, J, and K are radar-dark haloes that are apparently ice-free regions in the exposed upper and outer rims of these craters. These dark haloes only show up because they contrast with the diffuse, low-level brightness seen around the general vicinity of these craters. This so-called "diffuse patch", consisting of unresolved or barely resolved features, was discussed earlier by Harmon et al. (1994, 2001) and ascribed to ice deposits in secondaries from the D–K impacts or in shaded depressions in the hummocky ejecta blankets of these craters.

4.1.3 Large-Scale Distribution and Low-Latitude Features

The Goldstone-VLA images of the north pole had virtually all of the radar-bright feature contained within the 80° latitude circle (Butler et al. 1993; Butler 1994), and the bright features in the pre-upgrade Arecibo image were all north of 79°N Latitude (Harmon et al. 1994). However, the early (1998–1999) post-upgrade Arecibo images (Harmon et al. 2001) revealed a number of small bright spots at still lower latitudes, including several below 75°N. The lowest latitude ice feature in the MIH, labelled X2 (74.5°N), was traced to a tiny region under the south rim of crater Tung Yuan (see Table 2). The remaining three features (A3, B3, C3) found by Harmon et al. (2001) below 75°N were all located in the MUH (Table 2). Harmon et al. noted a tendency for the low-latitude features to avoid the so-called "hot poles" (0°W, 180°W), which are the two (alternating) subsolar longitudes at Mercury perihelion.

Arecibo observations since 1999 have confirmed all of the low-latitude ice features found by Harmon et al. (2001) and have identified several additional features at latitudes below 75°N (Table 2). Many of these, including one at a latitude of 70°, can be seen in the image in Fig. 27, which was obtained from Arecibo observations on August 14–15, 2004. A more recent image from August 5–6, 2005 (not shown) revealed an even more southerly spot at 67°N (Table 2). The recent images (e.g., Fig. 27) confirm the earlier suggestions that the low-latitude ice features tend to avoid the hot-pole longitudes.

Table 2 Low-latitude polar features

Long.	Lat.	Label[a]
298	+67.0	
286	+70.3	
255	+71.6	A3
293	+71.6	B3
303	+72.0	C3
315	+72.8	
254	+73.6	
282	+73.8	
281	−74.3	
62	+74.5	X2 (Tung Yuan)

[a]Label from Harmon et al. (2001)

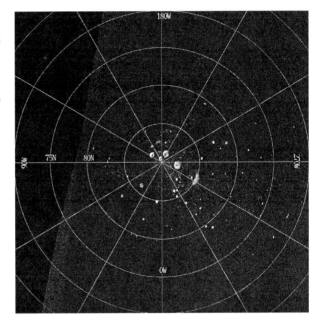

Fig. 27 Arecibo SC radar image (1.5-km resolution) of the north polar region of Mercury. This is a larger-scale image from the same 2004 observations as in Fig. 26. The radar illumination is from the right (281°W Long.) and the radar horizon is visible on the *left*

4.1.4 Radar Scattering Properties

The Mercury polar features are characterized by high radar reflectivities and inverted circular polarization ratios. The circular polarization ratio is defined as $\mu_c \equiv \sigma_{sc}/\sigma_{oc}$, where σ_{oc} and σ_{sc} are the OC and SC radar cross sections, respectively. Normally one sees $\mu_c < 1$ for conventional backscatter from rough surfaces. The three icy Galilean satellites of Jupiter all show inverted ($\mu_c > 1$) polarization ratios (Campbell et al. 1978; Ostro et al. 1992), as does the south polar ice cap of Mars (Muhleman et al. 1991). The polarization inversion of Mercury's north polar echo was apparent immediately upon discovery of the feature (Slade et al. 1992; Butler et al. 1993). Harmon et al. (1994) computed polarization ratios for all of the north polar crater features in the pre-upgrade Arecibo image, finding μ_c values in the range 1.0–1.6. Harmon et al. (2001) measured ratios for the larger features (D, E, H, J, K) from the first (1998) post-upgrade Arecibo image, finding a mean value of $\mu_c = 1.25$. For comparison, the icy Galilean satellites have μ_c values in the range 1.15–1.5 at both S-band and X-band (Ostro et al. 1992).

Although the polar features can be considered bright by any measure, quantifying that brightness in a meaningful way requires that one account for incidence-angle effects. Harmon et al. (1994, 2001) adopted the "equivalent full-disk albedo" $\hat{\sigma}$ as a useful brightness measure for any given feature observed at incidence angle θ. This is the radar albedo (normalized by planet projected area) that the entire planet would have if that particular feature were to cover the entire visible disk and were to conform to an assumed scattering law $\sigma°(\theta)$ (defined as the radar cross section per unit surface area at incidence angle θ). If one assumes the usual "cosine law" of the form $\sigma°(\theta) \propto \cos^n \theta$, then the equivalent full-disk albedo is given by $\hat{\sigma} = 2\sigma°(\theta)/(n+1)\cos^n(\theta)$. Assuming $n = 1.5$, Harmon et al. (2001) found a mean depolarized albedo $\hat{\sigma}_{sc} = 0.89$ for the five major north polar features (D, E, H, J, K). Harcke (2005) found similar albedos for these same features from Goldstone X-band observations in 2001; specifically he found that one could equalize the Arecibo S-band and Goldstone X-band albedos by varying n between 1.3 and 2.1 (depending on the feature), which spans the plausible range of scattering-law exponents. For comparison, Ostro et al. (1992) measured full-disk depolarized albedos in the range 0.4–1.6 for the icy Galilean satellites, with (like Mercury) very little wavelength dependence between S and X bands. Galilean satellite observations with the Arecibo 70-cm radar (Black et al. 2001a) showed weak echoes indicating a strong drop in radar albedo between S-band and the longer wavelength. A strong drop in the radar brightness of the south polar ice cap of Mars was also noted by Harmon et al. (1999) between X-band and S-band. By contrast, Arecibo 70-cm radar observations of Mercury (Black et al. 2002) showed a strong echo from the north pole with a cross section similar to that at S-band. The conclusion is that Mercury's north polar ice feature has roughly the same intrinsic radar reflectivity over a wide wavelength range (3–70 cm). Possible implications of this wavelength-independent scattering behavior are discussed later in Sect. 4.3.

4.2 South Pole

Arecibo observations in 1991–1992 (Harmon and Slade 1992) identified a south polar bright feature that appeared to be mostly concentrated in the floor of Chao Meng-Fu, a large (150-km-diameter) crater whose rim lies tangent to the pole. The improved 1992 image published in Harmon et al. (1994) confirmed a large circular feature occupying most of the floor of Chao Meng-Fu and showing inverted polarization ($\mu_c = 1.12 \pm 0.09$). Several other smaller bright spots were also found, three of which were located in the MIH and could be identified with known craters in the Mariner-based maps. As with the north pole, corrections to the Mariner-based south pole position had to be made in order to line up the radar features with their host craters. This adjustment amounted to sliding the NASA/USGS map grid (Davies et al. 1978) by 1.2° (51 km) to the right, i.e., along the 90°W longitude meridian.

The first X-band detection of a south polar bright feature was made in 1994 from Goldstone-VLA imaging by B.J. Butler and M.A. Slade (Fig. 3). The first X-band delay-Doppler image of the south pole was obtained at Goldstone in 2001 using the long-code method (Slade et al. 2001; Harcke et al. 2001). The most recent version of this image (Slade et al. 2004; Harcke 2005) shows all of the bright features seen in the pre-upgrade Arecibo image, as well as many new ones (at 6-km resolution). Nearly all of these features show circular polarization inversion. This Goldstone image confirmed the Arecibo-based pole

Fig. 28 Arecibo SC radar image (1.5-km resolution) of the south polar region of Mercury, from observations on March 24–25, 2005. The radar illumination is from the bottom (349°W Long.) and the radar horizon is visible at the top. The original Mariner-based pole is located at 276°W, 88.7°S on this more accurate grid

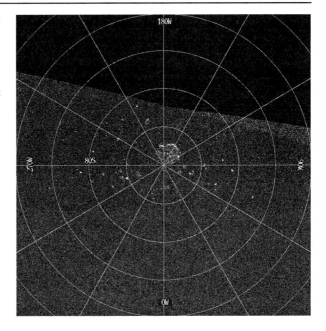

position of Harmon et al. (1994). Furthermore, Harcke (2005) found excellent agreement between the locations of the ice features and the revised Mariner-based grid of Robinson et al. (1999).

The south pole was out of view from Arecibo when the upgraded telescope came back on line, and would not return to view until 2004. An Arecibo image of the pole from that year was of poor quality owing to telescope pointing problems, and the first decent image was not obtained until March of the following year (albeit with only half the nominal transmitter power). This image, from March 24–25, 2005, is shown in Fig. 28. The image shows many of the same bright spots as the 2001 Goldstone image, although not all (owing to a different sub-Earth longitude aspect). Note the Chao Meng-Fu feature, which shows highlights from the north (pole-facing) rim wall and a complex ring of central peaks. The radar grid and inferred south pole position agree closely with the pre-upgrade imaging of Harmon et al. (1994) as well as with the recent Goldstone image (Harcke 2005) and the revised Mariner-based grid (Robinson et al. 1999). The original Mariner-based pole is located at 276°W, 88.7°S on the new grid in Fig. 28; conversely, the new pole is located at 96°W, 88.7°S on the old Mariner-based map of Davies et al. (1978).

4.3 Discussion: Polar Imaging

A strong argument in favor of the Mercury ice hypothesis is the fact that the radar scattering properties of the polar features are similar to those of icy surfaces on other bodies. The resemblance between the north polar spot in the 1991 Goldstone-VLA image of Mercury (Slade et al. 1992) and the south polar spot in the 1989 Goldstone-VLA image of Mars (Muhleman et al. 1991) was particularly striking. However, the actual prototypes for this enhanced radar backscatter from ice were the three icy Galilean satellites (Europa, Ganymede, Callisto), whose surprising radar characteristics were discovered back in 1975–1976 (Campbell et al. 1978). This discovery spawned a number of theoretical papers proposing diverse scattering mechanisms to explain the phenomenon. It is generally conceded that the first satisfactory explanation was the "coherent backscatter" theory proposed by Hapke (1990) and developed further by Peters (1992) (see also Black et al. 2001b). This theory treats the radar backscatter as a volume (subsurface) scattering process involving multiple forward scatterings off of inhomogeneities (voids, cracks, lenses, etc.) in the ice. Since clean ice has low electrical loss, a deep ice layer can support long scattering paths with minimal attenuation, resulting in most of the incident radar wave being scattered back out. Because individual scattering events are mostly in a forward direction, the helicity of a circularly polarized wave tends to be preserved, giving the observed polarization inversion. An important feature of this mechanism is a coherent addition of wavefronts traveling along time-reversed versions of the same scattering path. This coherent effect not only boosts the backscatter cross section but also enhances the helicity retention effect. An alternative theory based on refraction scattering has been proposed (Hagfors et al. 1997), although it, too, invokes a subsurface mechanism with long path lengths in a low-loss medium such as thick deposits of clean water ice.

At the time of the discovery of Mercury's polar radar-bright features, enhanced backscatter from ice had been observed for the icy Galilean satellites and Mars's southern ice cap. Since then, the list has expanded to include Mars's northern ice cap (Harmon et al. 1999; Butler et al. 2000) and several icy satellites of Saturn (Ostro et al. 2006; Black and Campbell 2006). It is becoming increasingly clear, then, that enhanced volume backscatter is a common radar response for icy bodies in the solar system. This alone makes ice a strong candidate for the material responsible for Mercury's polar radar features. However, Mercury obviously presents a much more severe thermal environment than do the other planets and satellites mentioned above. One certainly does not expect water ice to survive on unprotected flat surfaces anywhere on the planet. On the other hand, it may be possible to maintain ice in permanently shaded "cold traps" at the poles, as was suggested years ago by Thomas (1974) and Kumar (1976). This idea was, in fact, borrowed from even earlier speculation about ice at the lunar poles (Watson et al. 1961). The floors of polar craters were considered to be the most likely havens for lunar ice, since the Moon's low ecliptic obliquity would maintain such regions in permanent shadow (Watson et al. 1961; Arnold 1979). Mercury has an even smaller (near-zero) obliquity, so permanently shaded polar craters should be common there as well. At the same time, Mercury has a very high (7°) orbital inclination to the ecliptic. As a result, the poles can present an earthward tilt of up to 12°, giving Earth-based radars access to many polar areas that never receive direct sunlight.

The initial discoveries of the Mercury polar radar features were followed immediately by theory papers evaluating the thermal stability of water ice at the mercurian poles (Paige et al. 1992; Ingersoll et al. 1992; Salvail and Fanale 1994). These papers reinforced the importance of permanent shadowing and gave some credence to the notion that water ice could be stable in polar craters. At that time, Chao Meng-Fu was the only known crater definitely identified with a radar feature, and this crater was specifically modeled by Salvail and Fanale (1994). Paige et al. (1992) also pointed out several likely host craters in the northern MIH that were consistent with the location of the north polar bright feature. The final Arecibo pre-upgrade images (Harmon et al. 1994) fully resolved both poles into crater-size features and provided definite host-crater identifications for all those features in the MIH. The post-upgrade Arecibo imaging (Harmon et al. 2001) revealed many additional features and identified more host craters. These images, including those shown in this paper, provided more detailed mapping of the radar feature locations within their respective host craters and gave additional support for the confinement of the radar-bright deposits to permanently shaded locales. They also revealed the presence of radar-bright spots at relatively low latitudes. Since these low-latitude features were all tiny spots or slivers, it was concluded that they probably arise from isolated shaded regions in some of the steeper or rougher pole-facing interior rim walls of craters and, thus, do not violate the requirement for permanent shading (even though the colatitude may be lower than the average rim-wall slope for mercurian craters). Hodges (1980) made a similar argument in support of the possibility of ice hiding in the pole-facing rim walls of low-latitude lunar craters.

An important point recognized by all of the early thermal modeling papers is that permanent shading, while apparently necessary, is not a sufficient condition for maintaining water ice at the mercurian poles. The reason for this is that a permanently shaded portion of crater floor will also be warmed by scattered sunlight and infrared reradiation, especially from the adjacent interior rim walls. Paige et al. (1992) and Ingersoll et al. (1992) both showed that medium-size craters cannot support ice at latitudes below 84°, using 112 K as the threshold temperature (i.e., the temperature at which sublimation would erode a water ice deposit at a rate of 1 m/Gyr). In their discussion of the pre-upgrade Arecibo image of the north pole, Harmon et al. (1994) cited the existence of two features just below 80°N latitude as posing a potential thermal problem for ice. This and later radar findings have shown ways in which the ice theory may be strained, and this has been seized upon as a rationale for alternative theories. The first of these was offered by Sprague et al. (1995), who suggested sulfur as an alternative to ice. This is a plausible idea, since sulfur is a volatile that might tend to concentrate in polar cold traps but, being more stable than water ice, could tolerate higher temperatures. Sulfur also has a low electrical conductivity and so could serve the same function as water ice, that is, as a low-loss matrix material for volume scattering. More recently, Starukhina (2001) rejected volatiles altogether and suggested that cold silicates could serve as the low-loss scattering medium.

The most thorough thermal modeling of the Mercury poles to date has been that of Vasavada et al. (1999). This refined model adopted more realistic flat floors for medium to large craters and also studied the protective effects of an insulating dust mantle. Although the study showed that the coldest parts of flat-floored craters could have $T < 112$ K at latitudes as low as 82°, it found that the diurnal maximum surface temperatures were too high to support exposed water ice deposits in the lower-latitude crater features (S, T, W) seen by Harmon et al. (1994), and also too high to account for the sizes of some of the radar features seen at higher latitudes. Vasavada et al. (1999) found that one could significantly reduce the diurnal range of crater floor temperatures by covering the ice with a thin (0.1–0.5 m) insulating layer of dust. Such a layer would not only reduce sublimation losses, but also protect the ice from sputtering and UV erosion. The detailed modeling results in Vasavada et al. showed that dust-insulated water ice could account for the radar features in all of the original host craters identified by Harmon et al. (1994).

Harmon et al. (2001) concurred with the Vasavada et al. (1999) conclusion that all of the radar features in the pre-upgrade Arecibo images were consistent with dust-insulated water ice. However, they also pointed out that there could be a problem explaining the radar features seen in some small (<10 km diameter) craters in the post-upgrade images of the north pole. Such small craters, because of their bowl-shaped floors and high depth/diameter ratios, are exposed to high levels of indirect radiation and thus have high floor temperatures. Vasavada et al. (1999) claimed that such craters could not support water ice at latitudes below 88°, even if dust-mantled, whereas the post-upgrade Arecibo images showed several small crater features in apparent violation of this latitude restriction. Harmon et al. (2001) suggested that these craters may simply be shallower, for some reason, than typical mercurian craters of their size, in which case the indirect radiation heating would be less. Vilas et al. (2005) made the very same suggestion, although their argument was reinforced by their own observation (from Mariner images) that the north polar host craters tend to have abnormally low depth/diameter ratios.

Another issue raised by Harmon et al. (2001) is the fact that, even if dust-mantled ice is thermally stable in a given crater, one still has to deposit the protective mantle on a time scale shorter than the erosion time for exposed ice. This is a difficult problem that raises fundamental issues regarding ice and dust deposition sequencing and presumes accurate knowledge of low-temperature sublimation rates and ice-loss processes. The thermal problems are most severe for the low-latitude craters because of their higher levels of indirect heating. Harmon et al. (2001) pointed out, for example, that virtually the entire floor of crater W (Despréz) should experience diurnal maximum temperatures in excess of 140 K, whereas a temperature of only 130 K is sufficient to erode ice at a rate of 1 m/Myr (Vasavada et al. 1999). The problem is even worse for craters S and T, which should experience maximum diurnal floor temperatures in excess of 150 K (owing to their smaller diameters and hot longitudes). Estimated dust deposition rates for Mercury range from 0.2 m/Gyr (Killen et al. 1997) to 4 m/Gyr (Crider and Killen 2005). Even using the faster rate, it would still take 50 Myr to accumulate a 20-cm dust mantle, by which time hundreds of meters of ice would have evaporated. If one accepts 20–30 m/Gyr as the maximum ice deposition rate (Moses et al. 1999; Crider and Killen 2005), then there is an obvious problem providing thermal insulation quickly enough to protect ice deposits in some of the lower-latitude craters.

The dust-mantle timing problem could go away if the deposited ice is dirty. In this case, sublimation of the top layer of ice could leave a self-sealing lag deposit of dust that would protect the remaining ice (Vasavada et al. 1999). Getting this to work with exogenic (impact) water sources seems problematic, since ballistic settling times for impact-generated dust should be shorter than the migration time for the water vapor to polar cold traps (Harmon et al. 2001; Butler 1997). However, it is possible that dust could be entrained with water vapor in a temporary post-impact atmosphere (Vasavada et al. 1999). Various other suggestions have been made for mitigating temperature effects on ice loss, most of which are discussed in Vasavada et al. (1999), Harmon et al. (2001), and papers cited therein. Vasavada et al. (1999) pointed out that the ice sublimation rates used in their modeling were based on extrapolations and, hence, might overestimate ice loss at a given temperature. Furthermore, sublimation may not result in permanent ice loss unless there is subsequent photodissociation and/or ionization of the sublimated vapor (Killen et al. 1997). Hence, one might envision a scenario in which some of the ice is continuously redistributed amongst polar cold traps. Killen et al. (1997) also suggested that a very thin (1–2 cm) dust mantle might suffice as a diffusion barrier to escape of sublimation vapor. Finally, there is the interesting possibility that the temperatures in many shaded crater floor areas are lower than one would predict on the assumption of flat surfaces. This could be the case if crater floors and rims contain embedded craters or topographic depressions that provide a "double shielding" against indirect radiation (Hodges 1980; Carruba and Coradini 1999).

That indirect radiation is, in fact, an important effect is suggested by the longitude distribution of low-latitude features. Harmon et al. (2001) noted a tendency for the lowest-latitude north polar features to align along the 90°–270°W cold longitudes and avoid the 0°–180°W hot poles. The more recent north polar images (e.g., Fig. 27) support this. The clustering is stronger along the 270° direction than the 90° direction, with the latter side showing only one feature (X2) below 75°N latitude. This may simply be due to the fact that the 90°W side is largely covered with the Borealis smooth plains (Grolier and Boyce 1984), and thus presents a dearth of suitable host craters (the X2 source crater, Tung Yuan, being an isolated exception). The lowest-latitude south polar features also show a preference for the cold-pole axis, as is apparent from Fig. 28 and the 2001 Goldstone image (Harcke 2005). The temperature of a permanently-shaded region in a low-latitude crater is governed by the amount of indirect heating, which is greater for craters at hot longitudes. Therefore, the dearth of low-latitude features at hot longitudes suggests that indirect heating plays an important role in the thermal environment of polar craters and is probably the limiting factor in the stability of volatiles at low latitudes.

Another observational constraint on polar ice theories is the polar features' radar wavelength dependence, or lack thereof. As discussed by Black et al. (2001b) in the context of the Galilean satellites, the wavelength dependence for volume backscatter from ice is influenced by the size distribution of the subsurface scatterers and the depth of the scattering layer relative to the attenuation depth. The fact that no obvious drop in radar cross section is seen for Mercury's north polar features as one goes from 13-cm to 70-cm wavelength (Black et al. 2002) indicates that the scattering layer is optically thick at both wavelengths. This implies that (unlike the Galilean satellites) there is no dearth of scatterers at the larger (meter) scales and also that the scattering layer is many wavelengths thick at the longer wavelength. The latter condition, which is required to build up sufficient multiple scattering, implies that the ice deposits are at least several meters deep, and possibly much deeper. The presence of a lossy surface dust mantle can also affect the wavelength dependence by preferentially attenuating the shorter wavelengths, a point which is of obvious interest for polar ice models invoking a thermal insulating cover. Comparisons between Arecibo S-band and Goldstone X-band north-polar images were reported that suggested there is a drop in the feature cross sections at the shorter wavelength that would be consistent with attenuation in a 20-cm-deep dust mantle (Slade et al., 2001, 2004; Crider and Killen 2005). However, as pointed out in Sect. 4.1.4, the recent analysis by Harcke (2005) found the evidence for an intrinsic S/X-band wavelength difference to be inconclusive when incidence-angle differences are taken into account. What the Harcke (2005) analysis does suggest is that any overlying dust layer is very unlikely to be more than 20 cm deep, which, if one accepts 0.2–4 m/Gyr as a plausible range for the dust deposition rate, gives an upper limit for the age of the top ice layer of between 50 Myr and 1 Gyr. Finally, it is important to note that our comments above could conceivably apply to other low-loss volume scattering media besides water ice, and that the inferences drawn from the wavelength behavior relate mainly to the structure and depth of the polar deposits rather than their composition.

If the polar radar features really are water ice deposits, then where does the water come from? Possible endogenic sources include release of water vapor through volcanism or other outgassing. Exogenic sources include delivery by impacts (comets, asteroids, meteors) and reduction of surface iron by solar wind implantation. Volcanic sources seem unlikely, given the lack of identifiable volcanoes and the fact that most of the smooth-plains effusion (with possible associated release of water vapor) predated the radar-bright c_4-class host craters (Harmon et al. 2001). The consensus seems to be that an exogenic source is more likely, with comet impacts being the dominant delivery mechanism (Killen et al. 1997; Moses et al. 1999; Barlow et al. 1999; Vasavada et al. 1999). Impact delivery of water by objects such as comets is attractive, as it would be consistent with the episodic deposition required by those scenarios incorporating protective dust mantling (Vasavada et al. 1999; Crider and Killen 2005). Impact deposition is also preferred over endogenic sources by those scenarios requiring some of the ice deposition to be quite recent (Crider and Killen 2005). Moses et al. (1999) consider Jupiter-family comets to be the most promising source of water, delivering as much as 60 m of ice to polar cold traps over the last 3.5 Gyr. They consider Halley-type comets to be a less important input, owing to the lower collision probabilities and higher impact velocities (which reduces water retention) associated with their high-inclination, long-period orbits. Sungrazing comets, which may be a dynamically evolved subset of the Halley-types (Bailey et al. 1992), would seem to be a plausible source of mercurian water and one which has not heretofore been considered by any of the published studies. Although most of these objects (including the Kreutz sungrazers being discovered with regularity by SOHO) are very small, they probably derive from one or more massive progenitor comets (e.g., Sekanina and Chodas 2004) whose total mass may represent a substantial fraction of the Mercury-crossing comet flux. However, the same orbital factors as apply to the conventional Halley-types may inhibit the water-delivery potential of the sungrazers.

It was natural that the putative discovery of polar ice deposits on Mercury would help to revive old ideas about ice on the Moon and to stimulate Earth-based and spacecraft searches for lunar polar ice. Although the neutron spectrometer on the Lunar Prospector orbiter detected hydrogen concentrations at the lunar poles suggestive of water ice (Feldman et al. 1998), all Arecibo radar searches for Mercury-like polar features have proved negative (Stacy et al. 1997; Campbell et al. 2006). This suggested that any lunar polar ice is likely to be in the form of a dilute frost mixed in the regolith, not the thick ice deposits of relatively clean ice required to explain the Mercury echoes (Feldman et al. 2000; Campbell et al. 2006). The obvious question then arises as to why the lunar ice is so sparse, if it exists at all. Several possible explanations are discussed in Vasavada et al. (1999) and Harmon et al. (2001). One suggestion is that the Moon and Mercury may have had very different obliquity histories (with the Moon settling late into its current spin state), although this presupposes that most ice deposition is primordial. Another possibility is that the Mercury features represent ice deposited in a massive (and possibly recent) comet impact, in which case the Moon's lack of ice may simply be due to chance. Finally, one should not discount the possibility that the mercurian water is endogenic and that the Moon's crustal chemistry and outgassing processes are less conducive to water release (Harmon 1997).

5 Conclusion

Earth-based radar images of Mercury reveal a variety of radar-bright features associated with (1) fresh impact ejecta, (2) smooth plains, and (3) probable volatile deposits in permanently shaded polar craters. These results raise some important questions. Are the polar features really water ice? Is the brightness of the smooth plains associated with volcanic composition or texture? What accounts for the diversity of radar-bright impact features and the apparent differences with lunar impact features? Although we have made an attempt in this paper to offer some possible answers to these and other questions, much additional work is needed.

The most likely source of new insights, as well as the next great advance in Mercury science, will come from the imagers and spectrometers on the upcoming MESSENGER and BepiColombo spacecraft. The neutron spectrometers on these spacecraft should be capable of detecting any hydrogen concentrations associated with near-surface water ice (Feldman et al. 1997). The γ-ray and UV spectrometers may be able to distinguish between volatile species in the mercurian surface and atmosphere, respectively (Solomon et al. 2001). Although the neutron and γ-ray spectrometers will have low surface resolution, it should still be possible to make coarse spatial correlations with the polar radar images. It will also be interesting to compare new spacecraft optical images of the poles with the polar radar images, especially on the Mariner-unimaged side. This should enable us to make detailed maps of ice features in their host craters and, in particular, to determine precise crater locations for the intriguing lower-latitude features (most of which lie in the MUH). If the spacecraft spectrometers do find evidence for water ice at the poles, this should stimulate renewed efforts to understand the problem of ice deposition and maintenance in this demanding thermal environment. Spacecraft imaging and spectrometry of the smooth plains should give us a better idea of the composition, surface texture, and geologic character of the smooth plains, which may provide some insight into the reason for their enhanced depolarized radar reflectivity. New opportunities will also open up for making detailed comparative studies of impact craters and ejecta using spacecraft and radar images, along the lines of work that has already been done for the Moon. This should give, among other things, some new insights into the crater maturation process on Mercury and may help to explain some of the Mercury-Moon differences.

Finally, continuing investigations should be carried out using the radar imaging data itself. In particular, there is a need to do more work on radar polarization, which could include making images of circular polarization ratio and degree of linear polarization. Such analyzes could provide more information on mercurian radar scattering processes and, specifically, on the role of subsurface volume scattering in the polar deposits and smooth plains.

Acknowledgements The author wishes to thank André Balogh, Rudolph von Steiger, Brigitte Fasler, and the staff of the International Space Science Institute (ISSI) for their hospitality during the 2006 Mercury Workshop in Bern. The author is affiliated with the Arecibo Observatory of the National Astronomy and Ionosphere Center, which is operated by Cornell University under a cooperative agreement with the National Science Foundation. The Arecibo radar observations shown in this paper were also made possible with support from the National Aeronautics and Space Administration.

References

J.R. Arnold, J. Geophys. Res. **84**, 5659–5668 (1979)
M.E. Bailey, J.E. Chambers, G. Hahn, Astron. Astrophys. **257**, 315–322 (1992)
N.G. Barlow, R.A. Allen, F. Vilas, Icarus **141**, 194–204 (1999)
J. Baumgardner, M. Mendillo, J.K. Wilson, Astron. J. **119**, 2458–2464 (2000)
G.J. Black, D.B. Campbell, *AAS Div. Planet. Sci. Conf. 38*, abstract no. 72.02, Pasadena, CA, 2006
G.J. Black, D.B. Campbell, S.J. Ostro, Icarus **151**, 160–166 (2001a)
G.J. Black, D.B. Campbell, P.D. Nicholson, Icarus **151**, 160–166 (2001b)
G.J. Black, D.B. Campbell, J.K. Harmon, Lunar Planet. Sci. Conf. 33, abstract no. 1946, Houston, TX, 2002
B.J. Butler, 3.5-cm radar investigation of Mars and Mercury: Planetological implications. Ph.D. thesis, California Institute of Technology, Pasadena, 1994
B.J. Butler, J. Geophys. Res. **102**, 19,283–19,291 (1997)
B.J. Butler, M.A. Slade, D.O. Muhleman, 2nd Intl. Conf. Mars Polar Sci. & Explor., abstract no. 4083, Reykjavik, Iceland, 2000
B.J. Butler, D.O. Muhleman, M.A. Slade, J. Geophys. Res. **98**, 15,003–15,023 (1993)
B.A. Campbell, *Radar Remote Sensing of Planetary Surfaces* (Cambridge University Press, Cambridge, 2002)
B.A. Campbell, B.R. Hawke, T.W. Thompson, J. Geophys. Res. **102**, 19,307–19,320 (1997)
D.B. Campbell, B.A. Burns, J. Geophys. Res. **85**, 8271–8281 (1980)
D.B. Campbell, J.F. Chandler, S.J. Ostro, G.H. Pettengill, I.I. Shapiro, Icarus **34**, 254–267 (1978)
D.B. Campbell, B.A. Campbell, L.M. Carter, J.-L. Margot, N.J.S. Stacy, Nature **443**, 835–837 (2006)
V. Carruba, A. Coradini, Icarus **142**, 402–413 (1999)
P.E. Clark, M.A. Leake, R.F. Jurgens, in *Mercury*, ed. by F. Vilas, C. Chapman, M. Matthews (Univ. of Arizona Press, Tucson, 1988), pp. 77–100
D. Crider, R.M. Killen, Geophys. Res. Lett. **32**, L12201 (2005). doi:10.1029/2005GL022689
M.E. Davies, R.M. Batson, J. Geophys. Res. **80**, 2417–2430 (1975)
M.E. Davies, S.E. Dwornik, D.E. Gault, R.G. Strom, *Atlas of Mercury* (NASA, Washington, 1978)
R.A. De Hon, D.H. Scott, J.R. Underwood, *USGS Map I-1233* (U.S. Geol. Surv., Denver, 1981)
W.C. Feldman, B.L. Barraclough, C.J. Hansen, A.L. Sprague, J. Geophys. Res. **102**, 25,565–25,574 (1997)

W.C. Feldman, S. Maurice, A.B. Binder, B.L. Barraclough, R.C. Elphic, D.J. Lawrence, Science **281**, 1496–1500 (1998)

W.C. Feldman, D.J. Lawrence, R.C. Elphic, B.L. Barraclough, S. Maurice, I. Genetay, A.B. Binder, J. Geophys. Res. **105**, 4175–4195 (2000)

D.E. Gault, J.E. Guest, J.B. Murray, D. Dzurisin, M.C. Malin, J. Geophys. Res. **80**, 2444–2460 (1975)

R.R. Ghent, D.W. Leverington, B.A. Campbell, B.R. Hawke, D.B. Campbell, J. Geophys. Res. **110** (2005). doi:10.1029/2004JE002366

R.M. Goldstein, Science **168**, 467–468 (1970)

R.M. Goldstein, Astron. J. **76**, 1152–1154 (1971)

J.A. Grier, A.S. McEwen, P.G. Lucey, M. Milazzo, R.G. Strom, J. Geophys. Res. **106**, 32,847–32,862 (2001)

M.J. Grolier, J.M. Boyce, *USGS Map I-1660* (U.S. Geol. Surv., Reston, 1984)

J.E. Guest, R. Greeley, *USGS Map I-1408* (U.S. Geol. Surv., Denver, 1983)

T. Hagfors, I. Dahlstrom, T. Gold, S.-E. Hamran, R. Hansen, Icarus **130**, 313–322 (1997)

B. Hapke, Icarus **88**, 407–417 (1990)

B. Hapke, G.E. Danielson, K. Klaasen, L. Wilson, J. Geophys. Res. **80**, 2431–2443 (1975)

L.J. Harcke, Radar Imaging of Solar System Ices. Ph.D. thesis, California Institute of Technology, Pasadena, CA, 2005

L.J. Harcke, H.A. Zebker, R.F. Jurgens, M.A. Slade, *Proceedings of Workshop on Mercury: Space Environment, Surface and Interior*, LPI Contr. No. 1097, Chicago, IL, 2001, p. 36

J.K. Harmon, Adv. Space Res. **19**(10), 1487–1496 (1997)

J.K. Harmon, *Proceedings of Workshop on Mercury: Space Environment, Surface and Interior*, LPI Contr. No. 1097, Chicago, IL, 2001, p. 38

J.K. Harmon, IEEE Trans. Geosci. Remote Sensing **40**, 1904–1916 (2002)

J.K. Harmon, *COSPAR Scientific Assembly 35*, Paris, France, 2004, p. 853

J.K. Harmon, D.B. Campbell, in *Mercury*, ed. by F. Vilas, C. Chapman, M. Matthews (Univ. of Arizona Press, Tucson, 1988), pp. 101–117

J.K. Harmon, D.B. Campbell, *Lunar Planet. Sci. Conf. 33*, abstract no. 1848, Houston, TX, 2002

J.K. Harmon, M.A. Slade, Science **258**, 640–643 (1992)

J.K. Harmon, P.J. Perillat, M.A. Slade, Icarus **149**, 1–15 (2001)

J.K. Harmon, D.B. Campbell, D.L. Bindschadler, J.W. Head, I.I. Shapiro, J. Geophys. Res. **91**, 385–401 (1986)

J.K. Harmon, R.E. Arvidson, E.A. Guinness, B.A. Campbell, M.A. Slade, J. Geophys. Res. **104**, 14,065–14,090 (1999)

J.K. Harmon, M.A. Slade, R.A. Vélez, A. Crespo, M.J. Dryer, J.M. Johnson, Nature **369**, 213–215 (1994)

J.K. Harmon, M.A. Slade, B.J. Butler, J.W. Head, M.S. Rice, D.B. Campbell, Icarus **187**, 374–405 (2007)

B.R. Hawke, D.T. Blewett, P.G. Lucey, G.A. Smith, J.F. Bell, B.A. Campbell, M.S. Robinson, Icarus **170**, 1–16 (2004)

R.R. Hodges, *Proc. Lunar Planet. Sci. Conf. 11th*, 1980, pp. 2463–2477

A.P. Ingersoll, T. Svitek, B.C. Murray, Icarus **100**, 40–47 (1992)

R. Jeanloz, D.L. Mitchell, A.L. Sprague, I. de Pater, Science **268**, 1455–1457 (1995)

R.M. Killen, J. Benkoff, T.H. Morgan, Icarus **125**, 195–211 (1997)

J.S. King, D.H. Scott, *USGS Map I-2048* (U.S. Geol. Surv., Denver, 1990)

L.V. Ksanfomality, Sol. Syst. Res. **38**, 21–27 (2004)

L. Ksanfomality, G. Papamastorakis, N. Thomas, Planet. Space Sci. **53**, 849–859 (2005)

S. Kumar, Icarus **28**, 579–591 (1976)

M. Mendillo, J. Warell, S.S. Limaye, J. Baumgardner, A. Sprague, J.K. Wilson, Planet. Space Sci. **49**, 1501–1505 (2001)

D.L. Mitchell, I. de Pater, Icarus **110**, 2–32 (1994)

J.I. Moses, K. Rawlins, K. Zahnle, L. Dones, Icarus **137**, 197–221 (1999)

D.O. Muhleman, A.W. Grossman, B.J. Butler, Annu. Rev. Earth Planet. Sci. **23**, 337–374 (1995)

D.O. Muhleman, B.J. Butler, A.W. Grossman, M.A. Slade, Science **253**, 1508–1513 (1991)

S.J. Ostro, 11 coauthors, J. Geophys. Res. **97**, 18,227–18,244 (1992)

S.J. Ostro, 18 coauthors, Icarus **183**, 479–490 (2006)

D.A. Paige, S.E. Wood, A.R. Vasavada, Science **258**, 643–646 (1992)

K.J. Peters, Phys. Rev. B **46**, 801–812 (1992)

G.H. Pettengill, R.B. Dyce, Nature **206**, 1241 (1965)

C.M. Pieters, J.B. Adams, P.J. Mouginis-Mark, S.H. Zisk, M.O. Smith, J.W. Head, T.B. McCord, J. Geophys. Res. **90**, 12,393–12,413 (1985)

R.J. Pike, in *Mercury*, ed. by F. Vilas, C. Chapman, M. Matthews (Univ. of Arizona Press, Tucson, 1988) pp. 165–273

M. Rice, J. Harmon, Bull. Am. Astron. Soc. **36**, 1162 (2004)

M.S. Robinson, M.E. Davies, T.R. Colvin, K.E. Edwards, J. Geophys. Res. **104**, 30,847–30,852 (1999)
J.R. Salvail, F.P. Fanale, Icarus **111**, 441–455 (1994)
G.G. Schaber, J.F. McCauley, *USGS Map I-1199* (U.S. Geol. Surv., Denver, 1980)
P.H. Schultz, in *Mercury*, ed. by F. Vilas, C. Chapman, M. Matthews (Univ. of Arizona Press, Tucson, 1988), pp. 274–335
Z. Sekanina, P.W. Chodas, Astrophys. J. **607**, 620–639 (2004)
M.A. Slade, B.J. Butler, D.O. Muhleman, Science **258**, 635–640 (1992)
M.A. Slade, L.J. Harcke, R.F. Jurgens, H.A. Zebker, Bull. Am. Astron. Soc. **33**, 1026 (2001)
M. Slade, J. Harmon, L. Harcke, R. Jurgens, *COSPAR Scientific Assembly 35*, Paris, France, 2004, p. 1154
M.A. Slade, L.J. Harcke, R.F. Jurgens, J.K. Harmon, H.A. Zebker, E.M. Standish, *Lunar Planet. Sci. Conf. 31*, abstract no. 1305, Houston, TX, 2000
S.C. Solomon, 20 coauthors, Planet. Space Sci. **49**, 1445–1465 (2001)
A.L. Sprague, D.M. Hunten, K. Lodders, Icarus **118**, 211–215 (1995)
P.D. Spudis, J.E. Guest, in *Mercury*, ed. by F. Vilas, C. Chapman, M. Matthews (Univ. of Arizona Press, Tucson, 1988), pp. 118–164
N.J.S. Stacy, D.B. Campbell, P.G. Ford, Science **276**, 1527–1530 (1997)
L. Starukhina, J. Geophys. Res. **106**, 14,701–14,710 (2001)
R.G. Strom, Space Sci. Rev. **24**, 3–70 (1979)
G.E. Thomas, Science **183**, 1197–1198 (1974)
T.W. Thompson, Earth Moon Planets **37**, 59–70 (1987)
T.W. Thompson, R.S. Saunders, D.E. Weissman, Earth Moon Planets **36**, 167–185 (1986)
T.W. Thompson, S.H. Zisk, R.W. Shorthill, P.H. Schultz, J.A. Cutts, Icarus **46**, 201–225 (1981)
A.R. Vasavada, D.A. Paige, S.E. Wood, Icarus **141**, 179–193 (1999)
F. Vilas, P.S. Cobian, N.G. Barlow, S.M. Lederer, Planet. Space Sci. **53**, 1496–1500 (2005)
J. Warell, S.S. Limaye, Planet. Space Sci. **49**, 1531–1552 (2001)
K. Watson, B.C. Murray, H. Brown, J. Geophys. Res. **66**, 3033–3045 (1961)
S.H. Zisk, G.H. Pettengill, G.W. Catuna, Moon **10**, 17–50 (1974)
S. Zohar, R.M. Goldstein, Astron. J. **79**, 85–91 (1974)

Earth-Based Visible and Near-IR Imaging of Mercury

Leonid Ksanfomality · John Harmon · Elena Petrova ·
Nicolas Thomas · Igor Veselovsky · Johan Warell

Originally published in the journal Space Science Reviews, Volume 132, Nos 2–4.
DOI: 10.1007/s11214-007-9290-3 © Springer Science+Business Media B.V. 2007

Abstract New planned orbiter missions to Mercury have prompted renewed efforts to investigate the surface of Mercury via ground-based remote sensing. While the highest resolution instrumentation optical telescopes (e.g., HST) cannot be used at angular distances close to the Sun, advanced ground-based astronomical techniques and modern analytical and software can be used to obtain the resolved images of the poorly known or unknown part of Mercury. Our observations of the planet presented here were carried out in many observatories at morning and evening elongation of the planet. Stacking the acquired images of the hemisphere of Mercury, which was not observed by the Mariner 10 mission (1974–1975), is presented. Huge features found there change radically the existing hypothesis that the "continental" character of a surface may be attributed to the whole planet. We present the observational method, the data analysis approach, the resulting images and obtained properties of the Mercury's surface.

L. Ksanfomality (✉) · E. Petrova · I. Veselovsky
Physics of Planets, Space Research Institute, 84/32 Profsoyuznaya str., Moscow 117997, Russia
e-mail: ksanf@iki.rssi.ru

E. Petrova
e-mail: epetrova@iki.rssi.ru

I. Veselovsky
e-mail: veselov@decl.sinp.msu.ru

J. Harmon
National Astronomy and Ionosphere Center, Arecibo Observatory, Arecibo, PR 00612, USA
e-mail: harmon@inaic.edu

N. Thomas
Physikalisches Institut, University of Bern, Silderstr, 5, 3012 Bern Switzerland
e-mail: nicolas.thomas@phim.unibe.ch

J. Warell
Astronomika Observatoriet, Uppsala Universitet, 751 20 Uppsala, Sweden
e-mail: Johan.Warell@astro.uu.se

Keywords Solar system · Unknown side of Mercury · Ground based observation · Resolved images · Regolith physical properties

1 Challenges of Ground-Based Observations of Mercury

Because of the small angular diameter of Mercury, about 7 arcsec in a favourable configuration, and its small angular distance from the Sun (on average not exceeding 22°), the task of obtaining distinct images of its surface from Earth-based observations is extremely difficult. Few astronomers drew maps of Mercury (e.g., Antoniadi's map of Mercury, 1934). In the middle of the 20th century, several attempts were made by French astronomers (Dollfus 1961). Figure 1 shows several drawings made from visual observations of the planet by B. Lyot in 1942 (Fig. 1a) and A. Dollfus in 1950 (Fig. 1c). Due to the orbital resonance of Mercury relative to Earth, at both inferior and superior conjunctions practically the same sides of the planet are observed in turn. At the time of their observations, Lyot and Dollfus did not know about the 3 : 2 resonance of Mercury; they assumed that the planet rotates synchronously (Dollfus 1961). Only those experts who have observed themselves by telescope the scarcely resolvable disk of Mercury can estimate the achievement of Lyot and Dollfus whose drawings are so similar and, as shown in the following, reflect a reality of the largest formations on Mercury, as shown in Figs. 1a and 1c. Both drawings (Fig. 1a,c) coincide exactly with the position of the planet shown in Fig. 1b, for which nowadays the mosaic of Mariner 10 photos is available. [Here and in the following the electronic globe generated by the RedShift 4 code is used (Red Shift 4 2000)]. In 1974–1975, three flybys of Mercury were made by Mariner 10. Imaging of the planet by Mariner 10 covered only 46% of the planet's surface (Fig. 2), so that after the Mariner 10 mission the other, unobserved side of the planet became the main object of interest.

Observing Mercury by classic astronomical methods is difficult due to well-known handicaps caused by the terrestrial atmosphere. These handicaps impede observation of very small objects close to the horizon. One can imagine the atmosphere's heterogeneity as a random set of poorly refracting lenses, each of them in its own way deforming the image. The distortions are individually insignificant, but they accumulate. Other kinds of difficulties are the jitter of images and their washout, caused by the atmosphere's turbulence. An obvious method of improving astronomical images (besides adaptive optics techniques) is the reduction of their exposure to a duration during which turbulence does not spoil a picture. If one then selects from the large number of pictures with very short exposures those which have small distortions, their processing makes it possible to approach the diffraction limit of the instrument. Experiments of this sort were carried out in 1960s–1970s, but without a noticeable success. The exposure should be very short, on the order of milliseconds for a telescope

Fig. 1 Drawings of Mercury made by B. Lyot in 1942 (**a**) and A. Dollfus in 1950 (**c**) at visual observation of the planet. The view of the planet in this phase, constructed from a mosaic of photos made by Mariner 10 is placed in the centre (**b**). The central meridian is 110°W

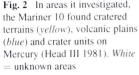

Fig. 2 In areas it investigated, the Mariner 10 found cratered terrains (*yellow*), volcanic plains (*blue*) and crater units on Mercury (Head III 1981). *White* = unknown areas

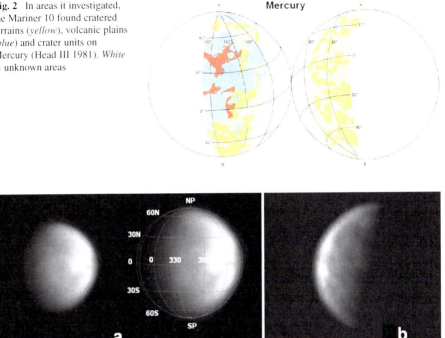

Fig. 3 The modern amateur's stacked photos of Mercury acquired by means of a small telescope and CCD matrix. (**a**) Image of Mercury obtained with a 10-inch telescope, October 1, 2006. Camera exposures were 1/50 s, a spectral range 700–900 nm, peaking at 800 nm. The image is a stack of 70 of the best single frames. North is up. Courtesy of S. Massey, Australia. (**b**) The image obtained with another 10-inch telescope using a red filter. The 100 images, each of 1/73 s, were stacked. Observation made during the daytime, November 22, 2006, with Central Meridian of 271°W. Courtesy of A. Allen, Santa Barbara, California

of moderate size (1.2–2 m); this is orders of magnitude shorter than the exposure needed by photographic emulsions. Nevertheless, a map of the planet was produced photographically (Murray et al. 1972). The millisecond exposures needed became possible when new, highly effective CCD detectors could be used in combination with computer programs that could process large volumes of observational data. The availability of CCDs nowadays allows us to obtain photographs of Mercury using even small telescopes, 20 to 36 cm. Several amateur astronomers are now obtaining excellent images of Mercury (Fig. 3).

The brightness of the sky is another problem. Observations of Mercury are inevitably made in twilight or even in daytime, against a bright sky. Brightness of the clean sky (produced by Rayleigh scattering) sharply decreases as the wavelength λ increases, as λ^{-4}. Shifting the centre of the spectral range, say, from 600 nm to 800 nm, decreases the sky brightness more than three times. The investigation of the surface of the atmosphereless celestial body yields usually the same results, whether they are grey or in colour (however, the colour hues may be of importance for mineralogy). That is why using near IR is very useful if the CCD spectral response is high enough in the IR range (see the discussion and Fig. 11). For the same reason, only observations at high mountain observatories are usable. Using the IR range is of importance in other ways, too. During twilight observations, Mercury is low above the horizon; this is the reason for a strong differential refraction and especially differential chromatic refraction. Selection of the IR range partly solves this problem. Due to

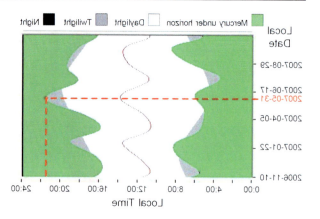

Fig. 4 Twilights observability of Mercury is limited by two or three short favorable periods each year. The diagram shows morning (*left*) and evening (*right*) elongations for one year, starting November 10, 2006. The deeper the grey area, the better conditions are

the reasons mentioned in this paper, the spatial resolution of astronomical images depends on the place of observation, time of day, density of aerosol component, zenith distance of Mercury (air mass, or sec z) and, primarily, on length of exposure.

As the angular distance of Mercury from the Sun in ground-based observations is always small, observatory administrators are reluctant to use large telescopes to observe Mercury, since there is a risk that the instrumentation of the telescope could be damaged by direct solar rays. Nevertheless, ground-based observations of Mercury by "classical" methods (without using short exposures) have been performed by many researchers. A detailed list of such observations can be found in the review by Warell and Limaye (2001). In comparison with the observations of other bodies in the Solar System, observations of Mercury at ground-based observatories are also subject to many restrictions. Since the greatest elongation of the planet does not exceed 28° (typically, no more than 22°), observations must be performed within less than 1 hour; in fact, the observing time in twilight rarely exceeds 20–30 min, as the planet is very low above the horizon when the large air mass (sec z) complicates the problem still further. More or less usable observations of Mercury are possible only at low-latitude mountain observatories. However, as is shown in the following, resolved images of the planet can still be obtained by ground-based technical and analytical means, although at the very limit of technical capabilities. There are three to four Mercury morning and evening elongations each year at which observations are feasible in principle. Figure 4 is calculated for Arhiz, the site of the Special Astrophysical Observatory in the Caucasus. Due to the inclination of planet's orbit to the ecliptic, not more than two of each set of elongations are really favourable with respect to the duration of visibility, indicated in gray in Fig. 4. As an example, the dashed lines indicate that on May 31, 2007, the year's most favourable elongation, Mercury sets on the mathematical horizon at 21:28, when sunset is 19:38. Keeping in mind that the planet is observable at least at few degrees above horizon, the duration of evening session for these favourable 10–12 days will each be less than one hour.

2 Advantages of a Short-Exposure Method and Achievable Surface Resolution

The characteristic time for which instant optical properties of atmosphere could change may be determined by researching frequency spectra of stellar scintillation and image jittering. It is seldom shorter than 30 ms (Ksanfomality 2003). Therefore, to use instant calm atmosphere, the exposure time should not be longer than that time. Advantages of the method

Fig. 5 Fast changes of quality in images of Mercury. Time increases with images numbered from down to up. Only images 2, 6 and 8 are precise (more or less). Image 1 is blurred; 3 is deformed. Image 4 is blurred along X axis and rotated; 5 is blurred along Y axis; 7 is deformed and appreciably differ from the distinct image 2. Intervals between photos are 99 ms. Expositions are 3 ms

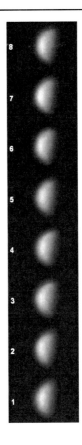

of short exposures were considered by Fried (1978), following Hufnagel (1966). In Fried's words: "In short-exposure imaging through turbulence, there is some probability that the image will be nearly diffraction limited because the instantaneous wave-front distortion on the aperture was negligible". Fried analyzed distortions created by atmospheric inhomogeneities and estimated that the possible gain in the resolution can be a factor 3.4 at short expositions. One of his conclusions was that optimum observations are before dawn, when atmospheric turbulence has not yet begun; for daytime observations (with moderate zenith distances) the relatively long exposures need the presence of strong winds. Experimental backing for these conclusions are available in both cases. We can attest to the correctness of these statements because three of the most successful series of our observations of Mercury were made in such conditions—two different sets were made during early autumn mornings in two different observatories, and another during spring evenings when observations at the Skinakas observatory were carried out in conditions of very strong winds. These results can be counted as a strong backing of Fried's (1978) conclusions. However, another of his conclusions, that "the probability of getting a good image" sharply decreases when diameter of telescope D grows, seems to be debatable, as he did not take into account, apparently, that exposure with a 1.5-m telescope can be about 10 times shorter than with a 0.5-m telescope, at the same signal-to-noise ratio.

In Fig. 5, part of a sequence of electronic pictures of Mercury (from observations made May 2, 2002) with an exposure of 1 ms, separated by intervals of 99 ms, shows how fast the quality of an image may change. Time increases with images from bottom to the top. Only

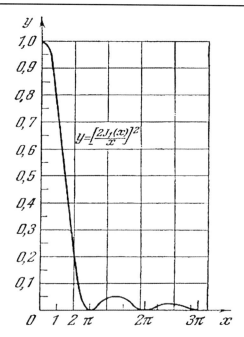

Fig. 6 Distribution of intensity in the image point. Three bright and dark rings are shown

images 2, 6 and 8 are precise, more or less. Other images are deformed and appreciably differ from the distinct image 2.

Steadier weakening or strengthening of atmospheric instabilities appear and disappear within several seconds (or tens of seconds). Another problem (besides blurring) is that in a series of a large number of electronic pictures, images sometimes appear rotated by a few degrees (Ksanfomality 2003) and that complicates their joint processing.

Returning to the potentially achievable resolutions and comparing them with the diffraction limit of an instrument, it is necessary to keep in mind that stacking very large numbers of primary pictures improves images substantially. As shown in the following, it is possible through this method to resolve details which are in size formally below a diffraction limit of the telescope. For the case of observations of Mercury in 90° phase (at geocentric distance 0.85–0.90 AU), in a near-IR range, as mentioned earlier, using the instrument with diameter, say, $D = 1.25$ m, at a wavelength $\lambda = 600$ nanometer, the formal diffraction limit is $1.22\,\lambda/D = 0.15$ arcsec. Thus, limiting resolution on the surface of the planet corresponds exactly to 100 km. Therefore the details of surface features of sizes of 40–80 km, considered below, should apparently be outside the capabilities of this method. However, this is only seemingly a paradox. As is known, distribution of intensity y on distance x from the centre of the image makes $y = [2J_1(x)/x]^2$, where J_1 is the coefficient of the first term in the polynomial expansion describing the ring pattern, and the factor 1.22π corresponds to position of the first minimum. If one chooses details at a level higher than 0.5 or 0.7 (Fig. 6), for example, at 0.9 level, their size, naturally, will be less. But it needs a very large number of suitable raw pictures (as is carried out in our program of stacking of images). Obtaining a resolution close to the diffraction limit of the instrument should make a resolution of, say, 0.12–0.15 arc seconds, using sharp optics, of course. At the same time, typical resolution for a 1.5-meter 1/10 reflector is seldom better than 1–1.3 arc seconds, being limited by atmospheric turbulence. Thus, it was impossible to obtain images resolved at the diffraction limit for Mercury, with a diameter in a quadrature phase on average is only

7.3 arc seconds, using photographic materials that do not allow reduction of exposure time to units of milliseconds. Such an image could be resolved typically to four to six lines only, as the photographic resolution is about 1.3 arc sec. Only with the advent of CCD detectors that have high quantum output, combined with fast image processing techniques that permit the combination of thousands of images chosen for high quality, does the high spatial resolution become achievable. It should be mentioned that a long-focus instrument has advantages in observing Mercury because that increases nominal resolution, which is important for the subsequent data processing.

3 Pioneering Short-Exposure Imaging of Mercury

Pioneering multicolour filter imaging of Mercury with rather short exposure times was made by J. Warell at the astronomical observatory Uppsala, Sweden (Warell and Limaye 2001). Warell began observations in 1995 at the 0.5-m Swedish Vacuum Solar Telescope (focus 22.35 m), on La Palma in the Canary Islands, at 11 wavelengths from 429 to 944 nm, using fast-readout CCDs. Two CCD cameras with 1.4 and 1.6 Mpixels were used with exposure times from 25 up to 360 ms. The outcome of a multiyear effort to map albedo variations of the global surface of Mercury at optical wavelengths was presented by Warell and Limaye (2001). They showed single (not stacked) pictures chosen from their large data set, using images of different parts of Mercury obtained during the elongations 1995–1998. Examples of successive electronic photographs of Mercury taken by J. Warell at the La Palma Observatory in the period 1995–1998 are shown in Fig. 7. Based on these sets, the images were produced that are shown in Fig. 8, (a) globes and (b) rectilinear map. The authors stated that a resolution ~200 km was realized. Despite the rather long exposure time for the single images, large details are defined and have permitted the authors a wealth of geologic

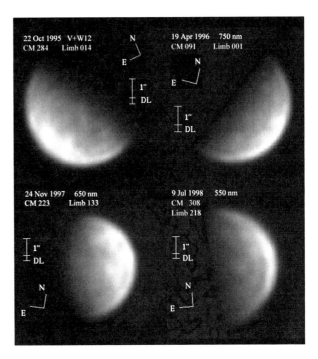

Fig. 7 Examples of electronic photographs of Mercury taken by J. Warell in period 1995–1998. Single Mercury images obtained with the 50 cm Swedish Vacuum Solar Telescope on La Palma observatory on four dates, from 1995–1998. The central meridian and limb West longitudes are indicated. Scale bars for 1 arc second and the diffraction limit are given for the used wavelength (Warell and Limaye 2001)

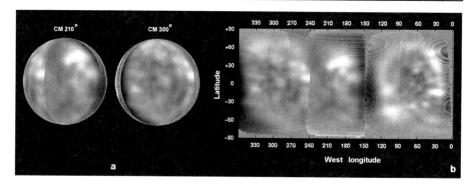

Fig. 8 (a) Globes and (b) global cylindrical projection of Mercury images created by J. Warell from the same ground-based observations as in Fig. 7. The central meridian is indicated as CM (Warell and Limaye 2001)

interpretations (see Sect. 10) based on multicolour analysis (Warell 2002, 2003; Warell and Valegård 2006).

Figures 8a,b are the first global map of Mercury's surface produced using CCD imaging. For this, the selected high-resolution raw images were individually reduced according to standard CCD procedures, sharpened by Wiener filtering to increase definition and contrast, and photometrically rectified and remapped for detailed analysis. The studies that were made of the morphologic and photometric properties of the detected albedo features, and their correlation with Mariner 10 image data of the well-known hemisphere, allowed identifications of probable bright ray craters on the poorly known hemisphere. Warell and Limaye (2001), as well as Warell and Valegård (2006), noted that in many ways the two hemispheres were found to be remarkably similar; no statistically significant differences could be detected in a comparison of the spatial distributions, number densities, photomorphologic variables or colour properties for the features on the two hemispheres. From a study of the Mariner 10 image data it was concluded that about 70% of the bright features on the poorly known hemisphere could be attributed to bright ray craters, while the rest were attributed to bright crater floors, concentrations of crater rays or diffuse patches which had been described based on Mariner 10 images. One feature corresponding to the location of the ray crater Kuiper (11.3°S, 31.1°W) discovered by Mariner 10, was found to be extraordinary in all its morphological parameters, with a contrast relative to the darkest albedo features of as much as 50%. Localized dark patches were found to correlate with surface of low radar backscatter signal (Harmon and Slade 1992; Slade et al. 1992; Butler et al. 1993). Bright spots were found to be less well correlated with radar signatures, but a bright albedo feature at 331°W, 3°N. The spots A and B detected in radar works were found to coincide with locations of less bright and more extended albedo features.

About this time papers were published (Ksanfomality, 2004, 2003) describing a new approach to the processing of Mercury's images. Subsequently, to produce a more detailed image of the planet, Warell passed the primary photographs to L. Ksanfomality to be processed by the software package (Ksanfomality, 2004, 2003; Ksanfomality et al. 2005). Unfortunately, because of the limited number of primary photographs, the software failed to perform a satisfactory correlation in matching the planetary images.

Another important pioneering work on short-exposure imaging of Mercury, simultaneously with Warell and Limaye (2001), was carried out by Dantowitz et al. (2000) and Baumgardner et al. (2000). On August 29, 1998, they electronically photographed Mercury in the

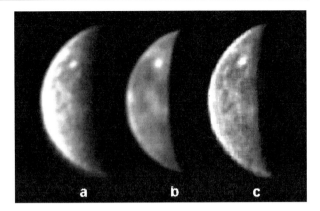

Fig. 9 Mercury image from 275° to 330°W longitude was produced by a short exposure technique and subsequent processing of a large number of data files by: (a) Dantowitz et al. (2000), (b) Baumgardner et al. (2000). A stack and co-add of (a) and (b) using the method of Ksanfomality (2004, 2003) is shown in (c). From Ksanfomality (2006)

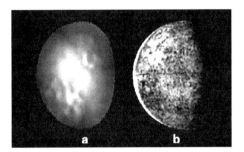

Fig. 10 Stacked image of Mercury from the observations at AbAO on December 3, 1999

longitude range that was not imaged by Mariner 10 (in 1974–1975), using an ordinary (general purpose) video camera mounted on the 1.5-m Mount Wilson Observatory telescope. The video camera was operated in standard mode, with a frame rate of 30 s^{-1} and an exposure of 17 ms. A total of 219,000 frames were taken, of which 1,000 photographs were chosen for further processing. Eventually two sets of 30 images with the highest contrast scores were used to portray the final results. Mercury was imaged at phase 106°, in the longitude range 265–330°W, and with the coordinates of the subterrestrial point 7.7°N, 254°W (Figs. 9a,b). This, we believe, was the first published result of stacking this large number of unprocessed electronic photos of Mercury; the author's stacked images were probably the best results of the works published in 2001–2002. Dantowitz et al. (2000) and Baumgardner et al. (2000) estimated the achieved resolution on the surface of Mercury to be 250 km. A review of all these results was published by Mendillo et al. (2001).

The results by Dantowitz et al. (2000) and Baumgardner et al. (2000) were processed additionally by the method presented in papers by Ksanfomality (2002, 2004) and Ksanfomality et al. (2005), for the reason given in Sect. 5. The processing improved appreciably the image of the planet (Fig. 9c) and permitted distinct large details to be defined. The stacked image Fig. 9c appears more detailed than the original images.

In December 1999, Ksanfomality et al. (2001) took a series of raw electronic photographs of Mercury at the Abastumany Astrophysical Observatory (AbAO) of the Republic of Georgia (42°50'E, 41°45'N), using a ST-6 CCD camera mounted at the Cassegrain focus of the AZT-11 telescope ($D = 1.25$ m, $F = 16$ m). Using these images, they performed tests of different kinds of processing by stacking software and produced a trial image in which some features could be distinguished (Fig. 10). The longitudes in the image were close to

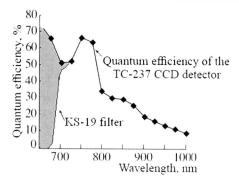

Fig. 11 The spectral range of the CCD camera and filter that were used for all images shown in the following

the electronic globe created from the published mosaics of Mariner 10, which allowed their comparison.

On November 3, 2001, the next series of observations of Mercury were performed at the AbAO again during morning elongation of the planet. The phase of the planet was 63°. The same AZT-11 telescope was used, but with a 700-nm filter that cut off the short wavelengths and with the STV CCD camera produced by SBIG Co (pixels are square 7.4 × 7.4 µm). The long-wavelength boundary was near 1 µm (Fig. 11) and was determined by the spectral properties of the CCD array. (It should be mentioned that the same camera and filter were used for all images shown in the following.) Since the object is bright, the integration time of several thousand signal units in each pixel even with a small-pixel CCD array still did not exceed a few milliseconds and short exposures, 3–10 ms, were used. On November 3, 2001, the disk of the planet was seen at an angle of 6.1 arcsec, which corresponded to a linear image size of 0.43 mm in the focal plane of the telescope. The observations were carried out under extremely favourable visibility conditions (Ksanfomality 2002). Several series of raw photos, including a series with the recording of the light polarization distribution over the planetary disk, were taken (Ksanfomality 2002). Based on the subsequent selection from the electronic raw photos, an image for the longitudes from 90° to 190° was produced (Fig. 12), using the first version of the data processing by means of correlational stacking software. In this experiment, when initial photos were stacked, a high resolution was achieved at the central part of the image that looked sharp. However, the image's periphery was blurred. The image of Mercury thus obtained was compared again with the electronic mosaic globe made from Mariner 10 photos, for the same phase and longitudes. This showed that features in the central part of the image with sizes of only 100–250 km are clearly identified with Mariner 10 data (Fig. 12). Later other features were also identified. At that time, this result was unexpected since the feature sizes were at the diffraction limit of the telescope, but it convincingly proved the actual capabilities of the new method. An example of coincident features in two image fragments is illustrated in Fig. 13, where 13a and 13c are fragments of Fig. 12; Figs. 13b and 13d are fragments of the Mariner 10 mosaic map, with coincident features indicated by arrows in b and c. A small, very dark area (22°N, 155°W) is marked as 1. Large features at (15°S, 150°W) and (18°S, 165°W) are indicated as 2 and 3. A minor bright dot 4 (10°S, 117°W) in Fig. 12a and Fig. 13c coincides with an unnamed bright crater 4 in Tolstoy, or Phaethontias Region (Fig. 13b). Two large, dark rounded regions 5 and 6 near the equator (centred on 5°N, 124°W and 1°S, 135°W) have the same form in Fig. 13b and c. These regions include the craters Mena (0.5°N, 125°W) and Lysippus (1.5°N, 133°W). Most convincing is a coincidence of shapes of features 4, 5, 6 in the 13 b and c images.

The features in other areas were also identified. Some details in Fig. 12a may be due to noise in view of the limited number of original images. Nevertheless, the similarity of the

Fig. 12 (a) Image of Mercury stacked from initial photos of observations made on November 3, 2001, and (b) its comparison with the globe of Mercury constructed from the Mariner 10 photo mosaic (from Ksanfomality and Sprague 2007, reworked)

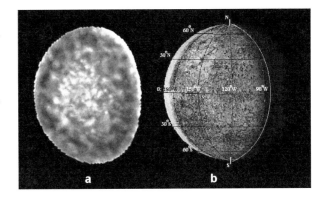

Fig. 13 Comparison of the fragments of the stacked image (a, c) and the Mariner 10 mosaic photomap (b, d)

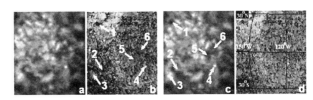

Fig. 14 Data counts from a trace across the image of Fig. 12a along the vertical line parallel to the polar axis and crossing the bright point (marked 4 in Fig. 13). From Ksanfomality and Sprague (2007)

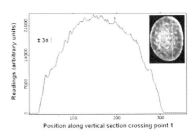

two images in Fig. 13 is obvious, despite the small size of the formations presented here (feature 3 is only 200 km in width). After all processing, the contrast of elements of features in the stacked image of Mercury in Fig. 12a is within the range 4–10%.

Figure 14 shows the distribution of contrast features in the section parallel to the polar axis and passing through the bright dot (10°S, 117°W) in Figs. 12 and 13. The position of features across the section is marked on the linear scale. Ksanfomality and Sprague (2007) estimated that the root mean square (rms) error for a signal obtained from one (single) pixel of the CCD for the case of Fig. 12 is about 9% for the absolute rms errors and 0.7% for the relative rms errors. The triple error $3\sigma R \approx 2\%$, that was quite comparable to the contrast of individual pixels (4–10%), as in Fig. 14. However, since the image is formed not by a single pixel, but by a large group of pixels (1,200 pixels in Fig. 13a), the probability of their erroneous combination is much lower.

Thus the first steps of the data processing by the correlational stacking software proved its efficiency. The data processing program was not perfect. Large image features were suppressed; the contrast of light and dark features was quite low; and the periphery of the image was blurred.

During subsequent years more effective codes were worked out, as described in the next section. In particular, the application of these codes successfully showed both the low spatial frequencies and a distinct periphery of the image.

4 Imaging Surface of Mercury in 2003 (Longitudes 90–190°W)

During the next few years, different data-processing programs were tested. Eventually it was found that less complicated codes yielded better results. The code AstroStack, created by Stekelenburg (1999, 2000), and the AIMAP code, created by Kakhiani (2003), were chosen to form the core of the processing system (Ksanfomality 2002). The technique of processing and confirmation of the results included about 10 other codes. The most important procedure was the correlation stacking of a large number of images. A short description of the processing technique was presented by Ksanfomality (2002) and Ksanfomality et al. (2005). Its newer version showed successfully both large image features and a distinct periphery of the image.

Together with the processing technique, a new, more effective method of observation was developed. To take the largest possible number of primary electronic photographs of Mercury, the STV camera mounted on the telescope was switched to fast image collection mode. It took 4 s (at a 1-ms exposure with 99-ms intervals) to acquire one frame (Fig. 15) composed of 40 photographs. It took approximately the same time to write the frame to a memory. After 15 frames were obtained, the entire set was transferred to a computer.

The STV camera is not the best solution. Only a serial (com) port is used in the STV camera for data transfer, which causes additional delays and is a shortcoming of the STV camera. Nevertheless, the fast collection of primary images of each frame in some measure compensates for these disadvantages. However, this requires stable seeing conditions, which are rarely met. The observational data acquired with the STV camera in its chosen mode had an 11- or 11.5-bit compressed format. The routine primary data processing operations included a precheck (the rejection of defective series), by which the best series of undistorted electronic photographs or single images were selected. For the case when visibility is poor even a single good image should be selected for processing.

A new series of observations of Mercury were made on November 2, 2003, under the same circumstances that were described about the observations made on November 3, 2001. The stacked and additionally processed image (Fig. 16) covers an area investigated by Mariner 10. The position of the planet was essentially the same as in Fig. 12 (64° is the phase of the planet, 120° is the central meridian). The longitudes 90–190°W are known after the Mariner 10 mission; however, the view of the planet in Fig. 16 differs much from the Mariner 10 images (e.g., Fig. 12b), by an abundance of dark and light areas. The stacked image in Fig. 16 much more closely resembles the far side of the Moon than the widely known and often published mosaic of Mariner 10 photographs. The main reason is that the Mariner 10 mosaic was created manually (Strom and Sprague 2003); the contrast of small features in the primary photographs was enhanced, and the boundaries of the photograph fields required alignment and were matched in contrast. This led to the suppression of the

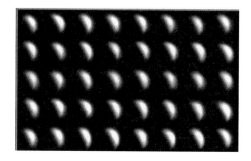

Fig. 15 Example of a frame from a group of primary electronic photographs of Mercury (November 24, 2006). The exposures are 3 ms each, the intervals are 99 ms; the size of each photograph is about 7 arcsec

Fig. 16 (a) Image of Mercury stacked from initial electronic photos of observations made on October 2, 2003, and (b) coordinate grid of Mercury for this date

low spatial frequencies, corresponding to the large structures lost in the Mariner 10 mosaic. However, it may well be that there is another reason. The onboard system memory is known to have malfunctioned during the operation of the spacecraft near Mercury. It is possible that this led to a mismatch of the video channel frequency properties of the Mariner 10 camera. In any case, the balance of spatial frequencies of the mosaic images was disrupted, due at least in part to the manual processing.

Though the resolution in Fig. 16 is not as high as in the central part of Fig. 12a, large details are seen better. The accuracy of the coordinate grid position is within 5°. In the left part of Fig. 16 the limb is placed a bit to the west, compared to the borders of imaging by Mariner 10 (Fig. 12b). In the Northern hemisphere, northward from the Mozart crater (8°N, 190°W) and westward from Caloris Planitia (30°N, westward from 180°W) a blurred large area is placed along the limb. This is the area of quadrangles Tolstoj H-8, Shakespeare H-3 and H-4 (Liguria), according to the *Atlas of Mercury* (Davies et al. 1978). A large, dark basin is placed at 45–60°N, westward from 180°W on the limb (10° on Mercury's equator corresponds to 426 km). Close to the North Pole there are two dark round basins, 300–500 km each. A very large dark area, extending for 1,300 km along the meridian, has an irregular shape. It coincides partly with the Basin Shakespeare. A bright crater at 40°N, 120°W could be its centre. Two distinct round craters are placed in its southern part 20°N, 115°W and 30°N, 110°W. Southward from them, a dark elongated feature about 400 km is centred at 20°N, 90°W. It is remarkable that many prominent dark or bright features, although they are well outlined (like a dark basin at the centre of the image), do not coincide with any known objects. The Basins Shakespeare, Van Eyck and Strindberg are difficult to identify. In the Southern hemisphere, two large dark basins are seen at 50°S, 110°W and 50°S, 150°W, about 400–500 km. An elongated bright area centred at 25°S, 170°W, is adjusted to a dark elongated valley, centred at the limb at 25°S, 190–200°W. The smallest features about 200 km (appearing as dark dots) are seen at 20°S, 170°W. A chain of dots about 200 km each is seen from 30°S, 140°W to 55°S, 130°W. For more reliable imaging one needs many more primary electronic photos.

As for initial photographs of Mariner 10, they are rich in bright and dark features (Fig. 17a). Here Caloris Planitia and the Mozart crater (marked CP and M) are seen at the morning terminator. The photo was made 12 hours after closest approach of Mariner 10. The image consists of two parts; they are joined at a line that crosses the Tolstoj basin (marked by T in Fig. 17a). Their boundary (just above mark "T") is well seen at the limb. Despite a minor phase difference, the upper photo shows the basin to be dark, the lower bright. This phenomenon is considered in Sect. 8; the 17b panel is considered in Sect. 7.

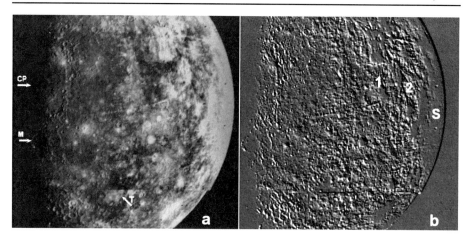

Fig. 17 (a) Unprocessed images of the Mariner 10 are rich in bright and dark areas. CP = Caloris Planitia, M = Mozart crater, T = Tolstoj basin. (b) Using high-pass filter provides quasi-relief information that may provide certain clues to the topographic expression. Craters *1* and *2* are prominent in the *right panel*, but are almost lost in original photo (a). By S marked saturated part of the photograph

5 The Surface of Mercury in the Longitude Range 210–285°W. "Skinakas" Basin

The stacked images of Mercury in the 200–290°W longitudes are based on several series of successful observations carried out in early May 2002 at the Skinakas Observatory of the Heraklion University (35°13′E, 24°54′N, Crete, Greece). The same STV CCD camera (with pixel size 7.4 × 7.4 μm) mounted on a Ritchey–Chretien telescope ($D = 1.29$ m, $F = 9.857$ m) was used. The disk of the planet on May 1 and 2, 2002, was seen, on average, at an angle of 7.75 arcsec with its linear size of 0.37 mm in the focal plane of the telescope and corresponded on the CCD array to only 50 lines in zoom mode, or 25 lines in normal mode (four pixels binned). The phases of Mercury were 93° and 97°. Short exposures, mainly 1 ms and up to 10 ms at large zenith distances, were used. Stable seeing conditions were at the Skinakas observatory on May 1 and 2, 2002; that was a key factor for successfully obtaining a very large number of raw electronic images, about 20,000. Since the size of each image (like those seen in Fig. 15) is only 7.75 arcsec, an image unsteadiness of only 1–2 arcsec would have brought part of the image outside the CCD fragment used.

Images of the surface range in Mercury's longitude 210–285°W were stacked in the course of processing the data set collected during the observations on May 2, 2002. The processing was continuously perfected and became more and more complex. The image resolution and the reliability of identifying the detected relief or albedo features of the planet increased accordingly. The first version of the stacked images from the processing of the 2003 observations was based on the 1,120 raw photographs; about 35% of these remained after manual selection (Ksanfomality 2003). This comparatively rough image (Fig. 18, phase 97°) immediately showed the largest structures in the longitude range 210–285°W, the most prominent of which is the "Skinakas Basin", considered in the following.

Further substantial progress in resolution was made in 2004. The second version of the image was based on a considerably larger number of primary photographs, 5,240; the photographs used to synthesize the first version of the image were excluded. Figure 19 with the coordinate grid shows the synthesized images of Mercury at phase 97°. It demonstrates how the images improved as increasingly more raw photographs were included in the processing and as the stacking technique itself was perfected. The coordinate identification error

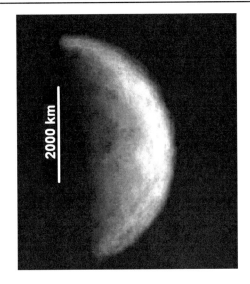

Fig. 18 The first version of the stacked image of the Mercury's surface in longitude range 210–285°W, based on the data set collected during the observations on May 2, 2002. This is a comparatively rough image, immediately showing Skinakas Basin, the largest complex structure in this longitude range. The phase is 97° (from Ksanfomality 2003)

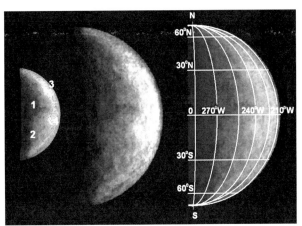

Fig. 19 Progress in the resolution of stacked images for Mercury at phase 97° (May 2, 2002). The central meridian is 286°W. Images (Fig. 18 and Fig. 19) were synthesized from independent raw data that number 1,120 and 5,240 primary electronic photographs, respectively. Details marked as 1–3 are considered in Sect. 8

does not exceed a few degrees, but it increases near the poles. The coordinate grid was constructed by taking into account the fact that the North Pole at the time of observations was displaced from the limb toward the observer by a few degrees.

Although all of the large image features are present in both Figs. 18 and 19, the version in Fig. 19 is appreciably sharper. The photometric function was removed more accurately. However, a contrast of large dark areas (like feature (1)) is lost to some extent. Both large and small structures can be distinguished in Fig. 19. An example is a large, about 1,200 km, pentagonal ray crater (2) at 36°S, 260°W, marked only in Fig. 18 but seen well in Fig. 19. Of the same size is a crater centred at 0°, 265°W. The sophisticated image processing resulted in Fig. 19, which allows us to discern similar features that are near the terminator. Details marked as 3–4 are considered in Sect. 8.

In the previous section, in the cases of Figs. 12 and 16, the reality of the results could be confirmed by their comparison with Mariner 10 images. However, what could be a proof that new formations found in the longitude range 210–290°W actually reflect the Mercury's geography? Since details of the planet's surface in this region were unknown, the first and

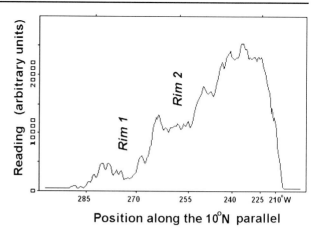

Fig. 20 Cross-section of rims of S Basin image along parallel 10°N

main criterion for the features being real was their presence in several synthesized images acquired independently and based on independent groups of primary electronic photographs (Ksanfomality, 2002, 2004). Using this criterion proved to be the most reliable method for confirming that these features are real. Nevertheless, in two cases, the detected features have been found in independent publications (Harmon et al. 2001; Dantowitz et al. 2000; Baumgardner et al. 2000). These are the formations near the North Pole and the Skinakas, or "S" Basin, respectively.

The most prominent feature (1) in Figs. 18 and 19 is a large, dark "basin" (according to the lunar terminology) near the terminator, centred approximately at 5–9°N, 270°W. The basin, the largest formation in the longitude range 210–290°W, has an inner diameter of about 1,000 km that is more or less regular in shape. The discernible irregular outer rim has twice the diameter of the inner part (Fig. 18). The structure of the inner rim of the basin, according to Fig. 18, is comparatively regular on the eastern side, but its regularity is distorted by an object centred at 35°N, 270°W in the north and by an extensive, lighter region located between 250° and 280°W, extending from 5°S to 25°S in the south. A large, right-angled area extends from 0 to 8°N, from 245 to 262°W and covers an area of 340 × 720 km; it is the darkest spot in this sector at 8°N, 250°W (Fig. 19). It could be a deep bottom-land (examined in Sect. 6).

When processing the observations, the nickname "Skinakas Basin" (after the Observatory where the data were obtained) was used for this formation, without having any pretension to making this legally binding; the IAU is known to name all objects on the surface of Mercury after writers, composers, artists, etc. Nevertheless, the name "Skinakas Basin" (or "Skinakas Mare") mentioned at several conferences (Ksanfomality 2004, 2005, 2006) has already been accepted in the literature. To avoid this problem here the name "S Basin" is used. In Ksanfomality et al. (2005), the S Basin and Caloris Planitia were compared and it was shown that S Basin is the largest basin on Mercury, about 1.3–1.5 times the diameter of the Caloris Planitia. The basin is probably a very old formation on Mercury, with broken rims that are actually overlapped by the boundaries of other, smaller basins. The morphology of the rims of the basin seems to be similar to that of Caloris Basin, or may resemble more or less the large old formation Aitken-South Pole on the far side of the Moon. The signatures of the rims are clearly seen in Fig. 20, which shows a distribution of brightness across Mercury's image shown in Fig. 18, along a parallel at latitude 10°, crossing the S Basin from terminator to limb. Distinct rims are seen along the 10° N latitude trace.

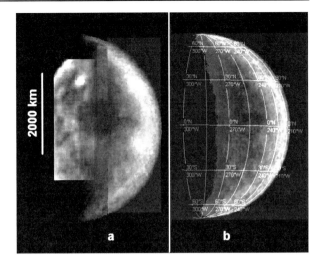

Fig. 21 (a) The composed view of S Basin from images in phases 29.08.1998 (Baumgardner et al. 2000; Dantowitz et al. 2000) and 2.05.2002 (Ksanfomality et al. 2005). (b) Location of coordinates grid

Only the eastern part of the Skinakas Basin is seen in Figs. 18 and 19. Nevertheless, it was possible to compose a full image of the S Basin. For this purpose, the eastern part of the basin was taken from Ksanfomality's data and its western part from the revised results of Dantowitz et al. (2000) and Baumgardner et al. (2000), shown above in Fig. 9c. As was noted in Sect. 3, Dantowitz et al. (2000) and Baumgardner et al. (2000) took numerous electronic photographs of Mercury at phase 106°, in the longitude range 265–330°W, on August 29, 1998, using the 60-inch Mount Wilson Observatory telescope. Each of the teams published their results and mentioned the bright object in the upper part of the image (35°N, 300°W). In addition, Dantowitz et al. (2000) pointed out that the observed features include an interesting ring object at the terminator, approximately at 10°S, 270°W. Baumgardner et al. (2000) mentioned it also and called this object a large, dark mare-like region at about 15–35°N, 300–330°W. In both cases, this position virtually coincides with the position of the centre of the S Basin. Due to a favourable coincidence, the terminator's position is in fact the same as evening terminator in Fig. 9 and morning terminator in Figs. 18 and 19. On this basis, the full image of the Skinakas Basin was constructed. The details are given in Ksanfomality (2006) and Ksanfomality et al. (2005). The favourable location of the S Basin in Figs. 9 and 18 allowed both its halves to be matched.

The results of joining the images of the western and eastern parts of the S Basin are shown in Fig. 21a. In Fig. 21b, the coordinate grid position is shown. Since matching was performed without coordinate transformation, it definitely cannot be exact. However, the limited resolution of the images makes the result quite acceptable even without coordinate transformation. As can be seen from Fig. 21a, the S Basin is a typical crater mare with a dark central part and a periphery of complex structure. The scale of the formation is shown on the left. As was estimated earlier, the sizes of its central and outer parts are about 1,000 and 2,000 km, respectively. The image was not sharp enough: only large features were identified reliably, but their detailed description was waiting for new observations. It is pertinent to note that the vague dark formation denoted by Mendillo et al. (2001) on their scheme coincides in position with the S Basin.

It is interesting to compare the S Basin sizes with the lunar forms. The inner part of the basin slightly exceeds in size the largest lunar Mare Imbrium, while its outer part has the scales of the lunar Oceanus Procellarum. In contrast to the S Basin and Caloris Planitia, the surface of Mare Imbrium is a lava field the formation of which dates back to the ancient

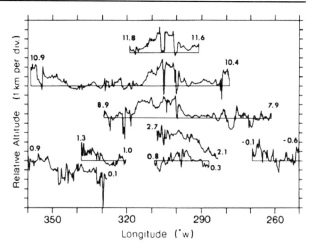

Fig. 22 Relief scans acquired by Arecibo radar imaging across Mercury crossing the area of S Basin. Latitude is indicated by the numbers at the ends of the scans. Relative altitude is given on vertical scales. From Harmon and Campbell (1988)

epoch of global lava flows on the Moon. However, it is not quite appropriate to compare the S Basin with lunar mare regions. In general, their albedo is a factor of 2–3 lower than that of the highland area. Since the data processing when Mercury's images were stacked included the "unsharp mask" operation, the albedo variations are difficult to estimate reliably. If the S Basin is assumed to be actually a broken impact crater, its sizes point to a very large impact event that dates back to the early history of Mercury.

The structure of S Basin is complex, and when studied in detail shows multiple impact features with degraded rims (Fig. 19). There are several complex structures and their relief cross-section may be traced in radar topographic tracks of Mercury's surface reproduced from Harmon and Campbell (1988) in Fig. 22. The longitude coverage of Fig. 22 includes the longitudes 250° to 300°W. The radar traces in Fig. 22 are labelled with latitude at the ends of the relative altitude traces. Of considerable interest is the trace at 10.4°N latitude which traces a deep depression of 2 km beginning at 277°W longitude and continuing to a maximum depth of 3 km at the rim near 300°W longitude. The same depression structure is traced at latitude 7.9°N at 260°W with an interruption of a degraded rim at 280°W continuing to the next rim at about the same longitude as the trace beginning at 10.4°N.

The dual-polarization, delay-Doppler upgraded Arecibo radar images clearly show the same large, multi-rimmed basin (Harmon 2007). In older radar images the bright region (that has been named "radar bright spot C", Butler et al. 1993) is an unusual, somewhat chaotic region with clusters of circular features. It coincides with a multi-rimmed basin in the Arecibo images of Harmon et al. (2007). Its northern part is shown in Fig. 23 and is in good agreement with the visible/IR images. There appears to be a large radar-dark albedo section extending from 210° to 300° and from 30°S to 40°N in latitude. The dark region is interrupted by various bright features. The bright semi-circular feature in the centre 8°N between 230° and 250°W could be the inner, less-degraded rim of the huge basin. The centre of the feature is 10°N, 277°W and the diameter of outer rims about 50° (corresponding to 2,100 km).

There are very dark regions in both the visible and radar images. The radar image albedo is highly dependent on terrain roughness on the scale of the wavelength of the observation (12.6 cm). The methods of obtaining visual/IR images of ultrahigh resolution have gradually improved. The best results were acquired at the end of 2006; S Basin was one of main goals. The observational data will need, naturally, more time to be processed, preventing their inclusion in this review. It was only possible to prepare a quite preliminary version of

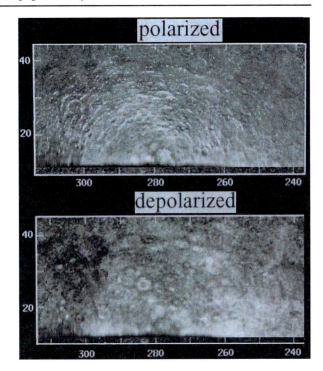

Fig. 23 Image of S Basin obtained by Arecibo radar imaging (Harmon et al. 2007). Northern half of S Basin

processed observational data collected on November 21, 2006. These observations are of a special interest as at that time the S Basin was on the sunlit side of Mercury.

In Fig. 24, the position of the S Basin is shown in the right panel by a black circle with the location of the coordinate grid. In the left panel in the centre of the image (and adjusted west to the S Basin centre) is a large crater, with a diameter 500–600 km. On the external rim, more than 1,600 km in diameter, numerous secondary formations are apparently imposed.

The S Basin may be one of the most interesting features in longitude range 210–290°W. Nevertheless, craters of complex structure are frequent on planets and satellites. Among the unique forms on Mercury, some unusual formations found in 2006 have been identified (Ksanfomality et al., in preparation).

6 Other Remarkable Features in Longitudes 210–285°W

Figure 25 shows the positions of the most prominent large craters discussed in the following. Evidently all of them are of impact origin. Large craters with extended ray systems are visible in many areas. The best conditions for discerning them are at the terminator of the planet, as contrasts are increased. In Fig. 25 large craters are marked by numbers 1–7. To improve their visibility we place in Fig. 25, together with the initial (b) and gamma-corrected image (c) for the phase 97°, contrasted versions of the fragments, taken from image (b). All craters but (5) have wide debris terraces. The centre of the largest and most noticeable crater (1) is placed at 30°S, 265°W. It has a pentagonal shape and a bright fresh rim of debris blanket. The whole rim formation is estimated to have sizes of 900–1,100 km. The outside rim is probably the excavation debris. About 5° northward of it there is another large crater (2), 700–800 km in size. Feature (3) is a large dark crater (or basin) at 1°S, 265°W at the extreme south border of S Basin with a large, about 1,000 km, rim of debris and a

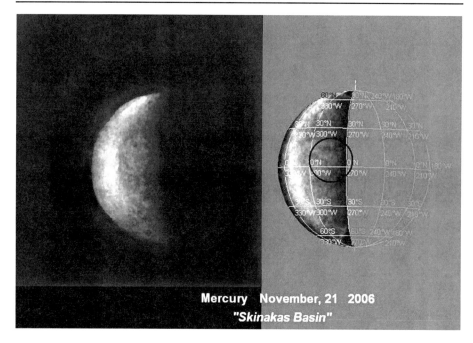

Fig. 24 Preliminary version of processing of experimental data of November 21, 2006, with S Basin (circled by *black ring* in location of coordinates grid) placed on lit side of the planet

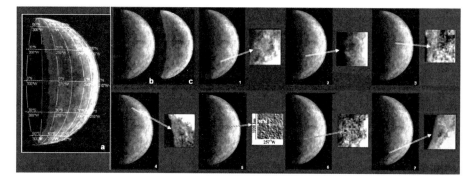

Fig. 25 Large impact craters, marked by numbers **1–7**, are shown, taken from the image (**b**) of May 2, 2002, for the phase 97°. The gamma-corrected image (**c**) presents details close to the terminator. Coordinates grid is shown in panel (**a**). The fragment (**5**) demonstrates rays from crater 5 acquired by means of high-pass filter. All the fragments are shown in 2:1 scale to the images (**b**) and (**c**)

central hill. The centre of feature (4) at 40°N, 275°W is a round, very dark 300-km crater of elongated shape with a size of destroyed outer rim of about 400–500 km. On the east periphery of the S Basin there are separate dark areas with a small dark object (5), about 80 km diameter, at 10°N, 257°W, surrounded with a light blanket of debris, about 200 km. It evidently possesses bright rays, discernable in the fragment 5 made by the high pass filter. The eastern light rays are traced up to 500 km in length. This is probably a young formation. The crater (5) is the smallest. To the west and south of it are extensive dark areas of 400–

Earth-Based Visible and Near-IR Imaging of Mercury 189

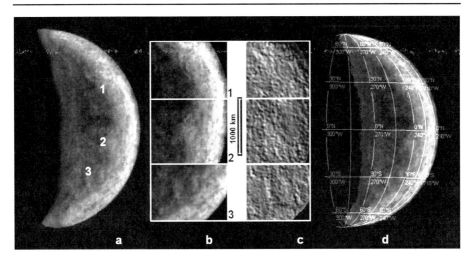

Fig. 26 (a) Large features at the limb in the longitude range 210–290°W, phase 97°. (b) The column of enhanced-contrast fragments. (c) The view of the same areas acquired using a special image processing code. (c) The location of the coordinate grid

600 km in size. The strangest and the largest (about 1,400 km) is the feature (6) centred at 13°S, 265°W. It is a chaotic relief area with few craters, no rim blanket, and the whole area is surrounded by a narrow, thin valley. The crater (7) at 45°S, 270°W has a common part of the rim blanket, with the whole area surrounded feature (1). It has a dark central crater of about 200 km and an outer rim of about 400 km.

A bright, large (about 900 km) and probably very young crater at 30°S, 210–220°W, with the extended blanket debris, is placed at the very limb while in 97° phase. It is seen much better in 93° phase image (considered in Sect. 8).

Some ray craters in the segment 210° to 290°W were considered by Ksanfomality and Sprague (2007). Close to the North Pole, there is a large impact crater at 85°N with an extensive debris blanket about 280 km in diameter (seen at the very north edge in Fig. 25c). Its central dark part is about 90 km in diameter. The size and position of the crater coincides with a crater K in the radar map (Harmon et al. 2001). The debris blanket on the radar map is seen as a dark belt without details, while in the optical image the debris blanket extends radially and has a higher albedo than the centre of the crater. These differences in albedo may have implications for the surface composition. It has been suggested that volatiles, for which a favoured candidate is water ice, produce coherent backscatter and bright depolarized radar albedo (Slade et al. 1992; Harmon and Slade 1992; Butler et al. 1993). Vilas et al. (2005) measured the depth-to-diameter ratios of the polar craters and found them to be abnormally shallow, supporting the notion that they have an infilling material. It may be that the albedo of the debris from these craters seen in the vis/IR images indicates the presence of unknown material.

Since the gradations of albedo or relief features are small, the possibilities for increasing the contrast are limited: for the same shape of the gamma function, increasing the contrast of some details leads to the suppression of others. Nevertheless the procedure of selective increases in contrast and using a high pass filter reveals numerous interesting features that become apparent at the limb of the segment 210° to 285°W (Fig. 26).

Very light regions near the limb (1) are placed at 20–35°N, 220–235°W. This is the brightest spot in the image of Mercury at phase 97° and at this part of the planet. At first

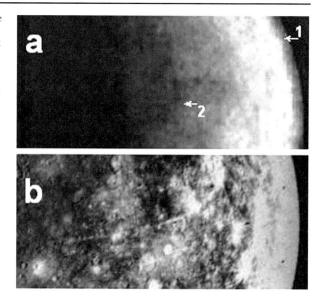

Fig. 27 (a) Thin arc-shaped (due to spherical projection) ray is emerging at the limb (*1*), at about 18°N, 210°W, crossing the equator at 260°W (*2*).
(b) Another thin arc-shaped ray is seen in Mariner 10 photograph (emerging at about 15°N, 120°W, terminating at equator, 160°W)

look it appears as a young impact crater, but no evidence of ray structure can be found. On contrasted fragment 1 (Fig. 26b) the area appears as a 500–700 km collection of chaotic high albedo sites with a round structure centred at 20°N, 245°W. Additional processing shows that the round feature is a 350-km basin of unusual shape, bordered by two large arcuate valleys of regular shape on both sides (Fig. 26c). The bright round feature (2) in Fig. 26b, centred approximately at 10°S, 230°W, after processing, appears to be a regular basin of lunar type, with a double, partially destroyed rim, about 800 km in diameter. Fragment (3) in Fig. 26b centred approximately at 30°S, 245°W is a huge, destroyed basin with a diameter of 700–800 km and a heavy eroded rim. The entire surface and periphery of the basin consists of a superposition of craters and basins with degraded rims that indicate the antiquity of the feature 3. This type of basin is called "relict circus", according to lunar terminology.

Besides the huge features described above, stacked images of Mercury (e.g., Figs. 19 and 25) show unusual features on Mercury's surface. One of them is very thin arc-shaped ray appearing at the limb at 18°N, 210°W, extending for a quarter of the planet and crossing the equator at 260°W. The spatial resolution of the image of the ray (as for a long, linear feature) is about 80–100 km, with its length between marks 1 and 2 being about 1,600 km. As for the arc-shaping, it is due to the spherical projection. The ray is seen well in contrasted images, like Fig. 26a. Another thin arc-shaped ray of this sort is seen in a Mariner 10 photo (Fig. 27). Similar features are known on the Moon, e.g., a ray crossing the Bessel crater at the centre of Serenitatis mare, or rays from the Tycho crater. However lunar rays are not this thin and do not reach this extent. Their origin supposes a huge impact. Keeping in mind that gravity acceleration on Mercury is twice the lunar value, the length of arc-shaped rays in Fig. 27 needs an explanation. Another interesting property of the ray in Fig. 27a is that when crossing the north outline of a large right-angled dark area at 8°N, 250°W, the ray is shifted a bit southward. It could be a manifestation that the dark area is a deep bottom-land. Its extents are 0 to 8°N, 245 to 262°W (8 × 17°), or 340 × 720 km.

Contrasting the images in Figs. 24, 25 and 26 for selection of interesting features in longitudes 210–285°W sector distorts inevitably the original distribution of bright and dark hues over the planet's surface. Their distribution in a less distorted way is presented in Fig. 19. The eastern part of the extended dark S Basin dominates in this sector.

7 High-Pass Filtering Images of Mercury: An Attempt at Restoring Relief Information

It is well known that restoring relief requires stereo images (cf. Cook and Robinson 2000). Is not possible to derive topographic information strictly from a single image, nevertheless, one may try to obtain useful information about topography of the area using a high-pass filter. Conversion of a single half-tone image to a so called "relief image" is a standard option of many image processing programs containing a task called "Relief". It is obvious that the brightness of a separate detail is defined by the conditions of illumination, the reflective properties (albedo) of the given surface, and its geometry. Retrieving quasi-relief information from the stacked images may provide certain clues to the topographic expression (Ksanfomality 2005). An example of conversion of a single half-tone image to quasi-relief form is illustrated by Fig. 17b (a conversion of a Mariner 10 photo shown in Fig. 17a). The irregularities of basin, craters and ray systems are clearly revealed. Relief of many features, like two large craters 1 and 2 placed close to the limb (not noticeable in Fig. 17a), becomes rich with details. The blind flat area S at the limb is a result of Mariner 10 data saturation. It is easy to see the thin, long arc-shaped ray across the photo, mentioned earlier.

The method was applied to the Mercury pictures using a specialized program. Conversion of the image stacked from primary images of Dantowitz et al. (2000) and Baumgardner et al. (2000), August 29, 1998, Fig. 9c, is presented in Fig. 28a. Limited topographic information appears, e.g., a long valley at the equator. More effective is the conversion of images stacked by L. Ksanfomality for data acquired on November 2, 2003, resulting in Figs. 28b and c. The direction of conditional illumination used for processing was in Fig. 28 from the southwest. Phase effects and shades of grey were suppressed. Comparison of stacked and converted images shows that fine features get underlined and became much more distinct than in the original pictures. More distinct are, for example, the giant round crater in the northeast of Fig. 28b and many other features. The results shown in Figs. 28a and c were presented for the first time at the 2nd AOGS conference (Ksanfomality 2005). In Fig. 28b and c, more detail is apparent than in Fig. 28a, because of the better quality and large number of the initial images. Converted pictures look more informative than original images.

The best result came from processing Fig. 19, May 2, 2002. Here many stacked images and their high quality played a positive role. The result does appear to give topographic information that delineates plausible features over the surface. It is impossible to determine whether explicit features are topographic highs or lows but by analogy to lunar cratered terrain and the side of Mercury imaged by Mariner 10, it is possible to recognize some relief

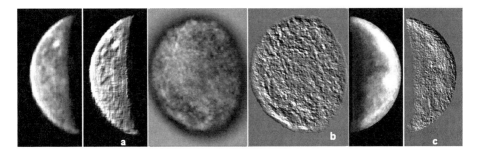

Fig. 28 It is impossible deriving topographic information from a single image. Nevertheless, one may try to acquire certain clues to the topographic expression using high-pass filter. Examples based on: (**a**) August 29, 1998 (Fig. 9c); (**b**) November 2, 2003 (Fig. 16); (**d**) May 2, 2002 (Fig. 19)

forms. In the left and right part of Fig. 28c the same regions are shown that were marked in Fig. 26. Figure 28c allows some of the estimates given above to be improved. The region of the S Basin is actually a superposition of 500–700-km broken old depressions, similar to crater maria, with central structures, hills or craters. Region 1 resembles a deformed 300–400-km crater mare of complex structure. The crater mare 2 (centred at 10°S, 230°W, diameter of about 700–800 km) with a double rim has the sharpest edges in the longitude sector 220–250°W. A broken large area 1,000 km in diameter with a centre at 13°S, 265°W and a broken double rim is adjacent to it on the west. Depression 3 and other areas are also distinguished. A more detailed geological interpretation of the longitude sector 215–285°W is needed.

8 Photometric and Polarimetric Properties of Mercurian Regolith Acquired in Experiments

It is well known from lunar observations that the view of an atmosphereless celestial body changes rapidly as it passes through its quadrature phases. It was interesting to trace the pattern of change in Mercury's view at this favourable phase, especially since, in contrast to the Moon, all sides of the planet are observable in principle at any phase. Unfortunately, only images at two neighbouring phases were acquired, since the observations in May 2002 were limited to only two evenings, May 1 and 2, due to meteorological conditions; the meteorological conditions on May 2 (phase 97°) were appreciably better. The image was stacked from the observations on May 1, 2002 (phase 93°) using more than 7,800 primary photographs. Despite their large number, the image at phase 93° is inferior in feature distinctness to that at phase 97° precisely because of the less favourable meteorological conditions. The observations themselves on May 1 and 2 were performed at the same time of the day. In one day, the subsolar point on Mercury is displaced, on average, by 6° toward the increasing west longitudes, while the phase angle changes, on average, by 4°. Thus, the rate of change in Mercury's phases is half that of the Moon. Two views of the planet, at phases 93° and 97°, are shown in Figs. 29a and b. Figure 29c shows the location of the coordinate grid. Although the phase difference is small and the subterrestrial point in Fig. 29a is displaced rightward, the objects at the terminator are seen much better owing to the earlier phase as, for example, the craters at 45°N, 264°W and 15°S, 270°W shown in Fig. 25. The structural features of the S Basin are seen much better. The large dark area that has not yet reached the limb at phase 93° is located northeast of region denoted 3 in Fig. 19. As regards region 3 itself, its brightness at phase 93° is appreciably lower than at phase 97°. It can probably be concluded from the sharp rise in the brightness of region 3 in Fig. 19, as the phase changes by only 4°, that region 3 is young. The young Tycho crater on the Moon is known to have the same properties. The image of Tolstoj basin (consisting of two photos in Fig. 17) showed the same effect, despite a minor phase difference between the upper and lower photos.

The lightest region near the limb, south of the equator (region 4 in Fig. 19), at this phase is the brightest object on the surface of Mercury. In general, the ring outlines of region 4 are the same as those at phase 97°; its entire structure reaching 1,000 km in diameter appears light, while its western part remains brightest at phase 97°. At phase 93°, region 4 is considerably brighter than region 3 which is the brightest at phase 97°.

One of the known powerful methods used to study physical properties of the regolith is polarimetry. In the observations of 2001 (in autumn morning elongation) an attempt was made to acquire data on the distribution of the degree of polarization of the light over the disk of the planet (Ksanfomality 2002). The data are the same that were used to produce

Earth-Based Visible and Near-IR Imaging of Mercury 193

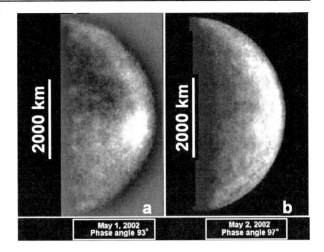

Fig. 29 Comparison of views of Mercury at longitude sector 210–285°W in two phases: (a) 93°; (b) 97°

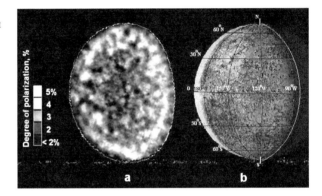

Fig. 30 Distribution of the degree of polarization of the light over the disk of Mercury at the longitude sector 90–180°W, at 63° phase

the image in Fig. 12, with the same limitation about blurring at the periphery and a high resolution only at the central part of the image. The result is shown in Fig. 30. The range of the degree of polarization was 2.5 to 5%, presented accordingly by dark and light tones.

9 Photometric Models and Physical Properties of Mercurian Regolith

Accurate photometric characterization of a surface is required for a number of important applications in planetary science. First of all, the reflectance function of a planetary surface provides information on the surface roughness and the regolith optical properties. It is also needed to accurately compare observations acquired under varying geometries and lighting conditions. In addition, for a space mission, knowledge of this function is required to establish instrument characteristics and to define operational parameters. For example, the bidirectional reflectance distribution function is needed to determine exposure times for the camera system while the zero phase angle reflectance is necessary to size the aperture–laser pulse energy product of a laser altimeter.

9.1 Models

In this section, we present the photometric functions most frequently used in interpretation of astronomical observations. If the surface is illuminated by a distant point source, the scattered radiance I in some direction is proportional to the incident solar irradiance πF

$$I = \text{RADF}^* F,$$

where RADF is the radiance factor (e.g., Hapke 1993), which is often denoted as I/F. If the surface is isotropic and flat, i.e., all directions in the surface plane are equivalent, I/F depends only on mutual orientation of the surface plane and the directions of incidence and observation, which is described by a set of three angles. They are the phase angle α (between the direction from the object to the light source and from the object to the observer) and the zenith angles of incidence i and emergence e. In astronomical observations, instead of i and e, another pair of angles is often used: photometric latitude φ (the angle between the normal to the surface and the scattering plane that contains the source, the object, and the observer) and photometric longitude λ (the angle in the scattering plane between projection of the normal and direction from the object to the observer).

These angles can vary in the following ranges:

$$0° \leq \alpha \leq 180°,$$
$$0° \leq i \leq 90°, \qquad \alpha - 90° \leq \lambda \leq 90°,$$
$$0° \leq e \leq 90°, \qquad -90° \leq \varphi \leq 90°.$$

And two sets of angles are connected with the following relations:

$$\mu_0 = \cos i = \cos \varphi \cdot \cos(\alpha - \lambda), \quad \tan \lambda = (\mu_0/\mu - \cos \alpha)/\sin \alpha,$$
$$\mu = \cos e = \cos \varphi \cdot \cos \lambda, \qquad \cos \varphi = \mu/\cos \lambda.$$

To describe the behaviour of the radiance factor as a function of these angles (α, i, and e or α, φ, and λ), a number of photometric functions were suggested.

(a) Lambert function

The Lambert function (Lambert 1760) is represented by a simple decrease of I/F with μ_0, and does not exhibit any specific dependence on α and μ:

$$I/F = A_\text{L} \cdot \mu_0.$$

The Lambert albedo, A_L, is the fraction of incident solar irradiance scattered by the surface into solid angle π (Hapke 1993).

(b) Minnaert function

The Minnaert function (Minnaert 1941) is the expansion of the Lambert function by introducing an additional term dependent on μ as well as the Minnaert coefficient, k, an empirical constant that sets the weighting between the incidence and emission angle contributions:

$$I/F = \pi A_\text{M} \cdot \mu_0^k \cdot \mu^{k-1},$$

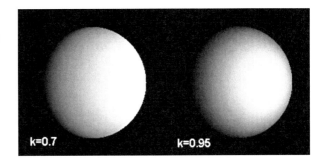

Fig. 31 The distribution of brightness over the disk of a planet with the surface reflecting according to the Minnaert law. Two cases of the Minnaert coefficient k are shown. The phase angle is 30°

where the Minnaert albedo, A_M, is the fraction of incident solar irradiance scattered normally from a normally illuminated surface. The coefficient k ranges from 0.5 (for a very rough Moon-like surface) to 1.0 (the Lambert law for a completely orthotropic reflecting surface). The model images below illustrate how the assumed value of k influences the view of the planetary disk ($\alpha = 30°$). The larger k (the closer the photometric function to the Lambert one), the stronger limb darkening is observed (Fig. 31).

Special tests of the Minnaert function in comparison with the Hapke function (see below) justified its use for limited angle ranges (Hapke 1981). However, it was shown that the "constants" A_M and k are both functions of phase angle.

(c) Lunar–Lambert function

A photometric function now known as the Lommel–Seeliger lunar function was derived by Chandrasekhar (1960) from the radiative transfer law to describe the scattering properties of the Moon:

$$I/F = (\omega/4) \cdot \mu_0/(\mu + \mu_0) \cdot f(\alpha),$$

where ω is the single-scattering albedo and $f(\alpha)$ is an arbitrary function describing the phase dependence of the surface radiance factor. For most surfaces observed in the Solar System, a linear combination of this function with the Lambert law, the so-called lunar–Lambert function,

$$I/F = A \cdot \mu_0/(\mu + \mu_0) \cdot f(\alpha) + B \cdot \mu_0$$

was found to provide a better agreement with the data than the Minnaert law and was used to study the photometric properties of various planetary surfaces and laboratory samples (e.g., Squyres and Veverka 1981; Buratti and Veverka 1983; Gradie and Veverka 1984; McEwen 1991; Pappalardo et al. 1998; Geissler et al. 1999). The fit which can be obtained with this function is as accurate as that provided by Hapke's photometric function considered below.

(d) Akimov empirical function

To approximate the measured brightness of the lunar disk, Akimov (1979, 1988) proposed an empirical expression, which contains two free parameters,

$$\frac{I}{F} = A \cdot \frac{\exp(-\eta\alpha) \cdot \cos(\alpha/2) \cdot (\cos\varphi)^{\nu\alpha+1} \cdot [(\cos(\lambda - \alpha/2))^{\nu\alpha+1} - (\sin(\alpha/2))^{\nu\alpha+1}]}{1 - (\sin(\alpha/2))^{\nu\alpha+1}}$$

$$\times \sec\varphi \cdot \sec\lambda,$$

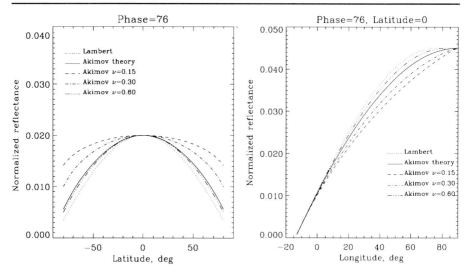

Fig. 32 The examples of the central meridian and luminance equator disk profiles calculated with the Lambert and Akimov photometric functions for the same phase angle as the profiles measured for Mercury (see below) and normalized to the same maximal value

where η determines the slope of the phase dependence and ν (the roughness parameter) averages approximately 0.16 to 0.31 for maria and highlands, respectively. This formula was also successfully used in the analysis of the images of different lunar regions obtained during the Clementine mission (Kreslavsky et al. 2000).

(e) Akimov theoretical function

Another approach in finding the reflectance law of the regolith surface is based on the condition that the loose particulate surface being slightly randomly undulated should have the same photometric function as before undulations (Akimov and Kornienko 1994). In this case, the formula obtained for the brightness of the area of the observed planetary disk contains only geometrical parameters and the surface albedo:

$$\frac{I}{F} = A \cdot \exp\left[0.5 \cdot (1 - 0.5\lambda_{00})\right] \cdot (\cos\varphi)^{\pi/\pi - \alpha} \cdot \cos\left[\frac{\pi(\lambda - \alpha/2)}{\pi - \alpha}\right] \cdot \sec\varphi \cdot \sec\lambda,$$

where A is the surface albedo and

$$\lambda_{00} = \frac{\pi}{\pi - \alpha} \cdot \left(\frac{\pi}{\pi - \alpha} + 1\right).$$

This formula, applied to the Clementine images of the Moon, produces a reasonable agreement with the data, though worse than that of the Akimov empirical formula (Kreslavsky et al. 2000).

Figure 32 illustrates the difference between the models and the influence of the roughness parameter of the Akimov empirical function with the examples of central meridian and luminance equator disk profiles. These profiles are calculated for the same phase angle as those measured for Mercury (see below) and normalized to the same maximal value. It is seen that the larger the roughness parameter the closer the profiles to the Lambert one. The smaller values corresponding to the lunar maria result in much flatter latitude cross-section and produce no limb darkening.

(f) Hapke function

The empirical photometric functions described above are useful for providing photometric corrections to data. However, photometric observations also provide important information about the nature of the surface. Extraction of this information requires a more sophisticated model whose parameters are more directly related to the properties of the surface. One such model is an approximated analytic radiative transfer model developed by Hapke (1981, 1993, 2002), which has been shown to accurately fit data from a wide variety of planetary surfaces and laboratory samples (e.g., Helfenstein and Veverka 1987; Hillier et al. 1994; Domingue et al. 1997; Guinness et al. 1997; Johnson et al. 2006).

Hapke's photometric model (Hapke, 1981, 1984, 1986, 1993) (without taking into account the coherent backscattering) is given by

$$\frac{I}{F} = \frac{\omega}{4}\frac{\mu_0}{\mu_0+\mu}\{[1+B(\alpha,h,B_0)]P(\alpha)+[H(\mu_0)H(\mu)-1]\}S(\alpha,\overline{\theta}),$$

where the factors $[1 + B]P$ and $[H(\mu_0)H(\mu) - 1]$ account for the singly and multiply scattered light, respectively. P is the single particle phase function; the function B accounts for the shadow hiding opposition effect; the parameters h and B_0 describe the angular width and amplitude of the opposition surge, respectively; H is Chandrasekhar's H functions, which approximate the multiple scattering as isotropic. Later, the assumption on isotropic multiple scattering was relaxed by a special term and the improved approximation for the H-functions was introduced (Hapke 2002). Finally, S is a function to account for shadowing caused by macroscopic roughness (Hapke 1984). It incorporates one parameter, $\overline{\theta}$, which describes the average slope angle of subresolution topographic relief.

Use of Hapke's functions has the advantage of producing a common framework in which to analyze observed data. On the other hand, it is becoming increasingly clear (e.g., Gunderson et al. 2006) that extrapolation of Hapke's function is unreliable and hence a complete data set is needed to produce a Hapke fit which can then be used as a predictor by interpolation. Nonetheless, several authors have fit Hapke's model to Mercury's visible phase curve data (e.g., Veverka et al. 1988; Domingue and Hapke 1989; Mallama et al. 2002; Warell 2004) although here too there are inconsistencies between the particular implementation of the Hapke relations used which make intercomparisons between results difficult.

9.2 Application of Models to Mercury

The scattering properties of Mercury's surface were previously addressed by Veverka et al. (1988), and many aspects of that work remain appropriate in the absence of additional data. Hapke, in his series of papers, has often used measurements of Mercury's integrated reflectance made by Danjon (1949) to illustrate the fit of his models to observational data (e.g., Hapke 1984).

Extraction of the scattering properties from the Mariner 10 imaging data is hampered by the lack of phase angle coverage (limited to $90 \pm 10°$) and the absence of accurate calibration. Hapke (1984, 1993), however, applied his model to the Mercury brightness profiles measured by Mariner 10 and estimated the model parameters fitting the observed longitude and latitude cross-sections of brightness. They are: $\omega = 0.25$, $\overline{\theta} = 20°$, $P(\alpha) =$

Table 1 Hapke parameters fit to the disk-integrated data from Mallama et al. (2002)

Symbol	Value
ω	0.20
h	0.065
B_0	2.4
b	0.20
c	0.18
$\bar{\theta}$	16°

$1 + 0.579 P_1 + 0.367 P_2$ (where P_1 and P_2 are the first and second Legendre polynomials, respectively). In this case the measurements were made far from opposition ($\alpha \approx 77°$), which allows the parameters accounting for shadow and coherent opposition effects to be assumed negligible. Whilst some work is continuing on the Mariner 10 data, there are unlikely to be major developments in the near future. Fortunately, however, there have been other observations (both ground-based and space-borne), laboratory work, and improved modeling work.

An important contribution from imaging observations in advancing our knowledge of Mercury is photometrically calibrated data, used in light-scattering models to derive information of macrophysical and optical properties of the surface. In this respect, both integral disk and disk-resolved image data are important separately and in conjunction, particularly if the phase angle coverage is extensive.

The most comprehensive disk-integrated data set is that of Mallama et al. (2002). Their data set was acquired in the V band with a small telescope and complemented with transformed magnitude observations from the SOHO LASCO spectrograph to cover a very extended phase angle range of 2–170°. It represents the first major achievement of its type since the work of Danjon (1949) who made visual measurements of Mercury's brightness for the range 3–123°. The Hapke parameter fit, using the Hapke (1986) formulation, derived by Mallama et al. is shown in Table 1.

Mallama et al. (2002) determined the geometrical albedo as 0.142 ± 0.005, which is at the upper end of previously published values, while the estimated average surface macroscopic roughness, $\bar{\theta} = 16°$, is at the lower end. Thanks to the coverage of small phase angles, it was possible to study the opposition effect with a higher precision than previously possible, and the brightness was found to surge by 40% from 10° to 2° phase angle. However, the phase curve still does not extend to small enough angles to determine the magnitude of the narrow peak of the opposition surge caused by the coherent backscatter effect. For this information, opposition geometry measurements are needed, which will likely be provided by the MESSENGER Mercury orbiter.

The photometric properties of Mercury compared with the Moon was studied by Warell (2004), based on the phase curve of Mallama et al. (2002) and the highest-resolution SVST images of Warell and Limaye (2001). The improved model of Hapke (2002) was used. Typical results of fitting the Hapke photometric model (lines) simultaneously to observational data for Mercury (red dots) in the form of central meridian and luminance equator disk profiles (left and central panels, respectively) and the Mallama et al. (2002) integral phase curve (right panel) for varying values of $\bar{\theta}$ are shown below. These profiles are for the 550 nm image obtained at the phase angle of about 76° (Fig. 33).

Similar to the work of Mallama, a small photometric roughness was determined from the integral phase curve data. A value of 8° fits best the observations at high phase angles, but considerably larger values were found to satisfy both integral and disk-resolved data if the

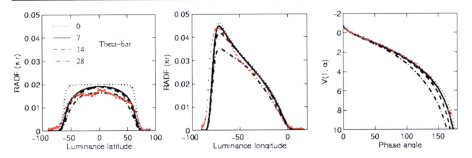

Fig. 33 Fitting the central meridian and luminance equator disk profiles (*left and central panels*, respectively) and the integral phase curve (*right panel*) observed for Mercury (*red dots*) with the Hapke photometric model (*lines*) for varying values of the parameter $\bar{\theta}$, which describes the average slope angle of subresolution topographic relief

phase angle coverage was restricted to moderate phase angles. This was explained as due to contribution of light from predominantly smooth plains units in the highest phase angle data from SOHO, which were determined to imply a surface roughness parameter of about 8°, about twice that of the average Mercurian surface with scattering from all types of geologic units combined. The single-particle scattering function was parameterized with the double Henyey–Greenstein representation, and it was found that the average Mercurian surface was more similar to lunar maria than to lunar highlands. These two observations indicate the possibility that Mercurian smooth plains units are more similar to lunar maria than lunar highlands in both textural and scattering properties. Note that a rather flat latitude cross-section and a weak limb darkening seen in the Mercury profiles comparing to the empirical model by Akimov also indicate a more probable similarity of the Mercurian regolith with that of the lunar maria.

The backscattering efficiency of the Mercurian surface is greater than that of the Moon, possibly due to a high density of internal scatterers in complex agglutinate-rich particles formed from strong maturation processes. In comparison to the Moon, the average brightness of Mercury was found to be 10–15% fainter in the V band based on the integral phase curve, which is supported with historical determinations of the geometric albedos of the two bodies.

While measurements of Mercury at zero phase are impossible from the ground, the results of this fit, and in particular the value of the B_0 parameter, indicate that the opposition effect for Mercury's surface is comparable to or even stronger than that of the Moon. This is critical for laser altimetry which observes, by definition, at zero phase. It is possible that the effect is even stronger than observed by Mallama et al. because the finite angular diameter of the Sun (1.1° to 1.7° in diameter) would tend to smear out the magnitude of the effect when compared to point source illumination (such as that provided by a laser). Extrapolation of the fit leads to a normal albedo ($= \pi I$ ($\alpha = 0, e = 0, i = 0$)/F) of 0.12 at V band (550 nm) wavelengths. Correlation between parameters in the Hapke fit suggests that interpretation and extrapolation of other parameters should be treated with caution (Gunderson et al. 2006).

Sprague et al. (2004) showed that the spectral slope at visible wavelengths increases linearly with wavelength to first order. Assuming that this finding can be extrapolated to small scales and that the shape of the bidirectional reflectance distribution function curve is not wavelength dependent over this limited range, a normal albedo of between 0.23 and 0.31 can be expected at 1,064 nm (the wavelength of the Nd:YAG lasers to be used for altimetry by the MESSENGER Laser Altimeter, MLA, and the BepiColombo Laser Altimeter, BELA). The constancy of the BRDF with wavelength is not obvious but has recently been verified over a limited wavelength range (Gunderson et al. 2005).

The surface contrast which will be evident to the MESSENGER and BepiColombo spacecraft has been estimated by Warell and Limaye (2001). Defining surface contrast as

$$\Gamma = \frac{I_B - I_D}{I_B} \cdot 100\%,$$

where I_B and I_D are the observed intensities of bright and dark areas, respectively, values of Γ of $\sim$50% at 550 nm were found.

Reviewing the above models for their applicability, it can be concluded that the selection of the optimum model to use in any particular case will require a further evaluation of the observations and their specific conditions.

10 Summary of Results from Imaging and Photometry: Composition and Structure of the Regolith

Ground-based imaging, photometric and polarimetric studies have made it possible to draw a range of conclusions regarding the composition and microphysical state of Mercury's surface.

In terms of the microphysical properties of the regolith, the small radar cross-section, the presence of a negative branch of polarization and the angular location of the maximum polarization value indicate a highly porous regolith similar to the Moon's (Mitchell and de Pater 1994; Dollfus and Auriere 1974). This is supported by photometric studies of the opposition effect of the integral phase curve (Mallama et al. 2002; Warell 2004), though observations at the smallest phase angles ($<2°$) indicative of the coherent backscattering effect are naturally lacking. Relative brightness variations across the disk in microwave imaging data indicate that the density of the regolith increases from about 1 g/cm^3 in the top few cm to about 2 g/cm^3 at a depth of about 2 m, with a concomitant increase in thermal conductivity (Mitchell and de Pater 1994). The optically active grain size is likely smaller than the case for the Moon (Dollfus and Auriere 1974), and values of around 30 µm have been determined (Sprague et al. 2002; Warell and Blewett 2004).

The scattering properties are generally lunar-like, but a stronger backscattering efficiency and dependence of colour on geometry may be related to a greater abundance of translucent complex glass-rich agglutinate particles (Warell 2004). The backscattering anisotropy, i.e., the amount of light backscattered to the observer, increases with wavelength both in the optical and near-infrared range and the throughout mm–cm wavelengths. In the latter case this effect is manifested in a decrease of microwave opacity with wavelength (Mitchell and de Pater 1994).

The average particle angular scattering function corresponds more closely to that of lunar maria than highlands which would indicate a surface macrophysically more similar to the former lunar terrain. An average photometric roughness of 20° has been derived from photometric modeling of VISNIR and thermal infrared spectra, but regional extents of smooth plains may have smaller surface roughness values of about 10° or less (Emery et al. 1998; Sprague et al. 2000; Mallama et al. 2002; Warell 2004). The absolute V-band brightness of Mercury is 10–15% less than nearside Moon, but considering the more strongly sloped Mercurian spectrum, the "bolometric" average VISNIR reflectances are probably similar

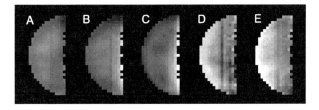

Fig. 34 Mercury's surface from 210° (limb) to 270° (terminator) W longitudes which is part of the poorly known hemisphere, analyzed with the Lucey method to map varying chemical composition. From left to right, images represent relative 750 nm optical albedo (**A**), optical maturity + ferrous iron content (**B**; brighter = less mature or lower iron), titanium abundance (**C**; brighter = more titanium), ferrous iron abundance (**D**; brighter = more iron), and degree of optical maturity (**E**; brighter = less mature). Adapted from Warell and Valegård (2006)

(Warell 2004). Optical photometric and polarimetric data indicate a refractive index of 2.1 compared to 1.8 for Moon, possibly due to the presence of metallic iron particles larger than the wavelength (Hapke 1993).

Compared to the Moon, the microwave dielectric loss tangent is about twice as small as for lunar maria, and 40% less than for the lunar highlands, indicating a surface low in opaques with FeO + TiO_2 around 1 wt% (Mitchell and de Pater 1994). This is supported by compositional modeling of the VISNIR spectrum (Warell and Blewett 2004). One possible explanation is effective secondary differentiation, aided by a high surface gravity, by density-driven plagioclase flotation in magma ocean resulting from a giant impact during the late phase of planetary accretion.

A subset of the images obtained by Warell and Limaye (2001) were analyzed by Warell (2002) in an attempt to study colour variations due to compositional variations on the surface. Individual sharp images of the poorly known hemisphere (longitudes 160–340 degrees) with a spatial resolution of 300 km or higher at four wavelengths from 550–940 nm were ratioed to each other to accentuate relative differences in flux. Of particular interest were the ratios of the 940 nm to the 550, 650 and 750 nm images, for which correlated variations might indicate abundance variations in ferrous iron-rich mineralogy. However, no such deviations were found, and a limit of 2% maximal differences in flux relative to the average was determined. Thus, the variations in mafic compositional abundances of the uppermost scattering surface of the regolith was found to be very small compared to the variation in albedo at a comparable spatial scale, though local variations greater than this limit, present at more local scales, could not be excluded.

A related study was performed by Warell and Valegård (2006) on the basis of new images of Mercury for the longitude range 210–270 degrees, obtained with the upgraded 1-m Swedish Solar Telescope. These images were acquired with basically the same system as used with the SVST. Substantial benefits were available with the new system: chromatic aberration and atmospheric dispersion were completely corrected. For each of six wavelengths from 450 to 940 nm, sharp images were aligned and averaged to increase the signal-to-noise ratio, and ratioed to each other to study variations of colour on the surface. Removal of the photometric function with the theory of Hapke (2002) and extension of methods developed for the Moon (Lucey et al., 2000a, 2000b) to determine abundances of FeO, TiO_2 and degree of maturation, allowed determination of the relative variations of these parameters across the observed hemisphere. With this data set, regional colour variations on disk was detected (Fig. 34) and were found to vary less than the case for the Moon, indicating a surface similar to the lunar farside. Highly reflective albedo features were found to be generally immature, have low to intermediate abundances of opaque minerals and intermediate

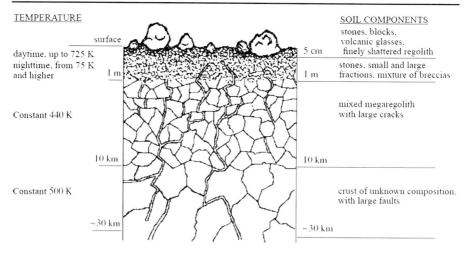

Fig. 35 Structure of Mercury's surface, according to Sprague et al. (1990)

abundances of ferrous iron. This is consistent with such features being bright immature ray systems of geologically young craters. The very darkest areas in reflectance have intermediate maturities, intermediate to high opaque mineral abundances and intermediate ferrous iron abundances. These may correspond to localized regions of more opaque materials than is typical for the general Mercurian surface, perhaps derived by local extrusive processes from crustal plutonic sources. Generally, Mercury's poorly known hemisphere and the hemisphere explored by Mariner 10 were shown to be very similar, which indicates that if the geologic units are of the same type as on the Mariner 10 hemisphere, they are also likely to have similar areal occurrence globally. Differences in the TiO_2-sensitive parameter across the disk were found to be substantially greater than the case for the FeO sensitive parameter, which may indicate that the abundance variations in opaque elements is greater than that of ferrous iron.

A schematic view of the physical structure of a layer of Mercury's regolith to a depth of 30 km is given in Fig. 35. This scheme was proposed by A. Sprague to explain the mechanisms of nighttime accumulation of Na and K in the regolith, with subsequent release of these elements into the atmosphere in the daytime (Sprague et al. 1990). On the surface, along with the finely shattered, sufficiently dry regolith, there are large blocks and stones. The share of glasses of volcanic origin should be low. By analogy with the Moon, it can be expected that the regolith of Mercury was also completely reprocessed in impact processes over geological time. Depending on the local features of the regolith composition, the mean density ρ falls within the range from 2.7 to 3.2 g cm^{-3}. The mean value of the thermal inertia $(k\rho c)^{1/2}$ is close to 0.0025 cal cm^{-2} s$^{-1/2}$ K^{-1}; i.e., it is much lower than the lunar value. The probing of the lunar crust has shown that, at depths from 1.5 to 30 km, the megaregolith represents a medium scarred by micro- and macrocracks. It is assumed that the megaregolith of Mercury has a similar structure. The maximum temperature (725 K) indicated in Fig. 36 refers only to the specific areas indicated above, while the night temperature (75 K) is observed only in the areas with the lowest thermal inertia, no more than 0.0012 cal cm^{-2} s$^{-1/2}$ K^{-1}, and only in the layer with a thickness less than 1 m. On the average, nighttime temperatures are close to 100 K and the diurnal/annual (176 days) thermal wave does not penetrate to depths greater than 1 m. Below this level, a constant temperature of about 440 K is established. The thermal gradient for Mercury's crust is estimated to be

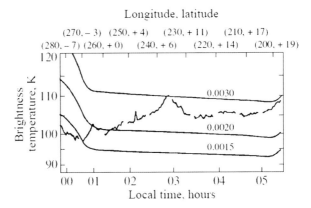

Fig. 36 Regions exhibiting differences in the coefficient of thermal inertia (and, perhaps, density) along the track of Mariner 10 measurements (Chase et al. 1974). The coefficients of thermal inertia, equal to 0.0015 and 0.0030 cal cm^{-2} s$^{-1/2}$ K^{-1} correspond, respectively, to lunar-like porous rocks and consolidated rocks

2 K/km. A temperature of about 500 K should correspond to a depth of 30 km. Due to the shattered surface and cracks going down to a large depth, the influx of Na and K to the surface should be stable. The mean value of the dielectric constant e at centimeter wavelengths comprises 1.8 in the surface layer (a few centimeters in thickness) and attains 2.9 at a depth of several meters.

According to the results of observations at centimeter wavelengths (from 1.3 to 6.2 cm) obtained by Mitchell and de Pater (1994), the temperature gradient of subsurface regolith layers can locally deviate upward or downward from the predicted value at equal values of the thermal inertia, calculated by a one-dimensional model of temperature diffusion into a depth x in the regolith layer. We should recall the specific features of the thermal model. The factor of thermal inertia for the regolith is defined as $I = (k\rho c)^{1/2}$. Here, k is the heat conductivity (which is a function of temperature), c is the heat capacity, and ρ is the density of the material. The model is built on the basis of the balance of three processes: (i) the absorption of solar radiation $E(1 - A)/a^2$ by a body with albedo A moving in the orbit with the semimajor axis a; (ii) the intrinsic thermal emission $\varepsilon\sigma$ from the surface having the temperature T_s and the emissivity e and the Stefan–Boltzmann constant σ; and (iii) the heat influx/outflux F_s in the vertical direction x. The balance is determined by the three equations (Ksanfomality 2001):

$$\frac{\partial}{\partial x}\left[k\frac{\partial T(x,t)}{\partial x}\right] = \rho c \frac{\partial T}{\partial t},$$

$$\varepsilon\sigma T_s^4 = E_\odot(1 - A)/a^2 + F_s,$$

$$F_s = -k\left[\frac{\partial T(x,t)}{\partial x}\right]_s.$$

A simultaneous solution of these equations enables one to find the value of the thermal inertia factor $I = (k\rho c)^{1/2}$ for the regolith. The brightness temperature T_b of Mercury's surface was estimated from the measurements in the range of 45 µm, which were made in 1974 along the Mariner 10 measurement tracks (Chase et al. 1974). Figure 36 shows the results of these measurements performed in the radiometer channels at 45 mm above the nightside of the planet in the equatorial region, from 23:40 to 05:30 of the local time. The brightness temperature T_b along the track varied from 98 to 110 K. The temperature T_b is determined by the thermodynamic temperature of the body T_s and by the emissivity ε as $T_b = T_s(\varepsilon)^{1/4}$. Since $\varepsilon < 1$, the temperature T_s is several degrees higher than T_b. The comparison of the

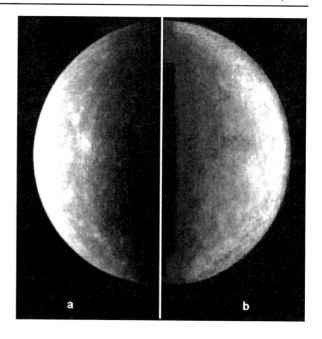

Fig. 37 The image taken by the Mariner 10 in 1973 from a distance of about 1 million km (a) and the part of Mercury for the longitudes 210–285°W by ground-based observations (b, phase 97°)

measured temperature with the model determined by three equations shown above made it possible to estimate the thermal inertia factor for the surface rock. In the case of Fig. 37, this factor was found from 0.0017 cal cm^{-2} s$^{-1/2}$ K^{-1}, i.e., somewhat higher than typical values for the Moon, up to 0.0030 cal cm^{-2} s$^{-1/2}$ K^{-1}; this value corresponds to a sufficiently consolidated rock whose density reaches 3 g cm^{-3}.

Sprague et al. (2000) obtained the first infrared images of Mercury's surface in the wavelength region of thermal emission from the surface, for the longitude range 210–250°W. They acquired data at 22 wavelengths from 8.1 to 13.25 μm with the Mid-Infrared Array Camera on the 2.2-m reflector at the Steward Observatory. A large number of 50 ms long exposures at each wavelength were coadded to achieve an angular resolution (0.9–1.4″) at or near the diffraction limit at all wavelengths. Individual coadded images, selected at wavelengths of known emission features in the spectra of terrestrial rocks (augite, diposide, Fe-rich olivine and labradorite), were ratioed to other wavelength images in order to search for contrast differences caused by rock abundance variations across the visible disk. At the angular resolution attained and the pixel scale (0.67″) used, however, it was not possible to detect any flux variations on the surface. Any such variations resulting from compositional variations on the disk were estimated to deviate less than 5% in flux from the surroundings at spatial scales of 700 km.

Concluding this overview of ground-based observations of Mercury, it is interesting to compare the obtained resolution of images of the planet with the Mariner 10 photographs as a whole. This comparison is shown in Figs. 37a and b. Although the longitudes are completely different, the similarity between the views of the surface is very convincing. The stacked image (Fig. 37b) roughly corresponds in resolution to the photographs taken by the Mariner 10 camera from a distance of 1 million km (Fig. 37a). The large dark areas in the new images cover an essentially larger area than can be distinguished in Fig. 37a. In general the view of the Mercurian hemisphere in the longitude range 15–190°W known from Mariner 10 photographs differs markedly from the planet at phase 97° for the longitudes 215–290°W (Fig. 19b), where larger dark regions are located. At the same time, the images of Mercury more closely resemble the first quarter of the Moon than the widely known and often published mosaic of Mariner photographs.

Ground-based telescopic observations ($D = 1.2$–1.5 m and $D/F = 1/10$–$1/20$) made with short exposures and with modern image processing, permit us to acquire previously unattainable resolution of Mercury's surface, about 80 km and less. For this, an extended data base, computer programs and high-quality observational material are required. The PSF (point spread function) should be used at a level of 0.85–0.90, instead of the usual 0.7. The image quality and presence (or absence) of small details permit choosing the level of processing. It is also worth mentioning that a technique to retrieve relief has been applied to different images, including the newly processed images in this paper. The result is promising with many features becoming visible and having the appearance of crater rims, central pits, scarps, depressions and hillocks.

11 Hypothesis on Possible Connection of the Magnetic Field of Mercury with S Basin

Here we suggest and discuss the hypothesis that the S Basin on Mercury (Ksanfomality 1998a, 1998b; Ksanfomality et al. 2005) can be associated with the measurable magnetic anomaly of crustal origin. The measured global magnetic asymmetry of the planet (Connerney and Ness 1988) can be related to the existence of the S Basin according to this hypothesis. We demonstrate that available knowledge about Hermean magnetic fields (Wicht et al. 2007) does not contradict this hypothesis. We indicate possible directions of search for correlations between magnetic fields and surface properties in complex multidisciplinary experiments onboard future spacecraft missions.

The Mariner 10 measurements (Connerney and Ness 1988) put only some upper limits on small-scale residual crustal magnetic fields, which appear to be of the orders of several nT to several tens nT on average on the length scales and of the orders of several tens to hundreds km on the charted Mercury surface beneath the Mariner's third flyby. It is not clear if such small residual crustal magnetization is typical for the whole planet or only for the investigated region. It is also not clear if stronger crustal magnetization exists and contributes to the global dipole and other harmonics. Only future measurements can clarify the situation.

11.1 Introduction to the Hypothesis

The main magnetic field of Mercury is poorly known. Our current knowledge is based on the interpretation of the limited in situ observations during two flybys of the Mariner 10 spacecraft on March 29, 1974 (I flyby) and March 15, 1975 (III flyby) on the nightside of the planet at altitudes above 707 and 327 km, respectively (Ness 1978; Ness et al., 1975a, 1975b). Flyby II was too distant from the planet for this purpose. Magnetic-field observations made during Mariner 10's third encounter with Mercury were interpreted such that an active dynamo is a more likely candidate than fossil magnetization for the origin of the field (Ness et al. 1976). Two possible source mechanisms for the observed intrinsic planetary magnetic field are evaluated: an active dynamo and a passive paleomagnetic field frozen into the planet's outer layers at an earlier epoch. It is noted that neither the available magnetic-field data nor models of the planetary interior permit one to distinguish uniquely between the two mechanisms (Ness 1978). Different approximations of the main magnetic field by dipolar, quadrupolar and octupolar components were suggested based on these measurements (Ness 1979). The results of the Mariner 10 mission about the Hermean magnetic field have not been fully explained in spite of many subsequent hypotheses (see, e.g., Strom and Sprague 2003 for their short description). Knowledge of the external magnetic field is the primary limiting factor in extracting reliable knowledge of the structure of Mercury's magnetic field from future observations (Korth et al. 2004).

Permanent magnetism and conventional dynamo theory are possible explanations for the magnitude of the Mercurian magnetic field, according to the point of view expressed by Stevenson (1987). A model was proposed and developed (Stevenson 1987; Schubert et al. 1988) in which thermoelectric currents driven by temperature differences at a bumpy core–mantle boundary are responsible for the (unobserved) toroidal field, and the helicity of convective motions in a thin outer core (thickness of about 100 km) induces the observed poloidal field from the toroidal field. The observed field of about 300 nT can be reproduced provided the electrical conductivity of Mercury's semiconducting mantle approaches 1,000/ohm per m. Tidal heating in the solid inner core plays the role of an additional heat source in this model. This model may be testable by future missions to Mercury. It predicts more complicated field geometry than conventional dynamo theories. However, it is argued that polar wander may cause the core–mantle topography to migrate so that some aspects of the rotational symmetry may be reflected in the observed field. Giampieri and Balogh (2002) reconsidered the thermoelectric dynamo model for the generation of the magnetic field of Mercury including its multipole expansion. Takahashi and Matsushima (2006) examined dynamo action possibly working in the fluid core of Mercury using numerical models in a thin spherical shell. Both dipolar and multipolar regimes are found for columnar flows inside and outside tangent cylinder. These results suggest that Mercury may have more complicated magnetic field than has been considered. Possibility was indicated of the robust testing of the further developed dynamo model by correlating future measurements of Mercury's magnetic and gravitational fields (Giampieri and Balogh 2002). Spohn and Breuer (2005), based on their calculations, found that the stagnant lid tends to thermally insulate the deep interior and fluid outer core shells for reasonable mantle rheology parameters even for compositions with as little as 0.1 weight% sulphur. The cooling time of the core can be much longer than naively assumed in the past. Radio measurements of the planetary rotation, using highly accurate radar speckle methods, are intended to find signatures of the existence of the molten interiors (Holin 1988; Van Hoolst et al. 2007). For an up-to-date review of the internal, dynamo-generated magnetic field of Mercury, see Wicht et al. (2007).

Stephenson (1976) suggested that the magnetic dipole moment of Mercury can be explained on the basis of thermoremanent magnetization acquired by an outer shell in an ancient Mercurian field produced by an internal dipole source such as a core dynamo which is now inactive. Such a shell will give rise to a dipole moment provided that there are differences of permeability between the shell and the interior, or the shell and free space. By assuming that the magnetic properties of the surface rocks of Mercury are similar to those of the moon it is shown that ancient fields of the order of 1 gauss and free iron concentrations of the order of a few percent are sufficient to produce the present dipole moment. We further develop this hypothesis and suggestion by Ksanfomality (1998a, 1998b, 2001) of possible similarity with the Moon and especially with Mars, where multiple crustal blocks exist with different remanent magnetization, but we do not exclude other hypotheses listed by Strom and Sprague (2003). In particular, liquid dynamo and solid magnets can coexist and contribute as in the case of Earth, but with different proportions between both contributors depending on specific evolutionary and structural details, which are not investigated now. Future measurements are needed to clarify the situation.

We have attempted to put some upper limits on the residual local magnetic fields, which could be tentatively associated with the local remanent magnetization of crustal blocks of Mercury's solid material based on available data.

11.2 Available Data and Method of Analysis

Available plots and numerical listings of 1.2-, 6-, and 42-s averages of magnetic field (Lepping R.P., Ness N.F., and Behannon K.W., http://nssdc.gsfc.nasa.gov/database/MasterCatalog?ds=PSFP-00072) can be used to calculate residual magnetic fields $B_{residual}$ which are defined as the difference between measured $B_{measured}$ and modelled main magnetic field B_{main} vectors $B_{residual} = B_{measured} - B_{main}$. The value B_{main} represents here some model field. Estimates of residual magnetic fields were obtained by the visual inspection of corresponding plots. They are of the order of several nT (up to several tens of nT) depending on the approximation and model used for B_{main} (actually, dipole or dipole plus one-two further lowest harmonics). Residual magnetic fields can contain contributions from internal and external magnetic fields produced by planetary and heliospheric 'sources of the magnetism' (this terminolgy is not good, but still used: magnetic field is divergenceless and has no sources by its definition!) not considered in the main field model and also by possible errors of measurements. Let us assume experimental errors being negligible and consider plausible models of the residual field.

The magnetic field around Mercury is produced by the internal electric currents and magnetization of the body of the planet and by magnetospheric electric currents. Mercury's internal magnetic fields are difficult to measure separately from magnetospheric contributions (see, e.g., Russell and Luhmann 1997). Magnetospheric currents strongly contribute to the measured field in the case, even at small orbital radii because of the small size of the magnetosphere. Giampieri et al. (2004) suggested exploiting the fact that the orbital and rotation periods of Mercury are nonsynchronous, in order to disentangle the internal field from external contributions. They simulated the magnetic field measurements onboard both BepiColombo orbiters (MMO and MPO) using for this purpose a Tsyganenko-type model for the Hermean magnetosphere, including contributions from the magnetopause and tail currents, as well as from reconnection. Using standard inversion techniques, these authors showed how to disentangle the magnetospheric field, periodically modulated by the orbital motion around the Sun, from the static internal field and concluded that collecting data from both orbiters over one or more orbital periods will give a good result. This procedure allows a fairly accurate measurement of the internal field multipoles, including higher order terms. Scuffham and Balogh (2006) developed a new empirical model of the Hermean field which is constrained wherever possible by the Mariner 10 data set. This model represents an appropriate rescaling of the Earth's magnetic field model to the Hermean conditions and can be used to fit a much larger spacecraft data set when it becomes available.

11.3 On the Models and Estimates of the Residual Magnetic Field

Mapping Mercury's internal magnetic field with a magnetometer in closed orbit around the planet will provide valuable information about its internal structure. By measuring magnetic field multipoles of order higher than the dipole one could, in principle, determine some properties, such as size and location, of the internal source. These expectations were quantified using conceptual models and simulation of actual measurement during the BepiColombo mission as well as analyzing the simulated data in order to estimate the measurement errors due to the limited spatial sampling. Ability to locate the field-generating current system within the planet and main limitation of the model, due to the presence of time-varying external magnetospheric currents was investigated (Giampieri and Balogh 2001). Our model assumptions differ from that of the paper by Giampieri and Balogh (2001) in several respects. In particular, we do not assume a priori an axial symmetry. We consider the residual field, tentatively, to consist of several or many blocks similar to magnetized strips and anomalies possibly associated with traces of ancient eruptions, impacts and dynamo action on Mars.

Balogh and Giampieri (2002) pointed out that it is unlikely that the Hermean magnetic field is dominated by the dipolar term; multipolar terms (quadrupole, octupole and higher order terms) are probably more important, when compared to the dipole, than in the case of the Earth. The main reason for this is that the location of the generation mechanism, at the vicinity of the core–mantle boundary, is much closer to the surface of the planet, due to the large size of the core, than at the Earth. Whether the field is generated by a hydrodynamic dynamo in a molten outer layer of the core, or by a thermoelectric dynamo at the core–mantle boundary, the mechanisms yield relatively significant higher order multipolar terms. Grosser et al. (2004), based on their two-layer model of Mercury on calculated contributions of magnetospheric or induced currents, depending on the core size.

The crustal magnetization can be observable or hidden depending on its strength and orientation. The horizontally oriented elementary magnets organized in linear or surface structures can be practically 'invisible' under appropriate geometry conditions. Only diffusion fields at the ends of such structures appear and can be measured above the surface of the planet. Vertically oriented magnetization is more easily observable from space because of the magnetic flux conservation and field line divergence from stronger fields to weaker ones. Divergence of the field lines with the altitude from 'local poles' at such places is smallest or could be even negligible in the case of large magnetized areas, when their linear size is larger than the altitude of the observation point. For the vertically placed thin sheet with the vertical magnetization along them one obtains rather slow decrease of the field with altitude above the edge. For linear structures which are thin in comparison with the altitude of observation and placed horizontally one obtains a two-dimensional situation with faster decrease of magnetic field strength as a function of the altitude. An even faster decrease follows for point-like 'poles'. Elementary considerations described in textbooks (Stratton 1941; Smythe 1950) give well-known results for different geometry.

The inverse mathematical problem that consists of determining unknown 'sources' (electric currents, magnetization fields near and beneath the surface) when the fields are given at one level above the surface is usually ill posed and has no unique solution when additional information is not known. This additional information is used practically for the reformulation of the inverse problem and to its 'regularization', i.e., to finally have well-determined solvable problems with unique and stable solutions. Corresponding numerical procedures are especially developed in the geophysics (Zhdanov 2002).

It is not useful to discuss and construct different quantitative, plausible and simplistic models at this stage, because of our limited input information about approximations for their selection. It is enough for our purposes to limit consideration by simple and very rough order of magnitude estimates taking into account a broad range of possible real situations behind scarce observations.

Based on these arguments, we performed the visual and qualitative inspection of data and models. The smooth dome-like shape of the magnetic field plot (Ness et al., 1975a, 1975b) obtained during the close approach (flyby III) with no large deviations from the main field indicates the absence of strong anomalies in the vertical magnetization on the ground beneath the spacecraft. As was also mentioned earlier, based on expert estimates of these residuals we obtained in this way upper limits between several nT and several tens of nT for the residual vertical fields on the ground, averaged over distances of several hundred km along the path of the Mariner 10 during its third flyby. We interpret this result as an indication of small anomalies, if any at the mentioned level of accuracy, in investigated regions around two lines—projections of the flyby trajectories on the surface of the planet. The interpretation is heavily dependent on the hypothesis that the main field can be considered separately and it is not of crustal origin.

11.4 On Possible Origins of Crustal Fields on Mercury

We propose that crustal magnetic fields on Mercury appeared after cooling of the magnetized fluid fraction below the Curie temperature. The magnetization of the fluid could be produced by electric currents (so-called MHD dynamo action). Some specific conditions are needed for the magnetic field self-generation, which can be fulfilled or not under given physical conditions and prehistory. Prehistory and 'memory' play a role in the nonlocal problem with many essential time scales for evolution and dissipation. There is no reliable scenario of the past MHD dynamo in Mercury at the moment and we can only speculate on its presence and characteristics.

We can suggest a possible scenario of the crustal magnetization process based on the available fragmentary information described above. The first step is the collision between comparable planetary bodies (formation of the Basin S could be the marker of the date and the size of this event). The relative velocity of the colliding bodies should be intermediate— not too large and not too small. For rather large relative velocities of colliding bodies (of the order of 10 km/s) their catastrophic destruction and subsequent disappearance of fragments and evaporation should be a consequence. For small relative velocities a slow, non-catastrophic gravitational merging process is not excluded. Such a process could lead to the formation of a jointed and rather cool solid body with no melting. At the intermediate relative velocities, an intermediate situation is conceivable: inelastic collision with the partial loss of mass in the shape fragments and partial melting and evaporation. Large, intermediate and small velocities are scaled by the characteristic thermal and gravitational escape velocities. The newly formed main body could be larger or smaller than initial colliding bodies. It could be solid, partially or completely melted depending on many unknown details of process. The proportion between different phases (solid, liquid, vapours, etc.) and fragments is difficult to estimate—it depends on many unknown details (composition, structure, velocity etc.).

We assume that the fluid fraction on Mercury was formed because of melting after its parent body collision with another celestial body of lesser, but comparable mass. We do not know when this hypothetical event happened—more studies are needed to clarify the age. A partial loss could happen of joint mass and partial melting of a newly formed jointed body. The hypothesis about a large collision is not new (see, e.g., Hartmann 1983). We find an indication of this event in the past: recent observations probably show huge (1,000–2,000 km) concentric structures on Mercury at longitudes 260–285°W (Ksanfomality 1998a, 1998b; Ksanfomality et al. 2005). These observations have been apparently confirmed by other methods (see, Harmon 2007, for the description of the radiolocation data and results, which initially did not show a clear confirmation). It is not excluded that Mercury, in the past, inelastically collided with another celestial body. We assume that the collision happened indeed and was not completely catastrophic for Mercury, which survived. The resulting mass of the newly formed main body (presumably present Mercury) after the collision can be larger or less than the mass of each initial colliding counterpart. There are also geological hints that the outer layers of mass were lost due to collisional sputtering of fragments and the remaining 'iron Mercury' was formed in this way.

We assume in our tentative scenario comparable masses of colliding bodies and anticipate the situation of an inelastic collision with appreciable melting of a large part of the newly formed planetary body. If masses are significantly different, the resulting mass of liquid is obviously not so large. Hence, we assume intermediate velocities, presumably, of the order of hundreds of meters per second or roughly, 1 km/s. Most elaborated numerical runs in the case of Mercury were performed only for higher relative velocities of about 20–30 km/s or so (Benz et al. 1988; Benz 2007). If the newly formed planet is partially or completely liquid, then standard dynamo action would work and produce magnetic fields. Subsequent cooling below the Curie point will produce crystal magnetization of solidified blocks, which is preserved up to present times and observed. Partial demagnetization of the crust can also occur due to heating by subsequent collisions with numerous smaller bodies (Artemieva et al. 2005), but these processes were not evaluated in the case of Mercury.

If our hypothesis is correct and the elapsed time is not enough to 'forget' about this collision and smear-out the memory of its consequences, we expect a rather strong internal asymmetry preserved inside the planet. A noncentred magnetic dipole and higher harmonics could be the natural manifestations of this memory. An actual dynamo can operate now or not; it is an open issue to be tested only in very long-term observations. Instantaneous measurements or short time episodes are not sufficient for any reliable conclusions. The measurements of the dynamical geology and dynamical magnetic fields are necessary for the separation of possible solid body and fluid contributions. The situation is similar in the geomagnetism. This point was correctly stressed by Connerney and Ness (1988).

The acquisition of thermoremanent magnetization (TRM) by a cooling spherical shell was studied for internal magnetizing dipole fields, using Runcorn's (1975b) theorems on magnetostatics (Srnka 1976). If the shell cools progressively inward, inner regions acquire TRM in a net field composed of the dipole source term plus a uniform field due to the outer magnetized layers. In this case, the global dipole moment and external remanent field are nonzero when the whole shell has cooled below the Curie point and the source dipole has disappeared. The remanent field outside the shell is found to depend on the thickness, radii, and cooling rate of the shell, as well as the coefficient of TRM and the intensity of the magnetizing field (Srnka 1976). In reality, Runcorn's theorems are not so essential and do not restrict the magnetization in the absence of spherical symmetry. We assume this situation.

The measured 'pole/equator ratio' for magnetic field on Mercury, i.e., the ratio of Mercury III to Mercury I maximal fields is rather large, about 3. It is larger than for the corresponding 'pole/equator ratio' in the case of the central dipole. This last value is equal to 2. Higher multipoles (quadrupole, octupole, etc.), large external magnetospheric fields (tail currents, Chapman–Ferraro currents, Birkeland currents, substorm currents, etc.) have been invoked as a possible explanation. We should remark that Fig. 5 in the paper by Ksanfomality (1998b) can be also interpreted as an indication of the possible strip magnetization of Mercury. Three to five strips extended along the meridian direction with a similar magnetization along them (or six to ten strips with opposite magnetization plus general field of larger scale) could provide the observed magnetic pattern for these two passes. In this case, the global dipole would be formed from the bundle of magnetic tubes or sheets presumably oriented close to the axis of the planet and placed at longitudes associated in some way with S Basin position at present time. The remaining magnetic field could be formed by the dispersion fields of finite length magnetic tubes and other sources. (In the case of Mars, this orientation of magnetic ribbons is along the parallels in the southern hemisphere.)

Ksanfomality (1998b) assumed two to three differently oriented blocks (500–1,000 km each in size) not specifying the shape and intensity. We speculate further about their elongated shape and field intensity, which depends on details of geometry and placement: tens to hundreds Gauss in the tubes for their compact position near the centre of the planet (this variant is not favourable for the explanation of the data) or more moderate and lower values up to hundreds of nanoTesla with different, 'more realistic' assumptions about the geometry and placement closer to the surface. The term 'realistic' is not quite correct in this context noting the lack of sufficient data to resolve the ambiguity and can be understood in the sense of a higher likelihood only.

The hypotheses of the combined (crustal and transient substorm-like) origin of the magnetic peculiarities is not excluded at the present scarce level of our knowledge after Mariner 10 about the Hermean magnetic field structure and its time variations, but this question needs further investigation. We can also speculate further that the situation can be partially similar in this respect to the Martian and terrestrial ones with their active magnetospheres, but not to the lunar case. Martian magnetic fields have relic crustal origins, nevertheless many important questions remain open (Acuna et al. 1999; Connerney et al., 2001, 2004a, 2004b; Kletetschka et al. 2005; Langlais et al. 2004). Lunar magnetic fields also have unclear origins (Runcorn 1975a, 1975b; Dolginov 1993; Richmond et al. 2003). The comparative study can bring new valuable constraints on the cosmogenic scenarios, which are far from being completed now.

References

M.H. Acuna, J.E.P. Connerney, N.F. Ness, R.P. Lin, D. Mitchell, C.W. Carlson, J. McFadden, K.A. Anderson, H. Reme, C. Mazelle, D. Vignes, P. Wasilewski, P. Cloutier, Global distribution of crustal magnetization discovered by the Mars global surveyor MAG/ER experiment. Science **284**(5415), 790 (1999)

L.A. Akimov, On the brightness distribution across the lunar disk and planets. Astron. Zh. **56**, 412–418 (1979) (in Russian)

L.A. Akimov, Reflection of light by the Moon I. Kinematika Fiz. Nebesnikh Tel. **4**(1), 3–10 (1988) (in Russian)

L.A. Akimov, Y.V. Kornienko, Light scattering by the lunar surface. Kinematika Fiz. Nebesnikh Tel. **10**(2), 15–22 (1994) (in Russian)

N. Artemieva, L. Hood, B.A. Ivanov, Impact demagnetization of the Martian crust: Primaries versus secondaries. Geophys. Res. Lett. **32**(22) (2005). doi:10.1029/2005GL024385

A. Balogh, G. Giampieri, The origin of Mercury's magnetic field and its multipolar structure. EGS XXVII General Assembly, Nice, 21–26 April 2002, abstract #5959

J. Baumgardner, M. Mendillo, J.K. Wilson, Digital high-definition imaging system for spectral studies of extended planetary atmospheres. 1. Initial results in white light showing features on the hemisphere of Mercury unimaged by Mariner 10. Astron. J. **119**, 2458–2464 (2000)

W. Benz, Space Sci. Rev. (2007, this issue)

W. Benz, W.L. Slattery, A.G.W. Cameron, Collisional stripping of Mercury's mantle. Icarus **74**, 516–528 (1988)

B.J. Buratti, J. Veverka, Voyager photometry of Europa. Icarus **55**, 93–110 (1983)

B.J. Butler, D.O. Muhleman, M.A. Slade, Mercury – Full-disk radar images and the detection and stability of ice at the North Pole. J. Geophys. Res. **98**, 15003–15023 (1993)

S. Chandrasekhar, *Radiative Transfer* (Dover, New York, 1960)

S.C. Chase, E.D. Miner, D. Morrison et al., Preliminary infrared radiometry of the night side of Mercury from Mariner 10. Science **185**, 142–145 (1974)

J.E.P. Connerney, N.F. Ness, Mercury's magnetic field and interior, in *Mercury* (University of Arizona Press, Tucson, 1988), pp. 494–513

J.E.P. Connerney, M.H. Acuña, P.J. Wasilewski, G. Kletetschka, N.F. Ness, H. Rème, R.P. Lin, D.L. Mitchell, The global magnetic field of Mars and implications for crustal evolution. Geophys. Res. Lett. **28**(21), 4015–4018 (2001). doi:10.1029/2001GL013619

J.E.P. Connerney, M.H. Acuña, N.F. Ness, T. Spohn, G. Schubert, Mars crustal magnetism. Space Sci. Rev. **111**(1), 1–32 (2004a)

J.E.P. Connerney, M.H. Acuna, N.F. Ness, D.L. Mitchell, R.P. Lin, H. Reme, A magnetic perspective on the Martian crustal dichotomy. Hemispheres Apart: the Origin and Modification of the Martian Crustal Dichotomy. LPI Contribution No. 1213. Proceedings of the conference held September 30–October 1, 2004, in Houston, TX, USA, 2004b, pp. 11–12

A.C. Cook, M.S. Robinson, Mariner 10 stereo image coverage of Mercury. J. Geophys. Res. **105**(E4), 9429–9443 (2000)

A. Danjon, Photometrie et colorimetrie des planetes Mercure et Venus. Bull. Astron. **14**, 315 (1949)

R.F. Dantowitz, S.W. Teare, M.J. Kozubal, Ground based-based high-resolution imaging of Mercury. Astron. J. **119**, 2455–2457 (2000)

M.E. Davies, S.E. Dwornik, D.E. Gault, R.G. Strom, Atlas of Mercury, in *NASA Scientific and Technical Report*, ed. by J. Dunn, NASA SP-42 (US Government Printing Office, Washington, 1978)

A.Z. Dolginov, Magnetic fields and nonuniform structures of the Moon. Twenty-fourth Lunar and Planetary Science Conference. Part 1: A-F, 1993, pp. 411–412

A. Dollfus, Pic-du-Midi visual and photographic observations of planets, in *Planets and Satellites*, ed. by G.P. Kuiper, B.M. Middlehurst (The University of Chicago Press, Chicago, 1961), pp. 482–485

A. Dollfus, M. Auriere, Optical polarimetry of planet Mercury. Icarus **23**, 465 (1974)

D. Domingue, B. Hapke, Fitting theoretical photometric functions to asteroid phase curves. The scattering properties of natural terrestrial snows versus icy satellite surfaces. Icarus **78**, 330–336 (1989)

D. Domingue, B. Hartman, A. Verbiscer, The scattering properties of natural terrestrial snows versus icy satellite surfaces. Icarus **128**, 28–48 (1997)

J.P. Emery, A.L. Sprague, F.C. Witteborn, J.E. Colwell, D.H. Kozlowski, R.W.H. Wooden, Mercury: Thermal modeling and mid-infrared (5–12 µm) observations. Icarus **136**, 104 (1998)

D.L. Fried, Probability of getting a lucky short-exposure image through turbulence. J. Opt. Soc. Am. **68**, 1651–1658 (1978)

P.E. Geissler, A. McEwen, L. Keszthelyi, R. Lopes-Gautier, J. Granahan, D.P. Simonelli, Global color variations on Io. Icarus **140**, 265–282 (1999)

G. Giampieri, A. Balogh, Modelling of magnetic field measurements at Mercury. Planet. Space Sci. **49**(14–15), 1637–1642 (2001)

G. Giampieri, A. Balogh, Mercury's thermoelectric dynamo model revisited. Planet. Space Sci. **50**(7–8), 757–762 (2002)

G. Giampieri, J. Scuffham, A. Balogh, BepiColombo measurements of Mercury's internal field. 35th COSPAR Scientific Assembly. Held 18–25 July 2004, in Paris, France, 2004, p. 2726

J. Gradie, J. Veverka, Photometric properties of powered sulfur. Icarus **58**, 227–245 (1984)

J. Grosser, K.-H. Glassmeier, A. Stadelmann, Magnetic field effects at planet Mercury. Planet. Space Sci. **52**(14), 1251–1260 (2004)

E.W. Guinness, R.E. Arvidson, I. Clark, M.K. Shepard, Optical scattering properties of terrestrial varnished basalts compared with rocks and soils at the Viking Lander sites. J. Geophys. Res. **102**, 28687–28703 (1997)

K. Gunderson, J.A. Whitby, N. Thomas, Visible and NIR BRDF Measurements of Lunar Soil Simulant, 36th Annual Lunar and Planetary Science Conference, abstract no. 1781, 2005

K. Gunderson, N. Thomas, J.A. Whitby, First measurements with the Physikalisches Institut Radiometric Experiment (PHIRE). Planet. Space Sci. **54**(11), 1046–1056 (2006)

B. Hapke, Bidirectional reflectance spectroscopy. 1. Theory. J. Geophys. Res. **86**, 3039–3054 (1981)

B. Hapke, Bidirectional reflectance spectroscopy. 3. Correction for macroscopic roughness. Icarus **59**, 41–59 (1984)

B. Hapke, Bidirectional reflectance spectroscopy. 4. The extinction coefficient and the opposition effect. Icarus **67**, 264–280 (1986)

B. Hapke, *Theory of Reflectance and Emittance Spectroscopy* (Cambridge Univ. Press, New York, 1993)

B. Hapke, Bidirectional reflectance spectroscopy. 5. The coherent backscatter opposition effect and anisotropic scattering. Icarus **157**, 523–534 (2002)

J.K. Harmon, Space Sci. Rev. (2007, this issue). doi:10.1007/s11214-007-9234-y

J.K. Harmon, D.B. Campbell, Radar observations of Mercury, in *Mercury*, ed. by F. Vilas, C.R. Chapman, M.S. Matthews (Univ. of Arizona, Tucson, 1988), pp. 101–117

J.K. Harmon, M.A. Slade, Radar mapping of Mercury: Full-disk images and polar anomalies. Science **258**, 640–642 (1992)

J.K. Harmon, P.J. Perillat, M.A. Slade, High-resolution radar imaging of Mercury's North Pole. Icarus **149**, 1–15 (2001)

J.K. Harmon, M.A. Slade, B.J. Butler, J.W. Head, M.S. Rice, D.B. Campbell, Mercury: Radar images of the equatorial and mid-latitude zones. Icarus **187**, 374 (2007)

W.K. Hartmann, *Moons and Planets* (Wadsworth Publishing Co., Belmont, 1983), Chap. 5, 510 p

J.W. Head, III, Surfaces of the terrestrial planets, in *The New Solar System*, ed. by J.K. Beatty et al. (Sky Publishing Corporation, London, 1981), pp. 45–56

I.V. Holin, Space–time coherence of signal scattered by diffuse moving surface in case of arbitrary motion and monochromatic illumination. Radiophys. Quantum Electron. **31**, 515–518 (1988) (in Russian)

P. Helfenstein, J. Veverka, Photometric properties of lunar terrains derived from Hapke's equation. Icarus **72**, 342–357 (1987)

J.K. Hillier, J. Veverka, P. Helfenstein, P. Lee, Photometric diversity of terrains on Triton. Icarus **109**, 296–312 (1994)

R.E. Hufnagel, Restoration of atmospherically degraded images. Proc. Nat. Acad. Sci. **3**, App. 2, 11 (1966)

J.R. Johnson, W.M. Grundy, M.T. Lemmon, J.F. Bell III, M.J. Johnson, R. Deen, R.E. Arvidson, W.H. Farrand, E. Guinness, A.G. Hayes, K.E. Herkenhoff, F. Seelos IV, J. Soderblom, S. Squyres, Spectrophotometric properties of materials observed by Pancam on the Mars Exploration Rovers. 1. Spirit. J. Geophys. Res. **111**(E02S14) (2006). doi:10.1029/2005JE002494

V.O. Kakhiani, Astronomical image processor AIMAP (2003, unpublished)

G. Kletetschka, N.F. Ness, J.E.P. Connerney, M.H. Acuna, P.J. Wasilewski, Grain size dependent potential for self-generation of magnetic anomalies on Mars via thermoremanent magnetic acquisition and magnetic interaction of hematite and magnetite. Phys. Earth Planet. Interiors **148**(2–4), 149–156 (2005)

H. Korth, J.B. Anderson, M.H. Acuna, J.A. Slavin, N.A. Tsyganenko, S.C. Solomon, R.L. McNutt, Determination of the properties of Mercury's magnetic field by the MESSENGER mission. Planet. Space Sci. **52**(8), 733–746 (2004)

M.A. Kreslavsky, Y.G. Shkuratov, Y.I. Velikodsky, V.G. Kaydash, D.G. Stankevich, Photometric properties of the lunar surface derived from Clementine observations. J. Geophys. Res. **105**(E8), 20281–20295 (2000)

L.V. Ksanfomality, Proper magnetic fields of planets and satellites (a review). Sol. Syst. Res. **32**(1), 31–41 (1998a)

L.V. Ksanfomality, The magnetic field of Mercury: A revision of the Mariner 10 results. Sol. Syst. Res. **32**(2), 115–121 (1998b)

L.V. Ksanfomality, Physical properties of the Hermean surface (a review). Sol. Syst. Res. **35**(5), 339–353 (2001)

L.V. Ksanfomality, High-resolution imaging of Mercury using Earth-based facilities. Sol. Syst. Res. **36**, 267–277 (2002)

L.V. Ksanfomality, Mercury: Image of the planet in the longitude interval 210–285°W obtained by method of short expositions. Sol. Syst. Res. **37**, 514–525 (2003)

L.V. Ksanfomality, A huge basin in the unknown portion of Mercury in the 250–290°W longitude range. Sol. Syst. Res. **38**, 21–27 (2004)

L.V. Ksanfomality, Global asymmetry of large forms of Hermean relief. The 2nd AOGS session, 2005, June, Singapore, Paper ID: 58-PS-A0974, 2005

L.V. Ksanfomality, Earth-based optical imaging of Mercury. Adv. Space Res. **38**, 594–598 (2006)

L. Ksanfomality, A.L. Sprague, New images of Mercury's surface from 210° to 290°W longitudes with implications for Mercury's global asymmetry. Icarus (2007). doi:10.1016/j.icarus.2006.12.009

L.V. Ksanfomality, V.P. Dzhapiashvili, V.O. Kakhiani, A.K. Mayer, Experiment on obtaining of Mercury's images by the short exposure method. Sol. Syst. Res. **35**, 190–194 (2001)

L. Ksanfomality, G. Papamastorakis, N. Thomas, The planet Mercury: Synthesis of resolved images of unknown part in the longitude range 250–290°W. Planet. Space Sci. **53**, 849–859 (2005)

J.H. Lambert, Photometria Sive de Mensura et Gradibus Luminis, Colorum et Umbrae. Detleffsen, Augsburg, 1760

B. Langlais, M.E. Purucker, M. Mandea, Crustal magnetic field of Mars, J. Geophys. Res. **109**(E2) (2004). doi:10.1029/2003JE002048

P.G. Lucey, D.T. Blewett, B.L. Joliff, Lunar iron and titanium abundance algorithms based on final processing of Clementine ultraviolet-visible images. J. Geophys. Res. **105**(20), 297 (2000a)

P.G. Lucey, D.T. Blewett, G.J. Taylor, B.R. Hapke, Imaging of lunar surface maturity. J. Geophys. Res. **105**(20), 377 (2000b)

A. Mallama, D. Wang, R.A. Howard, Photometry of Mercury from SOHO/LASCO and Earth. The Phase Function from 2 to 170 deg. Icarus **155**, 253 (2002)

Red Shift 4. Maris Multimedia Ltd, 2000. www.cinegram.com

A.S. McEwen, Photometric functions for photoclinometry and other applications. Icarus **92**, 298–311 (1991)

M. Mendillo, J. Warell, S.S. Limaye, J. Baumgardner, A. Sprague, J.K. Wilson, Imaging the surface of Mercury using ground-based telescopes. Planet. Space Sci. **49**, 1501 (2001)

M. Minnaert, The reciprocity principle in lunar photometry. Astrophys. J. **93**, 403–410 (1941)

D.L. Mitchell, I. de Pater, Microwave imaging of Mercury's thermal emission at wavelengths from 0.3 to 20.5 cm. Icarus **110**, 2 (1994)

J. Murray, A. Dollfus, B. Smith, Cartography of the surface markings of Mercury. Icarus **17**, 576–584 (1972)

N.F. Ness, Mercury—Magnetic field and interior. Space Sci. Rev. **21**, 527–553 (1978)

N.F. Ness, The magnetic field of Mercury. Phys. Earth Planet. Interiors **20**, 209–217 (1979)

N.F. Ness, K.W. Behannon, R.P. Lepping, Y.C. Whang, Magnetic field of Mercury confirmed. Nature **255**, 204–205 (1975a)

N.F. Ness, K.W. Behannon, R.P. Lepping, Y.C. Whang, The magnetic field of Mercury. I. J. Geophys. Res. **80**, 2708–2716 (1975b)

N.F. Ness, K.W. Behannon, R.P. Lepping, Y.C. Whang, Observations of Mercury's magnetic field. Icarus **28**, 479–488 (1976)

R.T. Pappalardo, J.W. Head, G.C. Collins, R.L. Kirk, G. Neukum, J. Oberst, B. Giese, R. Greeley, C.R. Chapman, P. Helfenstein, J.M. Moore, A. McEwen, B.R. Tufts, D.A. Senske, H.H. Breneman, K. Klaasen, Grooved terrain on Ganymede: First results from Galileo highresolution imaging. Icarus **135**, 276–302 (1998)

N.C. Richmond, L.L. Hood, J.S. Halekas, D.L. Mitchell, R.P. Lin, M. Acuña, A.B. Binder, Correlation of a strong lunar magnetic anomaly with a high-albedo region of the Descartes mountains. Geophys. Res. Lett. **30**(7), 48–51 (2003). doi:10.1029/2003GLO16938

S.K. Runcorn, An ancient lunar magnetic dipole field. Nature **253**, 701–703 (1975a)

S.K. Runcorn, On the interpretation of lunar magnetism. Phys. Earth Planet. Interiors **10**(4), 327–335 (1975b)

C.T. Russell, J.G. Luhmann, Mercury: Magnetic field and magnetosphere, in *Enciclopedia of Planetary Sciences*, ed. by J.H. Shirley, R.W. Fairbridge (Chapman & Hall, London, 1997), pp. 476–478

G. Schubert, M.N. Ross, D.J. Stevenson, T. Spohn, Mercury's thermal history and the generation of its magnetic field Mercury, in *Mercury* (University of Arizona Press, Tucson, 1988), pp. 429–460

J. Scuffham, A. Balogh, A new model of Mercury's magnetospheric magnetic field. Adv. Space Res. **38**, 616–626 (2006)

M.A. Slade, B.J. Butler, D.O. Muhleman, Mercury radar imaging—Evidence for polar ice. Science **258**, 635 (1992)

W.R. Smythe, *Static and Dynamic Electricity* (McGraw-Hill, 1950)

T. Spohn, D. Breuer, Core composition and the magnetic field of Mercury, American Geophysical Union, Spring Meeting 2005, abstract #P23A-01

A.L. Sprague, R.W.H. Kozlowski, D.M. Hunten, Caloris Basin: An enhanced source for potassium in Mercury's atmosphere. Science **249**, 1140–1143 (1990)

A.L. Sprague, L.K. Deutsch, J. Hora, G.G. Fazio, B. Ludwig, J. Emery, W.F. Hoffmann, Mid-infrared (8.1–12.5 μm) imaging of Mercury. Icarus **147**, 421 (2000)

A.L. Sprague, J.P. Emery, K.L. Donaldson, R.W. Russell, D.K. Lynch, A.L. Mazuk, Mercury: Mid-infrared (3–13.5 μm) observations show heterogeneous composition, presence of intermediate and basic soil types, and pyroxene. Meteorit. Planet. Sci. **37**, 1255 (2002)

A.L. Sprague, J. Warell, J. Emery, A. Long, R.W.H. Kozlowski, Mercury: First spectra from 0.7 to 5.5 μm support low FeO and feldspathic composition. 35th Lunar and Planetary Science Conference, 2004, abstract no. 1630

S.W. Squyres, J. Veverka, Voyager photometry of surface features on Ganymede and Callisto. Icarus **46**, 137–155 (1981)

L.J. Srnka, Magnetic dipole moment of a spherical shell with TRM acquired in a field of internal origin. Phys. Earth Planet. Interiors **11**(3), 184–190 (1976)

R. Stekelenburg, AstroStack manual (v. 0.90 beta), 1999, 2000. http://www.astrostack.com/

A. Stephenson, Crustal remanence and the magnetic moment of Mercury. Earth Planet. Sci. Lett. **28**(3), 454–458 (1976)

D.J. Stevenson, Mercury's magnetic field—A thermoelectric dynamo? Earth Planet. Sci. Lett. **82**(1–2), 114–120 (1987)

J.A. Stratton, *Electromagnetic Theory* (McGraw-Hill, 1941)

R.G. Strom, A.L. Sprague, *Exploring Mercury: The Iron Planet* (Springer, Chichester, 2003), 216 pp

F. Takahashi, M. Matsushima, Dipolar and non-dipolar dynamos in a thin shell geometry with implications for the magnetic field of Mercury. Geophys. Res. Lett. **33**(10) (2006), CiteID L10202

T. Van Hoolst et al., Space Sci. Rev. (2007, this issue). doi:10.1007/s11214-007-9202-6

J. Veverka, P. Helfenstein, B. Hapke, J. Goguen, Photometry and polarimetry of Mercury, in *Mercury*, ed. by F. Vilas, C. Chapman, M. Matthews (Univ. of Arizona Press, Tucson, 1988), pp. 37–58

F. Vilas, P.S. Cobian, N.G. Barlow, S.M. Lederer, How much material do the radar-bright craters at the mercurian poles contain? Planet. Space Sci. **53**, 1496–1500 (2005)

J. Warell, Properties of the Hermean regolith: II. Disk-resolved multicolor photometry and color variations of the "unknown" hemisphere. Icarus **156**, 303 (2002)

J. Warell, Properties of the Hermean regolith: III. Disk-resolved vis–NIR reflectance spectra and implications for the abundance of iron. Icarus **161**, 199–222 (2003)

J. Warell, Properties of the Hermean regolith: IV. Photometric parameters of Mercury and the Moon contrasted with Hapke modelling. Icarus **167**, 271 (2004)

J. Warell, D.T. Blewett, Properties of the Hermean regolith: V. New optical reflectance spectra, comparison with lunar anorthosites, and mineralogical modelling. Icarus **168**, 257 (2004)

J. Warell, S.S. Limaye, Properties of the Hermean regolith: I. Global regolith albedo variation at 200 km scale from multicolor CCD imaging. Planet. Space Sci. **49**, 1531 (2001)

J. Warell, P.-G. Valegård, Albedo–color distribution on Mercury: A study of the poorly known hemisphere. Astron. Astrophys. **460**, 625 (2006)

J. Wicht et al., Space Sci. Rev. (2007, this issue). doi:10.1007/s11214-007-9280-5

M.S. Zhdanov, *Geophysical Inverse Theory and Regularization Problems* (Elsevier, 2002)

Mercury's Surface Composition and Character as Measured by Ground-Based Observations

A. Sprague · J. Warell · G. Cremonese · Y. Langevin ·
J. Helbert · P. Wurz · I. Veselovsky · S. Orsini ·
A. Milillo

Originally published in the journal Space Science Reviews, Volume 132, Nos 2–4.
DOI: 10.1007/s11214-007-9221-3 © Springer Science+Business Media B.V. 2007

Abstract Mercury's surface is thought to be covered with highly space-weathered silicate material. The regolith is composed of material accumulated during the time of planetary formation, and subsequently from comets, meteorites, and the Sun. Ground-based observations indicate a heterogeneous surface composition with SiO_2 content ranging from 39 to 57 wt%. Visible and near-infrared spectra, multi-spectral imaging, and modeling indicate expanses

---

A. Sprague (✉)
Lunar and Planetary Laboratory, University of Arizona, Tucson, AZ 85721, USA
e-mail: sprague@lpl.arizona.edu

J. Warell
Department of Astronomy and Space Physics, Uppsala University, 75120 Uppsala, Sweden

G. Cremonese
INAF-Osservatorio Astronomico, vic.Osservatorio 5, 35122 Padova, Italy

Y. Langevin
Institut d'Astrophysique Spatiale, CNRS/Univ. Paris XI, 91405 Orsay, France

J. Helbert
Institute for Planetary Research, DLR, Rutherfordstrasse 2, 12489 Berlin, Germany

P. Wurz
Physics Institute, University of Bern, Sidlerstrasse 5, 3012 Bern, Switzerland

I. Veselovsky
Skobeltsyn Institute of Nuclear Physics, Moscow State University, 119992 Moscow, Russia

I. Veselovsky
Space Research Institute (IKI), Russian Academy of Sciences, 117997, Moscow, Russia

S. Orsini · A. Milillo
Istituto di Fisica dello Spazio Interplanetario (IFSI), via del Fosso del Cavaliere, 100,00133, Rome, Italy

S. Orsini · A. Milillo
Istituto Nazionale di Astrofisica (INAF), via del Fosso del Cavaliere, 100,00133, Rome, Italy

of feldspathic, well-comminuted surface with some smooth regions that are likely to be magmatic in origin with many widely distributed crystalline impact ejecta rays and blocky deposits. Pyroxene spectral signatures have been recorded at four locations. Although highly space weathered, there is little evidence for the conversion of FeO to nanophase metallic iron particles ($npFe^0$), or "iron blebs," as at the Moon. Near- and mid-infrared spectroscopy indicate clino- and ortho-pyroxene are present at different locations. There is some evidence for no- or low-iron alkali basalts and feldspathoids. All evidence, including microwave studies, point to a low iron and low titanium surface. There may be a link between the surface and the exosphere that may be diagnostic of the true crustal composition of Mercury. A structural global dichotomy exists with a huge basin on the side not imaged by Mariner 10. This paper briefly describes the implications for this dichotomy on the magnetic field and the 3 : 2 spin : orbit coupling. All other points made above are detailed here with an account of the observations, the analysis of the observations, and theoretical modeling, where appropriate, that supports the stated conclusions.

Keywords Mercury · Planetary surface composition · Mercury's surface composition · Remote sensing · Ground-based observations · Spectroscopy

1 Introduction

Mariner 10 made no direct measurements of Mercury's surface composition (for details see "The Surface of Mercury as seen by Mariner 10," by Cremonese et al. 2007, this issue). The vidicon cameras imaged 45% of Mercury's surface and showed it to be heavily cratered; in that sense, it resembles the Moon. Also, the crater morphology and tectonic expression of fault scarps and ejecta rays give the strong impression that the surface is covered with silicates. Ground-based imaging of the remaining 55% of Mercury's surface, in the visible and near-IR, microwave, and radar wavelengths, have found a surface similar to that found by Mariner 10—highly cratered, along with a huge basin and several locations of unusually bright visible and radar spots (for details see "Earth-based Visible and Near-IR Imaging of Mercury," by Ksanfomality et al. 2007, this issue). Ground-based telescopic spectroscopy has identified key spectral signatures that support the Mariner 10 impression of a heavily cratered, silicate surface. Several spectra have been interpreted in terms of Mercury's surface composition. The effects of space weathering and exogenic material on the spectral signatures are probably significant and must be accounted for in our spectral analysis, modeling, and interpretations.

Much work has occurred in the past decade, and this article focuses on these new results but also gives a review of older research, permitting the interested reader to find historical material. Also included are brief descriptions of laboratory spectral measurements of terrestrial, lunar, and meteoritic rocks, and minerals (at a variety of grain sizes) where necessary for complete description of the Mercury science. Laboratory spectral libraries are critical for the interpretation of spectra, especially from Mercury, a body for which we have no known "ground truth." All of the ground-based discoveries are discussed in the context of the challenges and opportunities of future ground-based observations and orbital measurements by the MESSENGER and Bepi Colombo spacecrafts.

2 Visible and Near Infrared Measurements of Mercury's Surface

Observational studies of visible and near infrared spectra of Mercury (wavelength range 0.4–2.5 µm) have been performed since the 1960s with a variety of techniques. Concise reviews of these efforts are given in, e.g., Vilas et al. (1988, 1984), Blewett et al. (1997) and Warell (2003). The techniques include filter photometry (Harris 1961; Irvine et al. 1968; Warell 2002), CVF photoelectric photometry (McCord and Adams 1972a, 1972b; Vilas and McCord 1976; Tepper and Hapke 1977; McCord and Clark 1979; Vilas et al. 1984), and low-resolution reflectance spectroscopy (Vilas 1985; Warell 2003; Warell and Blewett 2004; Warell et al. 2006). Papers since 2002 include disk-resolved data sets, while the earlier observations were made of the integrated disk.

Descriptions of the physics of absorption bands in the VISNIR have been reviewed by many authors (e.g., Burns 1993; Gaffey et al. 1993; Hapke 2001; Pieters and Englert 1993) and will not be repeated here. The two most sought-after bands in this wavelength region are the 1 μm crystal field band due to ferrous iron in crystalline silicates, and the very strong oxygen–metal charge transfer bands of iron and titanium oxides in the near-UV, below 0.5 μm. Historically, there has been much debate about the presence of the ferrous iron band in spectra obtained in the 1960s to 1990s, though it now appears to have been detected (Warell et al. 2006). The near-UV absorption edge has not been identified with certainty. This failure may be related to the strong maturity (highly space weathered) of the surface and to minute abundances of titanium in the crustal rocks.

The spectroscopic observations of Mercury obtained before the mid-1990s were rigorously analyzed by Blewett et al. (1997) in order to constrain Mercury's surface composition. The best spectra indicated an FeO content of 3% and a TiO_2 content of 1%, based on the color-ratio analysis methods formulated and calibrated for the Moon by Lucey et al. (1998, 2000).

Hapke (2001), using his improved model for light scattering in particulate semi-transparent media, summarized the present knowledge of Mercury based on Mariner 10 and ground-based spectroscopic studies to indicate the presence of a crust low in, but not devoid of, ferrous iron with an abundance of submicroscopic metallic iron that is similar to the Moon's, about 0.5 wt%.

Blewett et al. (2002) compared the spectra of a number of small, highly mature and iron-poor sites of the lunar farside (obtained during the Clementine Mission) with those of Mercury. A distinct difference between the lunar pure anorthosite (>90% plagioclase feldspar) regions compared to mercurian spectra is that the lunar spectra have a marked ferrous iron absorption band at 1 μm, clearly visible despite the very low abundance of FeO (2.8–3.3 wt%) and the fact that the Clementine photometry does not cover the full wavelength range of the band. Also, the lunar spectra were found to be slightly less sloped at wavelengths shorter than 0.75 μm than their mercurian counterparts. An explanation put forth by Blewett et al. (2002) is that Mercury's regolith has very little $npFe^0$.

The multicolor photometric imaging observations reported by Warell (2002) effectively covered the integral disk between 47°–236°E longitudes, on primarily the poorly known hemisphere. These data support the inferences from the first mercurian CCD spectrum obtained by Vilas (1985), covering 105°–227°E, indicating a strongly sloped and linearly shaped spectrum from 550–940 nm with no indication of the presence of a 1 μm ferrous absorption band. The disk-resolved nature of the imaging observations allowed the study of spectra of individual bright and dark albedo features, but the spectra of these two groups of features were not statistically different except for reflectance value. One finding was the presence of an inverse relation of the spectral slope with the emission angle, such that features at mid-disk have the steepest slopes. This effect was much stronger than the case for returned samples of the lunar regolith, for which it was proposed to be a result of wavelength-dependent backscattering efficiency—stronger than the case for (disturbed) lunar material. The spectral slope-photometric geometry relation was suggested as due to the presence of a greater abundance of complex, backscattering agglutinates (glassy regolith particles somewhat larger than average formed in the space weathering process) with absorption efficiencies smaller than those of their lunar counterparts.

Warell (2003) acquired 520–970 nm spectra from 60°–120°E longitudes in 1999 that support most of the previous work in terms of absence of the 1 μm absorption band and that verify the marked spectral slope-photometric geometry relation of Warell (2002). Spectral modeling with the light-scattering theory of Hapke (2001), which had previously not been rigorously applied to mercurian spectra, implied that the metallic iron abundance is likely less than 0.3%, and that modeled ferrous content (<3 wt%) in mercurian materials is lower than in lunar materials.

Noble and Pieters (2001) made a theoretical prediction of a spectral effect due to Ostwald ripening which might be operating on Mercury. This process involves an increase in the size of npFe^0 particles at elevated temperatures, and was postulated to be important at surface temperatures above about 470 K (Mercury's equatorial temperature exceeds this value by up to 250 K at local noon). The spectral result of coarsening metallic iron particles at equatorial and low mid-latitudes would be a general decrease in reflectance in the VIS–NIR range, in contrast to a strong reddening for the case of smaller (4–33 nm) particles which cause both darkening and reddening. The spectral manifestation of this prediction was checked by Warell (2003) using the disk-resolved data set of Warell (2002), and indications of the expected increase of spectral slope (reddening) with higher latitude was found.

A set of new CCD spectra of Mercury covering the blue range (0.40–0.65 μm) were combined by Warell and Blewett (2004) with previously obtained spectra and rigorously modeled with the improved Hapke (2002) theory to derive quantitative compositional information. The 0.6–2.5 μm CVF photometer disk-integral spectrum of McCord and Clark (1979), at the time the only published NIR spectrum of Mercury, and a synthetic 0.40–0.97 μm Mercury spectrum derived for standard observational geometry, were used to constrain the range of compositional models. Hapke-model physical parameters used in the analysis were based on modeling of well-characterized laboratory spectra of synthetically matured silicates, laboratory lunar regolith samples, and remotely sensed lunar pure anorthosite regions. Contrary to the case for the strongly spectrally sloped lunar pure anorthosite regions, the slope of the redder mercurian spectrum could not be explained without the assumption of a wavelength dependence in the amplitude of the backscattering lobe of the double Henyey-Greenstein single particle scattering function. For Mercury, the best-fit models consisted of a high reflectance labradorite plagioclase mixed with a minor amount of low-iron enstatite pyroxene, yielding an FeO abundance of 1.2 wt%. The abundance of submicroscopic metallic iron cannot be determined without knowledge of the average optically active grain size, which is unknown. However, grain sizes of the order of 30 μm or less (half the size of lunar particulate materials), were found to be the best fit of for all the modeled spectra. With that result, the wt% npFe^0 must be about 0.1 wt%, and likely to be from exogenic sources.

After an observational quest lasting about 40 years, it appears that an absorption band in the near-infrared spectrum of Mercury finally has been identified (Fig. 1). Warell et al. (2006) presented disk-resolved spectra for the 0.7–5.3 μm range with the SpeX instrument at the NASA Infrared Telescope Facility (IRTF) on Mauna Kea, HI. Two near-polar spectra showed the presence of broad shallow bands centered at about 1.1 μm. Their shapes and widths are consistent with the presence of Ca-rich clinopyroxene in various amounts in a highly matured regolith which is heterogeneous in composition on regional scales of 200 km or so. These results support the results of mid-infrared (3–14 μm) spectral studies of Sprague et al. (2002) who obtained Mercury emissivity spectra exhibiting spectral features similar to those of laboratory Ca-rich diopside.

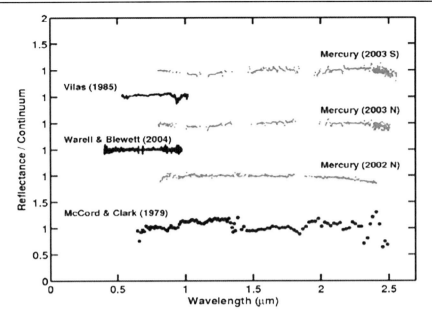

Fig. 1 First spectra to unambiguously show reflectance absorption near 1 μm (2003S, 2003N, 2002N) compared to previous efforts (Vilas, Warell, and Blewett; McCord and Clark). The strong red slope of the continuum has been divided out of these spectra to enhance the visibility of the absorption bands in the 2003S and 2003N spectra. From Warell et al. (2006)

2.1 General Summary and Implications of Vis- and Near-IR Spectroscopic Observations

The effective footprints of the mercurian spectroscopic targets are often large in ground-based observations (up to many hundreds of km) because of smearing by Earth's atmospheric turbulence. Thus, the sampled surface regions include a range of geologic units, many of which are probably highly space weathered. The spectroscopic signature of particularly interesting formations, possessing more crystalline crustal materials than the average surface, are present in the spectral footprints to smaller or larger extents depending on location. Such features are typically too small in areal extent to provide enough flux to dominate the observed spectra. Ideally, remote-sensing studies should be targeted to small and well-defined surface locations which are immature and contain a significant crystalline component. In addition, the exposed rocks should be excavated from depth to sample the native crust, and not merely represent the reprocessed local surface regolith such as breccias with a substantial exogenic component. With slit spectroscopy and the newer, sensitive instruments like SpeX, for 0.8- to 5-μm measurements, and the **MidInfraRedSpectrometer&Imager** (MIRSI), both at the IRTF, such smaller (200- to 300-km expanses) areal observations are possible. In this respect, it is interesting to note that the identification of absorption bands due to crystalline high-Ca pyroxene in two SpeX spectra from polar regions of Mercury (Warell et al. 2006) were made in surface regions where albedo maps indicate the presence of optically bright terrain, likely to have some exposed fresh material. Another site at which no band was detected was located in an optically dark surface region, likely to be highly space weathered.

Spectra from previously deep-seated crustal material make studies of the bulk material composition possible. Interior walls with mass-wasting morphologies, uplifted central peaks, and blocky ejecta blankets of young craters will be obvious targets for specific studies with MESSENGER and BepiColombo instrumentation. Also interesting will be other types of surface units likely to expose large amounts of crystalline material, such as lobate scarp faults.

3 Mid-Infrared Spectroscopy of Mercury's Surface

3.1 Laboratory Studies of Mid-Infrared Spectral Emission from Rocks and Minerals

Detailed mid-infrared laboratory studies of rock and mineral samples were begun by Logan and Hunt (1970), Salisbury et al. (1994), Henderson and Jakosky (1997), Wagner (2000), and others. More sophisticated laboratory study is now underway as part of the preparation for the Bepi Colombo mission to Mercury (Benkhoff et al. 2006; Maturilli et al. 2006a, 2006b) in the Planetary Emissivity Laboratory (PEL) at DLR. Currently these measurements are performed under ambient conditions, using a device built at DLR (Berlin) and coupled to a Fourier transform infrared spectrometer Bruker IFS 88 purged with dry air and equipped with a liquid-nitrogen-cooled MCT detector. In the near future measurements under vacuum conditions and in temperature ranges more applicable to Mercury are planned using a Bruker VERTEX 80v spectrometer and new detectors which allow measurement of the emissivity of samples in the wavelength range from 1 to 50 µm.

Figure 2 shows emissivity spectra obtained as part of the Berlin Emissivity Database (BED) under ambient conditions (Maturilli et al. 2006b). In general, spectra of the different mineral classes and subclasses can be easily distinguished from the high variability in positions and shapes of the major emissivity features. Three main spectral regions are: Christiansen feature (also the Emissivity Maximum-EM), the Reststrahlen bands (RB), and the Transparency Minima (TM). Hapke (1996) demonstrated that the actual wavelength of the Christiansen feature does not always coincide exactly with the EM which is the true diagnostic emissivity feature in silicates. In the spectra of silicates, The EM and TM shift to lower wave numbers (longer wavelengths) with decreasing polymerization degree of SiO_4 tetrahedra (Conel 1969). Therefore, these features are located at longer wavelengths in the emission spectra of olivines, compared to those of feldspars.

There is also a strong influence of particle size on the characteristics and shape of emittance spectra. This is well known but has often been neglected when studying the composition of planetary surfaces in the mid-infrared wavelength range. One reason for this neglect was the lack of a database with true emission spectra for a grain sizes less than 125 µm. The BED spectral library now contains entries for planetary analogue materials, separated in well-defined grain size ranges: <25 µm, 25–63 µm, 63–125 µm, and 125–250 µm. These grain sizes are ideal for application to thermal emittance spectra from Mercury because Mercury's regolith is highly comminuted and likely covered with fine dust of native and exogenic material.

The positions and shapes of RBs are specific for the chemistry and lattice structures of the particular material. The band depths and shapes are strongly affected by particle size and crystal orientation. In general an increase in the spectral contrast can be observed going from fine to coarse grain sizes. For size separates smaller than 25 µm, the contribution of volume scattering is significant. For these small particles, the spectral contrast within the Reststrahlen bands is strongly reduced and a separate emission minimum is mostly evident at about 800 to 870 cm^{-1} (11.5–12.5 µm). This emissivity minimum (TM), is a highly diagnostic spectral feature which is present as a maximum in reflectance spectra of fine-grained minerals and rocks. However, its wavelength position in the spectra of silicates

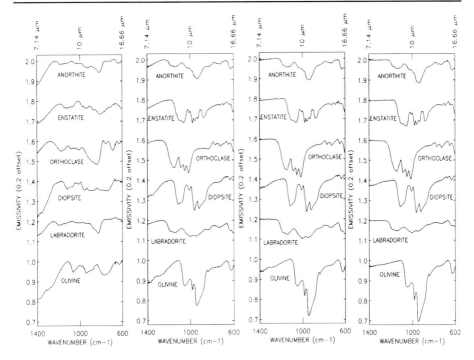

Fig. 2 True emissivity spectra of some of the MERTIS analogue materials (Benkhoff et al. 2006) obtained as part of the BED (Maturilli et al. 2006a, 2006b). Grain size fractions from left to right: <25 μm, 25–63 μm, 63–125 μm, and 125–250 μm. All spectra show some contribution from residual water vapor in the chamber at wave numbers higher than 1,300 cm^{-1}

is affected by particle size (Mustard and Hays 1997). The Christiansen feature, or more correctly the EM, is not affected by particle size, at 1 bar pressure (Cooper et al. 2002 and references therein). It should be noted that the EM is clearly seen as a maximum for grain sizes less than 25 μm or so. For coarser grain sizes it changes to a shoulder which is less diagnostic because the position is less well defined. Significant shift has been found when comparing the wavelength of the EM of solid and powdered samples (Cooper et al. 2002).

Thermal gradients in the sample influence the measured emission spectra in the case of an airless body (e.g., the Moon or Mercury), but not in the case of Mars which has an average atmospheric pressure of about 6 HPa (Henderson and Jakosky 1994, 1997). Detailed studies of heated samples in a vacuum environment and in controlled thermal conditions are badly needed and will be performed in the Emissivity Laboratory in the near future. The spectra will be available in the BED.

Figure 3 shows emission spectra of six plagioclase feldspars for the finest size separates (<25 μm). These minerals are members of the plagioclase solid solution series ranging from sodium-rich end-member albite (NaAlSi$_3$O$_8$) to calcium-rich end-member anorthite (CaAl$_2$Si$_2$O$_8$). As Ca^{2+} cation gradually substitutes for Na$^+$ in plagioclase structure, structural changes occur which affect frequencies of Si–O vibrations, as well as the related Christiansen frequencies. The spectral changes include a progressive shift of the EM and TM to shorter wave numbers (longer wavelengths) as highlighted by the lines in Fig. 3 which has vertical lines aligned with the position of these features for albite. Changes in relative contrasts of individual RBs are visible, although less pronounced for these spectra of fine-grained samples. The positions of the EM are unaffected by grain size at 1 bar pressure

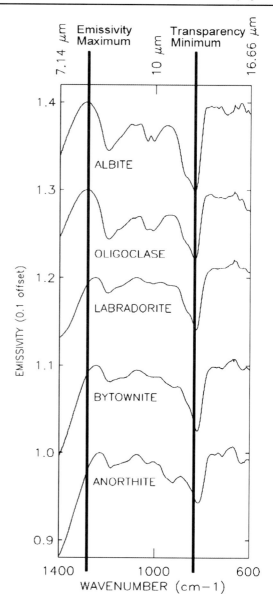

Fig. 3 Emission spectra of five plagioclase feldspars for the <25 μm grain size separate are shown offset for clarity. The trend for the Emissivity Maximum (EM) and the Transparency Minimum (TM) to appear at longer wavelengths as the chemistry changes from Na- and SiO$_2$-rich to Ca- and with diminished wt% SiO$_2$ is shown by the *two vertical lines* aligned to the respective features in the Albite spectrum. Data are from the BED (Maturilli et al. 2006a)

(Conel 1969; Cooper et al. 2002) and therefore are highly diagnostic as illustrated by an EM at 7.8 mm (or 1288 cm^{-1}) for albite and at 8.1 mm (or 1234 cm^{-1}) for anorthite.

3.2 Mid-Infrared Studies of Spectral Emissivity from Mercury's Surface

With the availability of sensitive mid-infrared detectors, spectral imaging data acquisition and analysis software, and large telescopes that can be used for daytime viewing when Mercury can be seen from the Earth's surface, a new wavelength region has become available for remotely sensing Mercury's surface composition.

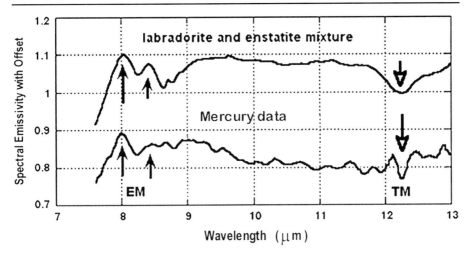

Fig. 4 A spectrum of Mercury's surface from ~240° to 250°E equatorial longitudes shows similarities to a mixture of Na-bearing feldspar (labradorite), and low-iron, Mg-rich orthopyroxene (enstatite). There is a good match at the EM for both minerals and at the transmission minimum (TM) for feldspar. Figure adapted from Sprague and Roush (1998)

Heterogeneous composition of Mercury's surface is indicated by mid-infrared spectra (Sprague et al. 1994, 2002); Sprague and Roush 1998; Emery et al. 1998; Cooper et al. 2001; Donaldson-Hanna et al. 2006). Qualitatively, most spectra match models including plagioclase feldspar and some pyroxene. The common solid solution series of plagioclase feldspar $(Ca,Na)(Al,Si)AlSi_2O_8$ is a strong candidate for a large component of Mercury's surface at several locations as seen in spectra of Figs. 4, 5, 6, 7, 8 and 9. Plagioclase feldspar is formed both at high and low temperatures. Na-rich plagioclase feldspar could have a designation such as $Ab_{95}An_5$ ($Albite_{95}Anorthite_5$) and be comprised of 95% $NaAlSi_3O_8$ (albite) and 5% $CaAlSi_3O_8$ (anorthite). However, as will be shown in the following, spectra from Mercury seem to be better matched by laboratory spectra in the Ab_{50} to Ab_{30} range, roughly labradorite. These spectral signatures may support the suggestion that Mercury has a thick anorthositic crust (Spudis and Guest 1988; Tyler et al. 1988; Sprague et al. 1994; Robinson and Taylor 2001). Members of the pyroxene group are closely related to one another and crystallize in two different systems. The most common orthorhombic pyroxenes are enstatite $(MgSiO_3)$ and hypersthene $(Mg,Fe)SiO_3$. The most common monoclinic pyroxenes are monoclinic forms of the common orthorhombic pyroxenes—diopside $(CaMgSi_2O_6)$, hedenbergite $(CaFeSi_2O_6)$, and augite (intermediate between diopside and hedenbergite with some Al).

Spectral measurements at Mercury from E longitudes 328° to 348°, 316° to 338°, 320° to 315°, 275° to 315°, 316° to 338°, and 240° to 250° are shown in Figs. 4, 5, 6, 7, 8 and 9. The areal extent of the spatial footprint is no smaller than 200 km by 200 km for the very best spatially resolved observations to date (Fig. 9), and as much as 1,000 km by 1,000 km for the least spatially resolved region (Fig. 8). Spectra exhibit emissivity maxima (EM) associated with the principal Christiansen frequency (Sprague et al. 1994; Emery et al. 1998). So far, all EM of spectra from the intercrater plains east of the crater Homer are indicative of intermediate silica content (~50–57% SiO_2). Figure 4 shows one example, a Mercury spectrum from 240° to 250°E longitude compared to a model spectrum created from laboratory spectra of a mixture of plagioclase feldspar and the Mg-rich pyroxene enstatite. Rocks composed of these two minerals might be anorthosite, leucogabbro, or leuconorite.

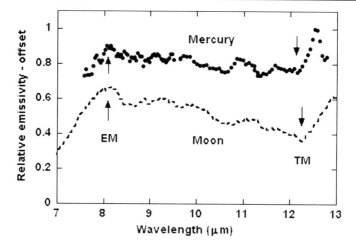

Fig. 5 A spectrum from Mercury's surface (equatorial ~335° to 340°E longitude) is compared with a laboratory spectrum from lunar mission Apollo 16 returned samples (particulate breccia No. 67031; 90% feldspar, 10% pyroxene). Figure courtesy of A.L. Sprague

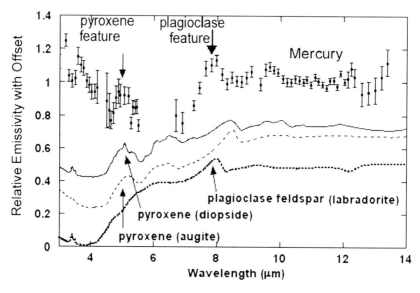

Fig. 6 A Mercury spectrum from the equatorial region of 275° to 315°E exhibits a 5 μm emission feature resembling laboratory spectra of clinopyroxene. The EM at 7.9 μm is indicative of intermediate SiO_2 content and is compared to plagioclase feldspar (Ab_{50}). Laboratory spectra are from Salisbury et al. (1987, 1991). Mercury spectrum is from Sprague et al. (2002)

Figure 5 shows a spectrum from an equatorial E longitudes near 335° to 340° on Mercury's surface and compares it to laboratory spectrum from a particulate breccia sample brought back from the Moon. The number of the sample, No. 67031, indicates that it was obtained at the Apollo 16 landing site. The lunar sample is ~90% anorthite (Ca-feldspar) and ~10% pyroxene. The EM for both spectra is centered close to 8 μm and marked with an arrow. These spectra may indicate the feldspar type on Mercury labradorite rather than

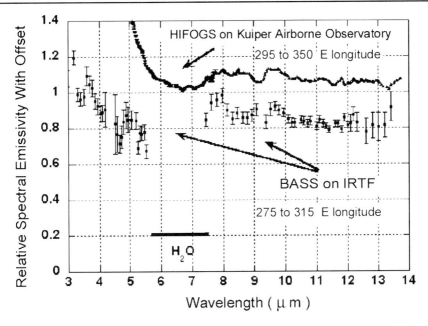

Fig. 7 Two Mercury spectra from adjacent locations on Mercury's surface. In both cases the principal EM is indicative of a feldspathic composition. Other features remain unidentified. Figure is from Sprague et al. (2002)

anorthite which is most common on the Moon. Other features in the Mercury spectrum are also similar to the lunar spectra and likely indicate the presence of pyroxene at this location.

Another spectral feature indicating the presence of clino-pyroxene in Mercury's regolith is a 5 μm emission feature in a spectrum from 275° to 315°E longitude (Fig. 6). The best fit model is to diopside ($CaMgSi_2O_6$). The low-FeO abundance indicated by near-IR reflectance spectroscopy also supports the presence of a low-iron bearing clino-pyroxene. The EM and TM present in the spectrum indicate ~45% SiO_2 and a mafic bulk composition. Unfortunately, the two pieces of information (pyroxene and weight %SiO_2) are not enough to uniquely identify the rocks that formed the regolith at this location. Powders of either low-iron basalt or anorthosite with about 90% plagioclase and 10% low-iron pyroxene could give the same features.

Based on the spectra shown in Figs. 4, 5, 6, 7, 8 and 9, it appears that Mercury's regolith has a high concentration of plagioclase feldspar, perhaps labradorite $(Na,Ca)Si_3O_8$. An alternative explanation for the apparent match to plagioclase feldspar is that the spectrum comes from the glassy soil on Mercury's surface that is very mature after aeons of meteoritic bombardment. Scientists have shown that if lunar soils are very mature, much of the FeO is removed from the glasses and they appear much more feldspathic in laboratory spectral measurements (e.g., Hapke 2001). However, the microwave observations show Mercury's regolith to be very transparent relative to the Moon so the elemental iron component cannot be very large (Mitchell and de Pater 1994).

A spectrum obtained using the High Resolution Faint Object Grating Spectrometer (HIFOGS) from the Kuiper Airborne Observatory (KAO) which flew at 35,000 ft., above much of the Earth's attenuating atmosphere (Fig. 7) has multiple EMs indicating a complicated bulk composition and or mixed mineralogy at 200° to 260°E (Emery et al. 1998). The spectrum from Fig. 6 is plotted again with the HIFOGS spectrum in Fig. 7 for comparison. While

Fig. 8 Spectra from Mercury's surface at six different locations show transparency minima (TM). Sub-Earth and sub-solar longitudes are given in the text. Wavelengths of TM are between 12 and 12.7 μm. Locations (spanning tens of degrees of longitude and equatorial latitude) are estimated from simple thermal models and centered at: from top to bottom, ~280°, 275°, 180°, 170°, 350°, and 215°E. Spectra are from Cooper et al. (2001)

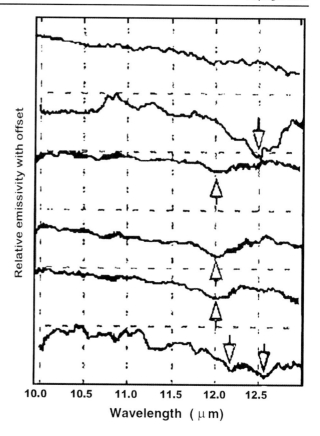

some spectral features appear to be repeated, more observations and modeling are required to make compositional identifications.

Figure 8 shows spectra exhibiting transparency minima (TM) measured from the Mc-Math Pierce Solar telescope on Kitt Peak. A circular aperture was placed over the entire Earth-facing disk of Mercury on six different days. The spectral signature comes from the hottest regions in the field of view. For the spectra shown (top to bottom in the figure), sub-solar longitudes are 280°, 275°, 180°, 170°, 350°, and 215°E, respectively. The sub-Earth longitudes on Mercury for the days of observation are: 350°, 345°, 104°, 94°, 280°, 131°E, respectively, and are the E longitude equivalent of the W longitudes given by Cooper et al. (2001). Wavelengths of the TM are between 12 and 12.7 μm. A simple model gives an estimate of the longitude responsible for the greatest flux at the detector. Spectra from ~180°, 170°, and 350°E longitude (3rd, 4th, and 5th from top) have probable transparency minima at 12 μm. The bulk composition associated with a transparency feature at this wavelength is intermediate to basic (45–57 wt%SiO_2). The spectrum from ~275°E (2nd from top) has a TM at 12.5 μm indicative of about 44 wt %SiO_2 or an ultra-basic composition. The spectrum (6th from the top) from ~215°E longitude has a doublet TM indicating two different dominant components in the regolith, one mafic and one ultra-mafic. For comparison, the Mercury spectrum from 240° to 250°E longitude (Fig. 3) has a clear and strong TM at 12.3 μm that is at the same location as the TM in a laboratory spectrum of labradorite powders.

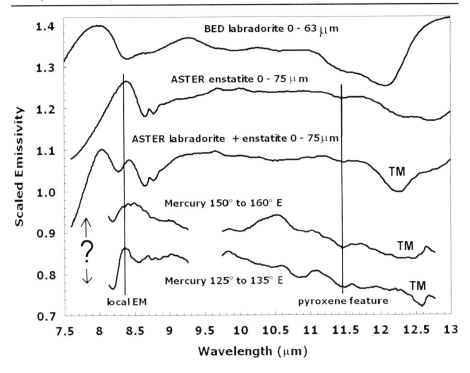

Fig. 9 Three laboratory spectra are plotted along with two spectra from Mercury's surface on the same wavelength scale. All spectra have been offset to distinguish similarities and differences. The ? indicates no data are available. From top to bottom: true emissivity spectrum of fine-grained (0–63 μm) labradorite (Ab_{50} to Ab_{30}) from the BED, enstatite.1F ($MgSiO_3$) reflectance spectrum, converted to emissivity using Kirchoff's relation, $E = 1 - R$ (Salisbury et al. 1991 and the ASTER data base; grain size 0 to 75 μm), model spectrum computed by linear mixing enstatite.1F ($MgSiO_3$) and labradorite.1F (Ab_{50} to Ab_{30}); grain size 0 to 75 μm, Mercury: region from Caloris Basin, Mercury: region west of Caloris Basin

Laboratory spectra of plagioclase feldspar (labradorite), pyroxene (enstatite), and a mixture of albite and enstatite are shown in Fig. 9 along with two Mercury spectra obtained April 7, 2006, at the NASA Infrared Telescope Facility (IRTF) using the Mid-InfraRed Spectrometer and Imager (MIRSI). The laboratory spectra span from 7.5 to 13.0 μm and thus exhibit the principal EM characteristic of feldspars between 7.5 and 8 μm. The Mercury spectra from Caloris Basin (150° to 160°E) and somewhat to the west of Caloris Basin (125° to 135°E) do not cover the 7.5 to 8.2 μm because MIRSI is not currently configured to permit measurements in that spectral region. Spectral features indicative of Mg-rich orthopyroxene (enstatite) are observed. Very suggestive are the local EM near 8.35 μm, a small emittance peak at 8.72 μm, and a local minimum centered at 11.45 μm as shown in Fig. 9.

Mafic spectral features are indicative of rocks that include olivine and pyroxene in their chemistry or are under-saturated in SiO_2 (feldspathoids). Spectral features indicative of clinopyroxene are apparent in ground-based spectra as discussed previously and shown in Figs. 1 (near-IR), 6, and 7. Orthopyroxene (enstatite) is indicated at other longitudes (Fig. 9). Olivine $(Mg,Fe)_2SiO_4$ forms a continuous solid solution of two end member components, Mg_2SiO_4 (forsterite) and Fe_2SiO_4 (fayalite). The low-iron end-member, Mg-rich variety, is a good candidate for Mercury because rocks of this type would not exhibit a strong FeO absorption band in their near-IR spectra. Picrite and dunite are olivine-rich, low-SiO_2 rocks. Tephrites have little olivine while basanites may have considerable olivine but

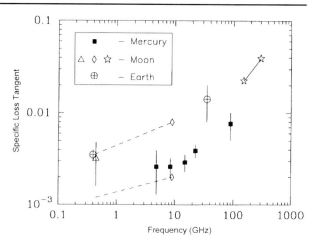

Fig. 10 Ground-based microwave imaging and modeling determined the specific loss tangent of materials in Mercury's regolith and found them to be systematically lower than those of the Moon and a suite of terrestrial basalts (Mitchell and de Pater 1994). *Solid and dashed lines* connect measurements made of the same sample at both 45 MHz and 9 GHz. The inference is that Ti and Fe are not as plentiful on Mercury's surface as on the Moon and that Mercury is not likely to have iron- or titanium-bearing basalt on its surface

both have $Na_2O + K_2O$ of 3 to 9 wt%. Feldspathoids like nepheline syenite are low in SiO_2 (<41 wt%) and have been suggested for Mercury on the basis of trends in some spectral data (cf. Sprague et al. 1994; Jeanloz et al. 1995 and Emery et al. 1998).

On the Moon, lava flows are mostly Fe- and/or Ti-bearing basalts. On Mercury there is little evidence for the Fe-bearing basalts or TiO_2 in significant abundance. Microwave observations and near-IR spectral modeling and comparisons to the Moon appear to rule them both out. In at least two cases the smooth plains overlie material that is bluer (higher UV/orange ratio) and enriched in opaque minerals relative to the hemispheric average (Robinson and Lucey 1997). An estimate of the FeO content of the mantle can be made as follows. Since the smooth plains are probably lava flows, and the FeO solid/liquid distribution coefficient is about 1 during partial melting, it is estimated that Mercury's mantle has a FeO content similar to lava flows and is <3%.

4 Microwave Emissivity of Mercury's Surface

The previous section shows how visible and near infrared spectral studies seem to indicate that there is no or little evidence for the Fe^{2+} charge exchange absorption band in Mercury's regolith. These observations do not preclude iron in nanophase particles or in sulfides. More ground-based observations may yet find some evidence for these. However, detailed measurements of Mercury's regolith from 0.3 to 20.5 cm wavelength indicate that Mercury's surface is more transparent to electromagnetic radiation at these wavelengths than the lunar surface and a suite of terrestrial basalts. These measurements demonstrated that Mercury's surface materials differ from those of the Moon by being significantly lower in materials opaque to electromagnetic radiation such as FeO and TiO_2, and the rocks and minerals containing them (Mitchell and de Pater 1994). This relationship is illustrated in Fig. 10, where the specific loss tangents as a function of frequency of Mercury's regolith are shown compared to those from samples of lunar maria, highlands, and terrestrial basalts.

5 What is the Expected Crustal Composition of Mercury?

The small number of papers that have been able to model the composition of Mercury (Mitchell and de Pater 1994; Sprague et al. 1994; Blewett et al. 1997; Emery et al. 1998;

Table 1 Oxide abundances (wt%) of bulk Mercury according to formation model

MC	ERR	Preferred	EVR	OC
Al_2O_3	16.6	3.5–7	3.3	
CaO	15.2	2.5–7	3.0	Some (in exosphere)
TiO_2	0.7	0.15–0.3	0.14	$\ll 1$ (>0)
MgO	34.6	32–38	32.1	
SiO_2	32.6	38–48	45.0	39–57
FeO	0	0.5–5	15.1	<3 (0.5–2)
Na_2O	0	0.2–1	1.4	Much? (in exosphere)
H_2O	0	Little	Much	Buried at poles? (late meteoritic)

MC = mantle crust system (bulk silicate fraction),
ERR = extremely refractory-rich,
EVR = extremely volatile-rich,
OC = observational constraints on oxide abundances

Sprague and Roush 1998; Sprague et al. 2002; Warell and Blewett 2004; Warell et al. 2006) give only rough bounds on the surface chemistry. Mid-infrared thermal emittance spectroscopy is particularly important with its potential to identify the presence of a range of rock types, specifically those which are iron-poor or iron-free which and are altered by space weathering processes.

5.1 Our Best Guess Models

The summary of present abundance estimates are not yet adequate to constrain Mercury's formation model. The details are given in Table 1 where the bulk content of oxide abundances in wt% for Mercury are shown according to formation model. The first four columns are from Goettel (1988). Observed abundance data are taken from Warell and Blewett (2004), Cooper et al. (2001), Sprague et al. (1998, 2002), Mitchell and de Pater (1994), Blewett et al. (1997), Moses et al. (1999), and Butler et al. 1993.

The extreme end-member models, with high abundances of either refractories or volatiles, are unlikely to have occurred because the circumstances leading to such compositions are highly restrictive. In the preferred model of Goettel (1988) the pure refractories (Al, Ca, Ti oxides) are fully condensed and thus not fractionated with respect to each other. They are moderately enriched above solar (chondritic) values because of accretion within the planetary feeding zone of some material in which Mg silicates were only partially condensed. The magnesium to silica ratio Mg/Si is greater than solar as Mg was enriched relative to pure refractories and thus depleted in Si. Most material, however, accreted from the region where both Mg and Si were fully condensed, and thus Mg/Si is close to the solar value. Significant amounts of (moderately) volatile components (alkalis, FeO, FeS, water) are predicted and are strongly dependent on the extent of the feeding zone. It is postulated that the Fe/Si bulk value of Mercury was increased following initial planetesimal accretion and differentiation by processes operating independently of composition of silicate component. Such processes might be aerodynamic fractionation (in the planetary nebula, larger, denser material is more subject to drag forces than smaller, less dense material) or giant impact(s).

The oxide abundances indicated by the available observations conform to those predicted by the preferred model, except perhaps for the high range of Si inferred from the measurements. However, none of the suggested formation models can be singled out as more or less likely. In terms of FeO, which is the only oxide for which the abundance has been possible to model based on VISNIR spectra, a bulk weight fraction of 1–2% is consistent with models predicting wide accretion zones (Goettel 1988; Lewis 1988; Wetherill 1988) and models with fractionation processes linked to preferential accretion of metal (0.5–6 wt%; Weidenschilling 1978). The value is greater than or equal to those predicted from the vapor fractionation model (0–3 wt%; Fegley and Cameron 1987), and compatible with mantle-stripping by late-stage giant impact (Benz et al. 1988). Not even the pure equilibrium condensation model (Lewis 1972) predicting zero FeO can presently be discarded on observational grounds, as the native crustal composition may have been fully iron-free and the iron contributed by chondritic infall.

Given these caveats, Mercury's surface appears to be heterogeneous in composition, trending toward feldspar with moderate Ca and Na. There may be feldspathoidal-rich terrains. Some mafic and ultra-mafic minerals such as clinopyroxene have been indicated by spectroscopy and olivine is likely present, but of the Mg-rich, low-Fe variety. None of the observations so far contradict the scenario of early core formation accompanied by contraction of the planet and retention of internal heat through closing off magma vents to the deep mantle (Solomon 1977; Jeanloz et al. 1995). Magma from the deep mantle would presumably be Fe-rich, which is contradictory to the spectral evidence from Mercury's surface.

Based on morphological and spectral observations, extrusive lavas are probably on the surface (Robinson and Taylor 2001; Strom and Sprague 2003), but it appears they may trend toward feldspathoids, alkali-rich alumino silicates with low (39–45 weight%) SiO_2 like tephrite or basanite (Jeanloz et al. 1995; Emery et al. 1998) and be enriched in Na, and possibly K. Such rocks are formed from plagioclase bearing lavas in which feldspathoids are present in greater abundance than 10 wt%.

A relatively large abundance of K in Mercury's exosphere has been reported over the longitude of Caloris Basin roughly in the mid-latitudes (Sprague et al. 1990). The K could be from impact melt associated with the impact that formed the basin, or from an episode of low temperature extrusive volcanism induced by the Caloris impact event. A well-developed fracture system in the basin may have K-rich minerals and serve to conduit K to the exosphere. This speculation requires testing by MESSENGER and Bepi Colombo.

Expansive regions near Homer and the Murasaki Crater Complex (315° to 330°E) appear feldspathic (Robinson and Taylor 2001), trending toward Ab70–Ab40, more Na-rich than the lunar anorthosites (Sprague et al. 1994). Bulk compositions are of intermediate silica content. Some mixed compositions of more basic silica content are present in equatorial longitudes from 200° to 292°E. Two candidates for the rocks at these locations are Na-rich anorthosites. Another possibility is low-Fe basalt, a rock with mixtures dominated by feldspar, minor pyroxene and olivine with a bulk rock chemistry of intermediate (52–57 wt% SiO_2) or mafic (46–51 wt% SiO_2) composition.

6 Augmenting Current Knowledge with Instrumentation on MESSSENGER and Bepi Colombo

The prospects of determining the bulk crustal composition and mode of formation will be very difficult or impossible with ground-based data. These issues will, we hope, be resolved with close-range spectroscopy and imaging with the upcoming orbiters, with information from the smallest resolvable scales (<10 m) playing a critical role in determining the composition of individual or clustered crystalline rocks or rock outcrops. Almost all of the observations have also raised new questions to be answered by the suite of instruments on the

MESSENGER spacecraft, scheduled to begin mapping Mercury's surface in 2011 (Solomon et al. 2001; Gold et al. 2001; Santo et al. 2001).

The ESA mission, Bepi Colombo, scheduled for launch in 2011, has instrumentation ideally suited to making surface composition measurements. If all goes as planned, those measurements will test the conclusions that have been presented above in this section. The MERTIS instrument is an IR-imaging spectrometer based on the pushbroom principle (Benkhoff et al. 2006). MERTIS uses a micro-bolometer detector which requires no cryogenic cooling and which makes it ideally suited for the hot environment around Mercury. MERTIS has an integrated instruments approach which will allow including a so-called μ-radiometer by sharing the optical entrance path, instrument electronics, and in-flight calibration components. The μ-radiometer uses highly miniaturized thermopile detectors and will be placed at the slit of the spectrometer.

MERTIS covers the spectral range from 7–14 μm at a high spectral resolution of up to 90 nm which can be adapted even along an orbital track, depending on the actual surface properties to optimize the S/N ratio. With this spectral coverage and resolution MERTIS will be able to detect the main features in this spectral region, such as the Christiansen wavelengths (emissivity maxima), the emissivity minima in the Reststrahlen region and the transparency features (Helbert et al. 2007).

MERTIS will globally map the planet with a spatial resolution of 500 m and a S/N of at least 100. For a typical dayside observation the S/N ratio will exceed 1,000 even for a fine-grained and partly glassy regolith. MERTIS will map 5–10% of the surface with a spatial resolution higher than 500 m. The flexibility of the instrumental setup will allow study of the composition of the radar bright polar deposits with a S/N ratio of >50 for an assumed surface temperature of 200 K. Depending on the exact orbit and therefore the distance from Mercury's polar regions, the spatial resolution could be as good as 1–2 km. In addition, by integration of the μ-radiometer MERTIS will be able to measure thermophysical properties of the surface like thermal inertia and internal heat flux and derive from this further information on surface texture and structure.

Another instrument that will be important for the study of Mercury's surface is the stereo camera (a channel of SIMBIOSYS—one of the payload instruments on board the Mercury Planetary Orbiter) of Bepi Colombo. It will perform 3-D global mapping of Mercury's entire surface and make multi-colored maps with a scale factor of 50 m/pixel at the periherm (400 km above the surface). It will allow the cataloging of the Digital Terrain Model of the entire surface and dramatically improve the interpretation of morphological features at different scales and topographic relationships.

The harsh environment of Mercury will strongly affect the functionalities and performance of the instruments, and even for the stereo camera we had to find a new solution and a new technique of acquiring the stereo pairs for generating the Digital Terrain Model of the surface. The instrument concept is based on an original optical design composed by two channels, looking at the surface at $\pm 20°$ from the nadir direction, converging on the same bidimensional focal plane assembly, with no mechanism (Da Deppo et al. 2006). The configuration of the focal plane assembly allows to apply the push-frame technique to acquire the stereo images, instead of the push-broom usually used by other planetary and terrestrial stereo imagers.

7 Space Weathering and Surface Maturation

Although the remote-sensing spectroscopic studies of Mercury are important, the results must be viewed in context of the exceptionally dynamic environment of the inner solar system. Space weathering processes and the contribution of exogenic chondritic material (see, e.g., Hapke 2001, 2007, this chapter; Langevin and Arnold 1977; Langevin 1997; Lewis 1988; Cintala 1992; Noble and Pieters 2003) strongly affect the chemistry and observed properties of the mercurian surface. Thus, no interpretation of the composition of Mercury's true crustal composition can be made without thoroughly accounting for space weathering, maturation and exogenic deposition on its surface.

According to the general maturation model, the VISNIR spectral properties of the regolith of an atmosphereless mafic silicate body are governed by three major components (Hapke et al. 1975; Rava and Hapke 1987; Hapke 2001): (1) ferrous iron as FeO in mafic minerals and glasses, increased abundance increases the depth of the near-infrared Fe^{2+} crystal field absorption band; (2) nm-sized $npFe^0$ particles formed by vapor deposition reduction in rims of mafic grains, increased abundance decreases the reflectance and increases the spectral slope ("reddening"); (3) spectrally neutral Ti-rich opaque phases such as ilmenite in minerals and glasses, increased abundance decreases the reflectance but decreases the spectral slope. The mean grain size is also a critical parameter for interpreting surface spectra, as band depths and band shapes strongly depend on this parameter for a given mineralogical composition. Nanophase metallic iron ($npFe^0$) is formed during the maturation process due to micrometeorite impacts and solar wind sputtering of the surface. The process is caused by condensation of Fe-bearing silicate vapors in silicate grain rims by selective loss of oxygen (Hapke et al. 1975; Hapke 2001 and references therein). Very small abundances of metallic iron as $npFe^0$ have drastic effects on VISNIR spectral shape and reflectance and must be understood and calibrated out of observed spectra in order to derive the composition of the unweathered material.

As discussed by Langevin and Arnold (1977) for the lunar regolith, maturity on an airless body such as the Moon or Mercury results from a competition between macroscopic gardening, which brings fresh material from the bedrock to the surface, and the space weathering processes which occur at or very close to the surface. The thickness of the debris layer (or regolith) grows as a function of time as new material is excavated. Over the last 4 billion years, the impact rate is expected to have remained stable, as it results from a steady state transfer of impactors close to resonances in the main belt (see e.g. Marchi et al. 2005). On the Moon, the mean regolith accumulation rate during this period has been in the range of 1 m/by, as demonstrated by the radiation exposure ages of deep-drill lunar cores. Given the stochastic nature of meteoroid bombardment, a soil sample can stay exposed at the surface for millions of years, then be covered by a new ejecta blanket, possibly to be exposed again several hundred million years later by an impact in the regolith. Nevertheless, the mean amount of time a given volume element of regolith spends within 1 mm of the surface on the Moon has to be close to 1 my. This mean residence time is controlled by the flux and velocity of large impactors which are able to dig down into the bedrock (size range of 1 m or more).

Freshly excavated ("immature") material is initially coarse grained (from the mm range to rocks, even m-sized boulders) and dominated by crystalline grains. A wide variety of space weathering processes modify the size distribution, mineralogy, and optical properties of the constituents of the regolith. A direct micrometeoroid impact can result either in a microcrater (if the grain is large) or in fragmentation. Conversely, macroscopic and microscopic impacts generate fused silicate droplets which can either solidify in flight (forming glass spherules) or weld together crystalline grains (forming glassy agglutinates). Eventually, a steady state grain size distribution is reached with a median size of $\sim$50 μm. The fraction of glassy material can reach 50% or more in highly mature lunar soil samples which have accumulated a lot of surface residence time. Nanophase metallic iron is formed during the maturation process due to micrometeorite impacts and solar wind sputtering of the surface. Other maturity effects such as the formation of particle tracks within the grains due to heavy ions in cosmic rays play a relatively minor role for optical and spectral properties.

Therefore, the most important space weathering parameters are the micrometeoroid flux (masses <1 mg) and velocity (which controls the production of glass and the fragmentation efficiency) and the flux of low-energy ions (solar wind or magnetospheric precipitation). When trying to assess the maturity state on Mercury, one has to assess these space-weathering factors as compared to the mean residence time of the surface which is controlled by large scale impacts. The well-documented situation on the Moon can then be considered as a reference case. This can be done on much firmer ground than in 1996 (date of the first Mercury workshop in London) as there are now reliable models of the relative impact rates on inner planets for various populations of meteoroids injected by resonances in the inner solar system. Marchi et al. (2005) concluded that the flux of large impactors on Mercury is one-tenth of that on the Earth. If one considers gravitational focusing, which enhances the impact rate on the Earth by a factor of 1.4, and the factor of 7 between the surfaces, the flux of large impactors per surface element, which controls regolith formation, is expected to have been similar on Mercury and on the Moon during the last 4 billion years. According to Marchi et al. (2005), there is a significant difference in terms of mean impact velocity (30 km/s on Mercury, 20 km s^{-1} at 1 AU). This will result in a cratering efficiency increase by a factor of 1.5 to 2. The mean rate of regolith formation on Mercury could therefore range from 1.5 m per by to 2 m per by.

One can expect a much higher flux of low-energy ions at 0.3 AU when compared to the Moon at 1 AU, even if the situation is made more complex by the shielding effect of Mercury's magnetosphere. The radial dependence of micrometeoroid fluxes is different from that of large-sized meteoroids, as the latter are dominated by asteroidal contributions while the small size range is much more complex (e.g. Divine 1993), with a large contribution of cometary material and zodiacal dust. From the evaluations of Müller et al. (2002), the dust flux near Mercury is enhanced by a factor of 1.5 to 3 when compared to that on the Moon. The impact velocity of the cometary component scale with the local orbital velocity, hence the available energy per dust impact is increased by a factor of $\sim$2.5 when compared to the lunar situation, with wide variations as a function of the true anomaly. The space-weathering efficiency of micrometeroids should therefore be 4 to 7.5 times higher on Mercury than on the Moon. The comparison between this factor and the factor of 1.5 to 2 between regolith accumulation rates suggests that the mean integrated maturity indices in the regolith of Mercury are at least twice higher than it would be in the lunar regolith. These new evaluations are therefore consistent with the conclusions of Langevin (1997).

Unlike at the Moon, the situation on Mercury is by no means equivalent for all surface elements. The solar wind sputtering flux is controlled by the 3:2 spin–orbit coupling, which causes an uneven irradiation pattern which is further modulated by magnetospheric screening and precipitation. Minor effects are implantation of solar wind species (C, N, H), and major effects include amorphication and/or chemical reduction of the outer rims of grains. The solar photon flux is also controlled by the spin–orbit coupling, and at the highest surface temperatures the effect is thermal annealing of radiation damage effects and an increase in npFe^0 particle size because of Ostwald ripening. Furthermore, impact velocities and rates are expected to strongly depend on true anomaly. Orbiter missions to Mercury may well reveal variations of the mean maturity indices between the two antipodal longitudes which are subsolar at perihelion and the intermediate longitudes which are subsolar at aphelion.

8 Exogenic Material from Comets, Meteorites and Interplanetary Dust Particles

8.1 Sungrazing Comets

Sungrazing comets have been observed for many hundreds of years. In the late 1880s and early 1890s, Heinrich Kreutz studied the possible sungrazing comets which had been observed and determined that some were sungrazers and some were not. He also found that those which were indeed sungrazers all followed the same orbit. That is, they were all fragments of a single comet which had broken up. It is probable that the original comet, and its fragments, have broken up repeatedly as they orbit the sun for a period of about 800 years. In honor of his work, this group of comets is named the *Kreutz sungrazers*. The Kreutz sungrazers come within about 50,000 km (perihelion distance of 0.005 AU) of the solar surface.

Beginning in 1979, coronagraphic observations from space allowed the detection of numerous additional Kreutz members that were completely vaporized as they grazed the Sun. Since 1996, the Solar and Heliospheric Observatory (SOHO) coronagraphs have revealed some 700 of these comets. This suggests that there is a constant stream of small members and that the break up occurred quite recently near aphelion. Break-up near perihelion requires that the observed dispersion of the orbital parameters would take many millennia. However, recent calculations have shown that the evolution can be substantially sped up by allowing fragments to be rotationally spun off at heliocentric distances of many tens of AU. Since the turn on of the LASCO instrument on December 30, 1995, through September, 2006, it has discovered 1,185 new comets of which about 8% belong to the Kreutz sungrazing group. The majority of the remaining comets belong to three new groups—the Meyer (5%), Marsden (2%), and Kracht (3%) groups—that were declared based solely on LASCO comet observations.

All the observed comets have clearly been very small, and the data suggest that few survived perihelion passage in any coherent fashion. The brightness distributions of the Kreutz comets detected with SOHO/LASCO indicate an increasing number of comets with decreasing size and the size distribution of nuclei is probably described with a power law. Based on a much smaller sample, observed from the end of 1996 until the end of 1998, Sekanina (2003) gave an apparition rate of 0.6 per Earth day. Then he estimated from his model that the initial diameters of SOHO sungrazers range from 17 to 200 m yielding the total mass of incoming comets of $M = 2 \times 10^{12}$ g arrived during the ~ 2 Earth years long interval, equivalent to 3.1×10^4 g s^{-1}.

Mann et al. (2005) tried to estimate the dust supply from the sungrazers assuming a spherical comet of 20-m radius fragmented into 10 µm spherical particles and distributed in a sphere of 10 solar radii. They found a number density of 10–17 cm^{-3}, which is below typical densities of 10–14 cm^{-3} for particles of this size range. They assumed this large size of dust grains, since for size distributions that are similar to that in the interplanetary dust cloud, the majority of mass would be contained in fragments of this size. Analysis of the dust tails of sungrazing comets shows that the sungrazers emit small particles of sizes $a = 0.1$ µm and that the dust in the sungrazing comets has a narrow size distribution (Sekanina 2000). When making this rough estimate for 0.1 µm spherical particles, the number density amounts to 10–12 cm^{-3}, which is comparable to typical dust densities in this size range. Moreover, the dust that is produced by sungrazers will quickly leave the solar corona. The Kreutz comets are in highly elliptic or hyperbolic orbits, their speed is approximately 230 km s^{-1} at 7 solar radii and can be described as bodies with initial speeds of zero at infinity that fall into the Sun. Dust grains released from the sungrazers are in similar orbits.

Furthermore, the perihelion distances and inclinations of the other three groups of sungrazers are 0.036 AU and 72.4°, 0.049 AU and 27.4°, and 0.048 AU and 13.6°, respectively, where each value is the mean weighted by the total observational time (Meyer 2003). The size of these comets is likely to be similar to the Kreutz family fragments SOHO detects, but because of the larger perihelion distances, the smallest members of the groups are probably not detected. The apparition rates of these comet groups and therefore the input to the near-solar dust cloud are clearly below those of the Kreutz group comets.

Mann et al. (2005) concluded that the dust supply from the frequently observed sungrazing comets is negligible, but they did not have a complete dynamical model of small particles released by these objects that may supply a large amount of dust. Considering that we do not have the size distribution released by the sungrazers, we cannot obtain a good estimate of the dust supply in the inner Solar System, and the particles smaller than 1 cm seems to be important for the Mercury's exosphere (Cremonese et al. 2005).

The dust supply from larger comets near the Sun can produce dust density enhancements that are comparable to the dust densities in the solar environment (Mann et al. 2000). The dust density may be raised many fold over a time span of weeks. Such relatively lower probability events are important for providing volatiles to the inner solar system.

As dynamical information on the sungrazers and their dust size distribution and production becomes better understood, we can apply this knowledge to how much dust is contributed to Mercury's surface and predict the depth to which the surface reflects more the exogenic source than the native crustal composition.

8.2 Meteoroids

In this section we report an estimate of the mass of meteoroids that has impacted the entire surface of Mercury during the last 3.8 Gy (from the end of the Late Heavy Bombardment to present). The calculations were performed using the model of Cremonese et al. (2005) and Bruno et al. (2006) in the size range of 10^{-8}–10^2 m radius. The size range has been limited to a radius of 100 m, since the differential velocity distribution of the meteoroids was determined by Marchi et al. (2005) only for 10^{-2}–10^2 m radius objects.

In the last 3.8 Gy the mass of meteoroids that has impacted the whole surface of Mercury has been 8.86×10^{18} and 2.66×10^{19} g, respectively. Only 2% of the mass comes from meteoroids with radius 10^{-2}–10^2 m. This is a rough estimate of the meteoroids' mass, obtained not considering the flux of objects with radius larger than 100 m and assuming a constant flux of the meteoroids with radius $\leq$100 m during all the 3.8 Gy. In any case, this estimate allows us to consider the variation of Mercury's surface composition owing to the contribution of meteoritic material. For the sake of simplicity, Bruno et al. (2006) assumed that their composition, for all the size range considered in their work, is equal to that of the S-type igneous asteroids (e.g., Krasinsky et al. 2002), which are the main constituents of the inner part of the Main Belt. In these rocks the olivine is the most abundant mineral (its composition varies from 0 to $\sim$30 mol.% Fe_2SiO_4), with subordinate amounts of low-Ca pyroxene, nickel-iron and troilite, and minor sodic plagioclase and diopside; the Fe contents ranges from $\sim$18 wt% to $\sim$30 wt%, the Ca content is of the order of $\sim$1–2 wt%, and the Na and K contents are $\sim$0.3–0.6 and 0.05–0.1 wt%, respectively (e.g., Dodd 1981).

Considering such meteoroids composition, Bruno et al. (2006) estimated that in the last 3.8 Gy meteoroids have supplied to Mercury: Fe = $(4.79$–$7.98) \times 10^{18}$ g, Ca = $(2.66$–$5.32) \times 10^{17}$ g, Na = 7.98×10^{16}–1.60×10^{17} g and K = $(1.33$–$2.66) \times 10^{16}$ g. It is clear that a percentage of this material has been ejected in the exosphere consequently to the meteoroid vaporization due to the impact with the planet surface.

Because there is very little evidence for iron on Mercury's surface, either as npFe⁰ or as FeO in surface rocks and minerals, the chemical model of Bruno et al. (2006) is not likely to be representative of the actual infalling material. This is an area that requires much more attention.

8.3 Delivery of Solar Material by The Solar Wind

The solar wind flux at the orbit of Mercury is much larger than at Earth orbit, typically about a factor of 10, which is of the order of 4×10^{13} ions m^{-2} s^{-1}. However, since Mercury has a sufficiently strong magnetic field it holds off the solar wind from reaching Mercury's surface most of the time. Since the Mariner 10 flybys through the magnetosphere were on the tail side, the solar wind stand-off distance had to be estimated from the particles and fields data. Early estimates based on scaling the terrestrial magnetosphere gave a stand-off distance of about 1.7 RM (Siscoe and Christopher 1975; Slavin and Holzer 1979). Later, computer modeling of Mercury's magnetosphere allowed more detailed investigations and it became clear that large fractions of Mercury's magnetosphere are open and solar wind ions can access the surface. These open areas are around the cusps and their size and exact location depend on the solar wind plasma parameters, which are mainly the speed, the density, and the magnetic field (Kabin et al. 2000; Sarantos et al. 2001; Kallio and Janhunen 2003; Massetti et al. 2003; Mura et al. 2005). These calculations give precipitating ion fluxes in the range from 108 to 109 cm^{-2} s^{-1} onto the surface at latitudes between 40° and 60° and longitudes ±60° from the subsolar point.

Figure 11 shows a map of precipitating protons calculated for four different cases of solar wind plasma parameters using a hybrid model (Kallio and Janhunen 2003). One can see large areas of proton precipitation in mid latitudes around the cusps (regions are high proton precitation are red, orange). In addition, one recognizes a band (yellow) of precipitation all around the planet arising from magnetospheric ions precipitating onto the surface (auroral precipitation). Again, from these model calculations the precipitating fluxes are in the range from 10^8 to 10^9 cm^{-2} s^{-1} for typical solar wind plasma conditions. The bottom right panel in Fig. 11 shows a situation of a plasma cloud from a coronal mass ejection (CME) interacting with Mercury's magnetosphere. Because of the much larger solar wind dynamic pressure in CME plasmas larger areas at Mercury's surface are open to interplanetary plasma, actually almost the entire dayside of the planet.

We have seen that there are large areas on Mercury's surface that get exposed to solar wind during regular solar wind conditions. Solar wind velocities are in the range of 300 to 800 km s^{-1} (slow and fast solar wind), which translates to energies of 0.5 to 3.3 keV/amu, and with a typical value of 1 keV/amu. In the solar wind, protons and alpha particles make up more than 99% of the ions, and heavy ions (from carbon to iron and up) together are about 0.1% of the solar wind ions in the number flux (Wurz 2005 and references therein). These ions and heavy ions are known to knock atoms bound in the surface regolith free and put them into Mercury's thin atmosphere. This process is called sputtering. For the lunar surface total sputter yields are about 0.07 per impinging ion for the typical mix of solar wind ions (Wurz et al. 2006). Note that the sputter yield is has a maximum around ion energies of 1 keV/amu. Heavier ions have sputter yields even larger than 1, but because their abundance in the solar wind is very low their contribution to the total sputter yield of the solar wind is negligible.

Since the total sputter yield for solar wind is significantly below 1 there is more material transported to Mercury's top-most surface by the solar wind than is sputtered away, and Mercury's surface is expected to be saturated with hydrogen and helium. However, for heavy

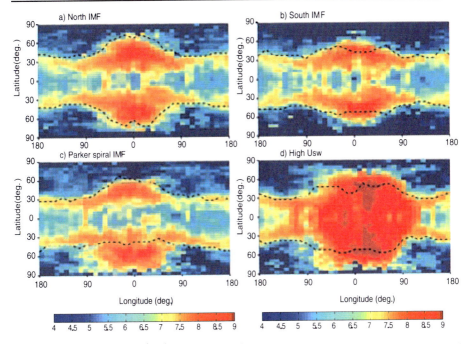

Fig. 11 The particle flux (cm^{-2} s^{-1}) of impacting H$^+$ ions (from Kallio and Janhunen 2003). The *dashed lines* show the open/closed field line boundary. The sub-solar point is in the *center of each figure*. The cases are: (**a**) pure northward IMF, (**b**) pure southward IMF, (**c**) Parker spiral IMF, and (**d**) high solar wind speed case (CME type)

atoms (from carbon to iron and up), the removal via sputtering exceeds the input from solar wind by 1 to 3 orders of magnitude depending on species, because the abundance of heavy ions in the solar wind is very low. Since Mercury's gravitational field is low, most of the sputtered atoms escape because their typical ejection speed exceeds Mercury's escape speed of 4250 m s^{-1}.

Koehn and Sprague (2006) argued that Mercury's surface is likely to be implanted with O and Ca from the solar wind. Using line-of-sight (LOS) O values from (Broadfoot et al. 1976) and an upwardly revised value (Hunten et al. 1988) of 1.0×10^{11} O atoms cm^{-2}, and a rough average of measured LOS Ca abundance (Bida et al. 2000; Killen et al. 2007) of 2×10^8 Ca atoms cm^{-2} gives a Ca/O ratio of 2×10^{-3} with a factor of 2 uncertainty owing to the O measurement and variability of several factors in the Ca measurement. The ratio of Ca/O in the solar wind is $1-4 \times 10^{-3}$ (cf. Wurz et al. 2003, and references therein). The similarity of the Ca/O ratio above the limb of Mercury to the Ca/O ratio in the solar wind compels the serious consideration that a primary source of Ca and O in Mercury's exosphere is the solar wind. If this is true, then spectroscopic measurements of Mercury's surface may be biased by the solar Ca and O abundances. However, within the error bars, using an exospheric surface density for oxygen of $N_O = 4.4 \times 10^4$ cm^{-3} (Hunten et al. 1988), which gives a column density of $N_C = 2.6 \times 10^{11}$ cm^{-2} and a model for thermal release by Wurz and Lammer (2003) and an average of the observed Ca abundance of $N_C = 1.1 \times 10^8$ cm^{-2} (Bida et al. 2000) the solar wind Ca/O ratio is at least a factor of ten larger than Mercury's. More observations of the exosphere are required to determine which scenario is correct.

The main loss process for exospheric atoms, whether they are native to Mercury's crust of recycled from exogenic material, is ionization. Once ionized, exospheric constituents

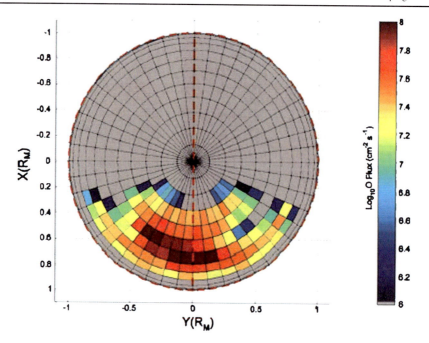

Fig. 12 Color-coded O-ENA flux generated by ion-sputtering on the northern surface of Mercury. Figure is from Mura et al. (2005)

become part of the magnetospheric system of Mercury and some of them are accelerated back to Mercury's surface with energies in the 10 keV range (Delcourt et al. 2003). Preferred places for the precipitation of ions are two belts at $+30°$ and $-30°$ latitude around the planet (the auroral precipitation region; Fig. 11) and the day-night perimeter of the planet (Delcourt et al. 2003). Na^+ fluxes of the order 10^6 cm^{-2} s^{-1} have been calculated for the auroral precipitation region. At these locations one expects enhancements of the top surface in Na and K (Sprague 1992a, 1992b).

Because the amount of exogenically contributed material is not well quantified, it is not possible to determine with certainty if the Ca-rich signatures in Mercury near-IR and mid-IR spectra are native or from a largely exogenic veneer. The search and discovery of true crustal sources of the elements in the exosphere by MESSENGER and Bepi Colombo will help to understand the surface remote sensing data.

9 Exospheric Clues to Surface Composition

9.1 Mercury'S Upper Surface

Mercury's upper surface is probably composed of: (i) interplanetary materials that is, by similarity with Earth's case, meteoritic material, interplanetary dust, and cometary materials; (ii) material originating in the planet's bulk. The relative abundance of these two types of material is unknown. On the basis of existing models, the fraction of meteoritic material is estimated to be in the range from 5–20% to 100% (Bentley et al. 2005) (virtually no release of regolith erosion products in this case). In the second hypothesis, the composition

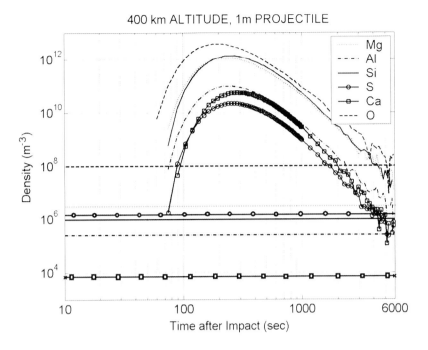

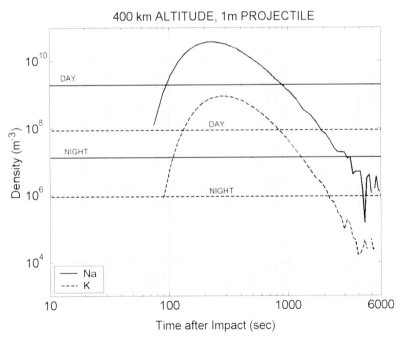

Fig. 13 Density versus time for an impacting object of 1 m radius, at 400 km altitude for the species whose mean density value does not change between day- and night-time (*top*); separately, for Na and K (*bottom*). Horizontal lines represent the exospheric background for each species according to the exospheric model of Wurz and Lammer (2003). Figure is from Mangano et al. (2007)

of Mercury's exosphere reflects the chemical composition of meteorites impacting Mercury, possibly mixed with solar-wind-added products, and no genetic link between regolith and exosphere exists. If only a small fraction of Mercury's exosphere is of meteoritic origin, the remaining comes from the regolith, more precisely from an upper superficial layer, in equilibrium with the exosphere (Killen et al. 2004). The composition of the upper superficial layer, eroded by energetic particles and radiation fluxes, probably significantly differs from the composition of endogenous layers, due to the different extraction rates of the species of different natures, undergoing various extraction mechanisms (Leblanc and Johnson 2003). In a steady state, the net escape fluxes of species at the top of the exosphere are equal to their net extraction fluxes at the surface, which reflect the chemical composition of the unperturbed subjacent regolith. Therefore, the composition of the bulk regolith should be better derived from the total escape rates of atoms and ions (balanced between inflow and outflow in the planet environment).

9.2 Remote Sensing of the Surface Composition Through Exospheric Fluxes

Generally, due to the strong link between the exosphere and the surface, by measuring neutrals and ions at relatively low altitudes it should be possible to get information on the upper surface composition. The subject of exospheric sources and sinks is discussed in detail in "Processes That Promote and Deplete the Exosphere of Mercury" by Killen et al. (2007, this volume). Therefore, we take only a brief visit to this topic.

Some release processes are non-stoichiometric, hence involving only selected species. For example, thermal evaporation and photo-sputtering are more effective for volatiles (like H, He, Na, K, S, Ar, OH). Conversely, ion-sputtering (Killen et al. 2001; Lammer et al. 2003) and impact vaporization (Cintala 1992; Gerasimov et al. 1998) are relatively stoichiometric in releasing species from the surface (also O, Ca, Mg, Si, and other refractory species). Nonetheless, the release efficiency, also for these more stoichiometric processes, depends on the mineralogy (the binding energy of the released atoms), and the altitude profile depends on mass and initial velocity distribution of each species. Therefore, in order to be able to infer a source composition from the in situ exospheric measurements, it is really important to know the mechanism of ejection and the surface properties. Hence, taking into account the effectiveness of the process in ejecting material, information on the surface composition can be deduced; conversely, if we know the upper surface composition, by observing the exosphere composition and the density altitude profiles, information of the active release process can be obtained.

From previous considerations, it follows that the processes that could provide more information on the surface composition are ion-sputtering and micro-meteoroid or meteoroid impact vaporization, since they are able to release the majority of the surface species.

The neutrals released via ion-sputtering can be discriminated from neutral populations released by other processes thanks to their specific characteristics. In fact, this process is highly dependent on local plasma precipitation rate, so that it is confined to a limited area and strongly variable (Kabin et al. 2000; Kallio and Janhunen 2003; Massetti et al. 2003; Mura et al. 2005). Furthermore, the released species are not only volatiles, as mentioned before. Since this process has a wide energy spectrum, that is, it has the bulk of neutral emission in the few eV range, with a not-negligible high-energy tail (hundreds of eV; (Sigmund 1969; Sieveka and Johnson 1984), these neutral atoms can be detected at both lower and higher energies. Finally, the altitude profile related to this process has a higher scale height with respect to other release processes (Mura et al. 2006). Since ion-sputtering also produces a release of charged particles from the surface, the detection of these planetary ions, associated with neutral particle emission, could be a signature of the efficiency of this process.

The micrometeoroid impact vaporization is expected to be the most efficient process in the night side and the only process constantly active over the whole surface of the planet (Killen et al. 2001; Wurz and Lammer 2003). In fact, other generally significant processes, such as ion- and photo-sputtering and thermal desorption, are mainly active in specific regions of the dayside. Anyway, any in situ detection of refractories at energy of a few eV with no signal at higher energies could be the signature of micrometeoroid impact vaporization.

Recently, Mangano et al. (2007) demonstrated that individual impact vaporization events of large meteoroids (about 10 cm or more in radius) on Mercury's surface have a high probability to be observed as elemental enhancements in the planet's exosphere during the Bepi Colombo mission. The material involved by the impact is vaporized within a volume proportional to the meteoroid mass (Cintala 1992). Hence, the analysis of such signals, taking into account the physics of the process and the particle transport in the exosphere, would provide crucial information of the planetary endogenous regolith composition, below the more superficial layers highly modified by space weathering action. In particular, vaporization of >1 m meteoroid impacts (with a frequency of occurrence comparable with the Bepi Colombo mission lifetime) could enrich the exospheric composition with endogenous material released from deeper layers (down to some meters, depending on density and porosity of the regolith). This possible detection could be the only way to obtain remote sensing information about the original composition of the planet surface.

9.3 Is There Enhanced Sodium in the Bright Visible and Radar Albedo Regions Known as Spots A and B?

A curious coincidence of enhanced exospheric Na has been observed at high radar and visible albedo regions (Fig. 14) centered at 25°S, 5°E and 55°N, 5°E, respectively. These have been labeled as spot B (north) and spot A (south) according to the scheme of Harmon (1997). These regions appear as rough in radar backscatter images (cf. Harmon et al. 2007) and presumably are relatively freshly excavated craters. Sprague et al. (1998) discussed several exospheric Na observations of enhanced column abundance over these two regions and Sprague and Massey (2006) showed other ground-based images. The exospheric connection is discussed further in this volume in the chapter by Killen et al. It may be that the crustal materials of Mercury are enhanced in these regions or it may be that the surface materials are enhanced by magnetospheric focusing onto these locations (cf. Potter and Morgan 1990). This puzzle will most likely not be solved until the MESSENGER and Bepi Colombo missions can acquire the data necessary to understand this phenomenon.

10 Radar Observations of Depolarized High Albedo Polar Deposits

The discovery of regions of depolarized radar signals backscattered from deposits near Mercury's north and south polar regions as shown in Fig. 15 (Slade et al. 1992; Harmon and Slade 1992; Butler et al. 1993; Harmon et al. 1994) astounded the planetary science community because the character of the material is similar to the radar backscatter signature from water ice at Mars' summer southern pole and the Galilean satellites. Thermal models (Paige et al. 1992; Butler et al. 1993; Vasavada et al. 1999) and Monte Carlo models of volatile distribution and storage in permanently shadowed regions (Butler 1997; Moses et al. 1999; Crider and Killen 2005) have shown water ice could be stable for a few million years following deposition even with Mercury's proximity to the Sun. It is likely that the stored volatile (if it is a volatile like water ice) is exogenic material from comets, meteorites, and

Fig. 14 Mercury image obtained by Frank Melillo, Holtzville, NY, using a red filter and small reflecting telescope. Both radar bright *spots* B and A are clearly seen as high albedo features in this image

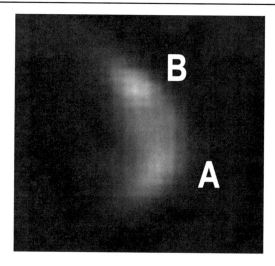

Fig. 15 High radar albedo and strong depolarization from radar backscatter indicates deposits of water ice or some other very radar-transparent material (sulfur, cold silicates, or other) in permanently shadowed craters. (**a**) north high latitudes, (**b**) south high latitudes. Figure is from Harmon et al. (1994)

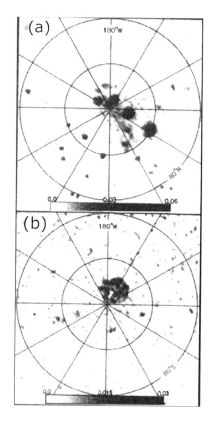

interplanetary dust particles as discussed in Sect. 8. An interesting alternative is that water molecules are formed by surface chemistry interactions with the solar wind impinging on the surface and followed by a systematic, long-time migration poleward (Potter 1995).

Subsequent, very high-resolution observations (Harmon et al. 2001) showed that the highly backscattered material is found at latitudes as low as 72°N making the water ice interpretation more difficult to explain but still possible because there are regions within craters that are permanently protected from insolation by steep walls and possible rocky debris.

The necessary physical property of any material responsible for the high albedo, depolarized radar backscatter is that it has low dielectric loss at temperatures where it is observed. Another substance with the necessary dielectric property is sulfur. Sprague et al. (1995) pointed out that sulfur is not as volatile as H_2O ice and that it should be abundant in impacting and vaporizing micrometeorites and/or may be sputtered from minerals such as sphalerite in the surface regolith. Migration and cold-trapping would naturally follow. They suggest that the discovery of S in Mercury's exosphere would be a strong indicator that S is the highly backscattered material in the permanently shadowed regions. Butler (1997) argued that the latitude range of stability for S is so great that there should be S polar caps if S is the backscattering material. Cold low-iron and low-Ti silicates also have the desired low-loss character at cold temperatures (Starukhina 2001). Because silicates are ubiquitous on Mercury's surface, the coherent backscatter may not require any volatile substance. This is one of the most exciting puzzles to be answered by MESSENGER and/or Bepi Colombo.

11 Surface Evidence for a Collisional Formation Scenario and Possible Implications

Ground-based images of Mercury in the sector 70° to 150°E longitude, a region not imaged by Mariner 10, show a large, double-rimmed basin whose inner portion extends 1,000 km across with a total dimension of the outer eroded rim slightly more than 2,000 km (extending over 55° to 115°E longitude and from 40°N to 20°S). Images of this basin can be seen in Ksanfomality et al. (2007, this volume). This basin includes and extends west and north of the dark albedo feature known as Solitudo Criophori and was unofficially called "Skinakas" after the observatory from which it was first imaged by Ksanfomality (1998). The inner basin extends from 73° to 97°E and from about 30°N to 5°S. Some regions within the basin have eroded circular rims that lack rayed structure or evidence of ejecta material (Ksanfomality and Sprague 2007 and the references therein) and apparently confirmed by recent radar imaging data (cf. Fig. 38 in Harmon et al. 2007).

The impact scenario has been studied in detail in vast numerical simulations by Benz (2007, this issue) for low relative velocities and can be invoked for explanation of the following important observations and their interpretations: the loss of a substantial part of the silicate crust (Benz et al. 1988).

Based on the impact hypothesis, we can speculate that specific properties of Mercury's surface morphology and magnetic field could be related in some way to its formation. In the course of a major collision during the accretionary period, perhaps the assymetric figure and remnant fields were formed (Ksanfomality and Sprague 2007). The measured global magnetic asymmetry of the planet is known (Connerney and Ness 1988). It is possible that future observations may reveal a global non-spherical anomaly in the gravitational field correlated with the planetary magnetic field dipole offset and the location of the basin. Any surface manifestations of this collision (roughness of the surface, composition, reflectance, color, etc.) have since been probably lost, however, because of the long time exposure to subsequent secondary bombardment processes which is thought to be in excess of 3.5 billion years (Strom and Sprague 2003).

The huge basin (called Skinakas by Ksanfomality 2002, 2003), and Basin S by Ksanfomality and Sprague 2007) is located on the short axis of the assymetric figure of the planet.

Fig. 16 The axial asymmetry of Mercury responsible for its 3:2 spin:orbit resonance is shown in exaggerated form above (adapted from Peale 1976). The shape is consistent with available data about the internal structure and inertia moments of the planet (van Hoolst et al. 2007, this issue)

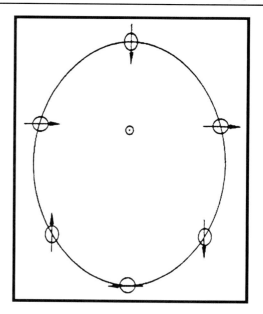

It has apparent dark-colored regions of undetermined composition and age. The hypothesis that the present-day Mercury is a result of collisions between comparable planet-sized bodies early in the history of the Solar system may be supported by the presence of this basin once its chronology is known following the MESSENGER and Bepi Colombo missions. Partial melting, fragmentation, evaporation, re-aggregation, and gravitational escape of material on such a body is conceivable and may have been responsible for the capture of Mercury in the 3:2 resonance between orbital and proper rotation periods as illustrated in Fig. 16.

12 Summary Statement

The spacecraft measurements promised by the MESSENGER and Bepi Colombo missions are badly needed to augment what is known about Mercury's surface. As we have described here, Mercury's surface is likely to have regions with rocks containing Na-rich feldspar with significant Mg-rich or Ca-rich pyroxene. FeO content is likely to be of the order of at most a couple of percent by weight on most of the surface, but may be greater locally at small scales. All evidence for npFe^0 iron is circumstantial and inferred from other planetary bodies. We do not know the composition of the material in the permanently shadowed crater regions. We do not know the extent of magmatic resurfacing or the age of surface units although cratering indicates that the original bulk surface is ancient, close to 3.5 billion years old. However, a constant addition of solar wind, meteoritic and cometary material has undoubtedly considerably altered the upper most layer.

References

J. Benkhoff, J. Helbert, the MERTIS team, Adv. Space Res. **38**(4) 647–658 (2006)
M.S. Bentley, A.J. Ball, M.D. Dyar, C.M. Pieters, I.P. Wright, J.C. Zarnecki, Lunar Planet. Sci. **XXXVI** (2005)

W. Benz, Space Sci. Rev. (2007, this issue). doi:10.1007/s11214-007-9284-1
W. Benz, W.L. Slattery, A.G.W. Cameron, Icarus **74**, 516–528 (1988)
T.A. Bida, R.M. Killen, T.H. Morgan, Nature **404**, 159–161 (2000)
D.T. Blewett, P.G. Lucey, B.R. Hawke, G.G. Ling, M.S. Robinson, Icarus **129**, 217–231 (1997)
D.T. Blewett, B.R. Hawke, P.G. Lucey, Meteorit. Planet. Sci. **37**, 1245 (2002)
A.L. Broadfoot, D.E. Shemansky, S. Kumar, Geophys. Res. Lett. **3**, 577–580 (1976)
M. Bruno, G. Cremonese, S. Marchi, Planet. Space Sci. (2006, in press)
R.G. Burns, in *Remote Geochemical Analysis: Elemental and Mineralogical Composition*, ed. by C.M. Pieters, P.A.J. Englert (Cambridge University Press, Cambridge, 1993), pp. 3–29
B. Butler, D. Muhleman, M. Slade, J. Geophys. Res. **98**, 15003–15023 (1993)
B.J. Butler, J. Geophys. Res. **102**, 19,283–19,291 (1997)
M.J. Cintala, J. Geophys. Res. **97**, 947–973 (1992)
J.E. Conel, J. Geophys. Res. **74**, 1614–1634 (1969)
B. Cooper, A. Potter, R. Killen, T. Morgan, J. Geophys. Res. **106**, 32,803–32,814 (2001)
B.L. Cooper, J.W. Salisbury, R.M. Killen, A.E. Potter, J. Geophys. Res. **107**(E4), 1-1 (2002). doi:10.1029/2000JE001462
J.E.P. Connerney, N.F. Ness, in *Mercury*, ed. by F. Vilas, C. Chapman, M. Matthews (University of Arizona Press, Tucson, 1988), pp. 429–460
G. Cremonese, M. Bruno, V. Mangano, S. Marchi, A. Mililo, Icarus **177**, 122–128 (2005)
G. Cremonese et al., Space Sci. Rev. (2007, this issue). doi:10.1007/s11214-007-9231-1
D.H. Crider, R.M. Killen, Geophys. Res. Lett. **32**(L12201), xx–xx (2005). doi:10.1029/2005GL022689
N. Da Deppo, V.G. Cremonese, E. Flamini, SPIE Proc. **6265**, 626527-1/9 (2006)
N. Divine, J. Geophys. Res. **98**(E9), 17029–17048 (1993)
R.T. Dodd, *Meteorites: A Petrologic-Chemical Synthesis* (Cambridge University Press, Cambridge, 1981), p. 368
K. Donaldson-Hanna et al. (2006). http://deneb.bu.edu/mercury/
D.C. Delcourt, S. Grimald, F. Leblanc, J.-J. Berthelier, A. Mililo, A. Mura, S. Orsini, T.E. Moore, Ann. Geophys. **21**, 1723–1736 (2003)
J.P. Emery, A.L. Sprague, F.C. Witteborn, J.E. Colwell, R.W.H. Kozlowski, D.H. Wooden, Icarus **136**, 104–123 (1998)
B. Fegley Jr., A.G.W. Cameron, Earth Planet. Sci. Lett. **82**, 207–222 (1987)
S.J. Gaffey, L.A. McFadden, D. Nash, C.M. Pieters, in *Remote Geochemical Analysis: Elemental and Mineralogical Composition*, ed. by C.M. Pieters, P.A.J. Englert (Cambridge University Press, Cambridge, 1993), pp. 43–77
M.V. Gerasimov, B.A. Ivanov, O.I. Yakovlev, Earth Moon Planets **80**(1/3), 209–259 (1998)
K.A. Goettel, in *Mercury*, ed. by F. Vilas, C. Chapman, M. Matthews (University of Arizona Press, Tucson, 1988), pp. 613–621
R.E. Gold et al., Planet. Space Sci. **49**, 1467–1479 (2001)
B. Hapke, J. Geophys. Res. **101**(E7), 16,817–16,831 (1996)
B. Hapke, W. Cassidy, E. Wells, Moon **13**, 339 (1975)
B. Hapke, J. Geophys. Res. **106**, 10039 (2001)
B. Hapke, Icarus **157**, 523 (2002)
J.K. Harmon, Adv. Space Res. **19**, 1487–1496 (1997)
J.K. Harmon, M.A. Slade, Science **258**, 640–642 (1992)
J.K. Harmon, M.A. Slade, R.A. Velez, A. Crespo, M.J. Dryer, J.M. Johnson, Nature **369**, 213–215 (1994)
J.K. Harmon, P.J. Perillat, M.A. Slade, Icarus **149**, 1–15 (2001)
J.K. Harmon, M.A. Slade, B.B. Butler, J.W. Head III, D.B. Rice, M.S. Campbell, Icarus **187**, 374–405 (2007)
D.L. Harris, in *Planets and Satellites*, ed. by G.P. Kuiper, B.M. Middlehurst (The University of Chicago Press, Chicago, 1961), pp. 272
J. Helbert, L. Moroz, A. Maturilli, A. Bischoff, J. Warell, A. Sprague, E. Palomba, Adv. Space Res. (2007, in press)
B.G. Henderson, B.M. Jakosky, J. Geophys. Res. **99**(E9), 19,063–19,073 (1994)
B.G. Henderson, B.M. Jakosky, J. Geophys. Res. **102**(E3), 6567–6580 (1997)
D.M. Hunten, T.H. Morgan, D. Shemansky, in *Mercury*, ed. by F. Vilas, C. Chapman, M. Matthews (University of Arizona Press, Tucson, 1988), pp. 562–612
W.M. Irvine, T. Simon, D.H. Menzel, C. Pikoos, A.T. Young, Astron. J. **73**, 807 (1968)
R. Jeanloz, D.L. Mitchell, A.L. Sprague, I. de Pater, Science **268**, 1455–1457 (1995)
K. Kabin, T.I. Gombosi, D.L. DeZeeuw, K.G. Powell, Icarus **143**, 397–406 (2000)
E. Kallio, P. Janhunen, Geophys. Res. Lett. **30**, xx–xx (2003). doi:10.1029/2003GL017842
R.M. Killen, A.E. Potter, P. Reiff, M. Sarantos, B.V. Jackson, B. Hick, B. Giles, J. Geophys. Res. **106**, 20509–20525 (2001)

R.M. Killen, M. Sarantos, A.E. Potter, P. Reiff, Icarus **171**, 1–19 (2004)
R.M. Killen et al., Adv. Space Res. (2007). doi:10.1007/s11214-007-9232-0
P.L. Koehn, A.L. Sprague, Planet. Space Sci. (2006, in press)
G.A. Krasinsky, E.V. Pitjeva, M.V. Vasilyev, E.I. Yagudina, Icarus **158**, 98–105 (2002)
L.V. Ksanfomality, Sol. Syst. Res. **32**, 133–140 (1998)
L.V. Ksanfomality, Sol. Syst. Res. **36**, 267–277 (2002)
L.V. Ksanfomality, Sol. Syst. Res. **37**, 514–525 (2003)
L.V. Ksanfomality, A.L. Sprague, Icarus **188**, 271–287 (2007)
L.V. Ksanfomality et al., Space Sci. Rev. (2007, this issue)
H. Lammer, P. Wurz, M.R. Patel, R. Killen, C. Kolb, S. Massetti, S. Orsini, A. Milillo, Icarus **166**, 238–247 (2003)
Y. Langevin, Planet. Space Sci. **45**, 31–38 (1997)
Y. Langevin, J.R. Arnold, Annu. Rev. Earth Planet. Sci **5**, 449–489 (1977)
F. Leblanc, R.E. Johnson, Icarus **164**, 261–281 (2003)
J.S. Lewis, Earth Planet. Sci. Lett. **15**, 286–290 (1972)
J.S. Lewis, in *Mercury*, ed. by F. Vilas, C. Chapman, M. Matthews (University of Arizona Press, Tucson, 1988), pp. 651–667
L.M. Logan, G.R. Hunt, J. Geophys. Res. **75**, 6539–6548 (1970)
P.G. Lucey, D.T. Blewett, B.R. Hawke, J. Geophys. Res. **103**, 3679–3699 (1998)
P.G. Lucey, D.T. Blewett, B.L. Jolliff, J. Geophys. Res. **105**(E8), 20297–20305 (2000)
V. Mangano, A. Milillo, A. Mura, S. Orsini, E. De Angelis, A.M. Di Lellis, P. Wurz, Planet. Space Sci. (2007, in press)
I. Mann, A. Krivov, H. Kimura, Icarus **146**, 568 (2000)
I. Mann, H. Kimura, D.A. Biesecker, B.T. Tsurutani, E. Grün, R.B. McKibben, J. Liou, S. Marchi, A. Morbidelli, G. Cremonese, Astron. Astrophys. **431**, 1123 (2005)
S. Marchi, A. Morbidelli, G. Cremonese, Astron. Astrophys. **431**, 1123–1127 (2005)
S. Massetti, S. Orsini, A. Milillo, A. Mura, E. De Angelis, H. Lammer, P. Wurz, Icarus **166**(2), 229–237 (2003)
A. Maturilli, J. Helbert, A. Witzke, L. Moroz, Planet. Space Sci. **54**(11), 1057–1064 (2006a)
A. Maturilli, J. Helbert, L. Moroz, Planet. Space Sci. (2006b, in press)
T.B. McCord, J.B. Adams, Icarus **17**, 585 (1972a)
T.B. McCord, J.B. Adams, Science **178**, 745 (1972b)
T.B. McCord, R.N. Clark, J. Geophys. Res. **84**, 7664–7668 (1979)
M. Meyer, Int. Comet Q. **25**, 115 (2003)
D. Mitchell, I. de Pater, Icarus **110**, 2–32 (1994)
J.I. Moses, K. Rawlins, K. Zahnle, L. Dones, Icarus **137**, 197–221 (1999)
M. Müller, S.F. Green, N. McBride, D. Koschnyb, J.C. Zarnecki, M.S. Bentley, Planet. Space Sci. **50**, 1101–1115 (2002)
A. Mura, S. Orsini, A. Milillo, D. Delcourt, S. Massetti, E. DeAngelis, Icarus **175**, 305–319 (2005)
A.A. Mura, S.A. Milillo, S. Orsini, S. Massetti, Planet. Space Sci. (2006, in press)
J.F. Mustard, J.E. Hays, Icarus **125**(1), 145–163 (1997)
S.K. Noble, C.M. Pieters, *Mercury: Space Environment, Surface, and Interior Workshop* (Lunar and Planetary Institute, Chicago, 2001), pp. 68–69
S.K. Noble, C.M. Pieters, Sol. Syst. Res. **37**, 31 (2003)
D.A. Paige, S.E. Wood, A.R. Vasavada, Science **258**, 643–646 (1992)
S.J. Peale, Icarus **28**, 459–467 (1976)
C.M. Pieters, P.A.J. Englert, *Remote Geochemical Analysis, Elemental and Mineralogical Composition* (Cambridge University Press, Cambridge, 1993)
A.E. Potter, Geophys. Res. Lett. **22**, 3289–3292 (1995)
A.E. Potter, T.H. Morgan, Science **248**, 835–838 (1990)
B. Rava, B. Hapke, Icarus **71**, 387–429 (1987)
M.S. Robinson, P.G. Lucey, Science **275**, 197–200 (1997)
M.S. Robinson, G.J. Taylor, Meteorit. Planet. Sci. **36**, 841–847 (2001)
J.W. Salisbury, B. Hapke, J.W. Eastes, J. Geophys. Res. **92**(B1), 702–710 (1987)
J.W. Salisbury, L.S. Walter, D.M. D'Aria, *Infrared (2.1–25 μm) Spectra of Minerals* (Johns Hopkins Univ. Press, Baltimore, 1991), 267 pp
J.W. Salisbury, A. Wald, D.M. D'Aria, J. Geophys. Res. **99**, 11897–11911 (1994)
A.G. Santo et al., Planet. Space Sci. **49**, 1481–1500 (2001)
Z. Sekanina, Astrophys. J. **545**, L69 (2000)
Z. Sekanina, Astrophys. J. **597**, 1237 (2003)
E.M. Sieveka, R.E. Johnson, Astrophys. J. **287**, 418–426 (1984)

P. Sigmund, Phys. Rev. **184**, 383–416 (1969)
M. Slade, B. Butler, D. Muhleman, Science **258**, 635–640 (1992)
M. Sarantos, P.H. Reiff, T.W. Hill, R.M. Killen, A.L. Urquhart, Planet. Space Sci. **49**, 1629–1635 (2001)
G. Siscoe, L. Christopher, Geophys. Res. Lett. **2**, 158–160 (1975)
J.A. Slavin, R.E. Holzer, J. Geophys. Res. **84**, 2076–2082 (1979)
S.C. Solomon, Phys. Earth Planet. Int. **15**, 135–145 (1977)
S.C. Solomon et al., Planet. Space Sci. **49**, 1445–1465 (2001)
A.L. Sprague, J. Geophys. Res. **97**, 18,257–18,264 (1992a)
A.L. Sprague, J. Geophys. Res. **98**, E1 1231 (1992b)
A.L. Sprague, T.L. Roush, Icarus **133**, 174–183 (1998)
A.L. Sprague, S.S. Massey, Planet. Space Sci. (2006, in press)
A.L. Sprague, R.W.H. Kozlowski, D.M. Hunten, Science **249**, 1140–1143 (1990)
A.L. Sprague, R.W.H. Kozlowski, F.C. Witteborn, D.P. Cruikshank, D.H. Wooden, Icarus **109**, 156–167 (1994)
A.L. Sprague, D.M. Hunten, K. Lodders, Icarus **118**, 211–215 (1995)
A.L. Sprague, W.J. Schmitt, R.E. Hill, Icarus **135**, 60–68 (1998)
A.L. Sprague, J.P. Emery, K.L. Donaldson, R.W. Russell, D.K. Lynch, A.L. Mazuk, Meteorit. Planet. Sci. **37**, 1255–1268 (2002)
A.L. Sprague, R.W.H. Kozlowski, F.C. Witteborn, D.P. Cruikshank, D.H. Wooden, Icarus **109**, 156–167 (1994)
P. Spudis, J. Guest, in *Mercury*, ed. by F. Vilas, C. Chapman, M. Matthews (University of Arizona Press, Tucson, 1988), pp. 118–164
L.V. Starukhina, J. Geophys. Res. **106**(E7), 14701–14710 (2001)
R.G. Strom, A.L. Sprague, *Exploring Mercury the Iron Planet* (Springer-Praxis, Chichester, 2003), 216 pp.
L. Tepper, B. Hapke, Bull. Am. Astron. Soc. **9**, 532 (1977)
A.L. Tyler, R.W.H. Kozlowski, L.A. Lebofsky, Geophys. Res. Lett. **15**, 808–811 (1988)
T. van Hoolst, F. Sohl, I. Holin, O. Verhoeven, V. Dehant, T. Spohn, Space Sci. Rev. (2007, this issue). doi:10.1007/s11214-007-9202-6
A.R. Vasavada, D.A. Paige, S.E. Wood, Icarus **141**, 179–193 (1999)
F. Vilas, Icarus **64**, 133–138 (1985)
F. Vilas, in *Mercury*, ed. by F. Vilas, C. Chapman, M. Matthews (University of Arizona Press, Tucson, 1988), pp. 59–76
F. Vilas, T.B. T.B. McCord, Icarus **28**, 593–599 (1976)
F. Vilas, M.A. Leake, W.W. Mendell, Icarus **59**, 60–68 (1984)
C. Wagner, in *Thermal Emission spectroscopy and Analysis of Dust, Disks, and Regoliths*, ed. by M.L. Sitko, A.L. Sprague, D.K. Lynch (Astronomical Society of the Pacific, San Francisco, 2000), pp. 386
J. Warell, Icarus **156**, 303–317 (2002)
J. Warell, Icarus **161**, 199–222 (2003)
J. Warell, D.T. Blewett, Icarus **168**, 257–276 (2004)
J. Warell, A.L. Sprague, J. Emery, R.W. Kozlowski, A. Long, Icarus **180**, 201–291 (2006)
S.J. Weidenschilling, Icarus **35**, 99–111 (1978)
G.W. Wetherill, in *Mercury*, ed. by F. Vilas, C. Chapman, M. Matthews (University of Arizona Press, Tucson, 1988), pp. 670–691
P. Wurz, H. Lammer, Icarus **164**, 1–13 (2003)
P. Wurz, P. Bochsler, J.A. Paquette, F.M. Ipavich, Astrophys. J. **583**, 489–495 (2003)
P. Wurz, in *The Dynamic Sun: Challenges for Theory and Observations*, ESA SP-600, 5.2, (2005), 1–9
P. Wurz, U. Rohner, J.A. Whitby, C. Kolb, H. Lammer, P. Dobnikar, Planet. Space Sci. (2006, submitted)

Processes that Promote and Deplete the Exosphere of Mercury

Rosemary Killen · Gabrielle Cremonese · Helmut Lammer · Stefano Orsini · Andrew E. Potter · Ann L. Sprague · Peter Wurz · Maxim L. Khodachenko · Herbert I.M. Lichtenegger · Anna Milillo · Alessandro Mura

Originally published in the journal Space Science Reviews, Volume 132, Nos 2–4.
DOI: 10.1007/s11214-007-9232-0 © Springer Science+Business Media B.V. 2007

Abstract It has been speculated that the composition of the exosphere is related to the composition of Mercury's crustal materials. If this relationship is true, then inferences regarding the bulk chemistry of the planet might be made from a thorough exospheric study. The most vexing of all unsolved problems is the uncertainty in the source of each component. Historically, it has been believed that H and He come primarily from the solar wind (Goldstein, B.E., et al. in J. Geophys. Res. 86:5485–5499, 1981), Na and K come from volatilized materials partitioned between Mercury's crust and meteoritic impactors (Hunten, D.M., et al. in Mercury, pp. 562–612, 1988; Morgan, T.H., et al. in Icarus 74:156–170, 1988; Killen, R.M., et al. in Icarus 171:1–19, 2004b). The processes that eject atoms and molecules into the exosphere of Mercury are generally considered to be thermal vaporization, photon-stimulated desorption (PSD), impact vaporization, and ion sputtering. Each of these processes has its own temporal and spatial dependence. The exosphere is strongly influenced by Mercury's highly elliptical orbit and rapid orbital speed. As a consequence the surface undergoes large

R. Killen (✉)
Department of Astronomy, University of Maryland, College Park, USA
e-mail: rkillen@astro.umd.edu

G. Cremonese
Osservatorio Astronomico-INAF, Padova, Italy

H. Lammer · M.L. Khodachenko · H.I.M. Lichtenegger
Space Research Institute, Austrian Academy of Sciences, Graz, Austria

S. Orsini · A. Milillo · A. Mura
Istituto di Fisica dello Spazio Interplanetario-CNR, Rome, Italy

A.E. Potter
National Solar Observatory, Tucson, AZ, USA

A.L. Sprague
Lunar and Planetary Laboratory, University of Arizona, Tucson, USA

P. Wurz
Physics Institute, University of Bern, Bern, Switzerland

fluctuations in temperature and experiences differences of insolation with longitude. Because there is no inclination of the orbital axis, there are regions at extreme northern and southern latitudes that are never exposed to direct sunlight. These cold regions may serve as traps for exospheric constituents or for material that is brought in by exogenic sources such as comets, interplanetary dust, or solar wind, etc. The source rates are dependent not only on temperature and composition of the surface, but also on such factors as porosity, mineralogy, and space weathering. They are not independent of each other. For instance, ion impact may create crystal defects which enhance diffusion of atoms through the grain, and in turn enhance the efficiency of PSD. The impact flux and the size distribution of impactors affects regolith turnover rates (gardening) and the depth dependence of vaporization rates. Gardening serves both as a sink for material and as a source for fresh material. This is extremely important in bounding the rates of the other processes. Space weathering effects, such as the creation of needle-like structures in the regolith, will limit the ejection of atoms by such processes as PSD and ion-sputtering. Therefore, the use of laboratory rates in estimates of exospheric source rates can be helpful but also are often inaccurate if not modified appropriately. Porosity effects may reduce yields by a factor of three (Cassidy, T.A., and Johnson, R.E. in Icarus 176:499–507, 2005). The loss of all atomic species from Mercury's exosphere other than H and He must be by non-thermal escape. The relative rates of photoionization, loss of photo-ions to the solar wind, entrainment of ions in the magnetosphere and direct impact of photo-ions to the surface are an area of active research. These source and loss processes will be discussed in this chapter.

Keywords Mercury · Exosphere · Surface composition · Particle release processes

1 Observations of the Mercury Exosphere

Early ground-based efforts to detect an atmosphere on Mercury were unsuccessful, leading only to successively smaller upper limits for the atmospheric density. The first real information about the Mercury atmosphere came with the Mariner 10 flybys (Broadfoot et al. 1976). Atomic hydrogen, helium and atomic oxygen were detected by the Mariner 10 ultraviolet photometers from sunlight scattered from these atoms.

A comprehensive discussion of the Mariner 10 measurements is given in "The Mercury Atmosphere" by Hunten et al. (1988), and the reader is referred to this source for details. Subsolar point densities were estimated at 6,000 atoms/cm^3 for helium, and 230 atoms/cm^3 for the thermal component of atomic hydrogen. A nonthermal component of hydrogen with a scale height of about 70 kilometers was observed near the limb above the subsolar point, providing a total number density of 23 atoms/cm^3 at the surface. Atomic oxygen was detected in the third flyby at a level of 44,000 atoms/cm^3, the number being dependent on the assumed scale height. This chapter is concerned with Mercury atmospheric observations made after the Mariner 10 flybys, all from ground-based observations.

1.1 Sodium in Mercury's Exosphere

In 1985, Drew Potter and Tom Morgan were at the McDonald Observatory of the University of Texas measuring the spectrum of light reflected from the Moon. They were looking for the "filling-in" of reflected solar lines caused by interactions of solar radiation with the lunar surface. It occurred to them that Mercury was similar in many ways to the Moon, and should show similar "filling-in" effects. The first solar lines they observed on Mercury were

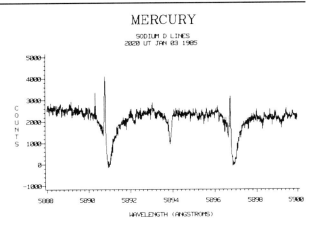

Fig. 1 Sodium discovery spectrum showing D$_1$ and D$_2$ sodium resonance emission lines within the solar Fraunhofer absorption lines reflected from the Mercury surface. Measured at the Harlan Smith telescope at the McDonald Observatory of the University of Texas

the pair of sodium Fraunhofer lines at 5,890 and 5,896 angstroms. They were astonished to find that the spectra of these features showed bright emission lines that extended well above the continuum. The wavelengths of the lines identified them as resonance scattering lines of sodium in the exosphere of Mercury. The discovery spectrum is shown in Fig. 1 (Potter and Morgan 1985).

The calculation of sodium column content from the spectra first requires that the emission intensity be extracted from the combination of surface reflection and sodium atom emission. The common practice has been to interpolate the solar Fraunhofer continuum from one side of the emission to the other, and subtract it to obtain the sodium emission line. Procedures that fit the Fraunhofer absorption line with a combination of Voigt profiles give the best results for defining the Fraunhofer background to be subtracted. However, Hunten and Wallace (1993) pointed out that this procedure underestimates the sodium emission, since about a fifth of the sodium light is hidden in a depression in the continuum caused from extinction by the sodium atoms.

Once the sodium emission intensity is determined, two methods have been used to calculate the column content of sodium atoms. The ratio of D$_2$ to D$_1$ emissions is a function of both temperature and the column content of sodium and the surface albedo (e.g. Killen 2006), so by assuming a value for temperature, the column content can be calculated from the ratio. This procedure was used to estimate the column content of sodium atoms in the discovery spectrum. However, this procedure is subject to large uncertainties. A better method is to use the planet itself as a "standard candle" to calibrate the sodium emission, since the solar continuum reflected from the planetary surface is available in each spectrum. The reflectance of Mercury can be calculated using the Hapke formulation for reflectivity with the optical constants for the Mercury surface. (The most recent set of optical constants was provided by Mallama et al. (2002).) It is then necessary to take into account the effect of atmospheric blurring of the Mercury image, which reduces the surface brightness as seen by ground-based observers.

Since the discovery of sodium emission, there have been a number of additional observations. Table 1 shows a summary of the observations that have been reported to date. These observations show that average sodium column densities computed from emission spectra seen above the planetary surface range between $1-10 \times 10^{11}$ atoms/cm^2. Column densities measured from absorption spectra seen during solar transit are smaller, in the range $1-3 \times 10^{10}$. However, these densities are measured at the evening terminator, where sodium densities are normally seen to be small.

Table 1 Column densities of sodium in the Mercury exosphere

Observer	Column content, atoms/cm^2	Remarks
	$8.1 \pm 1.0 \times 10^{11}$*	Slit spectra
Potter and Morgan (1987)	$1.9 \pm 1.4 \times 10^{11}$	Slit spectra
Potter and Morgan (1985)	$2.8–3.8 \times 10^{11}$	Slit spectra
Potter and Morgan (1997)	$4.4–6.0 \times 10^{11}$	5″ × 5″ image slicer, peak values
Sprague et al. (1997)	Min. $<1 \times 10^{10}$ equatorial	Slit spectra
	Max. 15.0×10^{11}	Slit spectra
	Avg. $1.20 \pm 1.49 \times 10^{11}$	Slit spectra
Sprague et al. (1998)	8.6×10^{11}	Northern spot, slit spectra
	8.0×10^{11}	Southern spot, slit spectra
Potter et al. (1999)	$0.84 \times 2.9 \times 10^{11}$	10″ × 10″ image slicer, average values
Barbieri et al. (2004)	$0.43–1.97 \times 10^{11}$	Slit spectra
Schleicher et al. (2004)	$3.4 \pm 0.1 \times 10^{10}$	Scanning Fabry Perot North pole
	$3.0 \pm 0.1 \times 10^{10}$	Scanning Fabry Perot South pole
	$1.5 \pm 0.1 \times 10^{10}$	Scanning Fabry Perot West equatorial
	$<0.2 \times 10^{10}$	Scanning Fabry Perot East equatorial
Leblanc et al. (2006)		Slit spectra, terminator low, solar limb high
Potter et al. (2007)	$1.5–3.5 \times 10^{11}$	10″ × 10″ image slicer, average values**

*Calibration from D_2/D_1 ratio. Other observations, except Schleicher et al. (2004), were calibrated from the Mercury reflectance signal

**Sodium seen in absorption. Column content derived from equivalent width

***Average values from 10 × 10 arc second slicer were scaled to the approximate area of planet

1.2 Temperature of the Sodium Exosphere

The velocity distribution of sodium atoms in the exosphere is an important indicator of the source of the sodium. Source processes such as photon-stimulated desorption and ion sputtering will produce non-Maxwellian velocity distributions (e.g. Madey et al. 1998). Subsequent interaction of the atoms with the surface can relax the distribution back towards Maxwellian. Thus, analysis of the velocity distribution can provide information on source

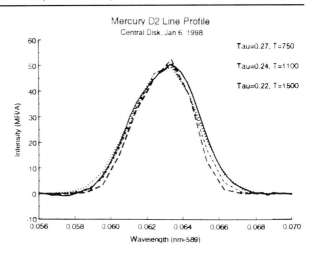

Fig. 2 The sodium D_2 line profile for the central disk of Mercury observed with the Anglo-Australian telescope on January 6, 1998 (*solid line*). A 1,500 K Maxwellian velocity distribution is plotted as a dotted line, an 1,100 K distribution is plotted as a *dot-dash line*, and a 750 K distribution is plotted as a *long-dash line*. The best fit to the line profile is given by the 1,500 K distribution (Killen et al. 1999)

processes and surface interactions. The measurable quantity that is related to the velocity distribution is the spectral profile of the emission line, which can be compared with line profiles computed assuming combinations of Gaussian velocity profiles for the atomic transitions. An initial effort was reported by Potter and Morgan (1987), who obtained a somewhat noisy line profile of the sodium D_2 emission at a dispersion of 7.2 mÅ per step. They concluded that the half-width of the line was consistent with a thermalized gas at a temperature of 500 K. A subsequent analysis that took into account the wings of the line emission suggested that the line profile was better fit by a temperature of about 1,000 K (Shemansky and Morgan 1991). Measurements at higher resolution and better signal-to-noise were reported by Killen et al. (1999). Line profiles were measured at the McDonald Observatory of the University of Texas at a dispersion of 1.98 mÅ per step, and at the Anglo-Australian Telescope (AAT) at Coonabarabran, Australia at a dispersion of 3.14 mÅ per step. The line profiles from the AAT were fit assuming a Chamberlain-type exosphere, summing the Gaussian line profiles for each of the hyperfine levels of the sodium resonance transition. An example showing three trial fits to the line profile of the central disk of Mercury observed is shown in Fig. 2.

The best fit to the central disk profile was given by assuming a 1,500 K Maxwellian velocity distribution. Line profiles from above the north and south pole were best fit by a 750 K gas. In all cases, the gas temperature appeared to be everywhere hotter than the surface temperature by 600–700 K. The importance of high signal-to-noise is evident in this plot, where the differences between profiles for different temperatures are small, becoming evident only in the wings of the emission profile.

Another approach to measuring the velocity profile is to measure the altitude distribution of the atoms. The observations by Schleicher et al. (2004) of Mercury in transit across the Sun measured the altitude distribution of sodium above the terminator. These data provide direct measures of scale heights and consequently temperatures. They also measured the width of the absorption line, finding that it corresponded to a Doppler temperature of 3,540 K. This apparent high temperature might be the consequence of sputtering over the polar regions and/or radiation acceleration, since the view of sodium atoms during a transit is such that velocity changes caused by radiation acceleration should be detectable (Potter et al. 2007). Their results are compared with the line profile estimates of scale height and temperature shown in Table 2. The temperatures from line profiles and estimated altitude distributions are in fair agreement near the center of the planet, but differ considerably at

Table 2 Temperatures and scale heights for the Mercury sodium exosphere

Observer/author	Location	Temperature, K
Shemansky and Morgan (1991)	Planetary average	1000
Killen et al. (1999)	North polar region	750
	Center	1500
	South polar region	750
Schleicher et al. (2004)	North polar region	1380 ± 307
	South polar region	1330 ± 307
	West edge	1540
	East edge	1540
	Planetary average	1540

the poles. The temperatures from altitude distribution are almost twice those derived from line profile data. The difference might be the result of slight differences in location of the observations near the poles.

The high temperature sodium absorptions in the polar regions reported by Schleicher et al. (2004) were at a maximum in regions offset towards the dawn terminator by about 15 degrees. In polar regions, 15 degrees *away* from the dawn terminator, the altitude distribution is much shorter, and the temperature is obviously much lower. Consequently, the AAT and transit observations may not have viewed the same regions of the exosphere near the poles, and temporal variations are certainly important. Recalling that the subsolar surface temperature of Mercury is about 700 K, and polar temperatures are much lower, the results summarized in Table 2 show that sodium is produced by an energetic process, and is not completely thermally accommodated. A planet-wide average temperature of 1,000 K for sodium corresponds to a scale height of about 97 km. For this scale height, the surface concentrations of sodium derived from the column densities of Table 1 would be in the range $1–10 \times 10^4$ atoms/cm^3.

1.3 Spatial Variation of Sodium Emission

One of the first things that investigators of the sodium emission noticed was that it was not uniform over the surface of Mercury and did not conform to the distribution expected for a classical exosphere. This is illustrated in Fig. 3 by profiles of the sodium emission across the planet taken with a slit spectrograph (Potter et al. 2006).

A similar set of profiles for the period April 3–6, 1988, were analyzed by Killen et al. (1990) using radiative transfer theory to make profiles of simple smooth models of a uniform exosphere. The observations could not be fit by these models. Compared to the models, the observed sodium emissions were concentrated in the sunward direction with polar enhancements. Several north–south distributions showed more sodium at the north polar regions than could be explained by a simple model. Furthermore, the distribution changed noticeably from one day to the next.

1.4 North–south Emission Peaks

Two-dimensional images of sodium distribution over the surface were obtained by Potter and Morgan (1990), using a 5″ × 5″ (arcsec) image slicer. Images from two different

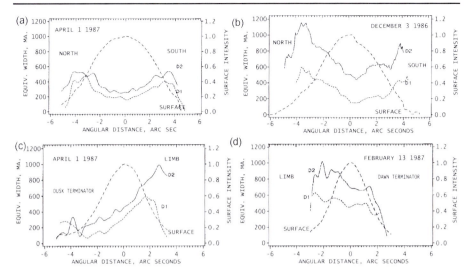

Fig. 3 Profiles of sodium emission (*solid lines*) and surface continua (*dotted lines*) across Mercury. The upper left profile (**a**) from April 1, 1987, shows symmetric limb brightening at the north and south polar regions. At the *upper right* (**b**), there is a north–south profile from December 3, 1986, that shows excess emission in the north polar region. At the *lower left* (**c**), there is a profile that starts at the dusk terminator and proceeds to the bright limb. Sodium emission is effectively nonexistent at the dusk terminator, in contrast to the *lower right profile* (**d**), which shows a terminator-to-limb profile for the case that the dawn terminator is in view. Here, the sodium emission decreases only about 30% as the terminator is approached

Fig. 4 Maps of the sodium D_2 emission from Mercury on July 7, and August 23, 1989

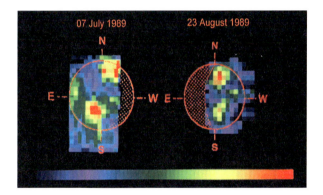

dates (July 7, 1989, and August 23, 1989) are shown in Fig. 4. These maps confirmed the long slit observations that the sodium is not distributed uniformly over the surface. Excess sodium emission was seen at both polar regions, plus some excess sodium near the subsolar point. Almost all the maps showed north–south excess sodium, and a sequence of three images from July 16–18, 1989, showed variations from one day to the next. The combination of north–south excess sodium emission and daily variability suggested to the authors that sodium was produced, at least in part, by sputtering at high latitudes by solar wind particles precipitated to the surface in the cusps of the magnetosphere.

The images shown in Fig. 4 view approximately opposite sides of the planet. Both images show emission peaks at north and south high latitudes and a smaller peak in the vicinity of the subsolar point. The peak values of sodium emission are 3.8 MR and 1.2 MR, respectively. The difference in peak intensity between the two images is due in large part to the difference of radial velocity. The July image was taken at a time when the radial velocity was about −4.7 km/sec, while the August image was taken near aphelion, when the radial velocity was about 0.69 km/sec. The peak values of atomic sodium density in the two images are similar. The images were made using a $5'' \times 5''$ image slicer having $0.5''$ slices to yield 5 arc second square images, each having 10×10 pixels of $0.5''$. To cover the whole planet, two such images offset from one another were combined, registered to one another by matching the surface reflection signals. The directions shown in this figure are those seen by a ground-based observer, and the terminator is the dusk terminator for the July image, and the dawn terminator for the August image.

The July image displays longitudes in the range from 270° (terminator) to 360° and 0° to 47° (limb). This range includes the high latitude radar- and optical-bright albedo spots B (north) and A (south) (reported by Slade et al. 1992; Butler et al. 1993; Harmon 1997), as well as the relatively freshly excavated Kuiper-Murasaki crater complex on the eastern limb. There is excess sodium in the vicinity of these features. The August image displays longitudes in the range 71° (limb) to 175° (terminator), approximately the opposite side of the planet from the July image. The sodium-bright spot pattern is similar to that seen for the July image, but there are no obvious anomalous geographic features in this case, other than the far eastern edge of Caloris Basin.

Sprague et al. (1997) made an extensive series of Mercury sodium measurements using a long-slit spectrograph. The slit provided three to five spatial resolution elements across the planet and, by stepping the slit across the planet, they were able to obtain low-resolution maps. They found that high-latitude enhancements of sodium emission were common, with the added characteristic that there was usually more at one hemisphere than the other. They observed notable sodium enhancements when the spectrograph slits were placed over the radar-bright spots B and A which are centered near at 355°W longitude and 55°N and 25°S latitude, respectively. In a repetition of the measurements on October 4, 1996, they again found sodium enhancements at the approximate locations of the radar-bright spots (Sprague et al. 1998). They also observed an exceptionally high emission value, corresponding to a column content of 15×10^{11} atoms/cm^2 at a location close to the Caloris Basin. Similar enhancements close to these locations were also observed by Potter and Morgan (1990), and the sodium map for July 7, 1989, shown in Fig. 4 includes the region of the radar-bright spots. Sprague et al. (1994, 1993) proposed that these sodium-bright spots were the result of a special quality of the radar-bright regions and Caloris Basin—most likely a surface composition rich in sodium as the result of geologic processes.

Potter et al. (2006) used a 10×10 element image slicer having $1''$ slices to produce $10'' \times 10''$ images with $1''$ pixels. Many of the images showed high latitude emission enhancements in one hemisphere relative to the other, and many showed enhancements in both hemispheres. The ratio of north to south hemispheric emission intensity was calculated for each image and the results are plotted in Fig. 5 as a function of the longitude of the center of the surface reflection image. The ratios appear to be random with respect to longitude, and there is no concentration of excess sodium emission in the northern hemisphere near the longitude of the Caloris Basin centered at latitude 22° north and longitude W 180°.

A correlation of high or low ratios shown in Fig. 5 with any specific longitude would suggest that specific sodium-rich areas of the planet were responsible for excess sodium at that longitude. However, there do not appear to be any strong correlations with specific longitudes. Images in which bright spots appeared in both hemispheres were of special interest in view of the possibility that the north–south radar-bright spots are sources of sodium

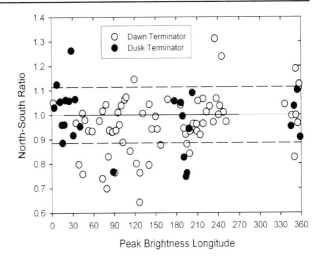

Fig. 5 Ratios of the northern hemisphere emission to the southern hemisphere emission plotted against longitude of peak brightness in the continuum reflection image

Table 3 Sodium images that show bright symmetric north–south emission peaks (Potter et al. 2006)

Date	Terminator	True anomaly, degrees	Radiation acceleration, cm/sec^2	Terminator longitude, degrees	Bright limb longitude, degrees
Aug. 31, 1999	Dusk	39.4	166.9	270.1	65.0
Feb. 6, 2000	Dawn	315.9	170.6	89.0	320.5
May 30, 2000	Dawn	81.5	157.8	281.1	181.0
Jun. 3, 2000	Dawn	98.4	135.1	289.9	200.7
Jan. 6, 2002	Dawn	297.7	172.2	85.3	325.0
Nov. 6, 2001	Dusk	68.8	173.1	96.2	225.7
Apr. 9, 2003	Dawn	113.4	113.4	269.3	162.6

emission. A selection of images from Potter et al. (2006) that showed bright spots in both hemispheres is listed in Table 3. For three of the observations listed in Table 3 (Aug. 31, 1999, Feb. 6, 2000, and Jun. 6, 2000) the radar-bright spots at 355° longitude are in view.

However, they are not in view for the remaining four observations. These images shared high values of solar radiation acceleration and, in fact, images measured at high values of radiation acceleration almost always showed bright spots in both hemispheres. Ip (1990) modeled the effect of radiation acceleration on the planetary sodium distribution, and showed that high values of radiation acceleration would push sodium towards the terminator. At the poles, the terminator and bright limb meet, so that at the poles, we expect to see limb brightening enhanced by sodium pushed to the terminator by radiation acceleration.

North–south excess sodium has also been observed in absorption against the Sun. Schleicher et al. (2004) observed the sodium D_2 line in absorption against the surface of the Sun during the transit of 2003. As shown in Fig. 6, sodium absorptions were observed at high northern and southern latitudes, in agreement with all the emission observations that find high latitude concentrations of sodium. This was the case despite the fact that radiation acceleration was not high, about 13.2% of surface gravity. Sodium absorption is observed along the western equatorial region, but not along the eastern equatorial region. It is puz-

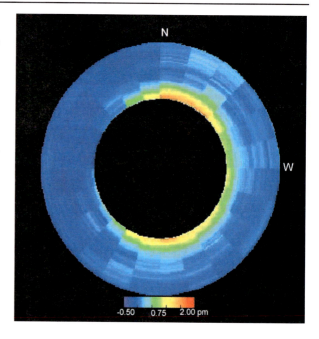

Fig. 6 Distribution of sodium D₂ absorption around the disk of Mercury seen in transit across the Sun. Concentrations of sodium vapor exist at high north and south latitudes. Sodium absorption is seen along the west equatorial limb (dawn), but not along the east equatorial limb (dusk). The true anomaly angle was 149.2° and radiation acceleration was 49.3 cm/sec², or about 13.2% of surface gravity. Figure from Schleicher et al. (2004)

zling to note that almost no sodium is observed over most of the eastern half of the planet. Observations of sodium emission do not show much, if any, deficit of emission on the dusk hemisphere of the planet. The temperatures derived from the altitude distribution of the sodium absorption are listed in Table 2.

Conclusions from this discussion are: (1) Excess sodium emissions often appear in either the northern or southern hemisphere, usually both. Sometimes they appear over the bright-albedo regions A and B. (2) Near the peak values for radiation acceleration, symmetric north–south spots tend to appear, which may be the result of the pile-up of sodium near the terminator pushed there by radiation acceleration. (3) North–south excess emissions sometimes appear even when radiation acceleration is relatively low (see Fig. 4). In addition to radiation acceleration effects, causes suggested for the appearance of sodium-bright spots include sodium-rich areas and high-latitude sputtering effects.

1.5 Diurnal Variation of Sodium Distribution

Sprague et al. (1997) reported a diurnal variation in sodium emission, such that the sodium abundance increased as the sunrise terminator was approached. This is illustrated in Fig. 7 (Fig. 4b from Sprague et al. 1997), where a bar chart shows the average sodium column abundance as a function of Mercury's local time. Bin 1 = early morning, 2 = mid-morning, 3 = mid day, 4 = mid afternoon, 5 = late afternoon. By mid- and late- afternoon the Na abundance is down by close to a factor of 3. Sprague et al. (1997) proposed that this effect results from evaporation of cold-trapped and ion implanted Na that is released as the sun rises and heats the surface of Mercury.

The ratio of terminator to limb emission over a range of true anomaly angles was reported by Potter et al. (2006). To calculate this ratio, sodium emission on the terminator side of the image was summed, and divided by the sum of sodium emission on the limb side of the image. For terminator-to-limb distributions where the emission falls to low values at

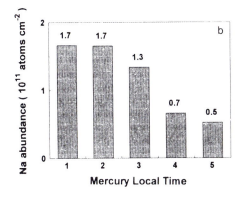

Fig. 7 The variation of sodium column content is shown at hourly intervals past sunrise (Fig. 4b from Sprague et al. 1997). This bar chart shows that the column content of sodium in bins of Mercury's local time: bin *1* = early morning, *2* = mid-morning, *3* = mid day, *4* = mid afternoon, *5* = late afternoon

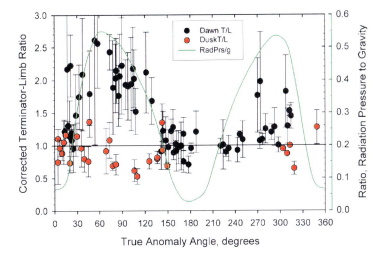

Fig. 8 The terminator-to-limb ratio variation with true anomaly angle is shown for cases when the dawn terminator is in view, and when the dusk terminator is in view Potter et al. (2006). The strength of solar radiation acceleration is plotted for comparison

the terminator (as seen in Fig. 3c), the terminator-to-limb ratio will be less than unity. For terminator-to-limb distributions where there is appreciable sodium at the terminator (as seen in Fig. 3d), the ratio will be unity or larger. Figure 8 shows a plot of this ratio against true anomaly angle. The ratio of radiation acceleration to surface gravity is also plotted.

For dawn-side observations on the "out" leg of the orbit, where true anomaly angles (TAA) < 180°, the ratios appear to be consistent with the hypothesis that dawn enhancement from evaporation of sodium occurs. The dawn-side ratio is larger than unity, indicating enhancement of sodium at the terminator. The dawn-side ratios are largest near maximum radiation acceleration. For values of TAA larger than 140°, the ratios for dawn-side observations drop down to near-unity values, coincident with the decrease of radiation acceleration. This change is inconsistent with evaporative enhancement of sodium near the terminator. It is reasonable to expect that the rate of evaporation of condensed sodium and hence the amount of sodium vapor near the terminator should depend on the rate of exposure of new surface at the terminator. In other words, the amount of excess sodium should depend on the rate of terminator advancement. The rate of terminator advancement is about −0.2 degrees/day at perihelion, rising to a maximum value of +3.38 degrees/day at aphelion. Consequently, one expects that the maximum amount of dawn enhancement should be found near aphelion, rather than near maximum radiation acceleration, but this does not occur. For the "in" leg of the orbit (TAA > 180°), the effect is much less pronounced. There are only a few dusk-side ratios. On average, the ratio values scatter around unity, with some below unity near maximum radiation acceleration.

Although some of the observations are consistent with the explanation that dawn enhancement occurs from evaporation of condensed sodium, it seems reasonable to expect that radiation acceleration could also play a role by pushing sodium towards the terminator, thus producing an excess of sodium there (see Ip 1990). This would explain why the effect is largest near maximum radiation acceleration.

The transit observations reported by Schleicher et al. (2004) (shown above in Fig. 6), found sodium emission clearly evident along the western equatorial region, which is the dawn terminator. Little or no sodium was visible along the dusk terminator on the eastern equatorial region. This observation is consistent with the existence of a dawn terminator enhancement of sodium vapor. Dawn/dusk asymmetries can also be caused by an asymmetric magnetosphere, with resulting asymmetric ion-sputtering and preferential re-implantation of photoions on the dawnside (Killen et al. 2004a).

1.6 Variation of Sodium Emission with Time

Sodium emissions sometimes vary in an apparently random fashion. An example was published by Potter and Morgan (1990), who observed the distribution of sodium over the planetary disk to change over a three-day period, from February 16 to 18, 1989. A bright spot appeared in the southern hemisphere on February 17, and disappeared the next day. The peak emission intensities rose from 2.2 MR on February 16 to 3.4 MR on February 17, and then dropped back to 2.2 MR on February 18.

Sprague et al. (1997) reported similar behavior. During the October 10–15, 1987 observing sequence, they saw a dramatic onset of north–south enhancement, followed by its total disappearance. The sodium abundance at high northern and southern latitudes exceeded equatorial values by a factor of 2 on October 14, while before and after this date, the abundance was similar at all latitudes.

Potter et al. (1999) reported observations of Mercury sodium over six days during the period November 13–20, 1997. Daily changes took place in both the total amount of sodium and its distribution over the planet. The cause of these variations is unknown, but their episodic nature suggests some connection with solar activity, or heating of the surface, both of which are known to change over short periods of time. Sprague et al. (1997) were unable to find any correlation between changes in sodium emission and the F10.7 cm solar radio flux. Likewise, Potter et al. (1999) found no correlation between the variations in their November 1997 data and F10.7 cm solar flux. They did note that there were a number of coronal mass ejection (CME) events at this time, some of which were directed towards the general area of Mercury, and suggested a possible connection with the changes observed on Mercury.

In addition to the random variations, there is a secular trend of sodium emission from one day to the next. This is illustrated in Fig. 9 (from Potter et al. 2007) which shows a plot of the planet-wide averaged emission intensity against true anomaly angle. The intensity is seen to change from day to day, mostly in a regular fashion, and is closely correlated to emission rate calculated for an individual sodium atom, caused by a variation in the continuum at the rest frequency of a sodium atom in the exosphere.

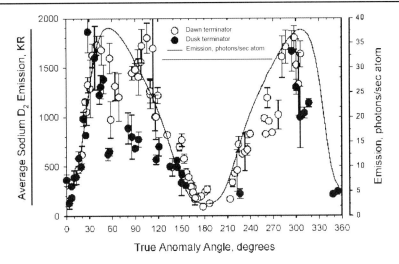

Fig. 9 The planetary average sodium emission was measured using a $10'' \times 10''$ image slicer over the period from 1997 to 2003, and is plotted here against the true anomaly angle. Depending on the true anomaly angle, the average sodium emission can change up to 20% per day. The calculated rate of emission from a single sodium atom is also shown in the plot (*solid line*)

The observed emission intensity shown in Fig. 9 tracks the calculated emission very well except for regions near maximum radiation acceleration. On the "out" leg of the orbit (TAA < 180°), there is a minimum just past the maximum acceleration at 50°. On the "in" leg of the orbit (TAA > 180°) there are deviations from the calculated curve both before and after the maximum radiation acceleration.

In large part, the secular changes of sodium emission intensity illustrated in Fig. 9 are the result of changes in the intensity of solar radiation in the rest frame of the sodium atoms as the radial velocity of Mercury changes.

However, changes in sodium column content also affect the emission intensity, and the deviations noted from the calculated emission curve may be due to changes in the sodium column content. By normalizing the emission intensities to a constant value of solar radiation intensity it is possible to separate the changes due to solar radiation from those due to sodium column content. Figure 10 shows the data from Fig. 9 normalized to the solar intensity seen by sodium atoms on Mercury at a true anomaly angle of 150.26°.

The maximum values of normalized emission, and hence of sodium column densities, occurs near perihelion and aphelion, where solar radiation acceleration effects are nearly negligible. In between perihelion and aphelion, there is a decrease of nearly a factor of three. The most likely cause for this decrease is the effect of radiation acceleration, which reaches a maximum as the normalized emission reaches a minimum, most clearly evident for the "out" leg of the orbit (TAA < 180°).

1.7 Effects of Radiation Acceleration on Sodium Emission

As noted in the previous paragraph, radiation acceleration appears to affect the amount of sodium visible on Mercury. Previous observations of the effect of radiation acceleration were inconsistent. Potter and Morgan (1987) found a decrease in column content as radiation acceleration increased. The average column content of sodium decreased by about 30% when the radiation acceleration relative to gravity increased from about 15% to 45%.

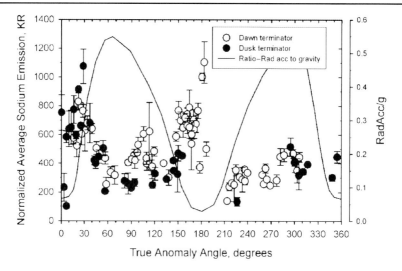

Fig. 10 Sodium emission intensities have been normalized to the values they would have at a constant value of solar radiation intensity (that seen by Mercury at a true anomaly angle of 150.26°). The normalized intensities are proportional to the average column content of sodium, in the absence of any secondary effects of radiation acceleration. The normalized emission shows maxima near perihelion and aphelion. The ratio of solar radiation acceleration to surface gravity (*solid curve*; axis RHS) reaches maximum values near true anomaly angles of 60 and 300 degrees

Fig. 11 The normalized sodium emission plotted against radiation acceleration. Dawn terminator observations are shown as *open circles* and dusk terminator observations as *filled circles*. There is a wide scatter of the data. A linear fit to the data yields an R-square value of 0.2

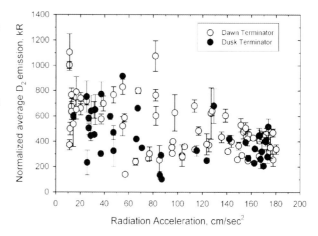

However, Sprague et al. (1997) found no correlation at all between column content and radiation acceleration. Recently, Potter et al. (2007) used the data shown in Fig. 10 to examine the effect of radiation acceleration. The data points from Fig. 10 are replotted against radiation acceleration in Fig. 11. There is no consistent trend with radiation acceleration. There are some sequences where the normalized emission intensity actually increases with increasing radiation acceleration.

Potter et al. (2007) found that most of the data scatter was not random, but could be explained by assuming that the sodium atoms on Mercury were exposed to the accelerating effects of sunlight for about 1,700 seconds. The effect of radiation acceleration was different whether the "out" leg or the "in" of the orbit was observed. There is a positive feedback loop

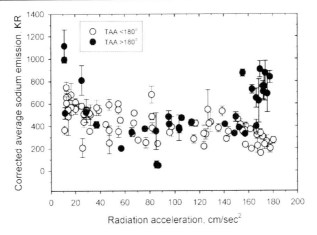

Fig. 12 The normalized sodium observations were corrected for the effect of solar radiation acceleration. The wide scatter of data points seen in the uncorrected emission values shown in Fig. 11 has diminished. A linear fit to the data yields an R-square value of 0.015 compared to 0.2 prior to the correction. An initial decrease is followed by a wide region of near-constant corrected emission, with changes near the end of the plot at maximum radiation acceleration. Near the end, there is a sharp increase of about 20% for observations on the "in" leg of the orbit. This effect may be real, or it might be an artifact of the simplified concept used for the correction

in the "out" leg of the orbit, such that radiation acceleration increases the solar continuum intensity seen by the atoms, and a negative feedback loop in the "in" leg of the orbit, such radiation acceleration decreases the continuum intensity. The result of this effect is that the emission intensity per sodium atom is increased over the non-accelerated value on the "out" leg of the orbit, and is decreased on the "in" leg of the orbit. The emission values corrected for this effect showed much less scatter, with a general trend of about 30% to lower values from minimum to maximum radiation as shown in Fig. 12.

After the correction for the secondary effects of radiation acceleration, the emission intensities should be truly proportional to the column content. To obtain column content values, the corrected emission intensities were multiplied by 9.712, which is the g-factor at a true anomaly angle of 150.26°. The results were plotted against true anomaly angle with the result shown in Fig. 13.

The results are compared with the theoretical predictions of Smyth and Marconi (1995) in Fig. 13. For true anomaly angles less than 180°, most of the data fall near the line predicted for perfectly elastic collisions with the surface ($\beta = 0$), suggesting that the interaction of sodium atoms with the surface is weak. The negative change in continuum with increasing radiation pressure at true anomaly angles greater than 180° causes the intensity per atom to decrease, so that the emission is actually under-corrected at true anomaly angles between 300 and 330 degrees of true anomaly angle. The curve should in fact be compressed, so that the atoms have a similar emission rate per atom in the entire region between 330 and 360 degrees of true anomaly angle.

Comparison of column densities at aphelion with those at perihelion is important, because at these two points, the effect of radiation acceleration is negligible. The column content at aphelion was larger than at perihelion by a factor of about 1.3, suggesting that the source processes for sodium do not quite keep up with loss processes as Mercury approaches the Sun, falling short by about 30%.

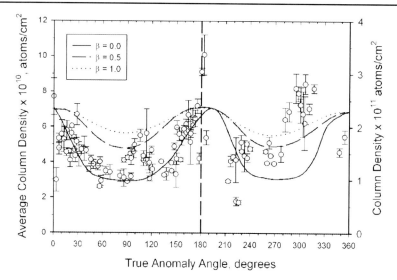

Fig. 13 Average column content data computed from the acceleration-corrected data of Fig. 12 is compared with predictions of Smyth and Marconi (1995). (Note that this average is taken over the area of the entire 10 × 10 arc second image slicer. The angular area of Mercury in view is about a third of this value, so densities referred to the Mercury surface would be about three times larger.) There are three overlays on this plot, taken from Fig. 15 of the Smyth and Marconi (1995) paper, in which they took into account radiation acceleration and the interaction of the sodium atoms with the surface

These results show that radiation acceleration can alter the emission intensity without any change in the column content of sodium. Consequently, the effect of radiation acceleration on emission intensities should be taken into account if column densities are to be calculated from emission intensities.

1.8 The Sodium Tail of Mercury

Models of the Mercury sodium exosphere published by Ip (1986) and Smyth and Marconi (1995) predicted that radiation acceleration could sweep sodium completely off the planet into a down-sun tail, depending on the energy of the atoms. Potter et al. (2002b) were able to observe the sodium tail of Mercury in twilight, mapping it downstream to a distance of about 40,000 km, as shown in Fig. 14. At that point, the velocity of the sodium had increased to about 11 km/sec as the result of solar radiation acceleration. The cross-section of the tail at 17,500 km downstream had a half-width of about 20,000 km, which implied transverse velocities of sodium in the tail of 2 to 4 km/sec. These velocities imply source velocities from the planet of the order of 5 km/sec. The total integrated flux of sodium in the tail was approximately 10^{23} atoms/sec, which corresponds to 1 to 10% of the estimated total production rate of sodium on the planet.

Considering that Smyth and Marconi (1995) estimated that atoms with velocities in excess of 2.1 km/sec could escape the planet into the tail, it appears the solar radiation acceleration can provide this velocity over a considerable range of the orbit. The radial velocity at which the observed tail disappears would be a measure of the initial velocity of the sodium atoms, but this would be different on the inbound and outbound legs of the orbit, on account of the fact that the effect of radiation acceleration is different for the inbound and outbound legs.

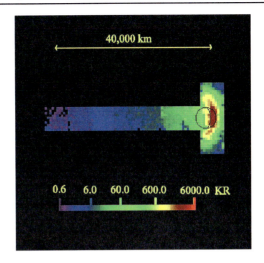

Fig. 14 The sodium tail of Mercury as observed May 26, 2001. A 10 × 10 arc second image slicer was used to capture square sections of the tail downstream from the planet

1.9 Potassium in the Exosphere of Mercury

Potassium in the exosphere of Mercury was reported by Potter and Morgan (1986). Sodium and potassium are chemically and physically very similar, so the appearance of potassium is not surprising.

The discovery spectrum is shown in Fig. 15, where it is seen that the emission line is very much weaker than for sodium. The abundance of potassium is much less than that of sodium. In the discovery spectra, the average column content of potassium was estimated to be about 10^9 atoms/cm^2, about 1% of the column abundance of sodium. Potassium observations reported by Sprague et al. (1990) found typical potassium abundances of 5.4×10^8 atoms/cm^2.

When they placed the slit over the longitudes of the "hot poles" where the Sun is overhead at perihelion, they found enhanced potassium abundance over the Caloris Basin and its antipode, as shown in Fig. 16.

Sprague et al. (1990) suggested that the potassium enhancements are consistent with an increased source of potassium from the well-fractured crust and regolith associated with the fractures in the basin floor and the hummocky terrain at the antipode. Killen et al. (1991)

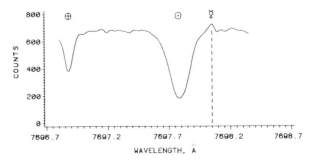

Fig. 15 Mercury reflectance spectrum showing emission from the potassium D_1 line at 7698.05 Å, Doppler-shifted from its rest wavelength of 7698.98 Å. The stronger D_2 emission line can only be observed when the Doppler shift is such as to move it out from under an atmospheric oxygen absorption line

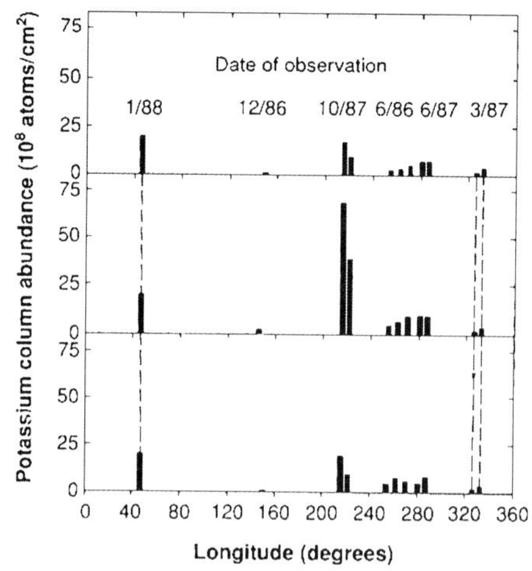

Fig. 16 Potassium observations (Sprague et al. 1990) showed an enhanced column abundance when the spectrograph slit was over the longitude of Caloris Basin and over the antipodal terrain 180° away. These longitudes are both under the Sun during perihelion

Fig. 17 Potassium observations of Sprague et al. (1990) show higher abundance in morning than in afternoon. This is explained in Sprague (1992) as a result of K ion implant into regolith materials during the long Hermean night with subsequent release to the exosphere the following morning. By afternoon, the source of implanted K ions is diminished

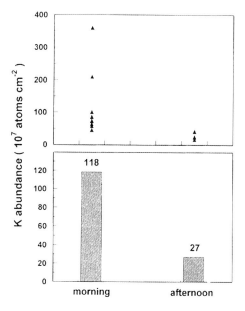

presented an alternate interpretation of the observation: because the localized sources of sodium could not be consistently correlated with specific locations on Mercury's, she noted that potassium sources may also not be consistently correlated with hermeographic locations either.

Sprague (1992) noted that potassium observations showed a marked morning/afternoon asymmetry (Fig. 17), similar to that observed for sodium, and presented a model in which potassium ions implant on the nightside of the planet where the footprint of the magnetosheath intersects the surface. In the morning the potassium is released by heating and thermal desorption.

Hunten and Sprague (2002) summarized all their data concerning the diurnal variation of potassium column content, and concluded that potassium would evaporate from the surface

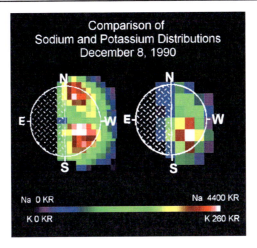

Fig. 18 Same-day maps of the sodium and potassium emission over the surface of Mercury. The sodium emission is on the left side of the image, potassium on the right. The resolution elements of the sodium image are 0.5″ square, while those of the potassium image are 1″ square. The weaker signal from the potassium exosphere required use of a lower resolution. Both sodium and potassium show excess emission in the southern hemisphere. Sodium also shows emission in the northern hemisphere equivalent to that in the southern hemisphere, while the potassium emission in the northern hemisphere is weak compared to that of the southern hemisphere. The phase angle was about 90°. The terminator is located at about 85°W longitude, and the limb at about 348°W longitude. The radar-bright spots are centered near 355°W longitude at about 55°N and 28°S, about 7° towards the center of the planet from the bright limb. True anomaly was about 284° and the radial velocity was near maximum at −9.8 km/sec. The enhancements for sodium are in the region of the radar- and albedo-bright spots A (south) and B (north). The potassium enhancement is over bright spot A

at temperatures of 400 K and higher. The chemical and physical similarity of sodium and potassium suggests that they should behave similarly on Mercury.

A comparison of sodium and potassium images supports this view. Same-day images of sodium and potassium emissions were measured by Potter and Morgan (1997) over a five-day period from December 6–10, 1990. Both the sodium and potassium emission were more intense at high northern and southern latitudes, and varied from one day to the next. The distributions of sodium and potassium emission over the Mercury surface were similar (but not identical), supporting the view that they are generated by similar processes. For this series of observations, the ratio of peak sodium intensity to peak potassium intensity was about 190. An example of the sodium and potassium images is shown in Fig. 18.

A series of sodium and potassium observations reported by Potter et al. (2002a) showed that the ratio of sodium to potassium was highly variable. This is illustrated in Fig. 19, where the ratios of planetary averages of sodium to potassium column densities are plotted against the potassium column densities.

Table 4 compares the Na/K ratio in different solar system bodies.

The ratio is found to vary from a low of 2 in the Earth's crust to a high of up to 190 in Mercury's exosphere. It is instructive to note that the highest values of this ratio are both seen in atmospheres (Earth) or exospheres (Mercury), indicative of differential loss. In the atmosphere of the Earth, the source of sodium is primarily sea salt, which is enriched in sodium due to differential solubility of Na/K in water. Thus a factor of 3 enrichment in the atmosphere must be attributed to removal of K, which is the heavier of the two. A low value of Na/K in the Earth's crust indicates dissolution of Na salt in the seawater, which is ultimately subducted to the mantle. In Mercury's case, we do not know the initial Na/K

Fig. 19 The ratio of sodium to potassium emission plotted as a function of potassium column content. Although the average ratio of sodium to potassium is about 100, there are significantly higher and lower values. The ratio decreases with increasing potassium density, suggesting that sodium and potassium densities can vary independently of one another

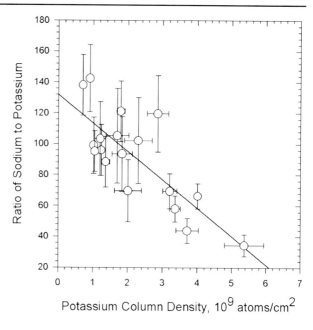

Table 4 Sodium to potassium ratios in the solar system

Object	Na/K	Source
Mercury exosphere	40–140	Potter et al. (2002a)
Io exosphere	10	Brown (2001)
Europa exosphere	25	Brown (2001)
Moon exosphere	6	Potter and Morgan (1988)
Lunar crust	7–9	Lodders and Fegley (1998, p. 177)
Meteorites	7–14	Lodders and Fegley (1998, p. 311)
Earth atmosphere	20–150	Gault and Rundle (1969)
Earth seawater	27	Lodders and Fegley (1998, p. 164)
Earth crust	2	Lodders and Fegley (1998, p. 143)
Solar system	15	Lodders and Fegley (1998, p. 80)

ratio in the crust. However, radiation acceleration is higher for potassium than for sodium, and the shape of the Fraunhofer lines are different. Consequently, the effect of radiation acceleration on emission intensity will be different for the two species, and that may be the cause of the variations of Na/K ratio observed. There also may also be ion/wave resonances that increase the loss rate of potassium ions. At present this is speculation, but measurement of wave modes by the upcoming missions should investigate these processes.

1.10 Calcium in the Exosphere of Mercury

Sprague et al. (1993) reported an unsuccessful search for calcium emissions at 4226.73 Å, estimating an upper limit for calcium of 7.4×10^8 atoms/cm^2. A tenuous calcium exosphere at Mercury at a density of 1–1.5×10^8 atoms/cm^2, principally seen in the polar regions, was first observed in July 1998, using the High Resolution Echelle Spectrograph (HIRES) at the W.M. Keck I telescope (Bida et al. 2000). Figure 20 shows results for calcium observations

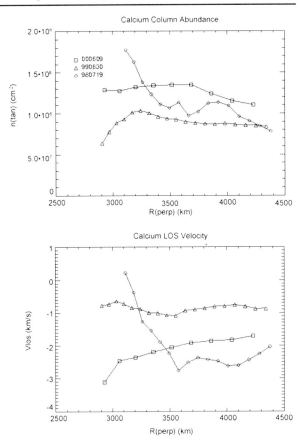

Fig. 20 Variation of column abundance and velocity for calcium as a function of radial distance from Mercury (Killen et al. 2005). Data are shown for three dates: July 19, 1998, June 30, 1999, and June 9, 2000

in 1998, 1999, and 2000. The emission was observed off the disk, but near the poles of Mercury.

Killen et al. (2005) summarized four years of observations of the calcium exosphere of Mercury. As seen in Fig. 20, the observations show a persistent but spatially variable blue shift, indicating an excess velocity toward the observer of up to 3 km s^{-1}, with an average excess velocity of 2.2 km s^{-1} above the south pole. In addition, the line profiles reveal a hot corona at the equivalent of 12,000–20,000 K in a thermalized atmosphere, indicating a large range of motion with respect to the observer. The calcium is not confined to the polar regions: rare and low Ca abundance is seen in the equatorial regions. Strong emission was seen anti-sunward on May 3, 2002. Apparent weak emission on the sunward hemisphere may be due to scattered light from the surface, or may indicate a high latitude source. Sputtering and impact vaporization could introduce calcium into the exosphere from Mercury's crust.

Killen et al. (2005) suggested that the likely source of the calcium is either impact vaporization in the form of CaO and clusters, which are subsequently photo-dissociated, or ion-sputtering of atoms, molecules and ions. The column abundance is somewhat, but not strongly, correlated with solar activity but data are sparse. Koehn and Sprague (2007) explore the possibility that the Sun is the primary source of O and Ca in Mercury's exosphere, a result of highly ionized atoms of O^{+6} and Ca^{+11} delivered by the Sun to Mercury.

1.11 Unknown Exospheric Components and Surface Composition

Despite several unpublished searches for visible and near-infrared emission from other elements in the Mercury exosphere, none have been found. Sprague et al. (1996) reported a search for lithium in the Mercury exosphere, with negative results. They estimated the upper limit for lithium abundance to be 8.4×10^7 atoms/cm^2. It is expected that the observed species represent only a small fraction of Mercury's exosphere, because at the surface the total pressure, derived from the sum of these known species, is almost two orders of magnitude less than the exospheric pressure of approximately 10^{-10} mbar, obtained by the Mariner 10 occultation experiment (Fjeldbo et al. 1976; Hunten et al. 1988). The surface composition plays a crucial role in the production and composition of Mercury's exosphere. To try to understand the importance of various processes that may be at work on Mercury one can initially use a suite of several possible surface compositions to bound the exospheric models.

The properties of Mercury's surface, especially its composition, age, origin and evolution, are not known precisely at the moment, due to the lack of sufficient remote and in situ investigations. Therefore, the surface composition can only be estimated, based on modeling and comparison with the Moon. Several scientific efforts have been made to estimate the surface composition of Mercury using spectral reflectance measurements in comparison with spectra of analog materials in laboratory studies.

Since the 1960s considerable optical and near-infrared spectra have been obtained, as discussed by Warell (2003). The measurements were improved with infrared detectors in the 1980s. The infrared spectra of Mercury, combined with laboratory studies of terrestrial, Lunar and meteoritic materials, indicate that the rock composition is dominated by feldspars and low iron pyroxene (Warell and Blewett 2004). Possible Mercurian surface compositions range from Lunar meteorites up to mixtures of Mercury analog materials such as labradorite and enstatite.

Burbine et al. (2002) used synthetic Mercury analogs to compare low-FeO anorthositic compositions with that of partial melts, derived from melting experiments on the EH4 chondrite Indarch (enstatite-rich chondrite). The goal of their work was to relate the compositions of basaltic partial melts and their residual aubritic materials to that of the Hermean crust and mantle, respectively. Blewett et al. (1997) in previous experiments used lunar anorthositic breccia MAC 88105, which is related to lunar meteoroid material, as analog to rocks of the Hermean crust. The synthetic Mercury composition used by Burbine et al. (2002) is depleted in FeO relative to the lunar anorthosite MAC 88105. However, to produce the observed spectral reddening of the surface by space weathering, surface soils should contain at least a few wt% of FeO in the bulk (Hapke 2001; Burbine et al. 2002). Mid-infrared spectral studies of Mercury's surface indicate Na-rich feldspars and pyroxene (Sprague and Roush 1998) and alkali basalt (Sprague et al. 1994). In addition, clino-pyroxene was identified (Sprague et al. 2002).

Mercury soil analogs which mirror the spectroscopic observations and range from Lunar meteorites up to mixtures of Mercury analog materials like labradorite and enstatite can be summarized as the following:

Synthetic Mercury: Burbine et al. (2002) compared low-FeO anorthositic compositions with that of partial melts, derived from melting experiments on the EH4 chondrite Indarch which is an enstatite-rich chondrite. The aim of their work was to relate the compositions of basaltic partial melts and their residual aubritic materials to that of the Hermean crust and mantle, respectively.

Lunar anorthositic breccia MAC 88105: This soil composition relates Lunar meteoroid material to rocks of the Hermean crust (Blewett et al. 1997). The Synthetic Mercury composition (Burbine et al. 2002) is depleted in FeO relative to Lunar anorthosite MAC 88105. However, to produce the observed spectral reddening of the surface by space weathering, surface soils should comprise at least a few wt% of FeO in the bulk (Hapke 2001).

*by*75 + *en*25, *la*75 + *en*25 (Bytownite–enstatite mixtures and labradorite–enstatite mixtures): Warell and Blewett (2004) investigated several mixtures of Mercury analog materials by means of visible near infrared (VNIR) reflectance spectroscopy. The best spectral fit was reached with a mixture of labradorite and enstatite (3:1) with about 0.1 wt% of submicroscopic metallic iron to simulate spectral reddening due to weathering. The USGS feldspar sample, which was used in the Warell and Blewett (2004) study.

*an*75 + *en*25, *ol*75 + *en*25 (Andesine–enstatite mixtures and oligoclase–enstatite mixtures): VNIR reflectance spectra show clear spectroscopic features related to electronic effects among transition elements, but is insensitive for thorough discrimination among plagioclase components. Therefore, the range was extended to andesine- and oligoclase-rich compositions.

VNIR reflectance spectra show clear spectroscopic features related to electronic effects among transition elements, but are insensitive for thorough discrimination among plagioclase components. Therefore, one cannot exclude andesine- and oligoclase-rich compositions. However, a mixture of andesine and enstatite adjusted to ratios of observed surface elements in the exosphere may be a good analog for Mercury's geochemical surface composition which contains most likely Si, Ti, Al, Fe, Mg, Ca, Na, K, Mn and O.

The presently observed and various expected exospheric species are shown in Table 5. Note that some of the reported species (e.g., N_2, O_2, CO_2 and H_2O) are just estimated as upper limits.

One should also note that the composition of Mercury's exosphere is non-stoichiometric with respect to the surface composition (Morgan and Killen 1997). The composition and temporal variability of the exosphere in part help to explain the weathering rate of the surface, and the volatile redistribution rate over both short and long time scales. What resides on the surface at this epoch is not the pristine surface, but a highly space-weathered surface that has been overturned by meteoroid bombardment, and desiccated of volatile content by many processes including photon-stimulated processes, ion sputtering and vaporization. The relative importance of these processes and their effectiveness at redistributing volatiles either to space or to high-latitude cold traps or to the nightside can be addressed by studying the exosphere (e.g., Leblanc and Johnson 2003).

2 Exospheric Sources and Their Models

2.1 Impact Vaporization

Impacting particles of small sizes (<100 μm) constantly rain onto Mercury's surface at a mean velocity of 20 km/s (Cintala 1992), churning the regolith and vaporizing the surface. Larger meteors impact more sporadically, but with higher mean velocity (Marchi et al. 2005). Many minor species, which are refractory during vaporization of silicates in vacuum, are highly volatile during hypervelocity impacts due to high temperatures and pressures during impacts (Gerasimov et al. 1998). Therefore the vapor ejected from impact vaporization will be the most representative of the surface composition.

Table 5 Expected abundances in Mercury's exosphere (Mililio et al. 2005)

Species	Surface abundance (cm^{-3})	Total Zenith Column (cm^{-2})
H	23; 230[a]	3×10^{9}[h]
He	6.0×10^{3}[a]	$<3 \times 10^{11}$[h]
Li		$<8.4 \times 10^{7}$[n]
O	4.0×10^{4}	$<3 \times 10^{11}$[h]
^{20}Ne	6×10^{3} day[c]	
	7×10^{5} night[c]	
Na	$1.7-3.8 \times 10^{4}$[a]	2×10^{11}[i]
Mg	7.5×10^{3}[d]	3.9×10^{10}[d]
Al	654[c]	3.0×10^{9}[d]
Si	2.7×10^{3}[d]	1.2×10^{10}[d]
S	5×10^{3}[d]	2.0×10^{10}[d]
	6×10^{5}[g]	2.0×10^{13}[g]
Ar	$<6.6 \times 10^{6}$[a]	$<9 \times 10^{14}$[b]
		1.3×10^{9}[k]
K	3.3×10^{2}[b]	2×10^{9}[b]
	5×10^{2}[h]	
Ca	387[d]	$<1.2 \times 10^{9}$[d]
	<239[f]	$<7.4 \times 10^{8}$[c]
		1.1×10^{8}[j]
Fe	340[d]	7.5×10^{8}[d]
H$_2$	$<1.4 \times 10^{7}$[p]	$<2.9 \times 10^{15}$[p]
O$_2$	$<2.5 \times 10^{7}$[p]	$<9 \times 10^{14}$[p]
N$_2$	$<2.3 \times 10^{7}$[p]	$<9 \times 10^{14}$[p]
OH	1.4×10^{3}[d,e]	1×10^{10}[d,e]
CO$_2$	$<1.6 \times 10^{7}$[p]	$<4 \times 10^{14}$[p]
H$_2$O	$<1.5 \times 10^{7}$[p]	$<1 \times 10^{12}$[c] $<8 \times 10^{14}$[p]

[a]Hunten et al. (1988): measurements or upper limits
[b]Potter and Morgan (1997)
[c]Hodges (1974): model abundance
[d]Morgan and Killen (1997): model abundances
[e]Morgan and Killen (1997): model abundances
[f]Sprague et al. (1993): measured upper limit
[g]Sprague et al. (1996, 1995): prediction
[h]Shemansky (1988): Mariner 10 measurements
[i]Killen et al. (1990): measured abundance
[j]Bida et al. (2000)
[k]Killen (2002): model abundance
[l]Killen and Ip (1999)
[m]Huebner et al. (1992) ionisation rates: experimental (e); theoretical (t) for quiet and active Sun
[n]Sprague et al. (1996): model abundance
[p]Broadfoot et al. (1976)
[q]Cremonese et al. (1997)

In addition, macro-meteors impact Mercury but at an unknown rate. Marchi et al. (2005) provided the distribution of impact probability as a function of impactor radius, up to objects of 100 m in radius. In particular, meteoritic impactors coming from the Main Asteroid Belt are expected to impact on Mercury. The contribution by these larger meteorites to the global Hermean exosphere is negligible; nevertheless, their impact is expected to produce strong, localized, but temporary increases in the exospheric density, enriched by material coming from deeper layers (Mangano et al. 2007). The impact frequency of such objects (especially in the lower size range) at Mercury is not negligible relative to the nominal duration of the BepiColombo mission (one year nominal plus one year extension).

Regardless of the size of the impactor, the initial ejecta from an impact will be high-temperature vapor (~5,000 K). This will quickly be followed by the "liquid and vapor" at a slightly lower temperature (2,500 K). However, only the vapor ejected from Comet Tempel I in the first milliseconds was hot, followed quickly by thermalized vapor. This suggests that impact vapor may be much cooler than previously supposed.

Impact events probe to a depth of several diameters of the impacting body. Because meteorite impacts probe much deeper than any process other than venting, and because the energy density of the process is very high, the exospheric products of this emission

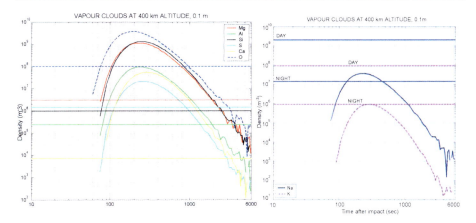

Fig. 21 Density versus time for an impacting object of 0.1 m radius, at 400 km altitude for the species whose mean density value does not change between day- and nighttime (*left*); separately, for Na and K (*right*). *Horizontal lines* represent the exospheric background for each species (from Mangano et al. 2007)

Table 6 Vaporization rates for sodium at Mercury due to micrometeoritic bombardment

Reference	Orbital position	Projectile → target	Total vaporization rate (g cm^{-2} s^{-1})	Vaporization rate Na (atoms cm^{-2} s^{-1}), with $f(\mathrm{Na}) = 0.005$
Morgan et al. (1988)	aphelion	combination	1.74×10^{-15}	2.27×10^5
	perihelion	combination	6.42×10^{-15}	8.40×10^5
Cintala (1992)	aphelion	diabase → regolith	8.18×10^{-15}	1.07×10^5
		regolith → regolith	7.25×10^{-16}	0.95×10^5
	perihelion	diabase → regolith	2.75×10^{-15}	3.60×10^5
		regolith → regolith	2.46×10^{-15}	3.22×10^5

process most closely represent the surface composition as a whole. Simulations performed by Mangano et al. (2007), analyze the effects in terms of the gaseous cloud produced by impacts of objects in the range 1 cm – 1 m. Particularly noticeable is the case of 10-cm meteor for which the enhancement, depending on the considered species, varies from 1 to 4 orders of magnitude higher than the mean exospheric background values (see Fig. 21). Figure 21 shows density versus time for an impacting object of 0.1 m radius, at 400 km altitude.

Durations are generally larger than 2,000 s, and their extension larger than 50° (calculated with respect to the center of the planet). Estimated vaporization rates from various groups are summarized in Table 6. A considerable variation in estimated rates results from uncertainties in the physical state of the surface as well as the impact flux as discussed in the following.

2.2 Interplanetary Space at Mercury's Orbit

The impact flux onto Mercury (as well as on other planets) is the consequence of several different physical processes which produce bodies on planetary crossing orbits. Detailed studies for the Earth have shown that the range of sizes impacting with our planet span over more than eight orders of magnitude: from μm up to hundreds of meters. There is no reason to doubt that the same is true also for the other terrestrial planets. Such a flux of material onto Mercury has several effects, like the formation of craters and the well-known "maturation" of the soils, for example. The role of the meteoroid flux on Mercury's exosphere is not well known.

A small fraction of volatiles released to the exosphere is thought to be produced by impact vaporization of meteoritic material. The composition of the Hermean exosphere thus reflects the chemical composition of the surface, and of meteorites impacting Mercury, mixed with traces of solar wind. Unfortunately, the meteoritic gardening and the impact history of the Mercury surface is presently unknown because it depends on variables related to the composition of the surface and the flux of meteoroids. The meteoroid flux used in literature for Mercury studies are roughly derived from estimate at the Earth's heliocentric distance, then extrapolated to the inner Solar System. It means we may not have a good estimate of the statistics on the number of impacts and the velocity distribution of the meteoroids. Cintala (1992) published very nice calculations on meteoroid impacts, but his work was restricted to sizes less than 1 cm, which are subject to Poynting–Robertson drag. His calculations cannot be directly extrapolated to larger bodies. Particles having a larger size follow a completely different dynamical evolution (Marchi et al. 2005).

The radiation pressure force deflects small particles directly antisunward. They may leave the Solar System after they are ejected from a comet or formed by collision, the exact condition depending on the initial orbital parameters. Particles for which solar gravity amounts to more than twice the radiation pressure may stay in bound orbits, and they form the main content of the interplanetary dust cloud. The radiation pressure influences the orbital evolution of this latter component mainly through the Poynting–Robertson effect. The momentum transfer caused by radiation falling onto a moving particle includes, when seen in the reference frame of the Sun, a small component anti-parallel to the particle's velocity that stems from the Lorentz transformation of radiation pressure force in the frame of the particle. This is the case for particles that move in orbital motion about the Sun and are exposed to the photon flux that is directed radially outward. The small deceleration that is induced by the anti-parallel component is denoted as the Poynting–Robertson effect. Thus a drift toward the Sun is superimposed on the motion in Keplerian orbits which limits the lifetime of the dust particles, just as collisions do.

Although the Poynting–Robertson effect may vary strongly with the size, composition, and structure of particles, the radial drift of the particles that it causes is small compared to the orbital velocities. The deceleration of particles by the Poynting–Robertson effect reduces the eccentricities and semi-major axes of their orbits. This leads to an increase of dust number density with decreasing solar distance.

The particles having size larger than 1 cm are not dominated by the Poynting–Robertson effect and they have to be studied following a different dynamical approach. Most of the meteoroids arriving on the terrestrial planets come from the asteroid main belt and the main delivery routes are the well-known resonances. Among them, the most efficient in ejecting material toward the inner Solar System are the 3 : 1 and $\nu 6$ (Morbidelli and Gladman 1998; Bottke et al. 2002).

2.3 Impacts of Meteoroids

Following the description of the meteoroid fluxes at the Mercury orbit we infer that we have to take into account two different populations of meteoroids based on their size and consequently different dynamical evolution.

In the case of the smaller particles Cintala (1992) provided a very good model of meteoroid impact vaporization considering objects with radii in the range 10^{-8}–10^{-2} m, that has been used by Cremonese et al. (2005, 2006) to calculate the differential number of impacts and the mass distribution to obtain the mass of the vapor produced by the impacts. Although the larger particles have been studied in a specific dynamical model realized for Mercury (Marchi et al. 2005), the size distribution of impactors on Mercury has been calibrated with the flux observed on the Earth, for which reliable data are available (Brown et al. 2002).

Indeed, in their numerical simulations, they estimated the ratio of impacts on Mercury versus the Earth for each projectile size, and they used this ratio to scale the impact rate with Mercury relative to that observed for the Earth. For this reason the dynamical model has been obtained for meteoroid in the size range of 10^{-2}–10^2 m, but we have to consider here only those impacts that are relevant for the daily production of the exosphere's elements that can affect the ground-based observations, limiting the size range to 10^{-2}–10^{-1} m.

The most important result concerning the velocity distribution of large particles is the wide range of impact velocities on Mercury: the mean impact velocity is about 30 km s^{-1}, but the tails span from about 15 to 80 km s^{-1}. For comparison, the Moon's impact distributions are much narrower with a maximum impact velocity of about 50 km s^{-1} (see Fig. 22).

Moreover, Mercury's impact distributions depend on the impactor sizes—that is, on the simplification $f(v, r) = f(v)$—used in some works (e.g., Cintala 1992), which does not hold in this case. To quantify the effects of the impactor sizes, we note that the percentage of high velocity impactors (defined as those having $v > 50$ km s^{-1}) are 25% and 19% for $r = 1$ and 10^{-2} m, respectively. Note that indeed on Earth it is possible to have impacts with velocities up to 80 km s^{-1}, but they are sporadic events related to retrograde swarm of fragments, presumably of cometary origin. Figure 23 shows $f(v, r)$ onto Mercury's surface as a function of the projectile velocity, averaged in the size range 10^{-2} and 10^{-1} m, in the average and perihelion cases.

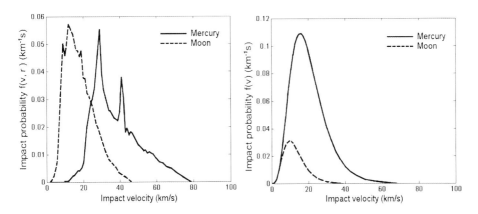

Fig. 22 *Left panel*: Impact probability $f(v, r)$, as a function of the projectile velocity, averaged over the large meteoroids range (10^{-2}–0.15 and 10^{-2}–0.10 m for Moon and Mercury, respectively). *Right panel*: Impact probability $f(v)$ as a function of the projectile velocity for small meteoroids (10^{-8}–10^{-2} m for Moon and Mercury, respectively)

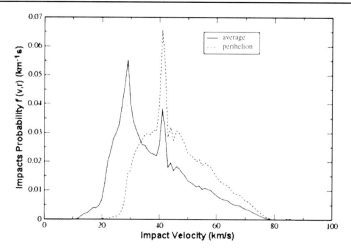

Fig. 23 Impact probability $f(v, r)$ onto Mercury's surface as a function of the projectile velocity, averaged in the size range 10^{-2} and 10^{-1} m, in the average and perihelion cases

Since the orbit of Mercury is quite eccentric ($e = 0.2$) there is some variation from the mean impact rate along its orbit. According to the model of Marchi et al. (2005) the distribution obtained for aphelion is almost the same as in the average case. On the contrary, for the perihelion case, the impact distribution is quite different and the relative number of high-velocity impacts is about 43% and 33%, respectively for $r = 1$, 10^{-2} m. Thus, impacts at perihelion happen at considerably greater velocity than the average case. Moreover, asymmetries in the rate of impacts onto planets or satellites have been widely studied for synchronous rotating bodies (e.g., see Horedt and Neukum 1984; Marchi et al. 2005) and for non-synchronous rotating bodies, like Mercury, the same considerations hold, but the asymmetry is related to the morning-evening (*am/pm*) hemispheres instead of to leading-trailing ones.

Following the model of Marchi et al. (2005) in the average case the ratio *am/pm* is greater than 1 except for $r < 13$ cm, while at perihelion the ratio is always *am/pm* > 1. The increase of the *am/pm* ratio with particle size is normal, as already pointed out by Morbidelli and Gladman (1998). It is due to the more numerous meteoroids having small semi-major axis, increasing the density of particles inside the Mercury's orbit, which typically tend to fall on the morning hemisphere. Also, it is normal that the *am/pm* ratio is larger for Mercury at perihelion, because the orbital velocity of the planet is higher, and the planet tends to catch up the meteoroids, rather than being caught up by them.

In the size range of 10^{-2}–10^{-1} m, the dynamical model by Marchi et al. (2005) estimates at perihelion *am/pm* = 1.2, while at aphelion *am/pm* = 0.8. Figure 23 shows the impact probability $f(v, r)$ in the average and perihelion cases.

The precise calculation of the amount of neutral atoms refilling the exosphere due to the impacts is not possible. Recent models that calculated the amount of vapor produced during the impact possibly represent an upper limit. The evaluation of the contribution to the exosphere requires the knowledge of the partitioning of the kinetic energy of the impact in the elements composing the vapor, allowing inference of the velocity distribution of the neutral atoms. The models considered by Cintala (1992) and Cremonese et al. (2005, 2006) treat vertical impacts of spherical projectiles into a regolith, with the shock behavior depending on the composition of the target and the meteoroid.

The quantity of melted and vaporized regolith produced by each impactor type is a function of the impact velocity and target temperature. An impact event of sufficiently high velocity creates what can be visualized ideally as a spheroidal volume, centered below the impact point, that grades from vapor through a liquid and vapor field into a completely liquid phase and on until a solid–liquid region merges into highly shocked, unfused target material. In this context it is clear that a smaller increase in internal energy will be required to initiate fusion in a hot target than a cold one, simply because the former is closer to its melting point. According to Cintala (1992) the effects of target temperature are not trivial, but they are secondary to the role played by impact velocity.

The above-mentioned models have been used to estimate the production rate of sodium, the main element observed with ground-based telescopes. That of potassium can be inferred from the ratio between the two atoms, in the composition assumed. It must be borne in mind that the cause of the high variability of Na/K observed in the exosphere is unknown. In the following we will report this calculation for the sodium.

Given the assumed velocity distribution of micro-meteors at Mercury, Cintala (1992) concluded that 1.2 projectile masses of vapor and 8.6 projectile masses of melt would be produced by a micro-meteor impact onto Mercury, whereas Morgan et al. (1988) calculated 5.4 projectile masses of vapor. As pointed out by Cintala Morgan et al. (1988) used a spatial density which was a factor or four greater than that used by Cintala, 1.8 g cm^{-3}, a velocity distribution with a higher mean impact velocity, and a lower sound speed than recommended by O'Keefe and Ahrens (1986). Without these differences, Morgan et al. (1988) would have obtained 0.63 projectile masses of vapor, only half that calculated by Cintala. Given these unknowns, the exact vaporization rate must be considered to be very uncertain. An additional factor that must be kept in mind when discussing the results of impact vaporization calculations is that, for micrometeoritic impact onto a regolith, the impactors are of the order of μm and the regolith particles are of the order of 100 μm in diameter. One might imagine that a certain fraction of the energy would go into angular velocity of the regolith particles, and in addition, that the vapor produced would be largely directed downward, and would fill voids in the regolith. Therefore our knowledge of the rate and temperature of the vapor ejected into the exosphere is rudimentary at best.

2.4 Physics of the Evaporation and Production Rate

The volume of target (regolith) material vaporized by a spherical projectile of radius, r, and impacting velocity, v, can be estimated using the relation of Cintala (1992)

$$V_{\text{vap}}(v, r) = \frac{4}{3}\pi r^3 (c + dv + ev^2). \tag{1}$$

The constants c, d (km^{-1} s) and e (km^{-2} s^2) depend on target temperature and projectile composition (Cintala 1992, Table 3, page 952). In these computations the constant values have been obtained for a diabase projectile and the target at 400 K. Equation (1) was derived by Cintala (1992) for meteoroids smaller than 10^{-2} m. Cremonese et al. (2005, 2006) assumed that (1) is valid up to 10^{-1} m, and for a regolith target with a modal composition similar to those of the basalt-derived soils from the Taurus Littrow floor (Ahrens and Cole 1974): pyroxene ∼60%, plagioclase ∼30%, olivine ∼5% and ilmenite ∼0.2%. In doing so the model is based on a regolith target with a modal composition having a higher plagioclase/pyroxene ratio. It follows that (1) should underestimate the amount of material vaporized with respect to the model of Cremonese et al. (2005, 2006) using a different composition. In fact, the energy needed to vaporize the plagioclase is lower than that needed to vaporize the pyroxene (i.e., Ahrens and O'Keefe 1972). Therefore, under the same conditions of stress, plagioclase-rich rocks produce more vapor than pyroxene-rich rocks.

The total mass of the vapor produced from the infalling of meteoroids on Mercury's surface can be obtained by using (1) given the flux of bodies impacting the planet, Φ, and their velocity distribution and size distribution

$$\Phi = \iint \phi(v,r) \cdot dr \cdot dv, \qquad (2)$$

where $\phi(v,r)$ is the differential number of impacts as a function of the meteoroid velocity and radius. Several authors (Cintala 1992; Cremonese et al. 2005; Marchi et al. 2005) used the following relation

$$\phi(v,r) = f(v)h(r), \qquad (3)$$

where $f(v,r)$ is the differential velocity distribution of meteoroids (km^{-1} s), and $h(r)$ is the differential number of impacts per year and per unit of impactor radius on the entire surface of the planet (year^{-1} m^{-1}). These functions can be taken from Cintala (1992), for small meteoroids (10^{-8}–10^{-2} m), as follows

$$f(v) = \kappa d^{0.2} \left[\frac{v}{\sqrt{d(v^2 - v_{MEe}^2) + v_{Ee}^2}} \right]^3 e^{-\gamma \sqrt{d(v^2 - v_{MEe}^2) + v_{Ee}^2}}, \qquad (4)$$

where $\kappa = 3.81$, $\gamma = 0.247$ (km^{-1} s), d is Mercury's orbital distance in AU, $v_{MEe} = 4.25$ km s^{-1} is the escape velocity at the surface of Mercury and $v_{Ee} = 11.1$ km s^{-1} is the escape velocity for the Earth at 100 km altitude; and

$$h(r) = -\frac{3 S_M T}{4 \rho_P \pi r^3 F_1} \left[\sum_{i=1}^{11} i c_i \ln\left(\rho_P \frac{4}{3} \pi r^3\right)^{(i-1)} \right] \exp\left[\sum_{i=0}^{11} c_i \ln\left(\rho_P \frac{4}{3} \pi r^3\right)^i \right], \qquad (5)$$

where $F_1 = 0.373$ and the constants c_i ($i = 0, 1, \ldots, 11$) were reported by (Cintala 1992, Table A1, page 968), S_M is the area of the planet surface, and T is the number of seconds in 1 year. ρ_P is the meteoroid density, which is assumed to be 2.5 g cm^{-3} consistent with the measurements of the densities of stratospheric cosmic dust particles (Rietmeijer 1998) and with densities data of S-type igneous asteroids (e.g., Krasinsky et al. 2002), which are the main constituents of the inner part of the Main Belt. Equation (4) simply assumes that the flux is governed by the gravity field.

In the case of the large meteoroids (10^{-2}–10^{-1} m), $h(r)$ was given by (Marchi et al. 2005; Cremonese et al. 2005) as

$$h(r) = \frac{a_1}{r^{a_2}} \left[1 - a_3 \exp\left(-a_4 r^{0.5}\right) \right], \qquad (6)$$

where $a_1 = 1.22$, $a_2 = 3.7$, $a_3 = 0.511$, $a_4 = 0.85$. Equation (6) is valid for a large size range, but the calculations have been limited to an upper limit of 0.1 m, because meteoroids with radius larger than 0.1 m are not relevant to the daily production of the exosphere. Figure 24 shows the differential distribution $h(r)$ of number of impacts per year and per unit of projectile radius in the case of large meteoroids.

The vapor composition is determined by the target compositions impact velocity (i.e., Flynn and Stern 1996), then the production rate (atoms cm^{-2} s^{-1}) of the neutral Na, S_{Na}, is

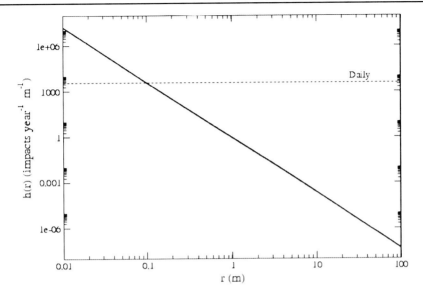

Fig. 24 The differential distribution $h(r)$ of number of impacts per year and per unit of projectile radius (on the entire surface of the planet) in the case of large meteoroids. The *horizontal line* corresponds to impacts that occur with a daily time scale

calculated by the following relation (Morgan and Killen 1997)

$$S_{Na} = M_{vap}\left(\frac{f_{Na}}{m_{Na}}\right)N_A, \qquad (7)$$

where M_{vap} is the vapor production rate (g cm^{-2} s^{-1}), m_{Na} is the atomic weight of Na, N_A is the Avogadro's number and f_{Na} is the mass fraction of Na in the regolith. M_{vap} has been calculated by solving the following equation

$$M_{vap} = \frac{\rho}{S_M T}\int_{v_{min}}^{v_{max}}\int_{r_{min}}^{r_{max}} \phi(v,r) V_{vap}(v,r)\,dv\,dr, \qquad (8)$$

where ρ is the target density (1.8 g cm^{-3}), $v_{min} = 4.25$ km s^{-1} and $v_{max} = 114$ km s^{-1}, $r_{min} = 10^{-8}$ m and $r_{max} = 0.1$ m.

Assuming that $f_{Na} = 0.038$, the Na production rate at mean orbit due to impact of meteoroids in the entire range considered (10^{-8}–0.1 m), is 1.82×10^6 atoms cm^{-2} s^{-1}, corresponding to 1.36×10^{24} s^{-1} reported in Fig. 25 as a function of the minimum meteoroid size (Cremonese et al. 2005).

The contribution of sodium to the exosphere due to large meteoroids, 10^{-2}–0.1 m, is less than 1%, and only the 7% of the Na comes from the impacts of meteoroids larger than 10^{-3} m. The number of impacts on the surface of Mercury are 6.69×10^{-7} impacts cm^{-2} s^{-1} in the size range of 10^{-8}–10^{-2} m and 2.59×10^{-21} impacts cm^{-2} s^{-1} in the range of 10^{-2}–10^{-1} m (Cremonese et al. 2005).

To estimate the contribution of the meteoroids to the production of Na the mass of meteoroids impacting the surface, for the entire size range, per unit of time and unit of surface

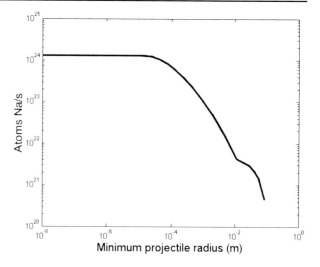

Fig. 25 Cumulative production rate of the neutral sodium atoms released in the vapor for the complete meteoroid size range, as a function of the minimum meteoroid radius. Each production rate value is due to all the meteoroids having a size larger than the corresponding radius in the x-axis. The rapid decline in the production rate at larger radius (at about 10^{-3} m) underlines the fact that the main contribution to the sodium production comes from this radius range

has been calculated as follows (Bruno et al. 2006)

$$F = \int_{v_{min}}^{v_{max}} \psi(v) \, dv, \qquad (9)$$

F is meteoritic flux in g cm^{-2} s^{-1}, and $\psi(v)$ is the differential meteoritic flux in g cm^{-2} s^{-1} (km/s)$^{-1}$

$$\psi(v) = \left(\frac{4\pi \rho_p}{3 S_M T}\right) \int_{r_{min}}^{r_{max}} r^3 \phi(v, r) \, dr, \qquad (10)$$

Figure 25 shows the cumulative production rate of the neutral sodium atoms released in the vapor for the complete meteoroid size range, as a function of the minimum meteoroid radius. Cremonese assumed that the atomic fraction of sodium in the regolith is $f_{Na} = 0.038$ (Cremonese et al. 2005) which yields a source rate of 2.91×10^{-16} g cm^{-2} s^{-1}. By assuming that the meteoroids are completely vaporized and have the same regolith composition ($f_{Na} = 0.038$), the Na derived from the meteoroid material is 5.3×10^5 atoms cm^{-2} s^{-1}. By adding this quantity to the previous one, they obtained a Na production of $\sim 2.3 \times 10^6$ atoms cm^{-2} s^{-1} if $f_{Na} = 0.038$, or 3.0×10^5 if $f_{Na} = 0.005$. This value of sodium production rate calculated by Cremonese et al. (2005, 2006) is higher than those reported by Hunten et al. (1988), 1.34×10^4 atoms cm^{-2} s^{-1}, but lower than that reported by Leblanc and Johnson (2003), 6.7×10^5 atoms cm^{-2} s^{-1}. It is in fair agreement with the upper limits given by Morgan et al. (1988), 1.9×10^6 atoms cm^{-2} s^{-1}. Killen et al. (2001) used $f_{Na} = 0.005$, and obtained a sodium source rate of 2.2×10^6 atoms cm^{-2} s^{-1} at perihelion and 1.15×10^5 at aphelion. Normalizing to the same composition, Cremonese et al. (2005) would have obtained a source rate for Na of 3.0×10^5 at mean orbit, in good agreement with the Killen et al. (2001) results.

3 Exospheric Sources and Their Models

3.1 Photon and Electron Stimulated Desorption of Surface Elements

In addition to the solar wind, plasma related to Coronal Mass Ejections (CMEs), cosmic rays and solar energetic particles, Mercury's surface environment is continuously bombarded by solar radiation and electrons. Solar radiation in the form of infrared and visible and energetic photons is absorbed by the surface and causes heating and alteration of the dayside surface area. This heating can reach dayside temperatures on Mercury of up to about 700 K at the planet's equator, which is relevant for thermal desorption. Moreover, solar photons with energies ≥ 4 eV can induce bond-breaking, which is important for photon-stimulated desorption (PSD) of absorbed elements. Lower energy electrons in the order of tens of eV are also important in addition to surface charging because they can cause electronic excitations that also lead to electron stimulated desorption (ESD) of adsorbed elements from the planetary surface. In the following subsections we describe these processes and their expected role in refilling Mercury's exosphere in more detail.

3.2 Photon-Stimulated Desorption

Photon-stimulated desorption (PSD) corresponds to the desorption of surface elements as a result of electronic excitation by a photon of a surface atom. Madey and Yakshinskiy (1998) showed that this process efficiently ejects alkalis from surfaces under laboratory conditions. Yakshinskiy and Madey (2004, 1999) found from their laboratory experiments with Na-covered lunar basalt samples that UV photons with energies of about 3–5 eV or greater than 5 eV cause desorption of "hot" Na atoms. This process acts through electronic transitions such as band gap excitation, valence electron excitation, or core excitation. Near-UV photons with energies ≤ 4 eV caused little or no detectable desorption of Na. Yakshinskiy and Madey (2004) estimated the PSD cross-section σ at photon energies of ≈ 5 eV to be about 10^{-20} cm^2, which is about seven times larger than that used by Killen et al. (2001). They found from their experiments that the desorbed Na atoms are suprathermal with a velocity peak in the PSD distribution of about 900 m s^{-1}. It was also discovered that desorption of Na varies with surface temperature and increased by a factor of 10 after the sample was heated from about 100 K to about 470 K (Yakshinskiy and Madey 2004). Cassidy and Johnson (2005) estimated that desorption from a regolith is reduced by about a factor of about three from that on a flat surface.

Bombardment of lunar silicates by UV photons ($\lambda < 300$ nm) was found to produce efficient desorption of Na atoms (Yakshinskiy and Madey 1999). The flux of atoms of species X desorbed by PSD can be given by

$$\phi_X = f_X N_S \int \phi_{ph}(\lambda) Q_X(\lambda) \, d\lambda, \tag{11}$$

$$\phi_{Na}^{PSD} = \frac{1}{4} \phi_{ph} Q_{Na} f_{Na} N_S, \tag{12}$$

where ϕ_{ph} is the solar UV flux at Mercury, $Q_X(\lambda)$ is the PSD cross-section for species X at wavelength, λ, f_X is the fraction of species X in the regolith, and N_s is the total regolith surface density in number of atoms cm^{-2}/mean free path. The experimental PSD cross section for Na has been given as $Q_{Na} = 1$–3×10^{-20} cm^2, integrated over the effective wavelength range, 250–400 nm (Yakshinskiy and Madey 1999). However, the actual yield in a regolith

Table 7 Solar UV flux at Earth's orbit compared with periherm (0.29 AU) and apoherm (0.44 AU)

Solar photons	1 AU [cm^{-2} s^{-1}]	0.44 AU [cm^{-2} s^{-1}]	0.29 AU [cm^{-2} s^{-1}]
UV-A (3.1–3.9 eV)	2×10^{16}	1×10^{17}	2.4×10^{17}
UV-B (3.9–4.4 eV)	2.5×10^{15}	1.3×10^{16}	3×10^{16}
UV-C (4.4–12.4 eV)	1×10^{14}	5.15×10^{14}	1.2×10^{14}

is reduced, perhaps by a factor of three (Cassidy and Johnson 2005) due to the possibility that ejected atoms will stick to grain surfaces before they can emerge from the regolith. The regolith surface number density is often assumed to be $N_s = 7.5 \times 10^{14}$ cm^{-2}/MFP, where MFP is the photon mean free path. The solar UV flux at Mercury integrated from 100–318 nm is $\phi_{ph} = 3.31 \times 10^{15}$ cm^{-2} s^{-1}/R_m^2, where R_m is Mercury's orbital distance from the Sun in AU. The sodium fraction in the lunar regolith is 0.0053, and this value has been extensively assumed for the sodium fraction in Mercury's regolith.

Photon-stimulated desorption is induced by electronic excitations rather than by thermal processes or momentum transfer. However, a temperature dependence in the yield was found and attributed to diffusion rates to the extreme surface (Yakshinskiy and Madey 2004). The velocity distribution of emitted atoms can be described by a Weibull distribution, which has a high-velocity tail. The energy distribution of the emitted atoms from electron stimulated desorption (ESD) has been given by Johnson et al. (2002) as

$$f_{\text{PSD}}(E) = \beta(1+\beta)\frac{EU^\beta}{(E+U)^{2+\beta}}, \tag{13}$$

where E is the energy of the emitted particle, U is the characteristic energy for PSD of a given species, and β is the shape parameter of the distribution. The shape parameters for Na and K were given by Johnson et al. (2002) as $\beta_{\text{Na}} = 0.7$ and $\beta_{\text{K}} = 0.25$. The peak of the Na velocity distribution from a PSD source is similar to that for a 1,100 K gas (Yakshinskiy and Madey 1999), corresponding to a $U = 0.0098$ eV in (13).

The temperature dependence for PSD, as given by the Yakshinskiy and Madey data, was fit by the following expression $(1.1448\text{E-}5 T^2 - 0.00163 T + 0.02128)/1.8$ (Killen, in preparation) where the maximum surface temperature used in this correction is 475 K. The desorption rate from a plane surface probably differs from that of a regolith due to the probability that a desorbed atom in a regolith will collide with a surface element before reaching the extreme surface (Cassidy and Johnson 2005). PSD is not effective in ejecting refractory species.

To investigate the PSD-induced release of atoms from the surface of Mercury, the variation of solar UV photons incident on the planetary surface over the planet's orbit have to be considered. The solar flux at Mercury's eccentric orbit differs substantially from the average condition present at 1 AU. Table 7 shows the solar UV-A, UV-B and UV-C flux at periherm of about 0.29 AU and apoherm at about 0.44 AU. One can see from Table 7 that the solar UV flux is about a factor 11 higher at the periherm and more than five times higher at apoherm than that at the Earth's orbit in 1 AU.

Lammer et al. (2003), studied the PSD-induced release of Na and K atoms along Mercury's orbit. The flux of desorbed Na atoms by incoming solar UV photons could be calculated by using (12) (Yakshinskiy and Madey 1999; Lammer et al. 2003). The study of Lammer et al. (2003) showed clearly that the largest PSD fluxes of released Na occur near equatorial latitudes at periherm, where the flux could reach values depending on used PSD

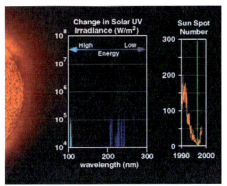 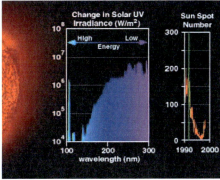

Fig. 26 The *left panel* shows the solar UV irradiance at Earth's orbit during low solar activity, while the *right panel* shows the solar UV irradiance at Earth's orbit during high solar activity where the irradiance could be up to 100 times higher than during quiet solar periods (M. Schoeberl and H. Mitchell, UARS/SUSIM, NASA-GSFC/SVS)

cross-section of about $4.5 \times 10^6 - 3.15 \times 10^7$ cm^{-2} s^{-1}. At apoherm the Na flux values at the equatorial regions are about three times lower.

The PSD-induced Na fluxes at latitudes higher than 75° are lower. Lammer et al. (2003) concluded that there should not be any noticeable PSD sources at Mercury's polar areas. The PSD-induced Na fluxes at the equatorial regions at apoherm and periherm of Lammer et al. (2003) are lower than the estimated fluxes of McGrath et al. (1986) of about $2.0 \times 10^7 - 2.0 \times 10^8$ cm^{-2} s^{-1} but larger than the estimated average flux value of about 2.0×10^7 cm^{-2} s^{-1} by Killen et al. (2001) and Killen and Ip (1999). A reason for the larger fluxes estimated by McGrath et al. (1986) could be that the fluxes estimated by these authors are overly optimistic since they were based on data for alkali halides (Killen and Morgan 1993).

By estimating the PSD-induced Na source rates one obtains between periherm and apoherm values in the order of about 1.0×10^{24} s^{-1}, which is in agreement with the observations by Killen et al. (2001) who found 7.6×10^{23} s^{-1} for November 13 and 1.4×10^{24} s^{-1} for November 20, 1997.

However, as one can see in Fig. 26 the solar UV flux increases up to about 100 times from low solar activity to active solar periods or during flare events. Therefore, one may expect that the Na PSD flux could reach values during high solar activity periods or flare interaction with Mercury of the order of about $\geq 10^8$ cm^{-2} s^{-1}.

A second element that can be desorbed from Mercury's surface due to solar UV radiation is K. Laboratory experiments by Madey et al. (1998) regarding the desorption of alkalis on oxide surfaces yield PSD cross-sections for K atoms which vary between $1.4 \pm 0.6 \times 10^{-20}$ cm^2 and $1.9 \pm 0.8 \times 10^{-21}$ cm^2 for wavelengths between 247.5 nm (5.0 eV) and 365 nm (3.5 eV). The most efficient cross-section in their experiments is about 1.8×10^{-20} cm^2 at 253.7 nm (4.9 eV). The exospheric observation by Potter and Morgan (1997) give an upper limit to Na/K ratio of about 200, which corresponds to PSD-induced K fluxes in the order of about 10^4 cm^{-2} s^{-1} for K atoms corresponding to the observed Na/K fractionation along a latitude strip that is directly facing the Sun. Because the Na/K ratio is extremely variable in the exosphere (Potter et al. 2002a) it is likely that loss rates play a role as well as source rates.

However, note that real regolith on Mercury's surface is different from the material studied by Madey et al. (1998). Since it has been irradiated, the alkali binding could be altered, the porosity of the material is unknown and sticking could be an efficient process. These factors can cut down the cross-section by more than an order of magnitude. Further, the surface layer can be depleted in alkali (Hapke 2001; Madey et al. 1998).

3.3 Electron-Stimulated Desorption

In addition to PSD experiments with adsorbed Na and K atoms on lunar silicates, Yakshinskiy and Madey (1999, 2004) also studied electron-stimulated desorption (ESD). They found that exposure of Na covered surfaces by low energetic electrons with energies from 3–50 eV causes also desorption of "hot" Na atoms. Generally there are intimate connections between ESD and PSD because similar electronic processes cause desorption of the atoms via electron or photon excitation. The release of Na via ESD is strongly temperature dependent. The observed average ESD cross-section for Na atoms and 10–50 eV electrons is about 1–2×10^{-19} cm^2 (Yakshinskiy and Madey 1999, 2004). The experimentally determined ESD cross-section for atomic Na has its initial threshold at about 4 eV, which is comparable with the PSD threshold. Furthermore, the desorption cross-sections have a similar magnitude for electron energies of about 5 eV and there is a resonance like feature at about 11 eV (Yakshinskiy and Madey 1999). If one assumes quasi-neutral solar wind plasma, then electron fluxes for electrons with energies of about 12 eV are about $5 \times 10^9 - 2 \times 10^9$ cm^{-2} s^{-1} at periherm and apoherm, respectively. The flux of adsorbed elements like Na or K due to ESD can be calculated by

$$\Phi_{\text{ESD}} = \Omega \Phi_e \sigma_{\text{ESD}} f, \qquad (14)$$

where Ω is the area where the electrons can reach the planetary surface divided by the whole planetary surface, Φ_e is the electron flux at Mercury's environment, σ_{ESD} is the ESD photon cross-section and f is the composition of the regolith abundance for instance of Na in Mercury's surface. Because Φ_e is several orders of magnitude smaller than the photon flux Φ_v, ESD during ordinary solar wind conditions will not be an efficient release process for adsorbed surface elements like Na or K compared to PSD, sputtering or micrometeoroid evaporization.

Leblanc et al. (2003b) considered a particular SEP event with particle energies larger than 10 keV. The event was observed in detail at the Earth's orbit (Reames et al. 1997) and rescaled to Mercury's orbit. Generally SEPs can reach Mercury before or few hours after the arrival of the shock and a magnetic cloud usually associated with a Coronal Mass Ejection (CME) that is for an unperturbed Mercury's magnetosphere for quiet solar wind conditions. Leblanc et al. (2003b) studied test particles representative of the energy flux distribution for each SEP ion species and for the electrons. They launched the particles from the magnetopause and followed them inside Mercury's magnetosphere and surface by using a magnetospheric model of Luhmann et al. (1998). They found that these particles can cause ESD of Na atoms and, because they penetrate more than the solar wind ions or UV photons, they can enhance the supply of Na atoms to the surface where it can be desorbed by thermal or PSD.

3.4 Particle Surface Sputtering

The impact of energetic ions on a solid surface (e.g., Mercury's surface) will cause the release of particles via momentum transfer, which is called sputtering, or more precisely physical sputtering. Particle sputtering will release all species from Mercury's surface into space, reproducing more or less the local surface composition on an atomic level. Preferential sputtering of the different elements of a compound will lead to a surface enrichment of those elements with low sputtering yields in the top-most atomic layers. However, the steady-state composition of the flux of sputtered atoms will reflect the average bulk composition. Thus, particle sputtering, when operative, will give us compositional information about the refractory elements of the bulk surface.

The normalized energy distribution for particles sputtered from a solid, $f(E_e)$, with the energy E_e of the sputtered particle, has been given as (Sigmund 1969)

$$f(E_e) = \frac{6E_b}{3 - 8\sqrt{E_b/E_c}} \frac{E_e}{(E_e + E_b)^3} \left(1 - \sqrt{\frac{E_e + E_b}{E_c}}\right), \tag{15}$$

where E_b is the surface-binding energy of the sputtered particle and E_c the cut-off energy for sputtered atoms. The cut-off E_c, which is the maximum energy that can be imparted to a sputtered particle by a projectile particle with energy E_i, is given by the limit imposed by a binary collision between a projectile atom, m_1, and the target atom, m_2, (to be sputtered) as

$$E_c = E_i \frac{4m_1 m_2}{(m_1 + m_2)^2}. \tag{16}$$

Figure 27 shows the normalized energy distribution for several elements. Note that the maximum of the energy distribution is at $E_{max} = E_b/2$, with E_b ranging from fractions of an eV to several eV depending on species and mineral/matrix. At higher energies the distribution falls off with E_e^2 until the energy E_e approaches the cut-off energy E_c.

The polar angle distribution of sputtered atoms, $f(\alpha)$, for polycrystalline surfaces is best described by a quadratic angular dependence, $f(\alpha) \propto \cos^2 \alpha$ for laboratory experiments (Hofer 1991). By modeling the details of sputtering in loosely packed regolith grains, Cassidy and Johnson (2005) found that for a fine-grained and porous regolith a better choice is $f(\alpha) = \cos \alpha$. For the azimuth angle a uniform distribution over 2π is a suitable description. Having the energy, the azimuth, and elevation angle one can calculate all three components of the initial particle velocity, **v**, and the trajectory of each sputtered particle in the exosphere. Using many such trajectories a vertical density profile $N_i(h)$ can be calculated (Wurz and Lammer 2003; Wurz et al. 2007). The density profile can easily be integrated to obtain the column content, which is the typical measurement obtained from telescopic observations of the exosphere. Either the exospheric density at the surface or the column content can be used to compare with observational data. The flux Φ_i of atoms sputtered from the planetary

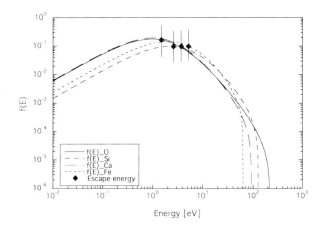

Fig. 27 Normalized energy distributions for sputtered O, Si, Ca, and Fe atoms by impact of protons of 1 keV energy using (1). The *symbols* indicate the energy corresponding to the escape velocity of each sputtered atom

surface can be calculated as

$$\Phi_i = \Phi_{\text{ion}} Y_i^{\text{tot}} = \Phi_{\text{ion}} Y_i^{\text{rel}} C_i, \qquad (17)$$

where Φ_{ion} is the energetic ion flux onto the surface and Y_i^{tot} the total sputter yield of species i; that is, the number of released surface atoms per incoming ion, from the surface with a given elemental composition. The total sputter yield Y_i^{tot} can be broken up into a relative sputter yield Y_i^{rel} and C_i the atomic abundance of species i on the surface. The total sputter flux of species i can also be written as

$$\Phi_i = N_i(0) \langle v_i \rangle, \qquad (18)$$

with $N_i(0)$ the exospheric particle density at the surface ($h = 0$), and $\langle v_i \rangle$ the average velocity of sputtered particles. Combining (17) and (18), the exospheric density at the surface resulting from the sputter process for species i is

$$N_i(0) = \Phi_{\text{ion}} Y_i^{\text{tot}} \frac{1}{\langle v_i \rangle}, \qquad (19)$$

$N_i(0)$ can be used in a calculation as a starting point to derive a quantitative density profile from the sputtering process for a given surface composition. The average release velocity is derived from the sputter distribution, (1), as

$$\langle v_i \rangle = \frac{3\pi}{4} \sqrt{\frac{2 E_{b,i}}{m_2}}, \qquad (20)$$

where $E_{b,i}$ is the binding energy of species i in the particular chemical/mineralogical mix of the surface (Wurz et al. 2007). Note that the most probable velocity is $v_{mp} = \sqrt{E_{b,i}/m_2}$, which is lower than the average release speed by a factor of about 3.3. These velocities have to be compared to the Hermean escape speed of 4.250 km s^{-1}. If we take oxygen as an example, with a binding energy of $E_b = 2.0$ eV, we get $\langle v_i \rangle = 11.57$ km s^{-1}, which exceeds the escape velocity considerably. The same is true for other elements. Thus, many sputtered atoms escape Mercury's gravity field. This can been seen in Fig. 27 where the energy corresponding to the escape speed is indicated in the energy distribution of sputtered atoms.

Therefore, if the flux of ions impinging the planetary surface, Φ_{ion}, is known one can calculate ab initio, with the sputter yields Y_i^{tot} for a particular surface composition, the sputtered flux, the surface density, the density profile, and the column content and compare these numbers with the observations. This has been done for the Moon recently (Wurz et al. 2007).

In addition to atoms, clusters of two and more atoms can be released from the solid surface via sputtering. For metallic surfaces the release of metallic dimers, trimers, etc. with yields of about $10^{-1}, 10^{-2}, \ldots$ with respect to the atomic sputter yield, has been observed in the laboratory (Gnaser and Hofer 1989; Wurz et al. 1990, 1991; Hansen et al. 1998). For oxide surfaces, which are more representative to the mineralogical surface of Mercury, one observes monoxides like SiO or CaO and larger oxide molecules, with yield ratios up to [Me/MeO] ~ 1 (Oechsner et al. 1978; Wucher and Oechsner 1986; Wurz et al. 1990). The sputter yield of such dimers (metallic and oxides) depends strongly on the chemical environment on the surface, mostly on the surface being oxidized or metallic. This has been investigated by performing sputter experiments on clean metallic surfaces that were oxidized in a controlled way (Wurz et al. 1990; Hansen et al. 1999).

It was found that the oxide molecule yields correlated strongly with the oxygen coverage, with the sputtered monoxide yield as large as the atom yield at maximum oxygen coverage, and the metal clusters anticorrelated with the oxygen coverage. The energy distribution of sputtered clusters is similar to the energy distribution of sputtered atoms, as given in (1) (Wucher and Oechsner 1986; Hansen et al. 1998), but the fall-off at higher energy of sputtered particles is $f(E_e) \propto E_e^{-n}$ with $n > 2$ (Coon et al. 1993; Betz and Husinsky 2004).

Since sputtering is a quite energetic process sputtered clusters have temperatures of several 1,000 K when released, which limits their stability against falling apart. So far clusters of sputtered atoms have not been observed directly in Mercury's exosphere, but it has been proposed that a significant contribution to the Ca exosphere arises from sputtered CaO molecules that fall apart at high altitudes (Killen et al. 2005).

A fraction of the sputtering atoms are positive or negative ions. The ion fraction of the sputtered atoms can be in the range between 10^{-4} and a few 10^{-1}, depending on element, matrix, and primary ion (Benninghoven et al. 1987). Sputtered ions were used for compositional analysis of ground-up sample simulants of mare and highland soils to mimic the sputtering behavior of lunar regolith (Elphic et al. 1993). Because of the lack of an atmosphere and ionosphere for Mercury, sputtered ions immediately are picked up by the electro-magnetic fields of Mercury's magnetosphere or escape with the solar wind.

Large areas on Mercury's surface are exposed to solar wind even during regular solar wind conditions, and solar wind ions are the most important ion population causing sputtering from Mercury's surface. Solar wind velocities are in the range of 300 to 800 km/s (slow and fast solar wind), which translates to energies of 0.5 to 3.3 keV/amu, and with a typical value of 1 keV/amu. Note that the sputter yield has a maximum around ion energies of 1 keV/amu. In the solar wind, protons and alpha particles make up more than 99% of the ions, and heavy ions (from carbon to iron and up) together are about 0.1% of the solar wind ions in the number flux (Wurz 2005, and references therein).

For the lunar surface total sputter yields are about 0.07 surface atoms per impinging ion for the typical H/He mix of solar wind ions (Wurz et al. 2007). Heavier ions have sputter yields even larger than 1, but because their abundance in the solar wind is very low their contribution to the total sputter yield of the solar wind is negligible. The He/H ratio in the normal solar wind is about 0.04 (Aellig et al. 2001), with He abundance varying from about 0.02 at solar minimum to 0.06 at solar maximum. For solar wind speed below 350 km/sec the average He/H is 1.8%. Increasing solar wind speed implies increasing He abundance in the solar wind, where He/H $= a + b v_{\text{sw}}$, where a and b depend on the solar cycle (Aellig et al. 2001).

During coronal mass ejections (CMEs) the He/H ratio can be enhanced to about 0.2 in a CME (Sarantos et al. 2007), which will increase the sputter yield accordingly. Moreover, the abundance of heavy elements can also be significantly increased in CMEs with respect to undisturbed solar wind (Wurz et al. 2001, 2003), which may become important for the sputter yields. Heavy ions are also highly enriched in SEP events (e.g., Wiedenbeck et al. 2005; Cohen et al. 2005); however, at these energies the sputter yields are very low since the particles penetrate far into the solid.

It is expected that the sputter yield on Mercury's surface will be about the same as that on the moon. The sputter yield has to be reduced by the porosity of the surface (Cassidy and Johnson 2005), thus we estimate that the actual sputter yield for Mercury's surface is between 0.02 and 0.025 surface atoms per impinging ion for the typical mix of solar wind ions.

The heavy ions in the solar wind are highly charged because of the million-degree hot solar corona. Oxygen, for example, is present in the solar wind with charge states of typically +6 and +7; iron is present with charge states in the range from +8 to +12. These high charge states mean that the ions have high internal energies (potential energies), for example, 295 eV for O^{6+} and 1055 eV for Fe^{10+}, as compared to singly or doubly charged ions. The charge state affects the sputtering yield due to the potential (ionization) energy available, hence the term "potential sputtering". However, these high internal energies (potential energies) have to be compared to their kinetic energies in the solar wind of typically 16 keV for oxygen and 56 keV for iron. It has been argued that the sputter yield for highly charged ions impacting on a planetary surface is increased by a factor of 10 to 1,000 as a result of their high internal energy (Shemansky 2003). The laboratory data on sputter yields for highly charged ions have been reviewed by Aumayr and Winter (2004), and we will briefly summarize their findings here. For metallic surfaces and semiconductors (Si and GaAs) no deviation of the sputter yield for highly charged ions from the sputter yield of singly charged ions was found, with the highest charge states investigated being Ar^{9+} and Xe^{25+}. Moreover, all the measured sputter yields agree with the TRIM calculations, a software package which considers only the kinetic energy of the impacting ion (Ziegler and Biersack 1985).

For ionic crystals (NaCl and LiF) a pronounced increase with ionic charge state was observed; for NaCl the sputter yield increased by a factor of 4 for Ar^{8+} ions compared to Ar^+ ions, for LiF the sputter yield increased by a factor of 25 for Ar^{14+} ions compared to Ar^+ ions. Note that Ar charge states in the solar wind range from +8 to +11.

For oxides, which are the best analog for Mercury's surface, a clear signature of a sputter yield increase for highly charged ions was observed for SiO_2 and Al_2O_3. For SiO_2 this increase was about 3 for Ar^{8+} ions compared to Ar^+ ions, and about 65 for Xe^{25+} ions compared to Ar^+ ions. Similar enhancements were found for the Al_2O_3 surface. Measured sputter yields of 1.5 keV Xe^{q+} onto Al_2O_3 show an approximately 40-fold increase in the sputter yield for Xe^{28+} over that of Xe^{9+}. Both of these materials appears to have a finite sputter yield at zero kinetic energy of the projectile. On the other hand, for a highly ionic oxide such as MgO, even though potential energy greatly increases the sputter yield, potential energy does not induce sputtering in the absence of kinetic energy of the projectile.

However, this enhancement is strongly depending on the ion dose the surface has been exposed to. After a removal of about a monolayer from the oxide surface the sputter yield for highly charged ions drops to about the values for singly charged ions. Removal of a monolayer of surface material corresponds to a heavy ion flux of a few 10^{13} ions $cm^{-2} s^{-1}$ at solar wind energies, which takes about two weeks at Mercury's orbit. This reduction in sputter yield is attributed to the very surface becoming reasonably conductive (by preferential loss of oxygen and the creation crystal defects) and thus the highly charged ions become decharged; that is, they lose their internal energy when they approach the surface.

To model ion sputtering it is important not only to model Mercury's magnetosphere but also to understand the composition of the solar wind at Mercury and its variability. These effects are being considered to correctly characterize the ion sputter source (Sarantos et al. 2007).

3.5 Surface Maps of Particle Fluxes and Energies

Sputtering of Na by solar wind ions impinging onto the surface of Mercury through the cusps of the magnetosphere was first suggested by Potter and Morgan (1990) to explain rapid variations in the observed Na exosphere, with high- to mid-latitude enhancements appearing and disappearing on intervals of less than a day. Such variations cannot be attributed to PSD which varies slowly with true anomaly angle of the planet, and which would be characterized by a sub-solar maximum in the distribution. Variations in the sodium exosphere during a week-long sequence in November 1997 were shown to correlate with possible variations in Mercury's magnetosphere such that increased ion sputtering correlated with opening of the cusp regions and an increased ion flux to the surface (Killen et al. 2001). The role of the precipitation of the solar wind plasma has received a noticeable interest because ion sputtering is proposed as a potential candidate to explain rapid temporal variations in Mercury's Na exosphere observed in Earth-based remote sensing measurements (e.g., Potter et al. 1999; Wurz and Lammer 2003; Leblanc et al. 2003b).

Sputtering occurs mostly due to solar wind precipitation, even if planetary ions may contribute as well (Delcourt et al. 2003). Solar wind particles are expected to precipitate in the dayside cusps (Massetti et al. 2003; Kallio and Janhunen 2003); during this motion, a large fraction (~90%) of the protons are bounced by the increasing magnetic field, the others (10%) reach the surface of Mercury and lead to ion-sputtering, or (1%) are neutralized due to charge-exchange effect (Mura et al. 2006a). During this motion, protons do not exactly follow magnetic field lines: particles are drifted northward by the $\mathbf{E} \times \mathbf{B}$ drift and westward by the grad-$\mathbf{B}$ drift (Mura et al. 2005).

The former is energy-independent; the latter is more efficient for the highest energies. Moreover, non-adiabatic effects may become of crucial importance for most of the ion magnetospheric transport (Delcourt et al. 2003; Massetti et al. 2007). Figure 28 shows an example of maps of surface precipitation for high- (1–10 keV) and low-energy (100 eV–1 keV) protons; high-energy protons (up to several keV), accelerated by the reconnection mechanism, precipitate at lower latitudes with respect to low-energy ones (see Massetti et al. 2003).

The intensity and the shape of the H^+ flux depend on the magnetospheric configuration which, in turn, depends on both the intrinsic magnetic field of Mercury and variable external parameters, such as the interplanetary magnetic field (IMF), and solar wind velocity and density (Kabin et al. 2000; Kallio and Janhunen 2003; Massetti et al. 2003).

Even if Mariner 10 measurements revealed the existence of an intrinsic magnetic field, this estimation has a considerable uncertainty, because it is difficult to separate the internal and external magnetic field components (Connerney and Ness 1988). Nevertheless, the di-

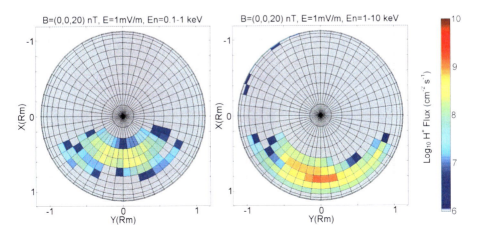

Fig. 28 Example of solar wind proton precipitating flux onto the north hemisphere, in an open magnetospheric configuration ($B_{IMF} = [0, 0, -20]$ nT). *Left panel*: low energy (<1 keV) protons; *right panel*: high energy (>1 keV) protons. Adapted from Mura et al. (2006c)

pole moment is probably between 284 and 358 nT R_M^3 (Ness et al. 1975); for comparison, the Earth's dipole is approximately 3×10^4 nT R_E^3. Peculiarities in Mercury's magnetosphere arise also from the specific conditions of the solar wind at Mercury's orbit (0.31–0.47 AU), which differ substantially from the average conditions present at 1 AU.

Parker spiral forms an angle of about 20° with the solar wind radial direction, while it is approximately 45° at the Earth. This implies a change of the relative ratio of the IMF components with respect to the near Earth conditions, and a modified solar wind–magnetosphere relationship. The average solar wind density is about ten times higher than at the Earth, and this value varies considerably due to the high eccentricity of the orbit of the planet, with average densities from 34 cm^{-3} at the aphelion, to 83 cm^{-3} at the perihelion. The dynamic pressure is, on average, approximately 16 nPa (Massetti et al. 2003). By applying all the above parameters, it has been estimated that the sub-solar point, where the internal and external pressures balance, is about 1.5 R_M out from Mercury's center (Siscoe and Cristofer, 1975; Goldstein et al. 1981), while at the Earth this value is 11 R_E. In this respect, the scale-length of Mercury's magnetosphere is 1/7 compared to the Earth's.

Ion precipitation at Mercury has been a subject of several analyzes motivated by Mariner 10 magnetic field and electron measurements made in 1974–1975. Unfortunately, there are no direct ion measurements available from the Mariner 10 flybys. Many studies of the solar wind-Mercury interaction have focused primarily on analyzing the role of the interplanetary magnetic field (Luhmann et al. 1998; Kabin et al. 2000; Killen et al. 2001; Sarantos et al. 2001; Ip and Kopp 2002; Kallio and Janhunen 2003; Massetti et al. 2003), and on the solar wind dynamic pressure (Sarantos et al. 2007), the motion of ions in the Hermean magnetosphere produced from the Hermean exosphere and emitted the surface (Ip 1987; Delcourt et al. 2003, 2002; Killen et al. 2003), and the motion of the solar wind protons injected from the tail (Lukyanov et al. 2001).

Both uncertain quantities and the large number of possible configurations make it very difficult to include all of them in a model of proton precipitation. Evaluation of H$^+$ flux on the surface may be obtained from MHD simulations (Leblanc et al. 2003a), quasi-neutral hybrid MHD simulations (Kallio et al. 2003), single particle models (Mura et al. 2005) and loss-cone estimations (Massetti et al. 2003); in the first two cases, the magnetic field is obtained from the model; in the other cases, it must be provided separately (for example, it may be reconstructed by adapting a magnetic field model of the Earth to Mercury's case (Massetti et al. 2003), assuming proper values for the IMF, the solar wind density and velocity). These precipitation models generally prescribe two areas of intense precipitation, roughly corresponding to the cusps regions, or a big area of precipitation located in the planetary dayside, depending on external conditions or simulation assumptions.

Table 8 summarizes some relevant quantities related to solar wind proton and planetary ion precipitation.

The H$^+$ flux onto the surface of Mercury may exceed values of 10^9 cm^{-2} s^{-1}, and the total integrated H$^+$ flux onto the surface of Mercury can be estimated as about 10^{25} s^{-1}. During solar energetic particle (SEP) events, high-energy integrated proton flux may be up to 10^{26}–10^{27} s^{-1} (Leblanc et al. 2003a). Alpha particles exhibit a smaller flux (approximately one tenth) but the yield for sputtering is considerably higher, so that they are expected to contribute to this process as well. The flux of the exospheric ions, like sodium, is much lower, and can reach 10^6–10^7 cm^{-2} s^{-1} (Delcourt et al. 2003; Leblanc et al. 2003b); those ions can precipitate and generate sputtering also in the nightside. The maximum sputtering flux due to this process has been estimated by Delcourt et al. (2003) and is around 10^4 cm^{-2} s^{-1}. The flux of ions to the Hermean surface depends strongly on solar wind dynamic pressure and on IMF (e.g., Kallio and Janhunen 2003; Delcourt et al. 2003), and strong disturbances such as SEP or CME events (e.g., Cohen et al. 2005).

Table 8 Flux related to solar wind proton and planetary ion precipitation

Quantity	Value	References, notes
Flux (cm^{-2} s^{-1})	1.5×10^8	McGrath et al. (1986) (H$^+$)
	4×10^8	Massetti et al. (2003) (H$^+$, IMF = $|0, 0, -10|$ nT)
	2×10^9	Massetti et al. (2003) (H$^+$, upper limit)
	2×10^9	Mura et al. (2005) (H$^+$, upper limit)
	$10^5 - 10^6$	Delcourt et al. (2003) (Na$^+$, perihelion-aphelion)
Integrated flux (total), (s^{-1})	8×10^{24}	Leblanc et al. (2003b) (SEP, H$^+$, 10 keV – 10 MeV)
	3×10^{23}	Leblanc et al. (2003b) (SEP, He2$^+$)
	1.1×10^{25}	Massetti et al. (2003) (H$^+$, IMF = $|0, 0, -10|$ nT)
	4×10^{25}	Mura et al. (2005) (H$^+$ IMF = $|0, 0, -20|$ nT)
	3.9×10^{25}	Kallio and Janhunen (2003) (H$^+$, IMF = $|0, 0, 10|$ nT)
	3.4×10^{25}	Kallio and Janhunen (2003) (H$^+$, IMF = $|0, 0, -10|$ nT)
	2.7×10^{25}	Kallio and Janhunen (2003) (H$^+$, IMF = $|32, 10, 0|$ nT)
	30×10^{25}	Kallio and Janhunen (2003) (H$^+$, High USW)
Fraction of precipitating ions	8%	Leblanc et al. (2003b) (SEP, H$^+$, 10 keV – 10 MeV)
	11%	Leblanc et al. (2003b) (SEP, other species)
	10%	Mura et al. (2005) (H$^+$)
Open field area (cm)	7.3×10^{16}	Killen et al. (2001) (H$^+$)
	2.8×10^{16}	Massetti et al. (2003) (H$^+$, IMF = $|0, 0, -10|$ nT)
	1.8×10^{17}	Mura et al. (2005) (H$^+$ IMF = $|0, 0, -20|$ nT)

One criticism of this source process is that the sputtering efficiency of protons, the dominant ion in the solar wind, is quite small and cannot account for significant sputtering (Koehn et al. 2003). When the IMF has either a negative B_z or a strong B_x component, magnetic reconnection occurs between IMF and the Hermean magnetic field. In this case, H$^+$ particles from the magnetosheath can cross the magnetopause, enter the magnetosphere and precipitate following open field lines. The shape of proton flux onto the surface (and, hence, of the ion-sputtering flux) approximately mimics the shape of the reconnection at the MP.

In a purely IMF-B_z case, reconnection occurs only if B_z is negative; however, the B_x component plays a substantial role in the magnetosphere solar wind coupling (Kabin et al. 2000; Sarantos et al. 2001; Kallio and Janhunen 2003; Massetti et al. 2007). In fact, the different angle of the Parker spiral at Mercury compared to the Earth suggests that the B_x component could be dominant. Non-zero B_x values introduce a north–south asymmetry in the H$^+$ surface precipitation; this feature is very important because it can easily explain the north–south asymmetries in the earth-based observation of some exospheric components (e.g., sodium).

An asymmetry in impacting ion flux translates into an asymmetry in the neutral atmosphere only when ion sputtering is a strong comparative source of atmospheric neutrals. This generally happens when the IMF is southward ($B_z < 0$). Since southward fields were observed roughly only half the time in Helios I and II data, a north–south asymmetry in impacting proton flux should be expected for only 16–28% of Mercury's lifetime. Still, visible asymmetries could sometimes result from a northward IMF when photon-stimulated desorption is weak—namely at aphelion. At a first look, this result implies that north–south asymmetries in impacting ion flux might be quite frequent at Mercury.

When the radial component of the IMF is dominant, most of the precipitating solar wind ions impact on one hemisphere: for a sunward-pointing component ($B_x > 0$) most ions reach the surface of Mercury's southern hemisphere; the opposite is true for an anti-sunward-pointing field ($B_x < 0$). This north–south asymmetry was first pointed out by Sarantos et al. (2001) who qualitatively showed that, for a negative B_x, solar ions have a velocity component parallel to Mercury's (open) magnetic field configuration in the northern hemisphere but antiparallel in the southern hemisphere, and vice versa. Sarantos et al. (2007) analyzed the Helios I and II magnetic field data collected while the spacecrafts were within Mercury's orbital range. A dominant B_x was defined as one being at least twice that of either of the other two components. Results show that the B_x was "dominant" according to the above definition between 32–57% of all time. (Each set of data is defined as one spacecraft pass between 0.31 AU to 0.47 AU. The lowest occurrence of a dominant B_x in a set was 32%, while the highest was 57%.)

To quantify this, Kallio and Janhunen (2003) tested one case of a Parker IMF ($B_x = 32$ nT, $B_y = 10$ nT, $B_z = 0$ nT). They found that in that case only 36% of the total proton flux impacted the northern hemisphere. This result implies that when there is a north–south asymmetry of the sputtering emissions in the atmosphere, an upper limit of a 2:1 ratio of total content in each hemisphere should be expected.

Figure 29 shows a numerical, single-particle simulation of the H$^+$ flux onto the dayside surface, as a function of IMF-B_x: in the figure, the $B_x > 0$ case is shown, and in this configuration the proton flux on the surface is expected to be higher in the southern hemisphere. To find how often the radial component of the IMF is dominant, Sarantos et al. (2007) implemented an analysis of the Helios data collected while the spacecrafts were within Mercury's orbital range. These authors used 40-second averages of the IMF and computed the probability density (normalized occurrence rate) estimate (see Fig. 30) At first look, this result implies that north–south asymmetries in impacting ion flux might be quite frequent at Mercury.

However, an asymmetry in impacting ion flux translates into an asymmetry in the neutral atmosphere only when ion sputtering is a strong comparative source of atmospheric neutrals. This generally happens when the IMF is southward ($B_z < 0$). Since southward fields were observed roughly only half the time in Helios I and II data, a north–south asymmetry in impacting proton flux should be expected for only 16–28% of Mercury's lifetime. Still, visible asymmetries could sometimes result from a northward IMF when photon-stimulated desorption is weak—namely at aphelion. Figures 31 and 32 show the precipitating flux of solar wind ions impacting Mercury's surface for likely aphelion and perihelion conditions. The relative recycling and escape rates of ions of Hermean origin has been a long-standing question (e.g., Ip 1987).

Global tracings of magnetospheric ions (Na$^+$ and K$^+$) in static magnetic field configurations given by the modified TH93 model (Killen et al. 2004a; Sarantos 2005) indicate that impacts dominate, while the escape rate of these species to the solar wind responds very slightly to external conditions: between aphelion and perihelion the escape ratio was seen to range from 30% to almost 40% (Sarantos 2005, PhD thesis). While long-term recycling is very strong, it is reduced by about a factor of 1.5 at perihelion. This prediction could help explain why the sodium atmosphere is observed to be denser at aphelion. In contrast, Delcourt et al. (2003) and Leblanc et al. (2003a) found significantly reduced relative recycling rates of ∼10%.

The sodium ion precipitation in Delcourt et al. was found to be similar to auroral precipitation at the Earth. We can only speculate about the reasons for these differences. The

Processes that Promote and Deplete the Exosphere of Mercury

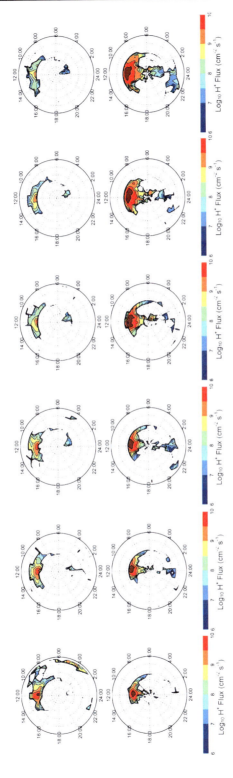

Fig. 29 Solar wind H$^+$ fluxes on the surface of Mercury, with different IMF-B_x external condition, from 5 nT (*left*) to 30 nT (*right*). *Top panels*: northern hemisphere; *bottom panels*: southern hemisphere. As a positive IMF-B_x component increases, higher fluxes are expected in the southern hemisphere. From Mura et al. (2006c)

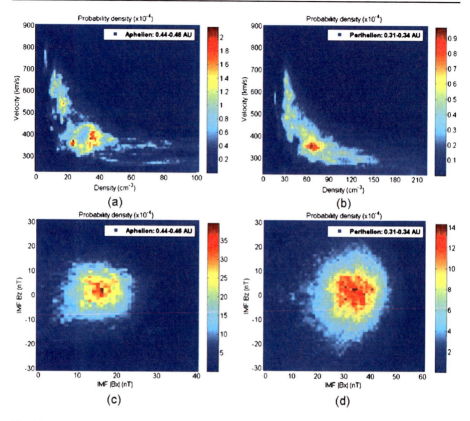

Fig. 30 Probability density estimates for important solar wind parameters from Helios 40-second data around the Hermean aphelion and perihelion (Sarantos et al. 2007)

models were run for different input conditions, with different resolutions and, most importantly, using different exosphere models which provided the initial ion distribution to be tracked. Possible explanations for these differences include pitch-angle and IMF B_x inclusion effects.

A pitch-angle effect may be related to the location of photoionization events by different exosphere models: the loss cone angle for particles launched below or at scale-height (e.g., Killen et al. 2004a; Sarantos 2005) is wider than for particles launched in the exoionosphere, so the latter may mirror instead of impact the surface.

On the other hand, the inclusion of IMF B_x in the TH93 model widens the open area, that is, the area of auroral precipitation, and makes it asymmetric. In conclusion, both investigations were limited in scope as they tested one case at aphelion and one at perihelion. These differences point to the need for more comparative simulations, with single input and boundary conditions.

Figure 33 (upper panel) illustrates the effect of likely solar wind conditions at perihelion versus aphelion on the fraction of surface area open to the solar wind.

At perihelion the solar wind is more dense than at aphelion, and the IMF is more radial, both of which act to create a more open magnetosphere. For the same B_z, the perihelion magnetosphere is more likely to have a larger surface area exposed to the solar wind. The likely precipitating solar wind flux (lower panel) is proportional to the open area and the

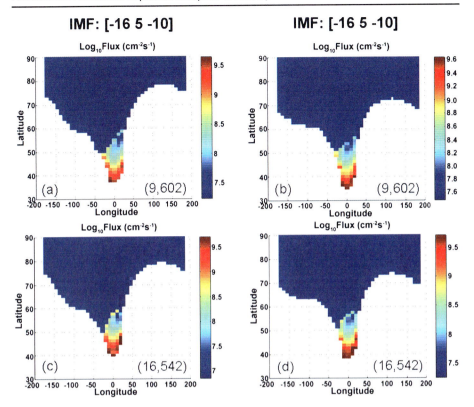

Fig. 31 Precipitating flux (*log scale*) of solar wind ions impacting Mercury's surface for likely aphelion conditions. These maps were produced with the modified TH93 model of the Hermean magnetosphere. Input conditions were selected by analyzing probability density estimates that are consistent with the Helios 40-second data in the 0.31–0.47 AU range. *Vertical columns* address the effects of increasing pressure on the cusp location and precipitation flux for the same IMF, while *horizontal rows* illustrate the effects of a more southward IMF for given pressure. The open-closed boundary exhibits a strong dawn–dusk asymmetry for the $B_z = -5$ nT cases as a result of the dominant B_x. In turn, the cusp becomes more symmetric as B_z grows comparable to B_x (cases with $B_z = -10$ nT). (Adapted from Sarantos et al. 2007)

dynamic pressure. For a precipitating integrated flux of 1.4×10^{26}, the upper and lower limits of ion sputter yield averaged over the solar wind species (0.05 and 0.15, respectively), give a yield averaged over the entire surface of the planet of 9×10^6 and 2.6×10^7 cm^{-2} s^{-1}, respectively. These yields are comparable to or greater than the PSD yields.

Heavy ions of planetary origin may be relatively energetic due to the centrifugal acceleration during $E \times B$ transport over the polar cap (Delcourt et al. 2003). Most of the Na$^+$ ions are lost into the dusk flank, but a localized region of energetic Na$^+$ precipitation develops at the planet's surface, and extends over a large range in longitude at mid-latitudes (30°–40°). These ions may sputter additional material. Characteristics of precipitating Na photoions are shown below in Fig. 34.

3.6 Extreme Solar Events

Because dense or fast solar plasma features could compress Mercury's magnetosphere so that more surface elements can be released due to sputtering into the exosphere, it is impor-

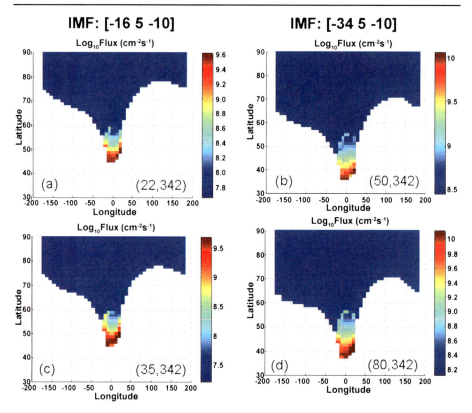

Fig. 32 Maps of the solar wind flux (*log scale*) that precipitates onto the Hermean surface for likely conditions at aphelion (**a, c**) and perihelion (**b, d**). Shown in *parentheses* are the density (cm^{-3}) and velocity (km/s) tested in each case. The open-closed boundary moves equatorwards by about 10 degrees for perihelion conditions. Responding to the denser plasma and stronger field magnitude, the precipitating flux at perihelion clearly increases both in the dayside and in the tail. The modeled integrated precipitating source (s^{-1}) increases fourfold, while the open area available to the solar wind doubles. (Adapted from Sarantos et al. 2007)

tant to study collisions with extreme solar particle events on Mercury's magnetospheric and surface environment.

Leblanc et al. (2003b) studied the interaction of a solar energetic particle (SEP) event reported by Reames et al. (1997) of protons with energy larger than 10 keV with Mercury's magnetospheric-surface environment. They expected that if a SEP encounters Mercury, a significant flux of energetic particles will reach Mercury's surface which may refill the exosphere. The simulations indicate that after the arrival of a SEP at Mercury, a population of quasi-trapped energetic ions and electrons is expected close to Mercury, which is stable for hours after their arrival. A significant dawn/dusk charge separation is observed and a fraction of about 10% of the initial energetic particles may impact the surface with a spatial distribution that exhibits north/south and dawn/dusk asymmetries. Furthermore, Leblanc et al. (2003a, 2003b) found that the flux of particles impacting Mercury's surface and the ability of a quasi-trapped population to be maintained near Mercury are highly dependent on the B_z sign of the interplanetary magnetic field. Leblanc et al. concluded that impacting SEPs can eject a non-uniform distribution of Na atoms into Mercury's exosphere,

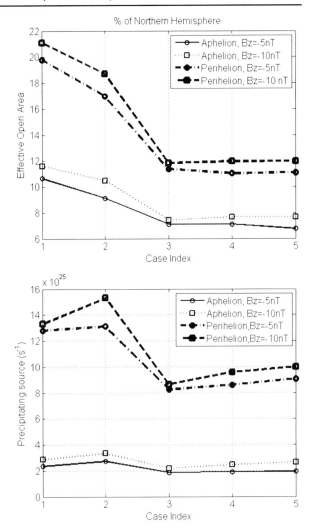

Fig. 33 Effective open area (*upper panel*) and precipitating flux (*lower panel*) for likely aphelion and perihelion conditions, respectively (Sarantos et al. 2007)

which may be the origin of several unexplained observed exospheric features, although they could not explain the total amount of Na atoms needed to reproduce the observations of Potter et al. (1999). However, much stronger SEP events than the one used in this work have been observed at the Earth (Mason et al. 1999) during the same month as the Potter et al. (1999) observations, and may have sufficient intensities. These authors proposed that the encounter of Mercury with CME or magnetic cloud (MC) events could cause such an enhancement.

Significant advances in the study of CMEs have been made by the Large Angle and Spectrometric Coronagraph (LASCO) on board of ESA's Solar and Heliospheric Observatory (SoHO), which observed more than 8,000 CMEs since January 1996. The observational data on CMEs are related to two spatial domains: the near-Sun region (up to 30 $R_{Sun} \approx 0.14$ AU) remotely sensed by coronagraphs; and the outer region, including the geospace and beyond, where in situ observations are made by spacecraft. At the larger distances such as at Mercury's orbit or beyond CMEs are traditionally called as Interplanetary CMEs or mag-

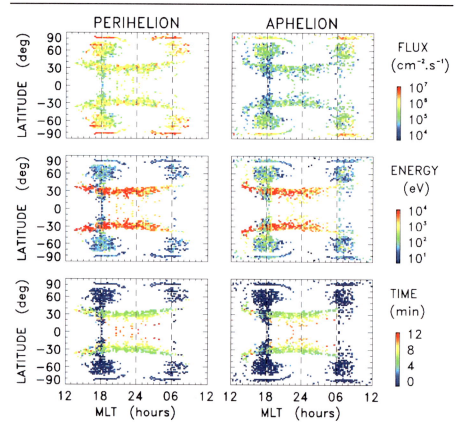

Fig. 34 Characteristics of precipitating Na$^+$ ions (*left*) at perihelion and (*right*) at aphelion. The panels from top to bottom show ion flux, average energy, and residence time in the magnetosphere (Delcourt et al. 2003)

netic clouds. CMEs are associated with flares and prominence eruptions and their sources are usually located in active regions and prominence sites. The basic characteristic of CME producing regions is closed magnetic structure. Recent studies on temporal correspondence between CMEs and flares provide arguments in favor of the so-called common-cause scenario, according to which flares and CMEs are different manifestations of the same large-scale magnetic process (Zhang et al. 2001). Although the details of this process still remain unclear, it definitely can be stated that an intensive flaring activity of a star should be accompanied by an increased rate of CME production. The probability of CME-flare association increases with the duration of a flare (Sheeley et al. 1983): $\approx 26\%$ for flare duration <1 h; and $\approx 100\%$ for flare duration >6 h. Multi-thermal structure of CMEs includes 1) coronal material in the front region (≈ 2 MK); and 2) possibly a core from solar prominence material ($\approx 10^4$ K), or hot flare plasma (≈ 10 MK).

The basic and widely considered characteristic of CMEs is their velocity, determined by tracking a CME feature in coronagraph image frames taken with a certain time cadence. According to the data from SoHO/LASCO, the velocity of solar CMEs ranges from tens of km s^{-1} to $>2{,}500$ km s^{-1} near the Sun, with an average value of about 490 km s^{-1} (e.g., Gopalswamy 2004). Due to the relatively large statistics ($>8{,}000$) of the considered SoHO/LASCO CMEs, the average ejecta velocity value can be considered as a representative quantity. It is also consistent with the results of other velocity measurement techniques applied to separate CME phenomena (Lindsay et al. 1999; Gopalswamy et al. 2001a, 2001b, 2004; Lara et al. 2004). Halo CMEs, a subclass of CMEs propagating toward the Earth, are currently believed to be the main drivers of space weather disturbances at the Earth. Compared to the general population of CMEs, halos have much higher average speeds of 1,004 km/s (Yashiro et al. 2004).

Besides the velocity of CMEs an additional important parameter for studying the plasma interaction with Mercury's magnetospheric-surface environment is the plasma density. Estimates of CME plasma density from white light (Vourlidas et al. 2002), radio (Gopalswamy and Kundu 1993), and UV observations (Ciaravella et al. 2003) give similar values of about 10^6 cm^{-3} at the distances $\approx 3-5 R_{Sun}$ which are consistent with the assumption of the coronal value of density of a CME material at the moment of ejection. At larger distances >30 R_{Sun} (i.e., >0.14 AU), the density and duration of ICMEs and associated MCs are measured in situ by spacecraft (Henke et al. 1998; Lepri et al. 2001). For the plasma density in MCs observed between 0.3–1 AU by the Helios satellites, Bothmer and Schwenn (1998) found a power law

$$n_{MC} = n_{MC}^0 \left(\frac{d}{d_0}\right)^{-2.4 \pm 0.3}. \tag{21}$$

where d is the radial distance to the Sun in units of AU, quantity $n_{MC}^0 = 6.47 \pm 0.95$ cm^{-3} is the MC plasma density at the near-Earth orbit, and d_0 is at 1 AU (Bothmer and Schwenn 1998). By using this power law one gets CME number density values at periherm of about 80–260 cm^{-3} and about 45–130 cm^{-3} at about 0.38 AU.

Finally, another important characteristic of CME activity and its influence of Mercury's environment is the CME occurrence rate. The data from Skylab, SMM, Helios, Solwind, and SoHO indicate a correlation between sunspot numbers (SSN) and the CME occurrence rate (Hildner et al. 1976; Howard et al. 1986; Webb and Howard 1994; Cliver et al. 1994; Cyr et al. 2000; Gopalswamy et al. 2003). At the same time, SoHO/LASCO observations found that although there is an overall similarity between the SSN and the CMEs occurrence rates, there are some differences in details. The most recent SoHO/LASCO observations give a CME occurrence rate ≈ 0.8 CMEs/day for solar minimum and ≥ 6 CMEs/day for solar maximum.

These numbers are consistent, but a bit higher as compared to previous estimations (Hildner et al. 1976; Webb and Howard 1994; Cliver et al. 1994), which is attributed to the better sensitivity and the high dynamic range of the LASCO coronagraphs. Focusing on halo and fast-and-wide CMEs may help estimate the occurrence rate of ICMEs at Mercury. First, full halo CMEs account for about 3.5% of all CMEs detected by SOHO. Second, the fraction of fast-and-wide CMEs ranges from 2% (1996) to 6% (2003, past solar maximum) of the total SOHO count (Gopalswamy 2004). Combining the observed occurrence rates of all CMEs originating at the Sun with the aforementioned fraction of fast-and-wide CMEs, and assuming that a typical halo has angular width of 120 degrees, we should expect 2–3 ICMEs/year at solar minimum and about 44 ICMEs/year (one every nine days) at solar maximum to impact Mercury.

Kallio and Janhunen (2003) studied the precipitation of the protons related to a MC or CME event with a proton density of about 75 cm^{-3} and a velocity of about 800 km s^{-1} with a self-consistent quasi-neutral hybrid model. In their simulation the particle flux of the plasma protons was calculated self-consistently with the kinetic model. By increasing solar wind dynamic pressure to about 10^{-5} Pa it was possible to push the magnetopause toward Mercury's surface, in agreement with MHD model runs of Kabin et al. (2000). In such a

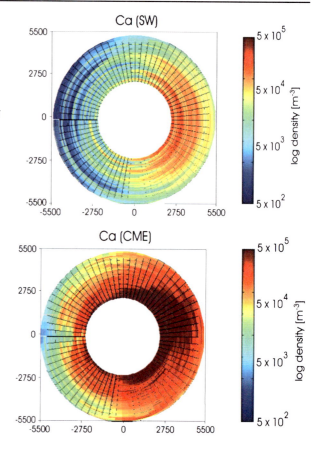

Fig. 35 3D exosphere calculation by using the test particle model for sputtered Ca atoms as a function of planetary distance in units of km for ordinary solar wind conditions (*left panel*) and during a simulated MC collision with Mercury with a plasma density of 75 cm^{-3} and a velocity of about 800 km s^{-1}

case an intense H$^+$ ion precipitation with fluxes up to about 10^9 cm^{-2} s^{-1} was found near the subsolar point.

Figure 35 shows an equatorial cut of sputtered exospheric Ca atoms during ordinary solar wind conditions where larger fluxes of solar wind protons only precipitate to the surface around Mercury's cusp regions (Massetti et al. 2003; Kallio and Janhunen 2003) and the MC event discussed earlier and studied by Kallio and Janhunen (2003). One can see that due to the large affected surface area the exosphere should be refilled by Ca atoms during the MC collision.

One can see from Fig. 35 that the results which give much denser and more distributed exospheric number densities of sputtered surface elements during expected collisions with CMEs or MCs are pertinent to future measurements on the Messenger and BepiColombo missions, which will be instrumented to observe the released exospheric particles.

Kinetic energies of solar wind ions are, on average, 1 keV/amu, where the sputtering efficiency peaks. Sputtering by H$^+$, which accounts for 85% of the total kinetic energy carried by the solar wind, is relatively inefficient. On average, He^{2+} accounts for about 13% of the kinetic energy carried by the solar wind, and is generally assumed to account for most of the space weathering effects. However, although heavy ions ($Z > 6$) account for only about 2% of the kinetic energy carried by the solar wind, they also carry ~1 keV each in potential energy due to ionization into a high charge state. The charge state of the impinging ion has little effect on the sputter efficiency of highly conducting targets (conductors and semiconductors), but has considerable effect on some insulators as discussed earlier (Aumayr and Winter 2004).

3.7 Sputter Effects on Surface Chemistry

Sputtering of the surface by any of the mechanisms discussed earlier may affect the topmost layer of the regolith. Ion implantation of solar wind ions occurring for billions of years may change the predominant apparent chemistry to more reflect the solar wind than the pristine composition from Mercury's origin. In addition, new chemical compounds may be made. Production of sodium and water by proton sputtering of sodium-bearing silicates was considered by the following mechanism (Potter 1995)

$$2H + Na_2SiO_3 \rightarrow 2Na + SiO_2 + H_2O. \tag{22}$$

The supply rate of water molecules is half the supply rate of Na by this process. Since the free energy of this reaction is -4.7 kcal/mole, it will proceed spontaneously; however, the activation energy is unknown. Chemical reactions, either on the surface or in the atmosphere, can enhance the loss rates of the reaction products if the reaction products are created with enough energy to escape. Very little work has been done to quantify rates for chemistry as a source or loss process in the context of planetary sciences.

In laboratory research such processes are investigated and are referred to as plasma-assisted etching, or chemically enhanced physical sputtering (e.g. Winters et al. 1983; Winters and Coburn 1992). These processes are used for a large variety of industrial and laboratory applications. Energetic ions and electrons of the plasma induce strong changes in the surface chemistry. Plasma-assisted etching involves the interaction of a plasma discharge with a solid surface to produce a volatile product. Laboratory research aims to produce product molecules at the surface that are weakly bound to the surface to enhance the sputter yield for micro-fabrication. The reaction can be divided into three steps: 1) the adsorption/implantation of external particles, 2) the product formation, and 3) the removal of the product from the surface.

Physical sputter theory predicts that the sputter yield is inversely proportional to the binding energy of the species to the surface (Sigmund 1969). Energy distributions under the conditions of chemical sputtering have been measured in the laboratory (Haring et al. 1982) and two components in the energy distribution have been identified (Winters and Coburn 1992). The first component is very well described by the energy distribution for physical sputtering (see (15)). The second component, however, peaks at energies between 0.1 and 0.5 eV (indicative of a low surface binding energy) and can be described by a Maxwellian distribution, and shows higher yield. Thus, the chemical alteration increased the total sputter yield from the surface. In laboratory experiments the interaction of halogens is studied mostly, but there are a few investigations where the increased sputter yield of silicon by hydrogen ion impact is reported (Winters and Coburn 1992, and references therein).

Such a low binding energy means that the created species can also be released via thermal desorption, as is the case for the example of Na and water given above. Thus the chemical alteration of Mercury's surface by precipitating ions and the later release of volatiles (e.g., Na and water) can also occur with considerable delay. A possible scenario is ion implantation when the surface is cold (e.g., at the nightside). At a later time, when the surfaces warms up because of solar insolation, the thermal release of the such created volatiles may occur.

Chemical reactions induced by solar wind ions impinging on the surface undoubtedly occur on Mercury. Earlier, we discussed the example of a source for water molecules and atomic sodium. The study of chemistry on the surfaces of bodies exposed to the solar wind is in its infancy, and more so the effects this has on the sputter yields.

4 Thermal Vaporization of Alkali Atoms

The vapor pressure of a gas in thermal equilibrium with a liquid is given by

$$P = P_0 \exp\left(-\frac{l}{RT}\right), \tag{23}$$

where l is the latent heat per mole, R is the gas constant and T is the temperature in Kelvins. In the case of vaporization of an adsorbed volatile from a solid surface, the rate constant for thermal desorption is described as

$$k_{des} = A_{des} \exp\left(-\frac{E_{des}}{kT}\right), \tag{24}$$

where A is a pre-exponential vibrational frequency, and E_{des} is the desorption energy. The value of the pre-exponential, the vibrational frequency, most often used is 10^{13} s^{-1} (e.g., Hunten et al. 1988) but in fact it can vary from 10^4 to 10^{23} s^{-1} (Holmlid 1998). The extremely large range for the preexponential term is due to the large number of physical processes actually involved in thermal desorption: including diffusion to and from the bulk rock or grains, surface diffusion between sites with different desorption energies, electronic excitation and de-excitation, jumps during the near-desorption in excited states.

Spatially resolved studies of iron-rich minerals show the complexity of real metal oxide surfaces: for instance, the desorption energy of alkali atoms is changed radically by the addition of other atoms in low concentrations. The thermal barrier for K desorption from the same type of mineral can range from 0.83 eV to 2.35 eV by adding 2 wt% Mn to the material (Kotarba et al. 2004). Values quoted in the literature for desorption energy of Na atoms range from 1.1 eV (Hunten and Sprague 1997), 1.8 eV (Madey et al. 1998) and 2.7 eV for metal and metal oxide surfaces (Holmlid and Olsson 1977). In the Monte Carlo study of Leblanc and Johnson (2003) an average desorption energy of 1.85 eV was used with a preexponential of 10^{13} s^{-1}. Holmlid (2006) suggested that this pre-exponential is much too large, and that the desorption energy of alkali atoms from any real oxide mineral is unlikely to be much smaller than 2 eV.

The rate of thermal desorption from a surface is in fact rate limited by the slowest process acting in the chain of events leading to desorption (see Fig. 36). For the surface of Mercury, Killen et al. (2004b) concluded that the rate-limiting process for thermal desorption from the surface of Mercury is diffusion of atoms from the bulk of the grains. This conclusion is consistent with the conclusions of Leblanc and Johnson (2003) that thermal desorption rapidly depletes most of the sunlit surface of Mercury of adsorbed atoms.

The measured temperature for the sodium atmosphere is high, about 1,200 K, whereas a high adsorption energy would imply efficient sticking at the surface, and hence rapid thermal accommodation to the surface temperature (Holmlid 2006). However, observations of the variation of sodium D2 intensities with true anomaly angle imply that the sticking coefficient is quite small, on the order of 0.15 (Potter et al. 2006). This seems to be inconsistent with a high adsorption energy. It is not inconsistent with a lower adsorption energy, and an efficient

Fig. 36 (**a**) Shows the time required for the flux of solute from the grain interior to the grain surface to fall below a constant value of 10^7 cm^{-2} s^{-1} (the maximal required at the subsolar point to maintain the exosphere) as a function of temperature (diffusion coefficient) and grain radius. We show the limiting size grains (radius in cm, color coded) that can maintain the thermal fluxes until half the solute is depleted for (**b**) glass and (**c**) crystalline minerals: 100 cm (glass) and 1 μm (mineral) at perihelion, and 10^3 cm (glass) and <1 μm (mineral) at aphelion (Killen et al. 2004b)

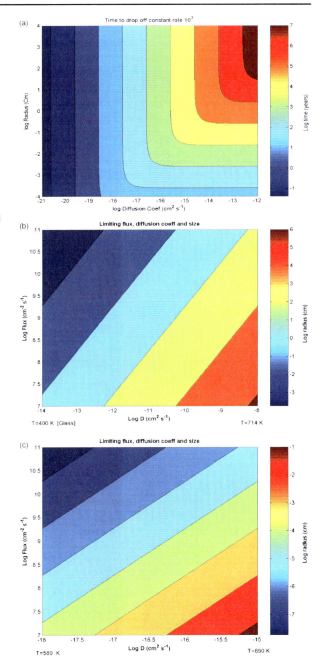

loss of adsorbed states, as described by Killen et al. (2004b) and by Leblanc and Johnson (2003).

In this case the source process for Na and K to the exosphere would be dominated by PSD, ion-sputtering and impact vaporization, which are all capable of ejecting the atoms from the bound state. The source rates for all processes other than impact vaporization, which accesses a depth equal to several impactor diameters, depends on the availability of atoms at the extreme surface. To maintain a long-term supply to the exosphere by PSD or ion-sputtering, atoms must diffuse from the bulk of the rock or grain to the extreme surface.

The long-term source rates are therefore limited by the diffusion rates. These are dependent not only on temperature, but grain size and lifetime of the grain on the surface. Thermal vaporization at the subsolar point on Mercury is very efficient, given a supply of adsorbed atoms at the extreme surface (e.g. Leblanc and Johnson 2003). Fluxes are calculated using (Crank 1975)

$$\frac{Dt}{a^2} = \frac{2}{\pi} - \frac{2}{\pi}\left(1 - \frac{\pi F}{3}\right)^{1/2} - \frac{F}{3}, \qquad (25)$$

where F is the fraction of solute lost at time t from a sphere of radius a. D is the diffusion coefficient and the solute is assumed to be initially uniformly distributed.

To obtain the flux per unit surface area of the planet at time, t, we solved (25) for F, took the time derivative of F and multiplied by the initial amount of solute and divided by the cross-sectional area of the grain. Thus the loss rate per unit cross-sectional area is given by

$$\frac{dn}{dt} = \frac{\frac{4}{3}\pi a^3 f_{Na}\rho \frac{dF}{dt}}{\pi a^2}. \qquad (26)$$

Once the adsorbed atoms have evaporated, the source rates drop to the values at which atoms can be supplied to the surface by diffusion. Results show that large thermal fluxes could be maintained by diffusion from hot glassy spheres as long as a mechanism exists for desorption of the atoms from the surface.

Any glass sphere smaller than the limiting size would be able to maintain the flux. For high fluxes, unreasonably small grains or high diffusion coefficients are required. Progressively smaller grains supply atoms to the surface at faster rates but for shorter periods of time, as the grains become depleted of volatiles (Killen et al. 2004b). The upper few μm of the Moon is turned over about once in 10^3 years (Heiken et al. 1991). We expect the turnover rate at Mercury to be about ten times that at the moon, since the impact flux is about ten times that at Earth orbit.

Thus a μm-sized grain will sit on the surface of Mercury for about 100 years. If the diffusion rate of a μm-sized grain drops to 10^7 cm^{-2} s^{-1} in one month, then a "thermal" vaporization rate of 10^{11} cm^{-2} s^{-2} is sustainable for less than 0.1% of the μm-sized grains on the surface, and none of the larger grains (Figs. 36a,b). It is concluded that the exospheric rate required to maintain the observed exosphere 10^7 cm^{-2} s^{-11}, is therefore the limiting value governed by diffusion of atoms to the surface of grains and by the gardening rate.

Radar bright spots near the poles of Mercury discovered in 1992 (Harmon and Slade 1992) were attributed to water ice. Numerous studies concluded that water ice is stable in permanently shadowed craters near the poles of Mercury (Paige et al. 1992; Ingersoll et al. 1992). It is intriguing that 13 radar bright features were found between 70° and 80° latitude (Harmon et al. 2001). All of these features are small, consistent with the limited amount of permanent shading expected at these latitudes. It has been suggested that the volatile in the cold traps is not water ice but in fact may be sulfur (Sprague et al. 1995) or cold silicate (Starukhina 2000).

We should ask ourselves the question: Is thermal vaporization a loss process for water? In other words, is evaporation a reversible or irreversible process? One must in fact show that vaporization is an irreversible process to rule out the possibility of its presence at a given latitude since atoms cannot be carried away by winds or other processes on an atmosphereless body. Unless the atoms are lost by an irreversible process such as ionization and entrainment into the solar wind, or chemical reactions, they must eventually return to the cold traps. Further, as pointed out by Holmlid (2006), the adsorption energy and preexponential terms are highly influenced by the local composition and charge state of the surface, which changes rapidly near the terminator for instance.

5 Loss Processes

Loss of atoms from the exosphere must be defined into categories of loss: reversible loss processes (sticking to the surface or ionization followed by impact with the surface and neutralization) and irreversible loss processes (Jeans escape and ionization followed by crossing the magnetopause boundary and entrainment into the solar wind). Reversible loss processes must be further characterized as long-term loss, such as burial or chemical reaction into a bound state with a large binding energy, or short-term loss such as adsorption on the surface of grains. Jeans escape is an irreversible loss process, but photoionization may or may not result in permanent loss. Sticking to the surface results in loss from the exosphere but not from the regolith. These different types of loss processes will be discussed in the following.

5.1 Jeans Escape

Only H and He have significant rates of Jeans escape at the ambient surface temperature on Mercury (Hunten et al. 1988). However, with the exception of thermal vaporization, the source processes populating the Hermean exosphere create non-thermal populations and the mean energy is species dependent. The mean velocities of atoms ejected by ion-sputtering may be above escape velocity, 4.25 km/s, and will depend on the binding energy of the atom in the substrate, and its mass (Table 9). As pointed out by Holmlid (2006) the binding energy of a given atom will depend not only on the particular mineral in which it is found, but also on the surrounding conditions. This is particularly important for a surface that is highly subject to space weathering.

The mean velocity of Ca atoms observed in the Hermean exosphere is about 3.5 km/s (Killen et al. 2005), consistent with the expected mean velocity for an ion-sputter source. If the main source of Mg and Al to the exosphere is ion-sputtering, then the bulk of the source would be lost by Jeans escape. Killen et al. (2005) suggested that dissociation of a calcium oxide could also account for the high velocity observed for calcium in the Hermean exosphere.

The temperature of meteoritic vapor is often described as being quite hot, perhaps as high as 6,500 K. However, only the initial vapor plume seen by the Deep Impact spacecraft in the first few seconds was hot (1,000–3,000 K), implying that a rapid cooling takes place in the dense fireball and possibly in the regolith (A'Hearn et al. 2005). The bulk of

Table 9 Mean velocity of an ion-sputtered source for selected atomic species

Species	v_{mean} [km/s]
Mg	4.5
Al	4.3
Ca	3.5
Fe	3.0

the vapor—that in the ejecta curtain—which is composed of excavated material, was at ambient temperature. The vapor resulting from micrometeoritic bombardment may cool by multiple collisions with the regolith as the micrometeoroids deposit their energy below the surface.

5.2 Photoionozation

Photoionization is an irreversible loss process if the ions cross the magnetopause and become entrained in the solar wind. Killen et al. (2004a) suggested that half of the photoions produced near the surface return to the surface where they are neutralized. An east-west electric field produces an asymmetric escape pattern such that those ions created on the dusk side reimpact on the dawnside, and those ions created on the dawnside escape across the magnetopause boundary or impact the nightside. However, the escape rate is higher as the ions are produced at successively higher altitudes.

Photoionization rates of many likely exospheric species are known. Rates for He, C, N, O, F, Na, S, Cl, Ar, Xe, OH were published by Huebner et al. (1992) and for Ca by Killen et al. (2005). Photoionization rates for most refractory species (e.g., Mg) have not been published. In any case, the energy with which refractory species are ejected by the two mechanisms capable of ejecting refractory species, ion-sputtering and meteoritic vaporization, will be energetic enough to eject most of the atoms at velocities exceeding escape velocity. Those refractories that do not directly escape will probably stick to the surface on impact. Thus their photoionization rates are not pertinent to their lifetimes in the exosphere.

The sodium ionization rate is controversial because of the discrepancy between the experimental and theoretical cross-sections, and further because there are theoretical calculations that agree with the experimental value. The Combi et al. (1997) value is quoted because it is commonly used in the cometary community, and the observed cometary abundances agree with calculations when this value is used for the ionization rate. However, the discovery of the sodium tail on comet Hale-Bopp (Cremonese et al. 1997) and the high-resolution spectroscopy performed along the tail was used to calculate a new value of photoionization lifetime for the neutral sodium atom that confirmed the value suggested by Huebner et al. (1992). The model used for the comet works also on the sodium tail of Mercury showing that it does not depend on the object observed.

The rates for sulfur are theoretical (Huebner et al. 1992). Rates from Kumar (1982) are 50% higher. The calcium ionization rate was calculated by Walter Huebner and communicated to the author. All other rates are from (Huebner et al. 1992). The photoionization rate is not the limiting rate if atoms stick to the surface.

5.3 Charge Exchange

A charge-exchange process may occur when an energetic ion collides with a exospheric neutral particle (target). In the interaction, an electron and a small amount of kinetic energy are exchanged between the neutral and the ion; the net result is an energetic neutral atom (ENA) and a thermal ion. The target (ionized) is scattered at an approximately perpendicular angle with respect to the projectile path; the newly created ENA retains approximately both the energy and the direction of the colliding energetic ion. The energy defect of the process is equal to the difference of the two atomic ionization potentials (for complete discussions see, e.g., Hasted 1964). Charge-exchange is a resonance process, which can be symmetrical (if the species of the ion and the neutral are the same) or accidental (Stebbings et al. 1964; Hasted 1964). The cross-section varies with species and energy and is of the order of 10^{-14}–10^{-17} cm^{-2}.

Table 10 Solar photo ionization rates at 1 AU

Species	Rate coefficient (quiet sun) [s^{-1}]	Rate coefficient (active sun) [s^{-1}]
H	7.26×10^{-8}	7.1×10^{-8}
He	5.25×10^{-8}	1.51×10^{-7}
O(^{3}P)	2.12×10^{-7}	5.88×10^{-7}
O(^{1}D)	1.82×10^{-7}	5.04×10^{-7}
O(^{1}S)	1.96×10^{-7}	5.28×10^{-7}
Na	$1.62 \times 10^{-5\,(e)}$	$1.72 \times 10^{-5\,(e)}$
	$5.92 \times 10^{-6\,(t)}$	$6.42 \times 10^{-6\,(t)}$
	$5.40 \times 10^{-6\,(a)}$	
S (^{3}P)	1.07×10^{-6}	2.44×10^{-6}
S(^{1}D)	1.08×10^{-6}	2.46×10^{-6}
S(^{1}S)	1.05×10^{-6}	2.31×10^{-6}
Ar	3.05×10^{-7}	6.90×10^{-7}
K	2.22×10^{-5}	2.36×10^{-5}
Ca	$7.0 \times 10^{-5\,(b)}$	7.8×10^{-5}
OH → O(^{3}P) +H	6.54×10^{-6}	7.17×10^{-6}
OH → O(^{1}D) +H	6.35×10^{-7}	1.51×10^{-6}
OH → O(^{1}S) +H	6.71×10^{-8}	1.64×10^{-7}
OH → OH+	2.47×10^{-7}	6.52×10^{-7}

[e] experimental
[t] theoretical
[a] Combi et al. (1997)
[b] Huebner (personal communication, unpublished)

Charge exchange at Mercury may occur due to solar-wind plasma (Mura et al. 2005) as well as due to planetary ions. Generated neutrals have typical energies of 1 keV or more; hence, such neutrals are no more trapped in ballistic orbits and the result is a net loss from the planet. This process mostly occurs in the dayside and dawnside regions close to the planetary surface (Mura et al. 2005); in general, the H-ENA production rate for unit length reaches values up to 10 (cm^{-2} s^{-1} sr^{-1} m^{-1}), close to the dayside planetary surface. It has been estimated that, approximately, less than 1% of the solar wind plasma circulating inside the magnetosphere of Mercury experience charge-exchange (Mura et al. 2005); the related CE loss rate, cumulated on all neutral species, is between 10^{22} and 10^{24} s^{-1}, which is, on average, small if compared to other loss mechanisms.

Figure 37 shows simulated fluxes of ENA coming from different directions (ENA images), as "seen" from a vantage point in the nightside, in a "fish-eye" projection. The fluxes are integrated between 1–10 keV. The intense fluxes coming from the dawnside of the planet (right side of the picture) are generated from westward drifting solar wind protons.

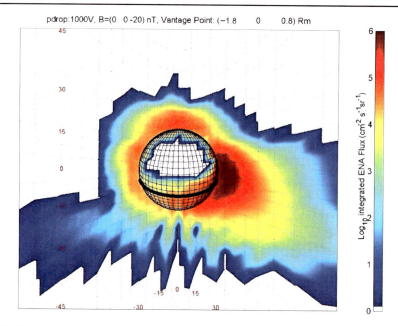

Fig. 37 Simulated ENA images, from a vantage point in the nightside (1.8, 0, 0.8 R_M). Color is coded according to ENA flux, integrated between 1 and 10 keV. From Mura et al. (2005)

5.4 Gardening

Gardening, fragmentation and burial of the regolith, has been studied extensively for the Moon (Heiken et al. 1991). Repetitive impacts agitate the surface by fragmenting, tumbling, burying and exhuming individual grains. The layer called the regolith is continuously churned. Although numerous small impacts homogenize the upper part of the regolith, a major role is played by the larger impacts in excavating previous sedimentary layers, fragmenting rock layers and depositing fresh material onto the surface. The number of times the lunar regolith has been turned over versus depth was calculated using results of laboratory cratering experiments in fine grained unconsolidated targets (Gault et al. 1974), and is given as a turnover versus depth as a function of time.

On the Moon, a depth of almost 1 cm is overturned once in 10^6 years with 50% probability (Heiken et al. 1991, p. 87). If the meteoritic influx scales as $1/R^2$ with distance from the Sun, then the rates at Mercury are on average 6.7 times those at the Moon, and the turnover times would scale as 0.15. Thus a layer of regolith 1 cm deep on Mercury should be overturned in 1.5×10^5 years with 50% probability. Turnover rates are important for long-term renewal of material at the extreme surface. Survival times versus grain size were estimated for the Hermean surface by Killen et al. 2004b, and compared with the time at which the diffusive flux of sodium from the grain interior to the surface would drop to 50% of its initial rate at a given prescribed rate (Fig. 38). This figure shows that the observed sodium exosphere would be sustainable, even if all of the volatile loss were irreversible, because fresh material will be supplied by regolith turnover before the volatiles are removed.

Processes that Promote and Deplete the Exosphere of Mercury

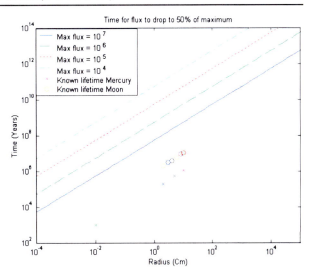

Fig. 38 Time it takes for a grain of a given radius to lose 50% of its solute at a given rate between 10^4–10^7 cm^{-2} s^{-1}. The *open circles* are lifetimes of grains on the Moon, and the *crosses* are the estimated lifetimes on the surface of Mercury, with the assumption that rates are approximately an order of magnitude faster on Mercury. This plot is a two-dimensional cut through a three-dimensional therefore temperature may vary across the plot

6 Space-Based Observations and Expected Results

The ESA cornerstone BepiColombo/MPO is planned to fly around Mercury in a polar orbit, with 400 km periherm and 1,500 km apoherm; the orbital period is about 2.3 hours. Two instruments in the MPO payload are mainly devoted to observing the exosphere: the FUV-EUV spectrometer, PHEBUS, and the comprehensive suite for particle detection, SERENA. The combination of these two experiments on BepiColombo will be an unprecedented opportunity to perform a detailed analysis of the exosphere composition and vertical profiles.

6.1 The BepiColombo/MPO/PHEUBUS UV Spectrometer

The BepiColombo/MPO/PHEBUS UV spectrometer PHEBUS is a dual FUV-EUV spectrometer (see the ESA BepiColombo webpage: http://www.rssd.esa.int/index.php?project=BEPICOLOMBO) working in the wavelength range from 55 to 315 nm plus a small two-channel detector used to measure potassium and calcium at 404 nm and 422 nm, respectively. This instrument is devoted mainly to the characterization of Mercury's exospheric composition and dynamics. In addition, some ionized species have emission lines in the spectral window; hence, this instrument will potentially contribute to the identification and characterization of the exo-ionosphere.

Thanks to the remote sensing of the exosphere, PHEBUS will provide measurements of density of many species (see Fig. 39) and their profiles at altitudes below the spacecraft periherm. Observations of the low altitude exosphere are particularly important for the characterization of the heavier atoms, such as N, C, Ne, Si, Fe, Mg and molecules like CO, that are unlikely to arrive at the BC/MPO orbit and that can be observed, most probably, solely by a remote-sensing instrument. PHEBUS will observe Na at 268 and 285.3 nm.

Since the observable emission lines are generated by the interaction between the solar photons and the exospheric atoms, the PHEBUS can only observe by looking toward the dayside or the terminator. In this sense, the two instruments, PHEBUS and the mass spectrometer, SERENA/STROFIO (see next subsection), are completely complementary. In fact, the combination of the PHEBUS remote sensing and the SERENA/STROFIO in situ measurements will provide a complete view of the day–night exosphere for almost all the foreseen components in a wide range of altitudes and Hermean conditions.

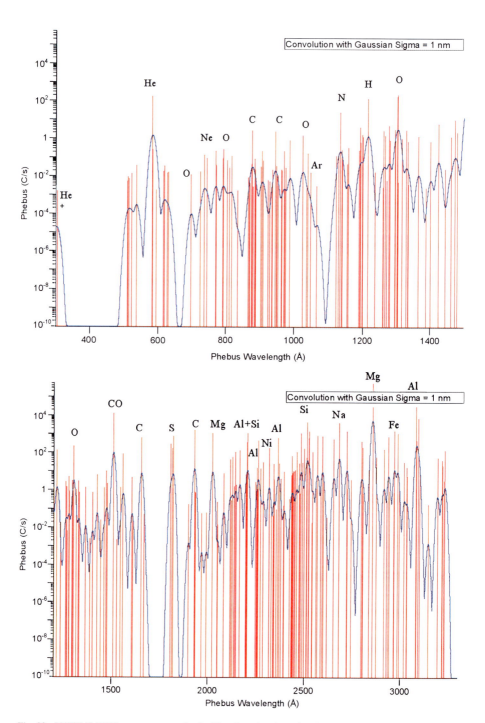

Fig. 39 PHEBUS EUV spectrum convolved with a Gaussian ($\sigma = 1$ nm)

6.2 The BepiColombo/MPO/SERENA Neutral Particle Detectors

A comprehensive suite for particle detection in the Mercury's environment, the SERENA instrument (see the ESA BepiColombo webpage: http://www.rssd.esa.int/ index.php?project= BEPICOLOMBO), is included in the MPO payload of the mission. This package consists of four units: STROFIO and ELENA will detect the neutral particles and measure their energies in the range from fractions of eV to a few keVs; MIPA and PICAM will measure and analyze ionized particles of planetary and solar wind origin from tens eV to tens of keV. ELENA (Emitted Low-Energy Neutral Atoms) is a neutral particle camera that investigates neutral gases escaping from the surface of Mercury, their dynamics and the properties of the related source processes. The ELENA sensor is a Time-of-Flight (TOF) detector, based on state-of-the-art choppers and mechanical gratings. The new development in this field allows unprecedented performances in timing and angular discrimination of low-energy neutral particles. STROFIO is a mass analyzer able to measure the neutral composition of the non-directional, thermal component of the exosphere.

The ELENA FOV (see Fig. 40, left panel) is always nadir-pointing, and most sectors look towards the planetary surface. In this way, the ELENA sensor will be able to detect the neutral flux coming from two different sources: charge-exchange ENAs, resulting from the interaction of solar wind and planetary plasma with the neutral exosphere, and ion-sputtering ENAs resulting from the precipitation of solar wind and planetary plasma onto the surface of Mercury. In general, it is possible to discriminate the neutral flux coming from those sources, since they have different typical energies and different generation regions.

In fact, instrumental simulations show that ELENA has both energy and angular resolution able to discriminate between those sources. The STROFIO FOV (see Fig. 40, right panel) points in the ram direction, so that low-energy exospheric particles will enter the sensor head due to the spacecraft motion. Regardless of the source generation process (provided that particles will have enough energy to reach the instrument location, see Fig. 41), STROFIO will count and identify the local exospheric components along the BepiColombo MPO orbit.

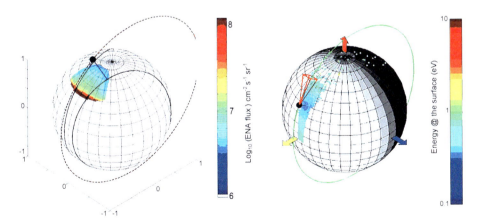

Fig. 40 ELENA and STROFIO field-of-views along the BC-MPO trajectories. *Left panel*: MPO orbit (*red dashed curve*); example of ELENA FOV (*blue lines*); boundaries of the projection of ELENA FOV onto the planetary surface (*green curves*); sample of the sputtering flux, due to s/w precipitation inside cusps, emitted from the surface (*color-coded map*). *Right panel*: MPO orbit (*green curve*; example of the STROFIO FOV (*red lines*) and source location of observable particle with the related energy are indicated

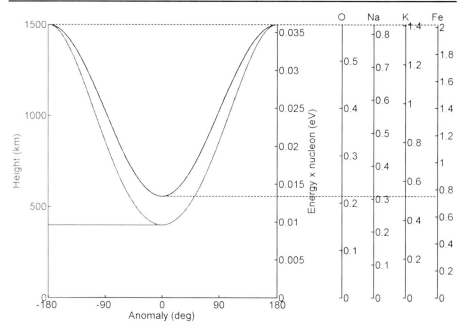

Fig. 41 MPO height function of anomaly angle (*red scale*, LHS). *Black scales* (RHS): minimum release energy necessary for a particle to get from the surface up to MPO orbit, for different species (O, Na, K, Fe), function of the anomaly angle

The major processes causing surface emission at Mercury are listed in Sect. 2. Among these processes, ion sputtering is the most effective in transporting exospheric gas up to BC-MPO orbit, and hence it is worth simulating the signal measured by SERENA. The ion-sputtering energy distribution function (f_s) (15) of the ejection energy usually peaks at few eV (Sigmund 1969; Sieveka and Johnson 1984), with a high-energy tail, up to approximately one hundred of eV. The value of the binding energy E_b determines the position of the spectrum peak, which is located at $E_b/2$; the projectile impact energy E_i determines the spectrum cut-off energy.

The spectrum differs from species to species but, in general, most of the ejected particles are able to reach the BC-MPO altitude. For sodium, as an example, the energy needed to reach the satellite altitude goes from 0.3 eV at S/C periherm, to 0.8 eV at S/C apoherm. If we assume the distribution given by (15), then a fraction above 90% of the ejected sodium is able to reach such altitudes. The detection of ion-sputtering ENAs deals with several scientific objectives of BC-MPO/SERENA (Milillo et al. 2005):

(1) particle loss-rate from Mercury's environment and surface emissivity;
(2) analysis of ion-sputtering process, since its yield, at high energy, is not well known; and
(3) analysis of proton precipitation, as a function of external conditions, performable in addition to other ion measurements.

The SERENA-STROFIO unit, in principle, is able to reveal sputtered neutral particles, depending on the relative composition of the S/C and particle velocity (see Fig. 42). Since the process is, with some approximation, stoichiometric, the mass analysis of the flux coming from a certain region on the surface gives information about the composition of the soil in that region.

Processes that Promote and Deplete the Exosphere of Mercury 315

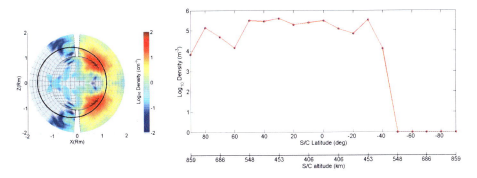

Fig. 42 *Left*: sample of ion sputtering density contours on the *x*–*z* plane (with trace of the MPO orbit). *Right*: actual STROFIO simulated data, according to instrument constraints. From Mura (2005)

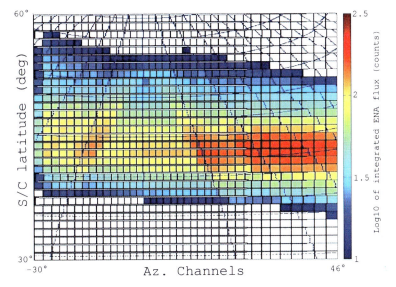

Fig. 43 2-D image of surface oxygen emission, obtained by the superimposition of 31 ELENA simulated measurements. Color is coded according to ELENA 30 s count rates, integrated over all energies above Oxygen escape energy. From Mura et al. (2006a)

On average, approximately 50% of the particles have more energy than the escape energy (few eV); those particles travel along quasi-linear paths, allowing a remote-sensing of the process via the SERENA-ELENA unit. The measurement of the total neutral flux, compared to the precipitating ion flux (observed by SERENA-MIPA), gives information about the effectiveness of the process (i.e., the process yield Y) and on the ion flux onto the surface, including its spatial and temporal distribution (see Fig. 43).

The 2D image of surface oxygen emission is obtained with the help of S/C motion, as a superimposition of 31 ELENA simulated measurements as the S/C is orbiting from 30°N to 60°N at 12:00 MLT (the periherm is at 0°, 12:00 MLT). The intense proton precipitation in the dayside, northern hemisphere results in an intense O-ENA production in the center of the global 2D image. In general, it is possible to have a global view of proton precipitation by means of ion-sputtered directional ENAs, with surface spatial resolution between 15 and

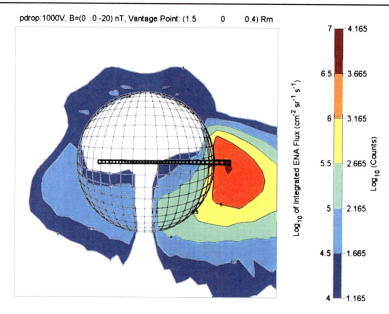

Fig. 44 Simulation of ELENA measurement, according to ENA flux in Fig. 37. FOV is shown by the *black bold grid*; each square represents an instrument sector ($2° \times 2°$). From Mura et al. (2006b)

50 km, depending on S/C altitude. The color is coded according to ELENA 30 s count rates, integrated over all energies above Oxygen escape energy.

Under different IMF conditions, the configuration of the Hermean magnetosphere changes, so that the area of Solar Wind proton precipitation (and subsequent neutral release) consequently drifts and changes in size (see, e.g., Massetti et al. 2003; Kallio and Janhunen 2003). According to the simulations shown, the ion-sputtering ENA signal from the dayside is high enough to be detected by ELENA. In fact, the spatial and time resolution is good enough to monitor instantaneous changes of the magnetospheric configuration; the spatial resolution permits discrimination of surface emissivity variations. The intensity of the directional ENA signal originating from ion-sputtering depends on both proton precipitation flux and surface properties (composition and yield).

However, our simulations show that it is possible to discriminate between these two factors. In fact, proton surface precipitation has a typical spatial scale factor of about 100 km (Mura et al. 2006b). Since ELENA spatial resolution is lower (between 15 and 50 km), any small-scale spatial change in ion-sputtering ENA signal would be probably due to surface property variations. Moreover, temporal variability in the ENA signal should be ascribed only to modifications in the proton circulation properties.

Charge exchange neutrals have typical energies of 1 keV or more (see Sect. 5.4); this process mostly occurs in the dayside and dawnside regions (Mura et al. 2005), up to altitudes of hundreds of km (see Fig. 37). The H-ENA production rate for unit length reaches values up to 10 ($cm^{-2} s^{-1} sr^{-1} m^{-1}$), close to the dayside planetary surface. This value, in some cases (depending also on the *line-of-sight* length), could lead to high values of H-ENA flux per steradian (up to 10^6 $cm^{-2} s^{-1} sr^{-1}$). To facilitate the detection of such H-ENAs, the ELENA central axis is tilted, with respect to the S/C nadir axis, by 8°. In this way, at least three sectors point away from the planet, if the S/C is in a ~15° orbital arc centered at the apoherm (see Fig. 44). In fact, the H-ENA signal is lower within sectors looking towards the planet, because the integration path is shorter. The S/C apoherm position will move, in longitude, during the MPO mission, thus allowing different, optimal vantage points.

As an example, from a nightside vantage point it is possible to detect the H-ENA signal generated from two different H^+ populations: the first one originates from protons that are precipitating into the cusp regions or circulating over the North pole. The second population originates from protons that have been drifted westward by the grad-B drift, from dayside to nightside through the dawn region. They reach low altitudes at their mirroring points, where exospheric densities are higher: they may produce an intense ENA signal that can be seen in the right side of the pictures. Charge-exchange ENA will be observed also by the other ENA sensor onboard the BC MMO satellite, with a more extended view due to the more distant apocenter (about 15,000 km).

The occurrence of impacts on Mercury by projectiles of the radius of 1 m seems to be not infrequent (2 events/year); such large projectiles would vaporize the regolith and reach a depth of meters, depending on the density and porosity of the regolith itself (Mangano et al. 2005). For the surface deeper layers are believed to be less contaminated by space weathering (Hapke 2001), the detection of the vaporized soil due to such impulsive events could be the only way to remote sense the real Hermean endogenous material.

One of the most interesting goals of the SERENA-STROFIO observations is the identification and detection of meteoroid impact vaporization process. In fact, as stated in Sect. 2, among the release processes active on Mercury, refractory species are released most efficiently by impact events (Gerasimov et al. 1998); hence, MIV could be a valid mechanism by which species like Mg, Al, and Si will be detected. Furthermore, larger projectiles (for instance, 1 m radius meteoroids that have a probability of two impacts/year) would vaporize the regolith reaching a depth of meters, hence layers less contaminated by space weathering (Hapke 2001), the detection of the vaporized soil due to such impulsive events could be the only way to remote sense the real Hermean endogenous material.

The frequent impact of meteoroids 10 cm in radius (more than two events/day, (Marchi et al. 2005)) makes them particularly interesting. In this case, the enhancement, depending from the considered species, varies from 1 to 4 orders of magnitude higher than the mean exospheric background values (Mangano et al. 2007). Durations are generally larger than 2,000 s, and their extension larger than $50°$ (calculated with respect to the center of the planet).

Such event will be observable by SERENA-STROFIO when the spacecraft flies over the impact zone during the event, and also by the UV spectrometer PHEBUS when the event happens in the daylight and along the instrument look direction. Figure 45 shows the simulated Na density observed by STROFIO as a function of time after impact and spacecraft position with respect to impact point. The signal is stronger when the impact is toward the ram direction (i.e., the look direction of the instrument); when the spacecraft pass the impact point fewer particles reach the instrument FOV.

In Fig. 46 the estimated counts rate of STROFIO for O, Mg and Na versus time in the case of a meteoritic impact under the spacecraft position are shown. These simulations seem to assure very high detectability, reaching almost the 100% of probability after only one month of monitoring from a probe in polar orbit and periherm at 400 km, apoherm at 1,500, as planned to be for the BepiColombo/MPO mission (Mangano et al. 2007).

Note that exospheric spatial inhomogeneities of refractory species with similar time scales could be generated by ion-sputtering. In this case the shape and intensity of the cloud would be significantly different as previously mentioned. Furthermore, the sputter-generated cloud would be characterized by an energy spectrum reaching higher energies (Wiens et al. 1997). The energy resolution of SERENA will allow discrimination between these two processes.

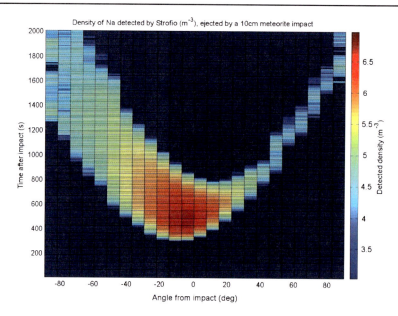

Fig. 45 Simulated Na density observed by STROFIO as a function of time after impact and spacecraft position with respect to impact point. The field of view of the instrument ($20° \times 20°$ in the ram direction) and the particle trajectories have been considered

6.3 3D Exospheric Modeling and Measurement Feedback

Mercury's exosphere consists of particles in ballistic orbits which originate from various release processes like thermal release, particle/photon sputtering and micrometeorite impact. The spatial distribution of a neutral exospheric component can be obtained by using a Monte Carlo single-particle model. In these models, it is assumed that the trajectory of a relatively small, but statistically representative, amount of test particles can reproduce the trajectories of all the real exospheric particles. A weight w is associated to each test particle, which takes into account the number of real particles that it represents.

For a given source process, the surface S where the process occurs is defined, then some number (N_{tp}) of test particles (usually some millions) are launched from randomly chosen starting points P_0 within S: $\Delta A = S/N_{tp}$, $w = j_0 \Delta A$, with j_0 being the initial flux through the surface element ΔA. The starting velocity $\mathbf{v}_0$ must be randomly chosen according to the velocity distribution function of the source; for an arbitrary velocity distribution function this can be done by using a Von Neumann (1951) algorithm. Alternatively, the surface of the planet can be divided into a number of elements and the initial flux of the sputtered particles can be prescribed for each element according to the composition of the soil and the sputter agents considered. In this case, several thousand particles are launched from each element corresponding to the initial conditions associated with the surface element and the contribution of each particle to the total number density is weighted according to the distribution function conforming to the release process considered. The initial elevation angle is determined via a given distribution while the azimuth angle is chosen randomly within the interval of 2π. Although the number of particles launched from each surface element is rather limited, their statistics are expected to approximately represent that of the many more particles forming the real exosphere of Mercury.

Processes that Promote and Deplete the Exosphere of Mercury

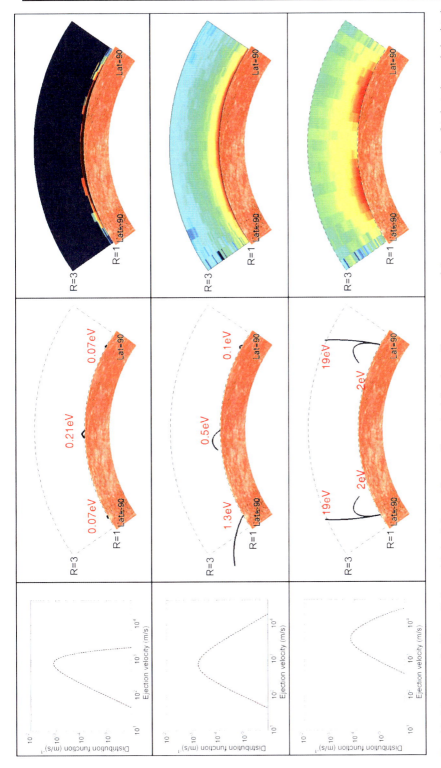

Fig. 46 Schematics of a Monte Carlo model simulation at Mercury. From the *left column*: Energy spectra of the source process; examples of trajectories, and simulated exospheres for different release processes (based on Mura et al. 2007). *Top row*: thermal desorption; *middle row*: photon-stimulated desorption; *bottom row*: ion sputtering. The planetary surface (*in brown*) represents a cut from $-90°$ to $90°$ of latitude; the curvature is not to scale

The equations of motion are solved in spherical coordinates (r, Θ, ϕ) and are given by:

$$\ddot{r} - r\dot{\theta}^2 - r\sin^2\theta\dot{\phi}^2 = -\frac{GM}{r^2},$$

$$\ddot{\theta} + \frac{2}{r}\dot{r}\dot{\theta} - \sin\theta\cos\theta\dot{\phi}^2 = 0, \qquad (27)$$

$$\ddot{\phi} + \frac{2}{r}\dot{r}\dot{\phi} + 2\cot\theta\dot{\theta}\dot{\phi} = 0,$$

where M is the mass of Mercury and G the gravitational constant. The trajectories of the particles are followed until they either leave the simulation box or hit the surface of the planet. Particles falling back onto the surface are assumed to be trapped in the soil and are therefore excluded from further calculations. If necessary, additional forces can easily be incorporated into the model, for example, radiation pressure is expected to significantly affect the trajectories of Na and K and should therefore be taken into account. On the other hand, due to the slow rotation period of Mercury ($P \sim 59$ days) the acceleration caused by the Coriolis force might be neglected. Finally, exospheric species may eventually become ionized and removed from the exosphere. Such loss processes can be taken into account both by removing test-particles, or by decreasing w:

$$\frac{dw}{dt} = w(t)\sum_i \frac{1}{\tau_i}, \qquad (28)$$

where τ_i is the lifetime of process i. To obtain the particle density, it is necessary to define a spatial grid Q. Each time a test particle crosses a grid cell, a quantity q is deposited there:

$$q = w(t)\Delta t, \qquad (29)$$

where Δt is the time elapsed inside the cell. After all trajectories have been simulated, the spatial dependent number density n can then be determined by dividing Q by the volume of the cells:

$$n(r, \theta, \phi) = \frac{Q(r, \theta, \phi)}{\Delta V}, \qquad (30)$$

further dimensions of the spatial grid can be used, as an example, to store the information about the energy of the particles.

At Mercury, several Monte Carlo models (e.g. Smith et al. 1978) have been proposed to reproduce the exospheric density due to relevant source processes such as thermal and photon-stimulated desorption, ion sputtering or micro meteoritic impact vaporization; as far as it concerns particle losses, it is necessary to include at least photo-ionization, being charge-exchange process negligible as a neutral loss mechanism Milillo et al. 2005; Mura et al. 2006b. Figure 46 shows an example of Monte Carlo modeling of the Hermean exosphere (based on Mura et al. 2007).

In Fig. 46, samples of exospheric vertical profiles related to various release processes are shown. TD produces a very dense exosphere, but at very low altitudes and only in the dayside. This process is not effective in global loss from the planet, since the fraction of particles that escape is well below 1% on average. In the case of PSD, the fraction of particles that are lost varies between 1% to 20% (Mura et al. 2007), while for IS it is found to be between 50% and 80%.

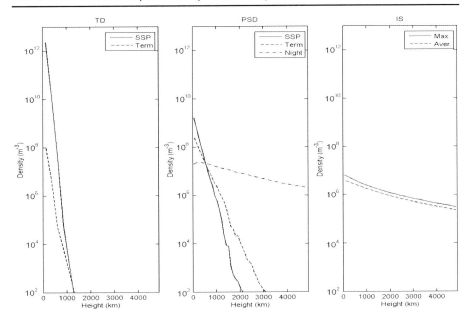

Fig. 47 Examples of Monte Carlo simulations of Na vertical profiles in the exosphere of Mercury. From the left: Thermal desorption (over sub-solar point (SSP) and over the terminator (Term)); Photon-stimulated desorption simulated using a Maxwellian-flux distribution with $T = 1,500$ K (over the sub-solar point, the terminator and over the anti-sun point (Night)); Ion-sputtering (over the point of maximum ion precipitation (Max) and averaged (Aver.) over all the dayside surface). Mercury–Sun distance is 0.38 AU, radiation pressure as estimated by Smyth and Marconi (1995) included. Adapted from Mura et al. (2007)

Figure 47 shows examples of Monte Carlo simulations of Na vertical profiles released from the surface due to thermal desorption, PSD and ion sputtering.

The Mercury exosphere simulations do provide significant background information, useful for the understanding of the "real" measurements which will be taken by the forthcoming in situ missions, like Messenger and BepiColombo. To deduce the physical meaning of these future observations, it could be worth deriving some functional forms—based on a best-fit approach of the available simulations—and link them to the basic physical exospheric parameters. In the future, such parameters could be properly tuned to best-fit the real data, and immediately provide a reliable signature of the planet's exospheric global characteristics.

In the simplified case of a surface with uniform concentration of a given component, the exospheric density generated from both PSD and TD has a cylindrical symmetry around the x axis (planet to Sun direction), and peaks at the sub-solar point. In this case, the vertical profile over this point can be reproduced by the simplified exospheric r-profile model (Rairden et al. 1986) by Chamberlain (1963) (first three terms of (28)). The angular dependence expressed by Rairden et al. (1986) for the Earth cannot be applied at Mercury since the Hermean exosphere is generated directly from the planet surface, and not from the exobase. Moreover, the radiation pressure acceleration can be very effective at Mercury (Smyth and Marconi 1995; Potter et al. 2002a), ranging from 0.2 to 2 m/s for Na, and from 0.3 to 3 m/s for K The remaining terms in (28) take into account the strong variation between day and night conditions, and describe the shape of the tail:

$$\log_{10}(n) = \underbrace{A + Be^{-C(r-1)} - \frac{r-1}{D}}_{\text{Raydern}} - \underbrace{\frac{F/r}{1+e^{-\frac{\alpha-\pi/2}{E}}}}_{\text{day/night modulation}} + \underbrace{G\left(1 - e^{-I(r-1)}\right)e^{-\frac{1}{2}\left(\frac{\alpha-\pi}{H}\right)^2}}_{\text{radiation pressure effect}}, \quad (31)$$

where n is the density, α is the solar zenith angle (from the sub-solar point), r is the planetocentric distance. A through I are free parameters depending on the process involved (PSD or TD) and on boundary conditions (Mura et al. 2007); their values have been derived by best-fitting the outcomes from Monte Carlo numerical model previously described.

The shape of the exosphere generated from ion sputtering strongly depends on the ion precipitation pattern onto the surface. Recently, numerical models have taken into account the contribution of protons (Sarantos et al. 2001; Massetti et al. 2003; Kallio and Janhunen 2003; Mura et al. 2007; Leblanc and Johnson 2003), minor s/w components such as alpha particles (Leblanc and Johnson 2003) and planetary ions (Delcourt et al. 2003). According to most authors, in the case of s/w protons, we expect a proton precipitation flux up to 10^9 cm^{-2} s^{-1} in the cusp regions, located in the dayside at mid latitudes. The flux, the shape and the relative size of these regions depend on the magnetospheric configuration. As stated before, the fraction of particles that escape from the planet is found between 50% and 80%.

Photoionization reduces, in general, the dayside exospheric density. Since the ionized particles are accelerated by the electromagnetic fields, this process produces a net loss from the planet. The estimated effect in exospheric profile, however, can be neglected.

Radiation pressure produces an increase of density in the nightside and a reduction in the dayside. This effect can be easily seen, for example, in the PSD exosphere, since particles generated by this release process have long residence time in ballistic trajectories (10^3–10^4 s); hence, they are efficiently accelerated in the anti-sunward direction.

7 Conclusions

The best-studied constituent in Mercury's exosphere, sodium, displays a rich variety in its spatial and temporal variability. Observations of this constituent have revealed a rapid variation in the ion sputtering component due to a variability in the magnetosphere, possible latitudinal and/or longitudinal variations in composition, long-term variations in photon-stimulated desorption and radiation pressure acceleration, long-term and short-term variations in meteoritic vaporization, and possible sequestration of volatiles on the nightside and at high latitudes. Observations of other species, both by Mariner 10 and by ground-based telescopes, are more sparse. A great wealth of information, both about the surface and about the interaction of high-energy radiation and particles with the surface, is expected to be gained with the advent of the two planned spacecraft, Messenger and BepiColombo, to the Hermean system. We expect to discover many more species in the exosphere in addition to the six currently known, but their expected relative abundances is widely debated. A comparison of simulated data with actual data from these spacecraft will allow us to test our current theories and revise them as appropriate. The abundances of the noble gases will tell us about processes from implantation of solar wind to the abundances in the deep interior. Asymmetries in the abundances of exospheric species will tell us not only how volatiles are sequestered and lost, but about the refractory species as well. We will undoubtedly be surprised by these discoveries, as all explorers have met with surprises in the past.

Acknowledgements A. Mura and H.I.M. Lichtenegger acknowledge support by the Europlanet project (http://www.europlanet-eu.org) for supporting working visits to the Space Science Institute of the Austrian Academy of Sciences in Graz, Austria and to the Istituto di Fisica dello Spazio Interplanetario-CNR, in Rome, Italy.

References

M.R. Aellig, A.J. Lazarus, J.T. Steinberg, in *Solar and Galactic Composition*, ed. by R.F. Wimmer-Schweingruber (2001)
M.A. A'Hearn, M.J.S. Belton, W.A. Delamere, J. Kissel, The Deep Impact Team, Science **310**, 258–264 (2005)
T.J. Ahrens, D.M. Cole, Proc. Lunar Sci. Conf. 5th, (1974), pp. 2333–2345
T.J. Ahrens, J.D. O'Keefe, Moon **4**, 214–249 (1972)
F. Aumayr, H. Winter, Phil. Trans. Roy. Soc. Lond. A **362**, 77–102 (2004)
C. Barbieri, S. Verani, G. Cremonese, A. Sprague, M. Mendillo, R. Cosentino, D. Hunten, Planet. Space Sci. **52**, 1169–1175 (2004)
A. Benninghoven, F.G. Rüdenauer, H.W. Werner, *Secondary Ion Mass Spectrometry: Basic Concepts, Instrumental Aspects, Applications and Trends* (Wiley, New York, 1987), 1227 pp
G. Betz, W. Husinsky, Phil. Trans. Roy. Soc. Lond. A **362**, 177–194 (2004)
T.A. Bida, R.M. Killen, T.H. Morgan, Nature **404**, 159–161 (2000)
D.T. Blewett, P.G. Lucey, B.R. Hawke, G.G. Ling, M.S. Robinson, Icarus **129**, 217–231 (1997)
V. Bothmer, R. Schwenn, Ann. Geophys. **16**, 1–24 (1998)
W.F. Bottke, A. Morbidelli, R. Jedicke, J.M. Petit, P. Levison, H.F. Michel, T.S. Metcalfe, Icarus **156**, 399–433 (2002)
A.L. Broadfoot, D.E. Shemansky, S. Kumar, Geophys. Res. Lett. **3**, 577–580 (1976)
M.E. Brown, Icarus **151**, 190–195 (2001)
P. Brown, R.E. Spalding, D.O. ReVelle, E. Tagliaferri, S.P. Worden, Nature **420**, 294–296 (2002)
T.H. Burbine, T.J. McCoy, L.R. Nittler, G.K. Benedix, E.A. Cloutis, T.L. Dickinson, Science **37**, 1233–1244 (2002)
B. Butler, D. Muhleman, M. Slade, J. Geophys. Res. **98**, 15,003–15,023 (1993)
M. Bruno, G. Cremonese, S. Marchi, Mon. Not. Roy. Astron. Soc. **1**, 1067–1071 (2006)
T.A. Cassidy, R.E. Johnson, Icarus **176**, 499–507 (2005)
J.W. Chamberlain, Planet. Space Sci. **11**, 901–960 (1963)
A. Ciaravella, J.C. Raymond, A. van Ballegooijen, L. Strachan, A. Vourlidas, J. Li, J. Chen, A. Panasyuk, Astrophys. J. **597**, 1118–1134 (2003)
M. Cintala, J. Geophys. Res. **97**, 947–973 (1992)
E.W. Cliver, O.C. St. Cyr, R.A. Howard, P.S. Mc Intosh, in *Solar Coronal Structures*, ed. by V. Rusin, J. Heinzel, C. Vial (VEDA Publishing House of the Slovak Academy of Sciences, 1994), pp. 83–89
C.M.S. Cohen, E.C. Stone, R.A. Mewaldt, R.A. Leske, G.M. Cummings, A.C. Mason, M.I. Desai, T.T. von Rosenvinge, M.E. Wiedenbeck, J. Geophys. Res. **110**, A09S16 (2005)
M.R. Combi, M.A. Disanti, U. Fink, Icarus **130**, 336–354 (1997)
J.E.P. Connerney, N.F. Ness, in *Mercury*, ed. by F. Vilas, C.R. Chapman, M.S. Matthews (University of Arizona Press, Tucson, 1988), pp. 494–513
S.R. Coon, W.F. Calaway, M.J. Pellin, J.M. White, Surf. Sci. **298**, 161–172 (1993)
J. Crank, *The Mathematics of Diffusion*, 2nd edn. (Oxford Univ. Press, Oxford, 1975)
G. Cremonese, H. Boehnhardt, J. Crovisier, H. Rauer, A. Fitzsimmons, M. Fulle, J. Licandro, D. Pollacco, G.P. Tozzi, R.M. West, Astrophys. J. Lett. **490**, L199–L202 (1997)
G. Cremonese, M. Bruno, V. Mangano, S. Marchi, A. Milillo, Icarus **177**, 122–128 (2005)
G. Cremonese, M. Bruno, V. Mangano, S. Marchi, A. Milillo, Icarus **182**, 297–298 (2006)
D.C. Delcourt, T.E. Moore, S. Orsini, A. Millilo, J.A. Sauvaud, Geophys. Res. Lett. **29** (2002). doi:10.1029/2001GL013829
D.C. Delcourt, S. Grimald, F. Leblanc, J.J. Berthelier, A. Millilo, A. Mura, S. Orsini, T.E. Moore, Ann. Geophys. **21**, 1723–1736 (2003)
R.C. Elphic, H.O. Funsten III, R.L. Hervig, Lunar Planet. Sci. Conf. Abst. **24**, 439 (1993)
G. Fjeldbo, A. Kliore, D. Sweetnam, P. Esposito, B. Seidel, T. Howard, Icarus **29**, 439–444 (1976)
B.C. Flynn, S.A. Stern, Icarus **124**, 530–536 (1996)
W.A. Gault, H.N. Rundle, Can. J. Phys. **47**, 85–98 (1969)
D.E. Gault, E.F. Horz, D.E. Brownlee, J.B. Hartung, Lunar. Planet Sci. Conf. **5**, 260 (1974)
M.V. Gerasimov, B.A. Ivanov, O.I. Yakovlev, Earth Moon Planet. **80**, 209–259 (1998)

H. Gnaser, W.O. Hofer, Appl. Phys. A **48**, 261–271 (1989)
B.E. Goldstein, S.T. Suess, R.J. Walker, J. Geophys. Res. **86**, 5485–5499 (1981)
N. Gopalswamy, in *The Sun and the Heliosphere as an Integrated system*, ed. by G. Poletto, S. Suess. ASSL Series (Kluwer, 2004), pp. 201–240
N. Gopalswamy, M.R. Kundu, Sol. Phys. **143**, 327–343 (1993)
N. Gopalswamy, S. Yashiro, M.L. Kaiser, R.A. Howard, J.L. Bougeret, J. Geophys. Res. **106**, 29,219–29,230 (2001a)
N. Gopalswamy, A. Lara, S. Yashiro, M.L. Kaiser, R.A. Howard, J. Geophys. Res. **106**, 29,207–29,218 (2001b)
N. Gopalswamy, A. Lara, S. Yashiro, R.A. Howard, Astrophys. J. **598**, L63–L66 (2003)
N. Gopalswamy, S. Nunes, S. Yashiro, R.A. Howard, Adv. Space Res. **34**(2), 391–396 (2004). doi:10.1016/j.asr.2003.10.054
C.S. Hansen, W.F. Calaway, B.V. King, M.J. Pellin, Surf. Sci. **398**, 211–220 (1998)
C.S. Hansen, W.F. Calaway, M.J. Pellin, B.V. King, A. Wucher, Surf. Sci. **432**, 199–210 (1999)
B. Hapke, J. Geophys. Res. **106**, 10,039–10,073 (2001)
R.A. Haring, A. Haring, F.W. Saris, A.A. de Vries, Appl. Phys. Lett. **41**, 174–176 (1982)
J.K. Harmon, Adv. Space Res. **19**, 1487–1496 (1997)
J.K. Harmon, M.A. Slade, Science **258**, 640–642 (1992)
J.K. Harmon, P.J. Perilat, M.A. Slade, Icarus **149**, 1–15 (2001)
G.B. Hasted, *Physics of Atomic Collisions* (Butterworths, London, 1964), p. 416
G. Heiken, D. Vaniman, B.M. French, *Lunar Sourcebook: A User's Guide to the Moon* (Cambridge Univ. Press, Cambridge, 1991)
T. Henke, J. Woch, U. Mall, S. Livi, B. Wilken, R. Schwenn, G. Gloeckler, R. von Steiger, R.J. Forsyth, A. Balogh, Geophys. Res. Lett. **25**, 3465–3468 (1998)
E. Hildner, J.T. Gosling, R.M. MacQueen, R.H. Munro, A.I. Poland, C.L. Ross, Sol. Phys. **48**, 127–135 (1976)
R.R. Hodges Jr., J. Geophys. Res. **79**, 2881–2885 (1974)
W.O. Hofer, in *Sputtering by Particle Bombardment*, ed. by R. Behrisch, R.K. Wittmaack (1991), pp. 15–90
L. Holmlid, J. Phys. Chem. **102**, 10,636–10,646 (1998)
L. Holmlid, Planet. Space Sci. **54**, 101–112 (2006)
L. Holmlid, J.O. Olsson, Surf. Sci. **67**, 61–76 (1977)
G.P. Horedt, G. Neukum, Icarus **60**, 710–717 (1984)
R.A. Howard, D. Michels, N.R. Sheeley, M.J. Koomen, in *The Sun and the Heliosphere in Three Dimensions*, ed. by R. Marsden, D. Reidel. ASSL, vol. 123 (Norwell, 1986), pp. 107–111
W.F. Huebner, J.J. Keady, S.P. Lyon, Astrophys. Space Phys. **195**, 1–294 (1992)
D.M. Hunten, A.L. Sprague, Adv. Space Res. **19**, 1551–1560 (1997)
D.M. Hunten, A.L. Sprague, Meteorit. Planet. Sci. **37**, 1191–1195 (2002)
D.M. Hunten, L.V. Wallace, Astrophys. J. **417**, 757–761 (1993)
D.M. Hunten, T.H. Morgan, D. Shemansky, in *Mercury*, ed. by F. Vilas, C.R. Chapman, M.S. Matthews (Univ. of Arizona Press, Tucson, 1988), pp. 562–612
A.P. Ingersoll, T. Svitek, B.C. Murray, Icarus **100**, 40–47 (1992)
W.H. Ip, Geophys. Res. Lett. **13**, 423–426 (1986)
W.H. Ip, Icarus **71**, 441–447 (1987)
W.H. Ip, Astrophys. J. **356**, 675–681 (1990)
W.H. Ip, A. Kopp, J. Geophys. Res. **07** (2002). doi:10.1029/2001JA009171
R.E. Johnson, Geophys. Monogr. **130**, 203–219 (2002)
K. Kabin, T.I. Gombosi, D.L. DeZeeuw, K.G. Powell, Icarus **143**, 397–406 (2000)
E. Kallio, P. Janhunen, Geophys. Res. Lett. **30**, (2003). doi:10.1029/2003GL017842
R.M. Killen, Meteorit. Planet. Sci. **37**, 1223–1231 (2002)
R.M. Killen, Publ. Astron. Soc. Pac. **118**, 1347–1353 (2006)
R.M. Killen, W.H. Ip, Rev. Geophys. **37**, 361–406 (1999)
R.M. Killen, T.H. Morgan, Icarus **101**, 294–312 (1993)
R.M. Killen, A.E. Potter, T.H. Morgan, Icarus **85**, 145–167 (1990)
R.M. Killen, A.E. Potter, T.H. Morgan, Science **252**, 474–475 (1991)
R.M. Killen, A.E. Potter, A. Fitzsimmons, T.H. Morgan, Planet. Space Sci. **47**, 1449–1458 (1999)
R.M. Killen, A.E. Potter, P. Reiff, M. Sarantos, B.V. Jackson, P. Hick, B. Giles, J. Geophys. Res. **106**, 20,509–20,526 (2001)
R.M. Killen, A.E. Potter, M. Sarantos, P. Reiff, Mercury, 25th Meeting of the IAU, Joint Discussion 2, Sydney, Australia, 2003. Meeting abstract
R.M. Killen, M. Sarantos, P.H. Reiff, Adv. Space Res. **33**, 1899–1904 (2004a)
R.M. Killen, M. Sarantos, A.E. Potter, P. Reiff, Icarus **171**, 1–19 (2004b)

R.M. Killen, T.A. Bida, T.H. Morgan, Icarus **173**, 300–311 (2005)
P.L. Koehn, A.L. Sprague, Planet. Space Sci. (2007). doi:10.1016/j.pss.2006.10.009
P.L. Koehn, T.H. Zurbuchen, K. Kabin, DPS Abstract, 35.2308, 2003
A. Kotarba, I. Kruk, Z. Sojka, J. Catal. **221**(2), 650–652 (2004)
G.A. Krasinsky, E.V. Pitjeva, M.V. Vasilyev, E.I. Yagudina, Icarus **158**, 98–105 (2002)
H. Lammer, P. Wurz, M.R. Patel, R. Killen, C. Kolb, S. Massetti, S. Orsini, A. Milillo, Icarus **166**, 238–247 (2003)
A. Lara, J.A. González-Esparza, N. Gopalswamy, Geofísica Internacional **43**, 75–82 (2004)
F. Leblanc, R.E. Johnson, Icarus **164**, 261–281 (2003)
F. Leblanc, D. Delcourt, R.E. Johnson, J. Geophys. Res. **108**, (2003a). doi:10.1029/2003JE002151
F. Leblanc, J.G. Luhmann, R.E. Johnson, M. Liu, Planet. Space Sci. **51**, 339–352 (2003b)
F. Leblanc, C. Barbieri, G. Cremonese, S. Verani, R. Cosentino, M. Mendillo, A. Sprague, D. Hunten, Icarus **185**, 395–402 (2006)
S.T. Lepri, T.H. Zurbuchen, L.A. Fisk, I.G. Richardson, H.V. Cane, G. Gloeckler, J. Geophys. Res. **106**, 29,231–29,238 (2001)
G.M. Lindsay, J.G. Luhmann, C.T. Russell, J.T. Gosling, J. Geophys. Res. **104**, 12515–12524 (1999)
K. Lodders, B. Fegley, *The Planetary Scientists Companion* (Oxford University Press, 1998)
J.G. Luhmann, C.T. Russell, N.A. Tsyganenko, J. Geophys. Res. **103**, 9113–9119 (1998)
A.V. Lukyanov, S. Barabash, R. Lundin, P. C:son Brandt, Planet. Space Sci. **49**, 1677–1684 (2001)
T.E. Madey, B.V. Yakshinskiy, V.N. Ageev, R.E. Johnson, J. Geophys. Res. **103**, 5873 (1998)
A. Mallama, D. Wang, R.A. Howard, Icarus **155**, 253–264 (2002)
V. Mangano, A. Milillo, S. Orsini, A. Mura, H. Lammer, P. Wurz, EGU Abstract, EGU05-A-01247, 2005
V. Mangano, A. Milillo, A. Mura, S. Orsini, E. De Angelis, A.M. Di Lellis, P. Wurz, Planet. Space Sci. (2007). doi:10.1016/j.pss.2006.10.008
S. Marchi, A. Morbidelli, G. Cremonese, Astron. Astrophys. **431**, 1123–1127 (2005)
G.M. Mason, J.E. Mazur, J.R. Dwyer, Astrophys. J. **525**, L133–L136 (1999)
S. Massetti, S. Orsini, A. Milillo, A. Mura, E. De Angelis, H. Lammer, P. Wurz, Icarus **166**, 229–237 (2003)
S. Massetti, S. Orsini, A. Milillo, A. Mura, Planet. Space Sci. (2007). doi:10.1016/j.pss.2006.12.008
M.A. McGrath, R.E. Johnson, L.J. Lanzerotti, Nature **323**, 696–696 (1986)
A.S. Milillo, S. Orsini, P. Wurz, D. Delcourt, E. Kallio, H. Rillen, R.M. Lammer, S. Massetti, A. Mura, S. Barabash, G. Cremonese, I.A. Daglis, E. De Angelis, A.M. Di Lellis, S. Livi, V. Mangano, K. Torka, Space Sci. Rev. **117**, 397–444 (2005)
A. Morbidelli, B. Gladman, Meteorit. Planet. Sci. **33**, 999–1016 (1998)
T.H. Morgan, R.M. Killen, Planet. Space Sci. **45**, 81–94 (1997)
T.H. Morgan, H.A. Zook, A.E. Potter, Icarus **74**, 156–170 (1988)
A. Mura, Neutral Atom Emission from Mercury Magnetosphere. AOG, Conference. Singapore, June 20–24, 2005
A. Mura, S. Orsini, A. Milillo, D. Delcourt, S. Massetti, E. De Angelis, Icarus **175**, 305–319 (2005)
A. Mura, S. Orsini, A. Milillo, A.M. Di Lellis, E. De Angelis, Planet. Space Sci. **54**, 144–152 (2006a)
A. Mura, S. Orsini, A. Milillo, D. Delcourt, A.M. Di Lellis, E. De Angelis, S. Massetti, Adv. Geosci. **3** (2006b). ISBN 981-256-983-8
A. Mura, D. Delcourt, S. Massetti, A. Milillo, S. Orsini, A. Di Lellis, E. De Angelis, Geophysical Research Abstracts, vol. 8, 06958, EGU General Assembly, Vienna (Austria), 2–7 April (2006c)
A. Mura, A. Milillo, S. Orsini, S. Massetti, Planet. Space Sci. (2007). doi:10.1016/j.pss.2006.11.028
N.F. Ness, K.W. Behannon, R.P. Lepping, Y.C. Wang, Nature **255**, 204–205 (1975)
H. Oechsner, H. Schoof, E. Stumpe, Surf. Sci. **76**, 343–354 (1978)
J.D. O'Keefe, T.J. Ahrens, Science **234**, 346–349 (1986)
D.A. Paige, S.E. Wood, A.R. Vasavada, Science **258**, 643–646 (1992)
A.E. Potter, Geophys. Res. Lett. **22**, 3289–3292 (1995)
A.E. Potter, T.H. Morgan, Science **229**, 651–653 (1985)
A.E. Potter, T.H. Morgan, Icarus **67**, 336–340 (1986)
A.E. Potter, T.H. Morgan, Icarus **71**, 472–477 (1987)
A.E. Potter, T.H. Morgan, Science **241**, 675–680 (1988)
A.E. Potter, T.H. Morgan, Science **248**, 835–838 (1990)
A.E. Potter, T.H. Morgan, Planet. Space Sci. **45**, 95–100 (1997)
A.E. Potter, R.M. Killen, T.H. Morgan, Space Sci. **47**, 1141–1148 (1999)
A.E. Potter, C.M. Anderson, R.M. Killen, T.H. Morgan, J. Geophys. Res. Planets **107** (2002a). doi:10.1029/2000JE014937
A.E. Potter, R.M. Killen, T.H. Morgan, Meteorit. Planet. Sci. **37**, 1165–1172 (2002b)
A.E. Potter, R.M. Killen, M. Sarantos, Icarus **181**, 1–12 (2006). doi:10.1016/j.icarus.2005.10.026
A.E. Potter, R.M. Killen, T.H. Morgan, Icarus **186**(2), 571–580 (2007)

R.L. Rairden, L.A. Frank, J.D. Craven, J. Geophys. Res. **91**(A12), 13613–13630 (1986)
D.V. Reames, S.W. Kahler, C.K. Ng, Astrophys. J. **491**, 414–420 (1997)
F.J.M. Rietmeijer, *Advanced Mineralogy* (Springer, Berlin, 1998), pp. 22–28
M. Sarantos, *Ion Trajectories in Mercury's Magnetosphere*, PhD thesis (Rice University, Houston, 2005)
M. Sarantos, P.H. Reiff, T.H. Hill, R.M. Killen, A.L. Urquhart, Planet. Space Sci. **49**, 1629–1635 (2001)
M. Sarantos, R.M. Killen, D. Kim, Planet. Space Sci. (2007). doi:10.1016/j.pss.2006.10.011
H. Schleicher, G. Wiedemann, H. Wohl, T. Berkefeld, D. Soltau, Astron. Astrophys. **425**, 1119–1124 (2004)
N. Sheeley, R.A. Howard, M.J. Koomen, D.J. Michels, Astrophys. J. **272**, 349–354 (1983)
D.E. Shemansky, Mercury Messenger **2**, 1 (1988)
D.E. Shemansky, *AIP Conf. Proc. 63, Rarefied Gas Dynamics 23rd Intl. Symposium*, 2003, p. 687
D.E. Shemansky, T.H. Morgan, Geophys. Res. Lett. **18**, 1659–1662 (1991)
E.M. Sieveka, R.E. Johnson, Astrophys. J. **287**, 418–426 (1984)
P. Sigmund, Phys. Rev. **184**, 383–416 (1969)
G. Siscoe, L. Christopher, Geophys. Res. Lett. **2**, 158–160 (1975)
M. Slade, B. Butler, D. Muhleman, Science **258**, 635–640 (1992)
G.R. Smith, D.E. Shemansky, A.L. Broadfoot, L. Wallace, J. Geophys. Res. **83**, 3783–3790 (1978)
W.H. Smyth, M.L. Marconi, Astrophys. J. **441**, 839–864 (1995)
A.L. Sprague, J. Geophys. Res. **97**, 18257–18264 (1992)
A.L. Sprague, T.L. Roush, Icarus **133**, 174–183 (1998)
A.L. Sprague, R.W.H. Kozlowski, D.M. Hunten, Science **249**, 1140–1142 (1990)
A.L. Sprague, R.W.H. Kozlowski, D.M. Hunten, F.A. Grosse, Icarus **104**, 33–37 (1993)
A.L. Sprague, R.W.H. Kozlowski, F.C. Witteborn, D.P. Cruikshank, D.H. Wooden, Icarus **109**, 156–167 (1994)
A.L. Sprague, D.M. Hunten, K. Lodders, Icarus **118**, 211–215 (1995)
A.L. Sprague, D.M. Hunten, F.A. Grosse, Icarus **123**, 345–349 (1996)
A.L. Sprague, R.W.H. Kozlowski, D.M. Hunten, N.M. Schneider, D.L. Domingue, W.K. Wells, W. Schmitt, U. Fink, Icarus **129**, 506–527 (1997)
A.L. Sprague, W.J. Schmitt, R.E. Hill, Icarus **136**, 60–68 (1998)
A.L. Sprague, J.P. Emery, K.L. Donaldson, R.W. Russell, D.K. Lynch, A.L. Mazuk, Meteorit. Planet. Sci. **37**, 1255–1268 (2002)
L.V. Starukhina, Proc. Lunar Planet. Sci. Conf. **31**, 1301 (2000)
R.F. Stebbings, C.H. Smith, H. Ehrahardt, J. Geophys. Res. **69**, 2349 (1964)
O.C. St. Cyr, R.A. Howard, N.R. Sheeley Jr., S.P. Plunkett, D.J. Michels, S.E. Paswaters, M.J. Koomen, G.M. Simnett, B.J. Thompson, J.B. Gurman, R. Schwenn, D.F. Webb, E. Hildner, P.L. Lamy, J. Geophys. Res. **105**, 18169–18185 (2000)
J. Von Neumann, *Various Techniques Used in Connection with Random Digits*. National Bureau of Standard Applied Mathematics Series, vol. 12. (1951), pp. 36–38
A. Vourlidas, D. Buzasi, R.A. Howard, E. Esfandiari, in *Solar Variability: From Core to Outer Frontiers*, ed. by A. Wilson, ESA SP-506 (ESA Publication, Noordwijk, 2002), pp. 91–94
J. Warell, Icarus **161**, 199–222 (2003)
J. Warell, D.T. Blewett, Icarus **168**, 257–276 (2004)
D.F. Webb, R.A. Howard, J. Geophys. Res. **99**, 4201–4220 (1994)
M.E. Wiedenbeck et al., AGU Fall Mtg 2005, abstract SH11B-0267
R.C. Wiens, D.S. Burnett, W.F. Calaway, C.S. Hansen, K.R. Kykkem, M.L. Pellin, Icarus **128**, 386–397 (1997)
H.F. Winters, J.W. Coburn, Surf. Sci. Rep. **14**(3), 161–269 (1992)
H.F. Winters, J.W. Coburn, T.J. Chuang, J. Vac. Sci. Technol. B **1**(2), 469–480 (1983)
A. Wucher, H. Oechsner, Nucl. Instr. Methods **B18**, 458–463 (1986)
P. Wurz, in *The Dynamic Sun: Challenges for Theory and Observations*. ESA SP-600 (2005), pp. 5.2, 1-9
P. Wurz, H. Lammer, Icarus **164**, 1–13 (2003)
P. Wurz, W. Husinsky, G. Betz, in *Symposium on Surface Science*, ed. by J.J. Launois, B. Mutaftschiev, M.R. Tempère (La Plagne, France, 1990), pp. 181–185
P. Wurz, W. Husinsky, G. Betz, Appl. Phys. A **52**, 213–217 (1991)
P. Wurz, R.F. Wimmer-Schweingruber, K. Issautier, P. Bochsler, A.B. Galvin, F.M. Ipavich, *Composition of Magnetic Cloud Plasmas During 1997 and 1998*. CP-598 (American Institute Physics on Solar and Galactic Composition, 2001), pp. 145–151
P. Wurz, R. Wimmer-Schweingruber, P. Bochsler, A. Galvin, J.A. Paquette, F. Ipavich, in *Solar Wind X* (American Institute Physics, 2003), pp. 679, 685–690
P. Wurz, U. Rohner, J.A. Whitby, C. Kolb, H. Lammer, P. Dobnikar, J.A. Martín-Fernández, Icarus (2007). doi:10.1016/j.icarus.2007.04.034
B.V. Yakshinskiy, T.E. Madey, Nature **400**, 642–644 (1999)

B.V. Yakshinskiy, T.E. Madey, Icarus **168**, 53–59 (2004)
S. Yashiro, N. Gopalswamy, G. Michalek, O.C. St. Cyr, S.P. Plunkett, N.B. Rich, R.A. Howard, *J. Geophys. Res.* **109**(A7) (2004). doi:10.1029/2003JA010282
J. Zhang, K.P. Dere, R.A. Howard, M.R. Kundu, S.M. White, Astrophys. J. **559**, 452–642 (2001)
J.F. Ziegler, J.P. Biersack, *TRIM and SRIM Program Version SRIM-2003.26* (Pergamon, New York, 1985). http://www.srim.org/

Electromagnetic Induction Effects and Dynamo Action in the Hermean System

Karl-Heinz Glassmeier · Jan Grosser · Uli Auster · Dragos Constantinescu · Yasuhito Narita · Stephan Stellmach

Originally published in the journal Space Science Reviews, Volume 132, Nos 2–4.
DOI: 10.1007/s11214-007-9244-9 © Springer Science+Business Media B.V. 2007

Abstract Embedded in a large mass density and strong interplanetary magnetic field solar wind environment and equipped with a magnetic field of minor strength, planet Mercury exhibits a small magnetosphere vulnerable to severe solar wind buffeting. This causes large variations in the size of the magnetosphere and its associated currents. External fields are of far more importance than in the terrestrial case and of a size comparable to any internal, dynamo-generated field. Induction effects in the planetary interior, dominated by its huge core, are thought to play a much more prominent role in the Hermean magnetosphere compared to any of its companions. Furthermore, the external fields may cause planetary dynamo amplification much as discussed for the Galilean moons Io and Ganymede, but with the ambient field generated by the dynamo and its magnetic field-solar wind interaction.

Keywords Planets · Mercury · Magnetosphere · Magnetospheric currents · Electromagnetic induction · Dynamo action

1 Current Systems and Associated External Magnetic Fields

Mercury is an elusive planet. Due to its proximity to the Sun, its dayside surface temperature reaches maximum values of 700° K and the large rotational period of 58.64 days causes nightside temperatures as low as 100° K. Its equatorial radius of $R_P = 2{,}439.7$ km and mean mass density of $5.42 \cdot 10^3$ kg/m^3 indicates the existence of a large metallic core. The smaller than terrestrial gravity field implies a low escape velocity of 4.25 km/s and is

---

K.-H. Glassmeier (✉)
Institut für Geophysik und extraterrestrische Physik, Technische Universität Braunschweig,
Mendelssohnstrasse 3, 38106, Braunschweig, Germany
e-mail: kh.glassmeier@tu-bs.de

J. Grosser · U. Auster · D. Constantinescu · Y. Narita
Technische Universität zu Braunschweig, Braunschweig, Germany

S. Stellmach
Institut für Geophysik, Universität Münster, Münster, Germany

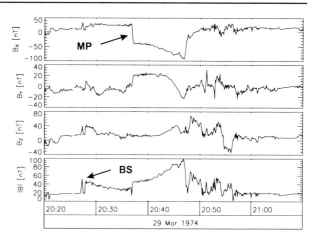

Fig. 1 Mariner 10 magnetic field measurements taken March 29, 1974. A Mercury-centered-solar-ecliptic coordinate system is used, where B_x, B_z, and B_y denote the magnetic field components along the Mercury–Sun line, towards the north ecliptic pole, and normal to the B_x–B_z plane; the value B gives the magnitude of the field. BS and MP denote the bowshock and magnetopause position, respectively

responsible for the absence of any significant atmosphere. A small planetary magnetic field with an equatorial surface strength of about 340 nT has been detected and causes a miniature magnetosphere in which planet Mercury is embedded (e.g. Ness 1979).

Solar wind properties outside this magnetosphere are also significantly different from terrestrial values. Mean solar wind density, flow velocity, and magnetic field magnitude are of the order 50–$70 \cdot 10^6$ protons/m^3, 500–700 km/s, and 50–70 nT, respectively (Russell 1989; Glassmeier 1997). The corresponding terrestrial values are $5 \cdot 10^6$ protons/m^3, 500–700 km/s, and 5–10 nT. Variations of these mean values are due to the large orbital eccentricity, $\epsilon = 0.2056$, of Mercury's path around the Sun. The large solar wind dynamic pressure and the low planetary magnetic field causes a mean magnetopause distance of 1,700 km ($0.7 R_P$) above the surface (Siscoe and Christopher 1975), which was traversed by Mariner 10 during its flybys in 1974–1975.

Figure 1 displays magnetic field measurements taken on March 29, 1974. The spacecraft entered the magnetospheric interaction region at the dawnside and traversed the bowshock at around 20:27 UT. The interplanetary magnetic field increased from about 20 nT in the upstream region to values of about 60 nT just behind the shock. After decaying in the magnetosheath to values of about 38 nT, a clear jump of about 24 nT across the magnetopause has been detected (Fig. 1). A Chapman–Ferraro sheet current density

$$\mathbf{j}_{CF} = -\frac{1}{\mu_0} \delta \mathbf{B} \times \mathbf{e}_N \quad (1)$$

causes such a jump of the tangential magnetic field component across a boundary assumed to be a plane sheet. Here $\delta \mathbf{B}$ denotes the change of the field across the boundary and $\mathbf{e}_N$ is its normal. The Chapman–Ferraro current system is a current system flowing in the magnetopause of any magnetosphere (see also Fig. 2). It serves to cancel the planetary field outside the magnetosphere. Inside it increases the field.

A 24 nT jump corresponds to a sheet current density of $1.9 \cdot 10^{-2}$ A/m. The extension of the current carrying layer in latitude is of the order of πR_{MP}, where $R_{MP} = 1.7 R_P$ is the average radial magnetopause distance. With this the total magnetopause current is of the order of $I_{CF} = 2.5 \cdot 10^5$ A.

Fig. 2 Schematic view of the magnetopause surface. *Solid lines* are magnetic field lines; *dashed lines* indicate the Chapman–Ferraro currents flowing on the magnetopause boundary (after Kivelson and Russell 1995)

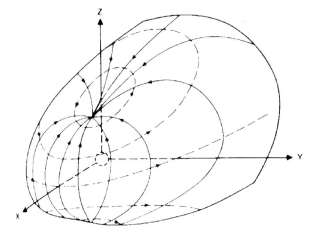

The magnetic field $\mathbf{B}_{CL} = (B_r, B_\phi, B_z)$ of a circular current loop in the plane of the loop is given by (e.g. Smythe 1968, p. 291)

$$B_r = 0; \qquad B_\phi = 0; \qquad B_z = \frac{\mu_0 I}{2\pi} \cdot \frac{1}{l+r} \left[K\left(\frac{r}{l}\right) + \frac{l+r}{(l-r)} E\left(\frac{r}{l}\right) \right], \qquad (2)$$

where I denotes the current, l the radius of the loop, and r the distance of the receiving point to the center of the current loop; K and E are complete elliptic integrals of the first and second kind. Approximating the magnetic effect of the magnetopause current at Mercury's dayside equator, B_{CF}, by the magnetic effect of such a circular current loop we put $I = I_{CF}$, $l = R_{MP}$, and $r = R_P$. With this we have $B_{CF} \approx 57$ nT. Apparently the magnetic field of the Chapman–Ferraro current significantly contributes to the magnetic field at the surface of Mercury.

Using the same approximation as above with $R_{MP} = 10 R_E$, where $R_E = 6{,}371$ km is the Earth radius, and a magnetic field jump of 62 nT across the terrestrial magnetopause, a magnetic field of about 17 nT at the Earth surface results. This value agrees well with earlier, more precise calculations by, e.g., Mead (1964), who found values of the order of 25 nT. This indicates that our simple ring current model for the Chapman–Ferraro current system is a sufficiently accurate approximation. We conclude that at Earth the ratio of the externally generated field to the dynamo generated field is of the order of 1/2000, while at Mercury this ratio is of the order of 1/6 and larger.

Three implications of such a large ratio of the external/internal generated fields are apparent. First, there are not only the Chapman–Ferraro currents flowing, but other magnetospheric currents may become very important as well. Second, the position of the magnetopause is not stable, but varying in time which causes significant temporal magnetic field variations at the planetary surface. Third, the mean external, magnetospheric field constitutes an ambient magnetic field in which any Hermean dynamo has to operate.

The first has implications for the analysis of magnetic field observations in the Hermean environment. In the terrestrial environment the magnetic field, $\mathbf{B}$, is described as the gradient of a scalar magnetic potential Ψ,

$$\mathbf{B} = -\nabla \Psi \qquad (3)$$

with Ψ usually expanded into spherical harmonics (e.g. Chapman and Bartels 1940, p. 639ff). Such a Gauss representation is merely correct if the description is applied to a current-free region of the magnetosphere. This condition is approximately fulfilled below the terrestrial ionosphere in the atmosphere or above the ionosphere in the magnetosphere, where the magnetic field of any current density can be neglected with respect to the dynamo-generated geomagnetic field, or where suitable models of the magnetospheric current systems exist to remove the corresponding magnetic field contribution. In the Hermean magnetosphere the situation is probably different. The Gauss representation is no longer a suitable tool to describe the magnetic induction field as measured onboard a satellite in space, but should be replaced by, for example, a Mie representation. For further discussion reference is made to Backus et al. (1996).

The second implication has far-reaching geophysical consequences. Siscoe and Christopher (1975) pointed out that, due to changing solar wind conditions, the magnetopause-to-surface distance may vary between 0.1 and $1.6R_P$, that is between 240 km and 3,900 km. Occasionally the magnetopause can even reach the Hermean surface. However, Hood and Schubert (1979), as well as Suess and Goldstein (1979), pointed out that magnetopause compression would be strongly opposed by induced currents in the planet's interior. But, as Slavin and Holzer (1979) showed, dayside reconnection would significantly increase the chances of the solar wind to impact the Hermean surface. Thus, the Hermean magnetosphere and planetary interior may be subject to rather severe solar wind induced variations. The effects of these will be studied in more detail in Sects. 2 and 3.

The third implication, the external field constituting an ambient field for the planetary dynamo, allows us to draw parallels between any dynamo action in Mercury's interior and dynamo action in Jupiter's moons Io and Ganymede, where the Jovian planetary magnetic field provides for an ambient magnetic field (Schubert et al. 1996; Sarson et al. 1997). The external field either causes induction effects in the rotating Hermean core as discussed in Sect. 4, or the Hermean dynamo is another example of a magneto-convection dynamo, details of which are discussed in Sect. 5.

Because little is known about the nature of the Hermean magnetic field, no firm conclusions can be drawn with respect to the above-mentioned processes. However, we feel that it is worthwhile to pursue these processes in future discussions.

2 Electromagnetic Induction in the Hermean Core due to Solar Wind Induced Field Variations

Due to changing solar wind conditions, the magnetopause is constantly in motion and changes its distance to the planetary surface. This causes temporal variations of the Chapman–Ferraro current generated magnetic fields at the surface and in the planetary interior. To elucidate these solar wind driven variations of the Hermean magnetospheric magnetic field, Fig. 3 displays actual observations of solar wind conditions scaled to the orbit of Mercury for February 20, 2006, 0–24 UT together with the induced variations of the magnetopause distance and the equatorial surface strength of the Chapman–Ferraro field. Significant variations are apparent with the externally generated field varying between 50 nT and 150 nT. Such large variations on time scales of minutes give rise to major induction effects in the electrically conducting planetary interior.

Induced currents in the Earth crust and mantle due to rapidly changing externally generated magnetic fields were and are subject of numerous scientific studies (e.g. Chapman and Bartels 1940; Rikitake 1966; Jacobs 1987). Inducing magnetic field variations in the terrestrial case are ULF pulsations, geomagnetic variations such as magnetic substorms, or Sq variations. As a rule of thumb one may conclude that induced magnetic fields are of

Electromagnetic Induction Effects and Dynamo Action

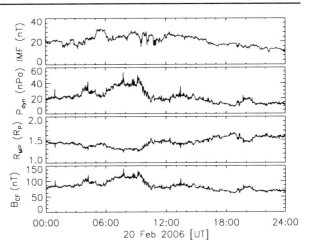

Fig. 3 Wind observations of the interplanetary magnetic field and the solar wind dynamic pressure scaled to the orbit of Mercury for February 20, 2006 (*upper two panels*) together with the Hermean magnetopause distance and the associated Chapman–Ferraro current generated magnetic field strength at the equatorial surface of Mercury (*lower two panels*). A magnetopause current of $I_{CF} = 2.5 \cdot 10^5$ A is assumed

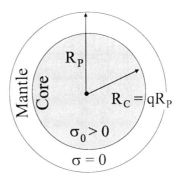

Fig. 4 Model of the electrical conductivity within planet Mercury. The parameter q gives the ratio of the core and planetary radii

minor importance for the structure of the terrestrial magnetosphere, but serve as an important tool to infer the electrical conductivity of the Earth crust and mantle.

Whether induction is of importance for the structure and dynamics of the Hermean magnetosphere can be discussed as follows (e.g. Glassmeier 2000; Grosser et al. 2004). For slightly different treatments, reference is made to Hood and Schubert (1979) as well as Suess and Goldstein (1979). Here we use the classical treatment of electromagnetic induction in a spherical body as developed by Lahiri and Price (1939) (see also Rikitake 1966), that is, we assume Mercury to be a spherical symmetric body with an electric conductivity distribution as shown in Fig. 4. Any dynamo-generated field is neglected for the time being. Only the induction effect of a time-varying external field is considered here. Dynamo-generated fields can be added later.

To describe the induced field $\mathbf{B}_{ind}$ in the planetary interior a vector potential $\mathbf{A}$ with $\mathbf{B}_{ind} = \nabla \times \mathbf{A}$ is introduced. Using the Coulomb gauge the induction equation reads

$$\frac{\partial \mathbf{A}}{\partial t} = \frac{1}{\mu_0 \sigma} \Delta \mathbf{A}, \qquad (4)$$

where $\sigma(\mathbf{r})$ denotes the electrical conductivity structure of the planetary body. For a spherically symmetric distribution $\sigma(\mathbf{r}) = \sigma(r)$ only the toroidal component of the vector potential becomes important, as its poloidal part gives rise to an induced toroidal magnetic field entirely confined to the planet. With $\mathbf{A} = \mathbf{r} \times \nabla u$, the induction equation then reads

$$\frac{\partial u}{\partial t} = \frac{1}{\mu_0 \sigma(r)} \Delta u. \tag{5}$$

A convenient Ansatz for its solution is (e.g. Rikitake 1966)

$$u(\mathbf{r}, t) = e^{i\omega t} \cdot \sum_{n=1}^{\infty} R_n(r, \omega) \cdot Y_n(\varphi, \theta), \tag{6}$$

where R_n describes the radial variation of $u(\mathbf{r}, t)$, Y_n are spherical harmonic functions, and ϕ and θ denote the azimuth and polar angle; ω is the frequency of the external field variation.

The magnetic induction in the planetary exterior may be described by a magnetic potential W, provided that this region is current free:

$$\Delta W = 0. \tag{7}$$

As noted earlier the magnetic field in the magnetosphere of Mercury is significantly influenced by local electric currents. This and the absence of an electrically isolating planetary atmosphere invalidates usage of (7). However, here we treat the Hermean mantle region as a non-conducting regime with $\sigma_0 = 0$ (see Fig. 4). An appropriate Ansatz for the exterior potential is $W = \sum_n W_n(\varphi, \theta)$ with

$$W_n(\varphi, \theta) = \left(g_{e,n} r^n + g_{i,n} r^{-n-1}\right) Y_n(\varphi, \theta), \tag{8}$$

where the first term in brackets denotes the external field due to, e.g., the magnetopause currents. The second term describes the magnetic field contribution of the electric currents induced in the Hermean core.

At the core-mantle boundary appropriate matching conditions are the continuity of the tangential magnetic field and of the normal component of the magnetic induction. These conditions lead to a relationship between the expansion coefficients of the internal induced field and the external induced field, $g_{i,n}$ and $g_{e,n}$, respectively, provided the radial function $R(r, \omega)$ is known. For the special case $\sigma_0 = \text{const}$, $R(r, \omega)$ may be represented using modified spherical Bessel functions, and we have at the planetary surface (e.g. Rikitake 1966):

$$\frac{g_{i,n}}{g_{e,n}} = \frac{n}{n+1} q^{2n+1} \cdot \left(1 - \left(\frac{k R_C}{2n+1} \frac{I_{n-1/2}(k R_C)}{I_{n+1/2}(k R_C)}\right)^{-1}\right). \tag{9}$$

Here $k^2 = i \mu_0 \sigma_0 \omega$ is the induction parameter, which is also a measure of the skin depth. Thus, if $g_{e,n}$ is known, the induced field can be determined for any σ_0 and ω. The parameter q denotes the ratio of the core to planetary radius.

To quantify the induction ratio (9) values for q and σ_0 are required. The precise radius of the Hermean core is unknown; values discussed range from 1,600 km ($q = 0.66$) (Ness et al. 1975; Gubbins 1977) to 2,100 km ($q = 0.86$). Here we adopt the value $R_C = 1,860 \pm 80$ km suggested by Spohn et al. (2001) and choose $q = 0.75$. Hood and Schubert (1979) assumed that the electrical conductivity of Mercury's core is comparable to that of the Earth's core, $\sigma_0 = 10^5$ S/m, while Gubbins (1977) argued in favor of a higher conductivity, $\sigma_0 = 10^6$ S/m. Here, we adopt this larger value.

For periods of the external variations of the order of 1 s and above, that is $\omega < 6$ Hz, the modulus of the argument of the modified spherical Bessel functions in (9) becomes large, $|kR_C| \gg 1$. In this case the induction ratio can be approximated as (cf. Grosser et al. 2004)

$$\frac{g_{i,n}}{g_{e,n}} = \frac{n}{n+1} q^{2n+1}, \tag{10}$$

and the ratio is mainly controlled by the ratio of the core and planetary radius.

To simplify matters the magnetic induction field $\mathbf{B}_{CF}$ of the magnetopause ring current is assumed here, to first order, as a uniform field, which in spherical coordinates reads (e.g. Smythe 1968, p. 295):

$$\mathbf{B} = \left(-\frac{\mu_0 I_{CF}}{2R_P} \cos\theta, 0, \frac{\mu_0 I_{CF}}{2R_P} \sin\theta\right). \tag{11}$$

Thus, at the planetary surface we have

$$g_{e,1} = \frac{\mu_0 I_{CF}}{2R_P}, \tag{12}$$

and the induction ratio approximates as

$$\frac{g_i}{g_e} = \frac{1}{2} \cdot q^3, \tag{13}$$

or, with our choice $q = 0.75$, $g_i/g_e \approx 0.21$. This implies that at the planetary surface induced magnetic fields of the order of 12 nT or about 4% of the global planetary field need to be taken into account. It should be noted that this is a conservative estimate.

The approximation used implies that the ratio (13) is a real number, which means that any phase lag between the inducing and the induced field can be neglected in the parameter regime treated here. This is the result of the large conductivity and the large radius of the Hermean core.

A comment is appropriate concerning the assumption of a non-conducting mantle region. Electrical conductivities of common Earth materials such as olivine and magnetite are about 10^{-4} Sm^{-1} to 10^3 Sm^{-1} at 300 K, respectively (Parkinson and Hutton 1989). For the lunar mantle Hobbs et al. (1984) estimated values of about 10^{-2} Sm^{-1}, which we adopt as a suitable value also for the Hermean mantle. The characteristic time for the external field to diffuse into the core, T_D, can be estimated by $T_D \approx \mu_0 \sigma_M L^2$ (e.g. Suess and Goldstein 1979), where σ_M is the mantle conductivity and $L = 0.25 R_P$ the thickness of the crust, respectively. With the given values $T_D = 4{,}670$ s $= 1.3$ hours. This restricts the applicability of the approximation used to periods above about 1.5 hours. For higher frequencies a more elaborate approach, a multi-layer model of the Hermean interior, is required. Figure 3 indicates a variety of temporal scales on which the solar wind conditions change. The larger amplitude ones are at periods above 1.3 hours.

The above estimate is also limited in its application as the Chapman–Ferraro currents are assumed as a simple ring current while the real current system is a more complex one. To relax this ring current assumption, a more sophisticated magnetospheric model is required. Grosser et al. (2004) presented such a study. The magnetosphere model used was originally developed by Voigt (1981). The magnetospheric tail is represented by a semi-infinite cylinder with radius $R_T = 2.35 R_{MP}$; the dayside magnetosphere is modeled by a half-sphere with the same radius. Mercury is located within the half-sphere at a distance $1.35 R_{MP}$

from the subsolar point. The planetary field is represented by a centered dipole field with the appropriate Gauss coefficients derived from Mariner 10 observations (Ness 1979; Grosser et al. 2004). The dipole assumed here is slightly inclined with respect to the rotation axis. The Gauss coefficients used are first estimates and future space missions are necessary for a more precise determination.

External magnetic field contributions are caused by Chapman–Ferraro currents flowing in the plane of the assumed magnetopause. They are constructed in such a way that the normal component of the total magnetic field vanishes at the magnetopause. No other currents flow in this closed magnetosphere. The model is of sufficient complexity to describe the major characteristics of the Hermean magnetic field. More sophisticated magnetospheric models such as described by Korth et al. (2004) are available but of no concern here.

Figure 5 displays the total magnetic field magnitude at Mercury's surface for three different magnetopause distances: $R_{MP} = 1.6 R_P$ (top), $1.35 R_P$ (middle), and $1.1 R_P$ (bottom). The field is dominated by the tilted planetary dipole moment. However, the closer the magnetopause is located to the planetary surface the more is the surface field influenced by higher multi-pole contributions from the Chapman–Ferraro currents flowing in the model magnetopause. These higher multi-pole contributions clearly indicate that the ring current approximation used above is only a first approach to describing the external field.

As the magnetospheric model used is a static model of the magnetosphere temporal variations can be enforced by harmonic variations of both, the magnetopause location as well as the tail radius. Following Grosser et al. (2004) the induced magnetic field can be determined using (13) in a stepwise fashion. For each instant in time the spherical harmonic expansion of the external magnetic field at the Hermean surface needs to be determined. With the resulting Gauss coefficients of the induced field, the coefficients of the external contribution, and the dynamo part the total field at any point of the magnetosphere can be calculated.

The orbit of the Mercury Planetary Orbiter (MPO) element of the BepiColombo mission to Mercury (Grard 2006; Schulz and Benkhoff 2006) will be a polar orbit with a periherm of about 400 km and an apoherm height of about 1500 km. For several orbits Fig. 6 displays the various magnetic field contributions expected to be measured by the magnetometer onboard the BepiColombo MPO spacecraft. The amplitude of the magnetopause variations assumed is $\pm 0.25 R_P$, and the period chosen is four hours, which is about the orbital period of the MPO s/c. It is apparent that the field is not dominated by the internal field, but is characterized by the external and induced parts as well. Any future magnetic field observations at Mercury, such as those planned onboard NASA's MESSENGER s/c (Solomon et al. 2006) or onboard BepiColombo need to take this into account when heading for a detailed determination of the internal field structure.

3 Induction Effects and the Magnetopause Position

Variations of the solar wind dynamic pressure close to Mercury are rather severe. As exhibited in Fig. 3 this solar wind buffeting has an important influence on the magnetopause position and raises the following questions: To what extent is the Hermean magnetosphere compressible and do induction effects have any influence on this compressibility as first discussed by Hood and Schubert (1979) and Suess and Goldstein (1979)? Those authors concluded that, due to induction effects, solar wind impingement on the Hermean surface is inhibited.

Glassmeier et al. (2004) introduced the concept of a magnetospheric compressibility parameter κ. Assuming a typical length scale of a magnetospheric system the subsolar standoff

Fig. 5 Total magnetic field at Mercury's surface using the Voigt (1981) magnetic field model for three different magnetopause distances: $R_{MP} = 1.6 R_P$ (*top*), $1.35 R_P$ (*middle*), and $1.1 R_P$ (*bottom*). In each panel an internal dipole Gauss coefficient based on Mariner 10 observations is used. The center panel is the subsolar point at the Hermean surface. Field values are given in nT. *Cross* and *times* symbols denote the Hermean magnetic south and north pole, respectively

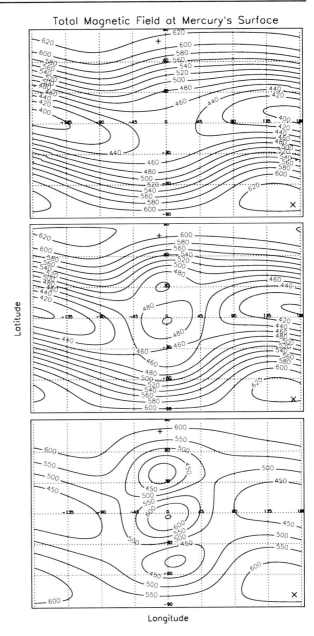

distance of the magnetopause R_{MP}, given by the pressure equilibrium between the dynamic solar wind pressure $p_{\text{ram}} = \rho_{SW} v_{SW}^2$ and the magnetic pressure of the planetary magnetic field via

$$R_{MP} = \left(\frac{B_0^2}{\mu_0 p_{\text{ram}}} \right)^{1/6}. \tag{14}$$

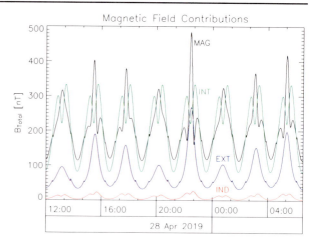

Fig. 6 Total magnetic field along modeled polar orbits around Mercury, calculated using the Voigt (1981) magnetic field model with a time-varying magnetopause distance and taking into account induction effects to first order. The *green line* exhibits the internal, dynamo-generated field magnitude; the *blue line* shows the external, magnetospheric contribution; the *red line* displays the induced field; and the *black line* shows the total magnitude as observed along the s/c polar orbit

where B_0 is the magnetic field strength at the planetary surface, the bulk modulus K of the magnetosphere is defined as the ratio of the solar wind dynamic pressure change required to obtain a specific relative change in the magnetopause position (Glassmeier et al. 2004):

$$K = -R_{MP} \cdot \frac{dp_{\text{ram}}}{dR_{MP}}. \tag{15}$$

With the above definition of the stand-off distance we have $K = 6 \cdot p_{\text{ram}}$, and the magnetospheric compressibility κ is given as

$$\kappa = \frac{1}{K} = \frac{1}{6 \, p_{\text{ram}}}. \tag{16}$$

We conclude that the compressibility of a planetary magnetosphere depends only on the solar wind dynamic pressure, not on the planetary magnetic field. Thus, any induced planetary magnetic field does not have an influence on the compressibility. It should be noted that the solar wind mass density along Mercury's orbit around the Sun is much larger than that along Earth's or Jupiter's orbit by about a factor of 10 and 200, respectively. Therefore the compressibility of the Hermean magnetosphere is much less than that of Earth or Jupiter. Based on the proposed definition of the compressibility, the Hermean magnetosphere is much stiffer than that of any other magnetosphere in our solar system.

However, the feedback of any induced planetary magnetic field on the magnetopause position needs a different approach. Here we tackle the problem in a slightly different manner than Hood and Schubert (1979) or Suess and Goldstein (1979). In case of severe induction, the surface field strength to be used in (14) is given by $B_0 = B_D + B_I$, where B_D is the field generated by any planetary dynamo action and B_I is the induced field strength. The corresponding field at the magnetopause location than reads

$$B_{MP} = B_0 \cdot \frac{R_P^3}{R_{MP}^3} = (B_D + B_I) \cdot \frac{R_P^3}{R_{MP}^3}. \tag{17}$$

The Chapman–Ferraro sheet current density j_{CF} is determined by the magnetic field jump across the magnetopause or, assuming a negligible solar wind magnetic field,

$$j_{CF} = \frac{B_{MP}}{\mu_0}. \tag{18}$$

If we approximate the Chapman–Ferraro current system as a ring current we need its electric current strength I_{CF}, which we approximate by

$$I_{CF} = \frac{B_{MP}}{\mu_0} \cdot \pi R_{MP}. \tag{19}$$

Here it is assumed that the Chapman–Ferraro currents flow over all of the magnetopause approximated as a sphere at distance R_{MP}. As noted earlier such an approximation yields a Chapman–Ferraro field at the Earth surface which is comparable to that of a more elaborate treatment. With (2) the Chapman–Ferraro current generated magnetic field at the surface of the Hermean core is approximated by

$$B_{CF} = \frac{\mu_0 I_{CF}}{2\pi} \cdot C(R_{MP}), \tag{20}$$

where

$$C(R_{MP}) = \frac{1}{R_{MP}+R_C}\left[K\left(\frac{R_C}{R_{MP}}\right) + \frac{R_{MP}+R_C}{(R_{MP}-R_C)} E\left(\frac{R_C}{R_{MP}}\right)\right]. \tag{21}$$

Assuming an induction ratio $g_i/g_e \approx 0.2$ (cf. (13)) the induced field strength can be approximated as

$$B_I = \frac{0.1 \cdot B_0 \cdot R_P^3}{R_{MP}^2} \cdot C(R_{MP}) = \frac{0.1 \cdot (B_D + B_I) \cdot R_P^3}{R_{MP}^2} \cdot C(R_{MP}), \tag{22}$$

or

$$B_I = \frac{01. \cdot R_P^3 \cdot C(R_{MP})}{R_{MP}^2 - 0.1 \cdot R_P^3 \cdot C(R_{MP})} \cdot B_D. \tag{23}$$

With this and $B_0 = B_D + B_I$ the magnetopause distance reads

$$R_{MP}^3 = R_{NoInd}^3 \cdot \frac{R_{MP}^2}{R_{MP}^2 - 0.1 \cdot R_P^3 \cdot C(R_{MP})}. \tag{24}$$

For any given non-induction influenced magnetopause distance $R_{NoInd} = \sqrt[6]{B_D^2/\mu_0 p_{ram}}$ the root of this cubic equation gives one the corresponding magnetopause distance R_{MP} taking into account induction effects. For a value $R_{NoInd} = 1.7 R_P$, that is for the average Hermean magnetopause distance, the ratio R_{MP}/R_{NoInd} is about 1.03. For other distances Fig. 7 displays the ratio in more detail. Based on our approximation of the induction problem we confirm that the induced magnetic field has a significant influence on the magnetopause distance. Figure 7 indicates that induction effects are probably sufficient to prevent the solar wind impacting the surface of Mercury: for a nominal magnetopause distance of $R_{NoInd} = R_P$ the actual, induction modified value is $R_{MP} = 1.15 R_P$.

Slavin and Holzer (1979) pointed out that reconnection driven erosion will have a profound influence on the magnetopause stand-off distance. However, induction effects may have possible influence on the erosion effect itself. Russell and Walker (1985) reported about flux transfer events at Mercury. The typical scale size of these events is of the order of 400 km, the typical time scale 1 s. Equation (10) can be used to estimate the induction effect on such an external, localized reconnection driven field variation. A scale of 400 km corresponds to a harmonic number 20 in a spherical harmonic analysis. From (10) we find $g_i/g_e \approx 10^{-5}$, which indicates that induction does not significantly alter physical processes

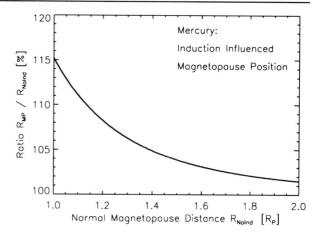

Fig. 7 The ratio of the magnetopause distance taking into account induction effects normalized to the non-induction influenced distance

in any Hermean flux transfer event taking into account induction in the core. Induction effects due to the low-conductivity mantle are difficult to estimate as no firm knowledge about the mantle electric conductivity is available.

4 Electromagnetic Induction in the Slowly Rotating Hermean Core

The discussions in the preceding chapters are based on assuming that the external magnetic field is changing in time with respect to a non-changing electrically conducting medium, the Hermean core in the case discussed here. A somewhat different situation emerges if the external field is static, but the conducting medium is rotating. Distortions in the magnetic field caused by differential rotation need to be considered. Any fluid motion in the interior of Mercury, such as thermo-gravitationally driven convection, causes complex fluid motions, part of which can be viewed at as differential rotation.

Early thermal evolution models (Stevenson et al. 1983) indicate that the core can remain at least partially molten if it contains a small concentration of a light element such as sulphur. More recent models that also take into account tidal heating, pressure and temperature-dependent rheology as well as the magmatic and tectonic evolution of Mercury indicate that outer liquid layers are easily possible for a wide range of parameters (Schubert et al. 1988; Conzelmann and Spohn 1999; Hauck et al. 2004). Thus, in an outer shell of the Hermean core we can expect differential rotation.

Two basically different situations can exist, a perpendicular and a parallel situation (e.g. Moffat 1978). In the perpendicular situation the magnetic field **B** lies in the plane perpendicular to the rotation axis Ω of the planet. Due to rotation the field is wound up into a spiral structure in the plane perpendicular to Ω. Finally the field is excluded from the differentially rotating regime. This form of field expulsion is related to the skin effect in conventional electromagnetism (Moffat 1978). Induction effects are involved as well in the rotating frame of reference though the inducing field is assumed static in the external, non-rotating frame of reference.

The second, parallel situation, where the field exposed to differential rotation lies in the plane containing the rotation axis Ω, is one well known from dynamo theory as the ω-effect (e.g. Gubbins and Roberts 1987). No expulsion of the field occurs, but an initially poloidal field winds up in a toroidal field configuration (Fig. 8). For a purely toroidal velocity field

Electromagnetic Induction Effects and Dynamo Action
341

Fig. 8 Sketch illustrating the distortion of lines of force of an initially uniform magnetic field if the field lines are parallel to the rotation axis of a differentially rotating body (after Moffat 1978)

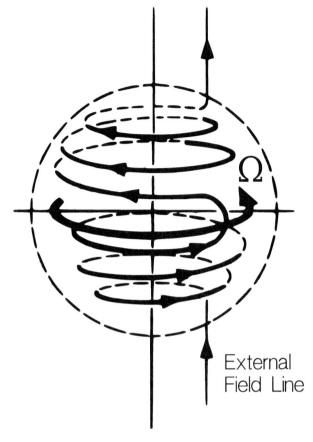

$\mathbf{u}_T = \mathbf{\Omega}(r, z) \times \mathbf{r}$ and a steady poloidal magnetic induction field $\mathbf{B}_P(r, z)$ the rotationally induced toroidal field evolves according to the induction equation (e.g. Moffat 1978)

$$\frac{\partial \mathbf{B}_T}{\partial t} = \nabla \times (\mathbf{u}_T \times \mathbf{B}_P) + \eta \nabla^2 \mathbf{B}_T, \qquad (25)$$

where r and z denote radial and axis-aligned coordinates of a cylindrical system, respectively; η is the magnetic diffusivity. The special situation here is that the initial poloidal field is not produced by some dynamo action, but by external fields, generated within the magnetosphere. The resulting toroidal field may well be viewed at as an induced field due to differential rotation in the electrically conducting Hermean core.

Magnetic diffusion finally limits the growth of $\mathbf{B}_T$ at a maximum value (Moffat 1978)

$$\max |B_T| = O(R_m) \cdot |B_P| \qquad (26)$$

with the magnetic Reynolds number $R_m = \mu_0 \sigma u L$. Here u and L denote typical values for the fluid speed and the system length scale. It should be noted that (26) only gives a first estimate.

For a dynamo to operate the Reynolds number needs to be large. Typical values will be of the order 100–1,000 for planetary dynamos. Using (2) the external, Chapman–Ferraro current generated poloidal field at the Hermean core-mantle boundary estimates at 47 nT. Thus toroidal magnetic fields due to differential rotation driven induction may be as large as 47,000 nT. This again demonstrates the possible importance of the external, magnetospheric magnetic field in the Hermean interior. Possible magneto-convective consequences need to be taken into account when discussing dynamo action in the Hermean core.

5 Magneto-Convection in the Hermean Core Induced by External Magnetospheric Currents

The Elsasser number $\Lambda = \sigma B^2/\rho \Omega$ (ρ is the mass density) describes the ratio between the Lorentz and Coriolis forces acting in a rotating planet and provides for a useful parameter to estimate whether the magnetic field directly impacts the dynamics of the planetary interior. If $\Lambda \approx 1$ a strong magneto-convection situation exists; the presence of the magnetic field alters the convection pattern or is even able to temporarily suppress convection (e.g. Weiss 1985; Wicht et al. 2007, this volume). For $\Lambda \ll 1$ planetary-scale convection is controlled by strong rotational effects. With $\sigma = 10^6$ S/m, $\rho = 5.42 \cdot 10^6$ kg/m^3, and $\Omega = 1.24 \cdot 10^{-6}$ s^{-1} (which corresponds to a rotation period of 58.64 days) the Elsasser number is of the order of $\Lambda_{CF} = 3.2 \cdot 10^{-7}$ based on the Chapman–Ferraro currents generated field, or $\Lambda_{tor} = 3.2 \cdot 10^{-4}$ using the field value possibly generated by differential rotation. These values do not suggest that a situation of strong magneto-convection exists in the Hermean core.

However, if the magnetic field considered is an imposed one, that is if the convective system is embedded in an ambient magnetic field other interesting consequences emerge: The imposed field destroys the invariance of the classical dynamo problem under the operation $\mathbf{B} \to -\mathbf{B}$ as is easily seen from the corresponding induction equation

$$\frac{\partial \mathbf{B}_{dyn}}{\partial t} = \nabla \times (\mathbf{u} \times \mathbf{B}_{dyn}) + \nabla \times (\mathbf{u} \times \mathbf{B}_{ext}) + \eta \nabla^2 \mathbf{B}_{dyn}. \tag{27}$$

Here indices dyn and ext denote the dynamo generated and imposed external field, respectively. Levy (1979) was amongst the first to point this symmetry breaking which is the result of the second term of the r.h.s. in (27).

This new term describes the interaction between the convection and the ambient magnetic field. It comprises a generator term opposing the action of the diffusion term. As pointed out by Levy (1979) this generator term, absent in classical formulations of the dynamo problem, also causes the absence of any threshold behavior of the dynamo. Dynamo action occurs without any vigorous convection of the fluid. The reason for this is particularly transparent when considering a poloidal external field in the plane of the rotation axis. Differential rotation as discussed above leads to the generation of a significant toroidal field.

Levy (1979) already pointed out that the Galilean satellites are embedded in the strong Jovian magnetic field. Magnetometer observations from the Galileo orbiter have shown that Io and Ganymede have significant magnetic fields of internal origin (e.g. Kivelson et al. 1996). Sarson et al., (1997, 1999) elaborated Levy's (1979) embedded dynamo action hypothesis performing numerical experiments of the magnetohydrodynamic processes in a rotating planetary fluid core under the presence of an ambient magnetic field. They convincingly demonstrated that at least Io operates what they call a magneto-convection dynamo. Conditions are different at Ganymede and Sarson et al., (1997, 1999) conclude that Ganymede is most probably operating a classical planetary dynamo.

It is tempting to compare the situations at Io and Ganymede with that one at Mercury. Table 1 gives the observed surface magnetic field strength, the ambient magnetic field magnitude, the corresponding external Elsasser numbers, and the putative process generating

Table 1 Comparing magnetic conditions of Io, Ganymede, and Mercury

	Surface field [nT]	Ambient field [nT]	Elsasser number due to B_{ext} [10^{-7}]	Dynamo process
Io	1300	1800	150	Magneto-convection dynamo
Mercury	340	47	≥ 3	Feedback dynamo
Ganymede	750	100	2	Classical dynamo

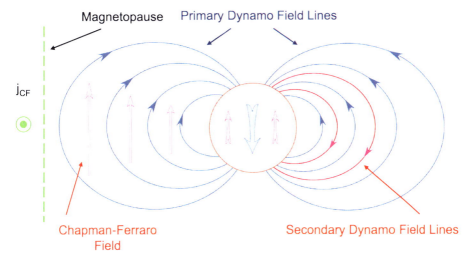

Fig. 9 Sketch illustrating the feedback situation of a possible Hermean dynamo. To the left the magnetopause is indicated with its Chapman–Ferraro currents. *Red arrows* display the additional magnetic field of these currents in the magnetosphere. *Blue field lines* represent the magnetic field of the primary dynamo action. Within the planetary interior, the *shaded circular region*, the *blue arrow* gives the direction of the primary magnetic moment while the *two red arrows* give the direction of the secondary moment, possibly generated by the magneto-convective action based on the Chapman–Ferraro field. On the right side *red field lines* opposing the *blue primary field lines* indicate the negative feedback situation

the planetary magnetic field. Obviously the external Elsasser number is large enough at Io to support a magneto-convection dynamo while at Ganymede the external field is too small to allow such a dynamo to operate. The situation at Mercury is intermediate. Though the external field, generated by Chapman–Ferraro currents in the Hermean magnetopause, is smaller than the Jovian field at Io's and Ganymede's orbit the Elsasser number is comparable at least to the one of Ganymede. The reason for this is the much smaller rotation rate of Mercury. This causes the Coriolis forces to be less significant. Mercury is therefore an interesting candidate for yet another type of dynamo action, neither classical nor magneto-convective.

This new type of dynamo action is of a feedback type. First, there is dynamo action in the planetary interior. The field generated causes the development of a small magnetosphere with a significant external Chapman–Ferraro field. This external field now can act back to the dynamo process. The question is whether this feedback loop is positive or negative. Levy (1979) as well as Sarson et al., (1997, 1999) point out that the orientation of the magneto-convective dynamo generated magnetic dipole moment is always such that it supports the ambient magnetic field in which the dynamo is embedded. Outside the dynamo region the dynamo field is, however, opposing the ambient field. Possible consequences of this are shown in Fig. 9: the magnetic field of the secondary dynamo action, based upon the work done by the planetary convection system against the external field, is expected to diminish the primary dynamo generated magnetic field. This causes an interesting possible feedback situation. As the feedback is not positive the process outlined here cannot explain the primary field. But one may speculate that the negative feedback has severe consequences for the primary dynamo action. We call this type of coupled magnetosphere-dynamo interaction a feedback dynamo.

6 Mercury: A Magnetically Strongly Coupled Planetary System

The magnetosphere of Mercury is unique in our solar system. Though the planetary magnetic field is rather small compared to those of Earth, Jupiter, or Saturn, it withstands the solar wind and gives rise to a small magnetosphere. Electric currents flowing in the Hermean magnetopause are comparable to Chapman–Ferraro currents flowing in the terrestrial magnetopause, but cause relatively larger magnetic fields at the planet's surface and its interior.

Temporal variations of these Chapman–Ferraro fields cause significant induced fields which, to a certain extent, support the primary planetary field in causing a magnetospheric system. Future magnetic field observations at Mercury and their geophysical interpretation need to take into account these induced fields as well as the large ratio of externally generated to internally generated magnetic fields. Separating all three contributions is a much more difficult task than at Earth.

Assuming that the magnetosphere has existed a long time implies that planet Mercury is embedded in a weak, but not negligible, external magnetospheric field generated by the interaction between the solar wind and the—probably dynamo action generated—planetary field. As viewed from the planet this external field is a poloidal field in which the planet rotates. If the planetary interior undergoes some differential rotation, significant generation of toroidal fields is possible.

If the internal convection system is more vigorous, the interaction of the external, ambient field with the fluid motion in the Hermean core needs to be considered. This results in an extra term in the classical induction equation, which not only serves to break symmetry with respect to the sign of the generated magnetic field, but also lowers the need for a very vigorous convection. However, in the Hermean situation, a more detailed consideration indicates that the magneto-convective dynamo action possibly causes a field opposing the primary dynamo field. As this primary field causes the Chapman–Ferraro field, a strongly coupled situation between the magnetosphere and the planetary dynamo is envisioned. One may speculate that the outlined feedback dynamo action is also a possible explanation for the rather weak planetary field. Whether such a coupled situation indeed can exist in the Hermean system needs further detailed theoretical and numerical studies as well as observations to be provided by the upcoming MESSENGER (Solomon et al. 2006) and BepiColombo (Schulz and Benkhoff 2006) missions to planet Mercury.

Acknowledgements K.-H. Glassmeier is grateful to Johannes Wicht for very illuminating discussions. The WIND plasma data and the WIND magnetic field data used in Fig. 3 were kindly provided by the MIT Solar Wind Group and Dr. R. Lepping (NASA), respectively. Financial support of the work of the BepiColombo MPO Mag Principal Investigator Team at the Technical University of Braunschweig by the German Ministerium für Wirtschaft und Technologie and the Deutsches Zentrum für Luft- und Raumfahrt under grant 50 QW 0602 is acknowledged.

References

G. Backus, R. Parker, C. Constable, *Foundations of Geomagnetism* (Cambridge University Press, Cambridge, 1996)
S. Chapman, J. Bartels, *Geomagnetism* (Oxford University Press, London, 1940)
V. Conzelmann, T. Spohn, Bull. Am. Astron. Soc. **31**, 1102 (1999)
K.H. Glassmeier, Planet. Space Sci. **45**, 119–125 (1997)
K.H. Glassmeier, in *Geophysical Monograph*, vol. 118 (American Geophysical Union, Washington, 2000), pp. 371–380.
K.H. Glassmeier, D. Klimushkin, C. Othmer, P. Mager, Adv. Space Res. **33**, 1875–1883 (2004)
R. Grard, Adv. Space Res. **38**, 563 (2006)
J. Grosser, K.H. Glassmeier, A. Stadelmann, Planet. Space Sci. **52**, 1251–1260 (2004)
D. Gubbins, Icarus **30**, 186–191 (1977)
D. Gubbins, P.H. Roberts, in *Geomagnetism*, ed. by J.A. Jacobs (Academic Press, London, 1987), pp. 1–184.
S. Hauck, A. Dombard, R. Philips, S. Solomon, Earth Planet. Sci. Lett. **222**, 713–728 (2004)
B.A. Hobbs, L.L. Hood, F. Herbert, C.P. Sonett, Geophys. J. R. Astron. Soc. **79**, 691–696 (1984)
L.L. Hood, G. Schubert, J. Geophys. Res. **84**, 2641 (1979)
J.A. Jacobs (ed.), *Geomagnetism* (Academic Press, London, 1987)
M.G. Kivelson, C.T. Russell (eds.), *Introduction to Space Plasma Physics* (Cambridge University Press, New York, 1995)
M.G. Kivelson, K.K. Khurana, C.T. Russell, R.J. Walker, J. Warnecke, F.V. Coroniti, C. Polanskey, D.J. Southwood, G. Schubert, Nature **384**, 537–541 (1996)
H. Korth, B.J. Anderson, M.H. Acuna, J.A. Slavin, N.A. Tsyganenko, S.C. Solomon, R.L. McNutt Jr., Planet. Space Sci. **52**, 733–746 (2004)
B.N. Lahiri, A.T. Price, Philos. Trans. R. Soc. Lond. A **237**, 509–540 (1939)
E.H. Levy, Proc. Lunar Planet. Sci. Conf. **10**, 2335–2342 (1979)
G.D. Mead, J. Geophys. Res. **69**, 1181–1197 (1964)
H.K. Moffat, *Magnetic Field Generation in Electrically Conducting Fluids* (Cambridge University Press, Cambridge, 1978)
N.F. Ness, K.W. Behannon, R.P. Lepping, Y.C. Whang, J. Geophys. Res. **80**, 2708–2716 (1975)
N.F. Ness, in *Solar System Plasma Physics*, vol. II, ed. by C.F. Kennel, L.J. Lanzerotti, E.N. Parker (North-Holland, Amsterdam, 1979), pp. 185–206.
W.D. Parkinson, V.R.S. Hutton, in *Geomagnetism*, vol. III, ed. by J.A. Jacobs (Academic Press, London, 1989).
T. Rikitake, *Electromagnetism and the Earths Interior* (Elsevier, Amsterdam, 1966)
C.T. Russell, Geophys. Res. Lett. **16**, 1253–1256 (1989)
C.T. Russell, R.J. Walker, J. Geophys. Res. **90**, 11067–11074 (1985)
G.R. Sarson, C.A. Jones, K. Zhang, G. Schubert, Science **276**, 1106–1108 (1997)
G.R. Sarson, C.A. Jones, K. Zhang, Phys. Earth Planet. Inter. **111**, 47–68 (1999)
G. Schubert, M.N. Ross, D.J. Stevenson, T. Spohn, in *Mercury*, ed. by F. Vilas, C.R. Chapman, M.S. Matthews (University of Arizona Press, Tuscon, 1988), pp. 429–460.
G. Schubert, K. Zhang, M.G. Kivelson, J.D. Anderson, Nature **384**, 544–545 (1996)
R. Schulz, J. Benkhoff, Adv. Space Res. **38**, 572–577 (2006)
G. Siscoe, L. Christopher, Geophys. Res. Lett. **2**, 158–160 (1975)
J.A. Slavin, R.E. Holzer, J. Geophys. Res. **84**, 2076 (1979)
W.R. Smythe, *Static and Dynamic Electricity* (McGraw-Hill, New York, 1968)
S.C. Solomon, R.E. Gold, J.C. Leary, Adv. Space Res. **38**, 564–571 (2006)
T. Spohn, F. Sohl, K. Wieczerkowski, V. Conzelmann, Planet. Space Sci. **49**, 1561–1570 (2001)
D.J. Stevenson, T. Spohn, G. Schubert, Icarus **54**, 466–489 (1983)
S.T. Suess, B.E. Goldstein, J. Geophys. Res. **84**, 3306–3312 (1979)
H.G. Voigt, Planet. Space Sci. **29**, 1–20 (1981)
N.O. Weiss, in *Solar System Magnetic Fields*, ed. by E.R. Priest (Reidel, Dordrecht, 1985).
J. Wicht, M. Mandea, F. Takahashi, U.R. Christensen, M. Matsushima, B. Langlais, Space Sci. Rev. (2007), this issue

Hermean Magnetosphere-Solar Wind Interaction

M. Fujimoto · W. Baumjohann · K. Kabin ·
R. Nakamura · J.A. Slavin · N. Terada · L. Zelenyi

Originally published in the journal Space Science Reviews, Volume 132, Nos 2–4.
DOI: 10.1007/s11214-007-9245-8 © Springer Science+Business Media B.V. 2007

Abstract The small intrinsic magnetic field of Mercury together with its proximity to the Sun makes the Hermean magnetosphere unique in the context of comparative magnetosphere study. The basic framework of the Hermean magnetosphere is believed to be the same as that of Earth. However, there exist various differences which cause new and exciting effects not present at Earth to appear. These new effects may force a substantial correction of our naïve predictions concerning the magnetosphere of Mercury. Here, we outline the predictions based on our experience at Earth and what effects can drastically change this picture. The basic structure of the magnetosphere is likely to be understood by scaling the Earth's case but its dynamic aspect is likely modified significantly by the smallness of the Hermean magnetosphere and the substantial presence of heavy ions coming from the planet's surface.

Keywords Mercury · Magnetosphere · Substorm · Solar wind

M. Fujimoto
Institute of Space and Astronautical Science, Sagamihara, Japan

W. Baumjohann (✉) · R. Nakamura
Space Research Institute, Austrian Academy of Sciences, Graz, Austria
e-mail: baumjohann@oeaw.ac.at

K. Kabin
University of Alberta, Edmonton, Canada

J.A. Slavin
Goddard Space Flight Center, NASA, Greenbelt, MD, USA

N. Terada
NiCT, Koganei, Japan

L. Zelenyi
Space Research Institute, Russian Academy of Sciences, Moscow, Russia

1 Shape and Dimensions of the Hermean Magnetosphere

A magnetosphere of a planet is formed as the super-sonic flow of ionized gas from the Sun, the solar wind, impinges upon the planet's intrinsic magnetic field. The two major factors that determine the shape and dimensions of a planet's magnetosphere are the ram pressure of the solar wind, which varies with distance from the Sun, and the magnitude of the planet's intrinsic magnetic field. Since Mercury is closer to the Sun and has a weaker magnetic field than Earth, one can easily infer that the size of its magnetosphere is smaller. As shown in Fig. 1, Mercury's magnetosphere is indeed quite small, with its intrinsic magnetic field standing off the solar wind at only about 4,000 km from the center of the planet. This distance should be compared with the planet's radius of $R_M \sim 2,500$ km (the solar wind is stopped at only $\sim 0.5 R_M$ above the surface), and it is only 0.1% the size of the largest magnetosphere in the solar system, the Jovian magnetosphere.

There are, however, numerous other factors, such as IMF orientation and dipole tilt angle, etc., coming into play to determine the magnetospheric shape and dimensions. The physics underlying some of these factors is well understood based upon theory and observation at Earth and can readily be estimated in the case of Mercury's magnetosphere. On the other hand, the arguments predicting Mercury's situation by some scaling are based on a magnetohydrodynamic (MHD) framework. There is a possibility that Mercury's magnetosphere is so small that non-MHD (kinetic) effects may be crucial in determining the macro-scale structure of the magnetosphere.

1.1 Solar Wind Conditions

The dimensions of planetary magnetospheres are determined by both internal conditions, such as intrinsic magnetic field properties and plasma sources, and the stresses exerted by the upstream solar wind. The pressure normal to the magnetopause is dominated by the solar wind dynamic pressure over the forward portion of the magnetosphere, where the wind makes a head-on collision with the obstacle. With increasing downstream distance it gradually switches over to domination by the solar wind thermal and magnetic pressure as the flaring of the magnetopause slowly goes to zero (Slavin et al. 1983).

Table 1 shows that the mean dynamic, or "ram," pressure of the solar wind at Mercury is expected to be between ~ 5 and 10 times the 1-AU value due to the $1/r^2$ increase in plasma density with decreasing distance from the Sun. The squares of the sonic and Alfvénic Mach numbers give the ratio of the dynamic pressure to the gas thermal and magnetic pressures, respectively. The decrease in these mean Mach numbers from 1 AU to Mercury's perihelion in Table 1 indicates that the relative strength of the dynamic pressure to the thermal plasma and magnetic pressures has decreased by factors of ~ 1.7 and ~ 5.8, respectively. Furthermore, the ratio of the magnetic to thermal pressure increases by a factor of 3~4 in going from Earth orbit to the highly eccentric orbit of Mercury. As discussed by Slavin and Holzer (1981), and confirmed by Helios observations (Russell et al. 1988), these important changes are due to the very different radial scaling of solar wind density, ion and electron temperature, and magnetic field intensity with distance from the Sun.

Helios data showed that, on occasion, the Alfvénic Mach number becomes comparable to 1, implying that there is higher probability than at Earth of observing low Alfvénic Mach number shocks, or even a slow mode or an intermediate bow shock at Mercury. The bow shock structure under these conditions may be significantly different from what is typically observed at Earth. The readers are referred to Kabin (2001) and DeSterck and Poedts (2000) for more details. The slow mode bow shock is expected to have an outwardly concave front.

Magnetospheric Hierarchy

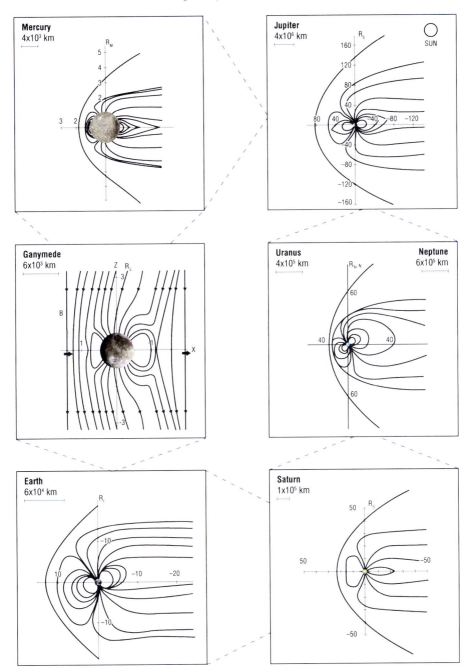

Fig. 1 Spatial dimensions of planetary magnetospheres

Table 1 Variation with distance from the Sun of the solar wind parameters which are important in determining planetary magnetosphere dimensions (cf. Slavin and Holzer 1981)

Planet	R (AU)	$P_{sw} \cdot 10^{-8}$ Dynes/cm^2	M_S	M_A	β	Spiral angle
Mercury	0.31	26.5000	5.5	3.9	0.5	17
	0.47	11.0000	6.1	5.7	0.9	25
Venus	0.72	5.0000	6.6	7.9	1.4	36
Earth	1.00	2.5000	7.2	9.4	1.7	45
Mars	1.52	1.1000	7.9	11.1	2.0	57
Jupiter	5.20	0.0920	10.2	13.0	1.6	79
Saturn	9.60	0.0270	11.6	13.3	1.3	84
Uranus	19.10	0.0069	13.3	13.3	1.0	87
Neptune	30.20	0.0027	14.6	13.3	0.8	88

In addition to this curious shape, there is ongoing debate regarding the slow shock's ability to accelerate particles. The combination of more radial interplanetary magnetic field that forms a quasi-parallel shock in the sub-solar region and low Alfvénic Mach number would make flow deflection by magnetic tension to be non-negligible and create dawn–dusk asymmetric pattern in the shocked flow. This may have some impact on subsequent solar wind–magnetosphere interaction processes. These arguments show that the Hermean environment is suitable to studying the physics of low Mach number collisionless shocks.

In addition to the normal stress (pressure) exerted on planetary magnetic fields by the solar wind, the tangential stresses also play critical roles in determining magnetospheric dimensions. In particular, the magnetic flux content of the Earth's tail region and the current systems that support it are determined primarily by the tangential stress of the solar wind. While this is still an area of active research, it is clear that the changes in the magnetic flux content of the tail lobes are determined by the magnitude and duration of the component of the interplanetary magnetic field oriented opposite to the magnetic field just inside the subsolar magnetopause (e.g., Milan et al. 2003). The underlying key process is reconnection on the front side surface of the magnetosphere (magnetopause). As a result of reconnection at the magnetopause, previously "closed" flux tubes, i.e. both ends rooted in the planet or its ionosphere, are "opened" when one end becomes connected to the interplanetary magnetic field. This new magnetic field topology has a kink at the magnetopause which creates a tailward "Maxwell stress" that transports the newly "opened" flux tubes and the plasma contained therein into the lobe region of the magnetospheric tail.

Mercury also has northward magnetic field just inside the front-side magnetopause. A southward component of the IMF that is required for reconnection at the dayside magnetopause arises due to the collision of parcels of solar wind with differing velocities, Alfvén waves, magnetic flux ropes, interplanetary shocks and the draping of interplanetary flux tubes about coronal mass ejections. Overall, this North–South, or B_z, component of the IMF has been found to scale with the intensity of the total IMF with distance from the Sun. That is, in going from the Earth to Mercury, it varies as $1/r$ (Slavin et al. 1983; Russell et al. 1988). In addition, the rate of reconnection at the dayside magnetopause is expected to scale inversely in proportion to the local Alfvén Mach number in the solar wind which may result in reconnection at the dayside magnetopause of Mercury being twice as "efficient" as at Earth (Slavin and Holzer 1979). This implies that even after the spatial scale factor has been normalized properly, Mercury's magnetosphere may react more quickly to changes in the solar wind than Earth's magnetosphere.

Since the Earth's magnetotail is not observed to dissipate when the IMF is northward, there are presumably additional sources of tangential stress, often termed "quasi-viscous" processes, which maintain the tail and replenish the slow sunward convection that persists in the central plasma sheet during these intervals. Furthermore, the Earth's magnetotail is known to change its structure significantly depending on the sign of IMF B_z. The plasma sheet, the region in the magnetotail sandwiched between the two lobes where plasma thermal pressure dominates over the magnetic pressure, is nominally filled by hot and tenuous plasma. It is considered that this hot plasma originates in the solar wind and is heated while they are transported via reconnection at dayside (opening of the field lines) and in the distant tail (re-closing of the field lines). Under extended northward IMF, however, when this mode of plasma transport becomes inefficient, a different mode of transport dominates and the plasma sheet becomes filled with cold and dense plasma (e.g., Terasawa et al. 1997). The transport mechanism that does not result in plasma heating is not yet clearly understood and it is interesting to see if the same switch in the state of the plasma sheet can be seen at Mercury.

It is well known at Earth that the dayside reconnection site moves poleward of the cusps under northward IMF (e.g., Song and Russell 1992; Lavraud et al. 2006). The open field lines are transported tailward and become closed without experiencing severe plasma depletion that takes place in the high-latitude part during southward IMF (e.g., Li et al. 2005). Since plasma depletion is the necessary element for plasma sheet heating, its absence implies little heating during the transport under northward IMF. This is proposed to be the formation mechanism of the cold and dense plasma sheet (e.g., Øieroset et al. 2005). While this is basically a large-scale MHD picture, other proposed mechanisms assume local plasma diffusion from the flanks using small scale turbulence and kinetic effects (e.g., Johnson and Cheng 1997; Nakamura et al. 2004). It is curious to see if the different setting at Mercury would result in different results or prefer a different mechanism from Earth.

1.2 Solar Wind Stand-off Distance

The best understood and most commonly studied magnetospheric dimension is the distance from the center of the planet to the sub-solar distance where the solar wind and magnetospheric pressures balance, generally termed the "solar wind stand-off distance." As expressed by Spreiter et al. (1966) this distance can be readily computed for a dipolar planetary magnetic field with a low plasma beta:

$$R_{ss} = \left[k f^2 M^2 / 2\pi P_{sw} \right]^{1/6}. \tag{1}$$

Here k is a gasdynamic "drag" coefficient, 0.88, and f is amplification of the internal subsolar magnetospheric magnetic field due to the curvature of the magnetopause current system, which has a value of ~ 1.22 at the Earth. Equation (1) is readily derived if the pressure balance in the Sun–Earth direction is considered.

Numerous applications of this relation to Mercury's magnetosphere (e.g., Siscoe and Christopher 1975; Slavin and Holzer 1979) have found that the range of magnetic moments and solar wind pressures inferred from the Mariner 10 magnetic field (see Giampieri and Balogh 2001; Korth et al. 2004) and plasma (Ogilvie et al. 1977; Slavin and Holzer 1979) measurements indicate a mean solar wind stand-off distance of $1.5 R_M$, or the distance from the surface to the subsolar magnetopause (thickness of the dayside magnetosphere) as small as $0.5 R_M$. Since this average altitude is rather low, it is intriguing to ask if the magnetopause will lie below the surface occasionally. A similar question regarding the Mars ionopause location under high solar wind dynamic pressure condition was cast. The concern was that the peak ionospheric pressure may not be large enough for the extreme times. Here we use the same scheme (Luhmann et al. 1992) for our case. From the above equation, one can see that a solar wind pressure 11 times higher than average is required to make this very strange situation reality. One may use the scaled 1-AU solar wind pressure values or use Helios 1 and 2 direct measurements at 0.3–0.5 AU to find that the subsolar magnetopause will rarely, if ever, lie at or below the surface of Mercury.

There are, however, two additional considerations that will affect the dimensions and dynamics of the dayside magnetopause. The first is the effect of dayside magnetic reconnection on the strength of the internal current systems and their effect on the dimensions of the dayside magnetosphere. At the Earth, the transfer of magnetic flux into the tail has been shown to reduce the distance to the sub-solar magnetopause by $\sim$10–20%. This conclusion is based on data obtained during a typical interval of southward IMF (Holzer and Slavin 1978; Sibeck et al. 1991). Indeed, analysis of the Mariner 10 boundary crossings by Slavin and Holzer (1979), after scaling for upstream ram pressure effects, indicated that the larger values of sub-solar magnetopause distance inferred from the individual Mariner 10 boundary crossing corresponded to IMF $B_z > 0$ and the smaller to $B_z < 0$. The physical interpretation for these effects is that the integrated effect of intensified tail current systems, and the Region 1 field-aligned currents that accompany them, is to produce a magnetic field on the dayside of the magnetosphere that is southward and opposes, or "bucks out," some of the magnetic flux that is standing off the solar wind. In other words, these nightside currents tend to reduce the effective magnitude of the planetary magnetic field that balances the ram pressure of the solar wind. These effects have been captured in many global models and MHD simulations of the Earth's magnetosphere. More recently, MHD simulations of Mercury have reproduced the effect (Kabin et al. 2000; Ip and Kopp 2002). The application of the MHD model for this problem would be appropriate as the interest here is in the integrated effect, in which total current rather than the spatial distribution or the peak value of current density matters.

The other process affecting the dimensions of the dayside magnetosphere is the rapid variation in solar wind dynamic pressure. Numerous calculations have now indicated that step-like increases in solar wind pressure associated with high-speed streams and interplanetary shocks will generate large-scale induction currents in the planetary interior that will aid the dayside magnetosphere in resisting these compressions (Hood and Schubert 1979; Grosser et al. 2004). Physically these induction currents are such that they will temporarily add closed magnetic flux to the dayside magnetosphere and effectively enhance the magnitude of Mercury's intrinsic magnetic field. Accordingly, the dimensions of Mercury's dayside magnetosphere may be governed by relations more complicated than one might expect simply from (1). The solar wind is not only colliding with a given obstacle which is fixed in time, but the temporal variation in the solar wind itself modifies the obstacle with which it interacts. This effect should be significant when the magnetopause is approaching the planet due to larger ram pressure of the solar wind, as in an Interplanetary Coronal Mass Ejection (ICME) event, which would be one of the most curious situations of the Mercury's magnetospheric dynamics.

1.3 Magnetotail Radius

The structure of the nightside magnetosphere are controlled by the flaring of the magnetopause downstream of the terminator plane, the magnetic flux content of the lobes and the thickness of the plasma sheet. As with the dayside, the physical issues all depend on the conditions in the solar wind and the magnetosphere and how they come into equilibrium at the magnetopause. On the nightside this pressure balance condition in the direction orthogonal to the tail-axis can be expressed as

$$P_{sw} \cos^2 \psi + nk(T_i + T_e) + B_{IMF}^2/8\pi = B_{lobe}^2/8\pi, \tag{2}$$

where ψ is the flaring angle between the direction of the Sun and the local normal to the magnetopause. At the sub-solar point where ψ is small, the left-hand side of this "Newtonian" pressure balance condition is dominated by the solar wind dynamic pressure and the effects of the thermal and magnetic pressure terms are negligible. However, these so-called "static" pressure terms become important near the inner edge of tail which is taken to be about one solar wind stand-off distance downstream of the center of the planet. This distance is generally taken to be the inner edge of the tail, X_0.

For Mercury, the Mariner 10 aphelion measurements show that the initial tail field strength, B_0, and radius, R_0, of the tail at $X_0 \sim 1.5 R_M$ are $\sim$40 nT and $\sim 3 R_M$ (e.g., Ogilvie et al. 1977). The sub-solar stand-off distance is $1.5 R_M$ at Mercury compared to $10 R_E$ at Earth, and there is a factor of seven difference when the respective planetary radius is taken as the unit for each case. Application of the factor of seven or eight scaling has been found to be quite successful in comparing the magnetosphere of Earth to Mercury (e.g., Slavin 2004). Applying this scaling to the Mariner 10 observations yields a tail diameter corresponding to the terrestrial case of 21–24 R_E at $X \sim -10 R_E$, which is in reasonable agreement with the observations at Earth (Sibeck et al. 1991).

Observations at the Earth have shown that tail magnetopause flaring ends very near the location of the distant neutral line (DNL) at $X \sim -100 R_E$ (Slavin et al. 1985; Nishida et al. 1995), where the field lines opened at the dayside reconnection are re-closed to form magnetospheric closed field lines. At that point the dynamic pressure term in (2) becomes negligible and the lobe magnetic field intensity and tail radius achieve their "terminal" values, B_T and R_T, respectively, and the lobe field ceases decreasing and the radius of the tail stops increasing with growing downstream distance. Assuming conservation of magnetic flux, and given the Mercury aphelion plasma beta value of $\sim$1 in Table 1, the terminal lobe magnetic field intensity, B_T will be $B_0/\sqrt{2}$ or $\sim$28 nT. Going one step further, and assuming that loss of magnetic flux from the lobes across the magnetopause and neglecting the volume of the cross-tail current sheet, the terminal diameter of Mercury's tail may also be computed using conservation of lobe flux:

$$B_0 R_0^2 = B_T R_T^2 \quad \text{or} \quad R_T = (B_0/B_T)^{1/2} R_0 \tag{3}$$

with the result that $R_T \sim 3.6 R_M$.

As remarked earlier, approximate correspondence was found at the Earth between the location of the DNL and the point where tail flaring ceased (the tail achieved its terminal radius). Furthermore, Slavin et al. (1985) found close agreement between the location of cessation of tail flaring, X_T, and the analytic tail model of Coroniti and Kennel (1972). The Coroniti–Kennel model uses as input only the distance to the inner edge of the tail, X_0, the initial and terminal radii of the tail, R_0 and R_T, and a fast mode Mach number equal to $M_F = [M_S^2 M_A^2/(M_A^2 + M_S^2)]^{1/2}$. Using the Mercury perihelion values from Table 1 yields $M_F \sim 4.2$. The Coroniti–Kennel model then predicts for Mercury

$$X_T = X_0 + M_F R_T \left[0.6 - (R_0/R_T)^3/3\right] \sim 7.7 R_M. \tag{4}$$

Note here that the convenient "factor-of-seven scaling" gives a different value ($\sim 14 R_M$) for X_T. This difference appears because the fast mode number M_F is different by a factor of ~ 2 between the two magnetospheres. Extrapolation blindly based upon the Earth's magnetosphere would suggest that the downstream distance at Mercury where the magnetotail would cease to flare and the distant neutral line would be located is $\sim 8 R_M$.

1.4 Polar Cap Size

The magnetic flux in the lobe region is emerging from the polar cap region. On the other hand, the intensity of the lobe field is determined by the pressure balance with the solar wind. Likewise the radius of the magnetotail is proportional to the sub-solar stand-off distance and is thus determined by the solar wind conditions. Noting that the intrinsic magnetic field intensity of Mercury is small, the question arises: Is there enough flux out of the polar cap to sustain the lobe region whose characteristics are determined by the external condition? This is equivalent to asking another interesting question: What controls the size of the polar cap of a planet?

Here let us develop a simple scaling argument to estimate what size Mercury's polar cap has to be.

(1) Assuming that the tail radius R_t is proportional to the sub-solar stand-off distance, one gets $(R_t/R_P) \sim B_d^{2/3} P_{sw,dyn}^{-1/3}$, where R_P is the planetary radius, B_d is the dipole field intensity at the equator of the planet's surface and $P_{sw,dyn}$ is the solar wind dynamic pressure.

(2) The lobe magnetic pressure has to balance the solar wind pressure. Assuming that the flaring angles of the magnetotail are the same for the two planets, one gets $B_{lobe} \sim P_{sw,dyn}^{1/2}$.

(3) The lobe field is emerging from the polar cap. Then one gets the relation $B_d \Omega_{pc} R_P^2 \sim B_{lobe} f_{lobe} R_t^2$, where Ω_{pc} is the solid angle of the polar cap and f_{lobe} is the area filling factor of the lobe in the cross-section of the magnetotail.

From these we obtain, $f_{lobe} \sim \Omega_{pc} B_d^{1/3} P_{sw,dyn}^{-1/6}$. To be specific, $f_{lobe,Mercury} \sim 0.15 (\Omega_{pc,Mercury}/\Omega_{pc,Earth}) f_{lobe,Earth}$, that is, if the lobe at Mercury is to fill the same fraction as at Earth of the cross-section of the tail, the solid angle of the Mercury's polar cap has to be 6~7 times that of Earth's.

It is interesting to inspect the global 3-D MHD simulation results from this point of view. For example, Kabin et al. (2000) reported the Hermean polar cap size to have an area of about $2 R_M^2$ which is indeed a factor of six larger in solid angle than the corresponding typical value for the Earth. In their simulation, the polar cap boundary latitude under typical solar wind conditions and Parker spiral IMF was at about 52 degrees on the dayside and at ~ 20 degrees on the nightside. Comparable sizes of the Mercury's polar cap area were obtained in the global 3-D MHD simulations of Ip and Kopp (2002) who studied several magnetospheric configurations with various IMF clock angles and zero B_x. Similar polar cap sizes were obtained except for the pure northward IMF case when, just like at the Earth, the Hermean magnetosphere is almost completely closed. Similar very large polar caps appear in the hybrid simulations (here ions are treated as particles while electrons are approximated by mass-less charge neutralizing fluid) of Kallio and Janhunen (2004), as well as in the semi-empirical model of Sarantos et al. (2001). Thus, the currently available models for the Hermean space environment predict that the adequate tail magnetic flux can be provided by the enlarged polar cap areas, however, this question certainly warrants further investigation because of the numerous simplifications inherent in these models.

1.5 Plasma Sheet Thickness

It is likely that the lobe fraction in the tail cross-section at Mercury is the same as at Earth. Then the plasma sheet fraction should be the same. At Earth, the plasma sheet half-thickness is $\sim 1/20$ of the tail radius. Then, with the tail radius being $\sim 4R_M$ at Mercury, one estimates the plasma sheet half-thickness to be ~ 300 km. At Earth the plasma sheet proton temperature is ~ 5 keV under southward IMF and is controlled by the solar wind kinetic energy (e.g., Terasawa et al. 1997). Then, since the solar wind speed is the same at Mercury, the plasma sheet proton temperature at Mercury would be also ~ 5 keV. This 5 keV proton has a Larmor radius based on the tail lobe field (~ 30 nT at Mercury) as large as the plasma sheet half-thickness.

The problem here is that such a "thin" current sheet may not be stable due to kinetic effects (Nakamura et al. 2006). The small thickness has been estimated based on MHD-like scaling argument, and when it comes to the small magnetosphere of Mercury, kinetic corrections may become as visible as to change the gross structure of a meso-scale structure, the plasma sheet in the present context, from what is expected by an MHD argument.

Indeed there is a curious discrepancy between MHD modeling and the observations. Along the orbit of Mariner 10 in the tail, the peak absolute value of the B_x component was 100 nT. Meanwhile a virtual spacecraft flying along the Mariner 10 orbit in the numerically simulated Hermean magnetosphere (global 3-D MHD model) shows the peak value of 140 nT. This number does not change significantly even when very strong southward IMF is adopted as the upstream condition. The same peak value of 140 nT which does not agree with the observations is also obtained when the magnetic field is taken from a scaled-Tsyganenko model, which is a MHD-like model. A possible solution is that the plasma sheet is thicker than what the MHD-like models predict and that this difference is due to kinetic effects. Forthcoming missions with appropriate plasma measurements will certainly resolve the issue.

2 Reconnection as the Basic Driver of Magnetospheric Activity

The discovery of the Mercury's planetary magnetic field by the Mariner 10 mission was one of the most significant (and surprising) achievements of the mission (Ness et al. 1975). The discovery, of course, excited the science communities interested in the origin and the present interior state of Mercury (e.g., Balogh 1997; Spohn et al. 2001), and it also encouraged the magnetospheric physicists to start developing ideas in the new parameter regime that had not been visited. Mercury's magnetic field is non-zero but weak. How large does it have to be to attract the strong interest of a magnetospheric physicist? As has been mentioned, the small magnetic field of the planet is yet strong enough to stand-off the solar wind. The boundary of the magnetosphere (magnetopause) is located at $0.5 R_M$ above the planet's surface in the subsolar region. This means that the planet can possess its own small magnetosphere, across whose boundary plasma and magnetic field can have different properties and different orientations. In the Earth's magnetosphere, which has the same setting as the Hermean one, reconnection of magnetic field lines can take place at the magnetopause. This field-line opening reconnection on the dayside and another subsequent field-line re-closing reconnection on the nightside are considered to be the basic driving mechanism of the magnetospheric activity, which drives the circulation of the field lines and the flow of the solar wind plasma into the magnetosphere (Fig. 2). There are differences, however, between the two planets, which likely lead to different consequences at Mercury.

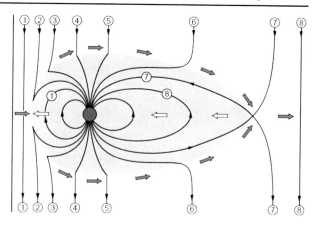

Fig. 2 Reconnecting magnetosphere: at Earth, it is well established that reconnection at two locations (dayside–nightside) is the fundamental driver of magnetospheric activity (after Baumjohann and Treumann 1996)

2.1 Reconnection

There are many outstanding questions regarding magnetic reconnection between the interplanetary magnetic field (IMF) and Mercury's intrinsic magnetic field. Substorm-like events in the Mariner 10 observations (Siscoe et al. 1975) indicate that the dayside–nightside magnetic reconnection cycle drives unsteady convection leading to the onset of the explosive phenomenon. However, there are several distinctions between the Hermean and the terrestrial magnetospheres, which may lead to the differences in the way reconnection works to vitalize the space surrounding the planets. Mercury's magnetosphere provides us with a unique setting to study this ubiquitous process in magnetized plasmas.

One of the curious features of the Hermean magnetosphere is that the planet's surface can be a substantial source of heavy ions to the magnetosphere. The ion supply is in two steps. First, atoms of volatile elements are ejected from the planet's surface by photon-stimulated desorption, and/or by thermal desorption, and/or by solar wind sputtering, and/or by micro-meteoroid vaporization. Then the intense radiation of the Sun photo-ionizes these particles quite efficiently.

Mercury's magnetosphere is much smaller than the Earth's in absolute size. The size comparison that is relevant to the underlying physics, however, should be done in the normalized form. To be specific, since our interest here is to see if MHD at the Hermean magnetosphere is as useful as it is at the Earth's, the scale length should be normalized by the ion inertial length or the ion Larmor radius, at which, roughly speaking, MHD approximation starts to break down. At Mercury, in addition to its small absolute size, because heavy ions of planetary origin such as Na^+ are considered to be abundant, the size of the magnetosphere in normalized units becomes even smaller. In addition to the small absolute size that causes the dynamic changes to occur faster (typical Alfvén transit times are of the order of seconds rather than minutes as in the terrestrial magnetosphere), the small size in normalized units should increase the underlying reconnection rate via the dynamics of easily demagnetized ions. For example, the gyro radius of a 1-keV Na^+ ion is as large as 1/16 of the typical scale of the Hermean magnetosphere.

On the front side, the conditions in the solar wind plasma at Mercury also act to increase the rate of reconnection because of the smaller plasma beta at 0.39 AU than at 1 AU. It is well known from the observations at Earth that smaller beta conditions lead to higher reconnection rates (e.g., Slavin and Holzer 1979; Paschmann et al. 1986).

In the terrestrial magnetosphere reconnection drives the flow in the magnetosphere. The field lines, which move together with the plasma, have their foot points anchored in an ionosphere that is less mobile. It is the field-aligned currents that transmit stresses from the magnetosphere to the ionosphere. The current loops have their low-altitude ends embedded inside the conducting ionosphere and the $J \times B$ force exerted by these low-altitude ends of the current loops on the dense ionospheric plasma accelerates the flow that eventually forms the ionospheric convection, the reflection of the counterpart in the magnetosphere. Field-aligned currents seem to exist in the Hermean magnetosphere (Slavin et al. 1997) as well. However, at Mercury, there is hardly any ionosphere and it is unclear at present how the field-aligned currents are closed at low altitudes (e.g., Glassmeier 2000; Ip and Kopp 2004; Baumjohann et al. 2006). What is clear, however, is that the Hermean ionosphere is too dilute to be an effective agent for mitigating magnetospheric circulation. Thus magnetospheric circulation and perhaps reconnection might also be enhanced compared to Earth due to the little anchoring effect of the Hermean ionosphere.

Substorms occur because the dual-reconnection driven convection cannot be steady, and loading as well as unloading phases are inevitably associated with it. Substorm interpretation of the events in the Mariner 10 data (Siscoe et al. 1975) implies the presence of dayside–nightside magnetic reconnection and unsteady convection driven by them. An alternative idea is that the events are simply due to instant responses to the IMF changes without loading and unloading phases (Luhmann et al. 1998). More discussion on the substorm issue is given later.

A magnetic cloud associated with a Coronal Mass Ejection (CME) often has strong southward magnetic field. At Earth, this leads to strong solar wind-magnetosphere coupling and finally to the most dramatic changes known as magnetic storms. During storms, heavy heating and outflow of ionospheric O^+ ions are often seen. These O^+ are so substantial that they can occupy a half of the plasma sheet number density (e.g., Kistler et al. 2006) and an equally substantial fraction of the energy density of the ring current population (e.g., Daglis 1997). At Mercury the high southward magnetic field is expected to be even higher and will lead to even more drastic effects as follows: Higher southward IMF will lead to more opened magnetic field lines. This will lead to a large supply of Na particles since more solar wind ions can access the surface for sputtering along the more opened field lines. After photo-ionization, the Na^+ ions find themselves in fluctuating electromagnetic fields under the influence of the magnetic cloud passage and are elevated in energy by the kick from the fields. Then the magnetosphere may become filled with energetic heavy ions, whose precipitation to the surface may set a positive feedback loop to the ion supply process. The enhanced reconnection at dayside under the strong southward field would also lead to enhanced reconnection at nightside, which would further contribute to energizing the heavy ions. It should be quite interesting to see how the energy is partitioned among the particle species in these extreme times and to compare that with the O^+-rich storm time events of the Earth.

2.2 Asymmetric Supply of Ions to the Magnetotail Current Sheet

The solar wind plasma is transported along the dayside reconnected field lines and forms the plasma mantle region. Numerous observations in the terrestrial magnetosphere show that the thickness and the occurrence of the plasma mantle depend on IMF direction (e.g., Sckopke et al. 1976). Furthermore non-zero B_x component of IMF causes difference between the northern and southern hemispheres. The IMF B_x effect is in many ways similar to the dipole tilt effects (e.g., Taguchi and Hoffman 1995) and the resultant asymmetries have been studied well.

At Mercury IMF is more radial than at Earth. Reconnection at dayside with this more radial IMF would result in stronger North–South asymmetry of the plasma content in the

Fig. 3 Asymmetric profiles of magnetic field: **a** Mariner 10 observations during current sheet crossing of the Mercury magnetotail; **b** results from a self-consistent theoretical model by Zelenyi et al. (2007). The dimensionless profile of the magnetic field component B_x is a function of the dimensionless coordinate z. The *red line* in Fig. 1b corresponds to $r = 0\%$ (total flow crossing from North to South), the *blue line* corresponds to a total flow reflection $r = 100\%$; the *black curves* are for $r = 10, 20, \ldots, 90\%$

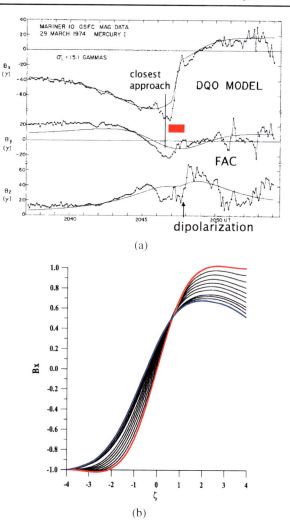

mantle region. Indeed the results of a global simulation shown by Kabin et al. (2000) exhibit strong density asymmetry between the two cusps. The plasma in the cusp will constitute the mantle region and then will flow into the current sheet in the tail. Here we show that when the plasma supply shows the North–South asymmetry, the force balance in the tail current sheet requires the magnetic field profile along the North–South direction to be asymmetric as well.

During the first Mariner 10 flyby across the nightside of the Hermean magnetosphere, the spacecraft entered the tail current sheet and measured the three magnetic field components (Whang 1977) as shown in Fig. 3. The B_x component of the magnetic field changed by $\sim$80 nT in 40 s, and the profile of the magnetic field component clearly exhibited a North–South asymmetry. The estimated thickness of the tail current sheet was about 150 km.

Whang (1977) developed the DQO model to describe these observations. The model has the following components:

(1) The planetary intrinsic field is represented by the sum of a dipole, a quadrupole and an octupole fields. The quadrupole is responsible for any of the North–South asymmetries of the planetary origin.
(2) The external field is modeled as a sum of an image dipole and a field arising from the magnetotail current sheet.

The influence of the interplanetary magnetic field outside the magnetosphere is not included. The minimization of nine free parameters brought good agreement with observations as shown in Fig. 3 (black points = observations; solid black line = the DQO model).

While the DQO model concludes that the non-dipolar component of the planetary origin is responsible for the observed asymmetry, we propose here another mechanism that is not considered in the model. The Mariner 10 observations suggest that the encountered current sheet was "thin," analogous to the one in the Earth's magnetotail (Nakamura et al., 2002, 2006; Runov et al. 2006). Indeed the Larmor radius for a 2 keV proton is about 100 km (with $B \approx 40$ nT), which is comparable with the aforementioned current sheet thickness of $\sim$150 km. Current density inside such a thin current sheet is supposed to be carried by ions impinging the neutral sheet from the northern and the southern mantle regions. At the same time the earthward tension of magnetic field line of the thin current sheet with a non-zero B_z component is balanced by the inertial force of ions (Zelenyi et al. 2004). If the plasma supply shows North–South asymmetry, then this would change the magnetic field structure as shown recently by Zelenyi et al. (2007). It is shown that diamagnetic currents of magnetized ions are stronger on the side with the more intense plasma source. This leads to a reduction of the B_x component in the corresponding lobe.

The results of a numerical calculation are shown in the right-hand panel of Fig. 3. For simplicity, this simulation considers only a single plasma source (at $z > 0$). Free parameter r denotes the coefficient of particle reflection at the current sheet plane, which non-monotonously depends on adiabaticity parameter κ (Chen 1992; Büchner and Zelenyi 1989). For typical conditions in a stretched current sheet ($\kappa \leq 0.1$–0.3) reflection coefficient may vary in a large range ($0 \leq r \leq 1$). The comparison between the observations and the calculated results show that qualitative agreement can be achieved when large r is taken for a highly asymmetric plasma source. The analogous results are obtained when the plasma temperatures or in-flow velocities of plasma sources are different in northern and southern lobes.

There is another interesting consequence of the asymmetric plasma supply. The center of the equilibrium current sheet is displaced towards the weaker plasma source side. This phenomenon might cause a vertical flapping motion of the thin current sheet when the plasma source is temporally variable, that is, for the Mercury situation, when IMF B_x is flipping its sign.

2.3 Substorms

In the Earth's magnetosphere, the fundamental energy conversion process in the magnetotail is the substorm. Due to the imbalance in magnetic flux transport from dayside to nightside and from nightside to dayside, energy is stored in the form of magnetic field and a thin current sheet is formed in the magnetotail. Instabilities in the thin current sheet lead to sporadic energy release which can be detected as changes in the electric current pattern in the magnetosphere-ionosphere system, enhanced particle precipitation (aurora) in the ionosphere, and enhanced plasma flows and particle acceleration in the magnetotail. At Mercury, where a particularly strong magnetosphere-solar wind coupling is expected but where an ionosphere is missing, it is yet a debated issue whether substorm-like processes can take place. This is why different interpretations of the scarce Mariner 10 observations exist.

Siscoe et al. (1975) was the first to suggest that the highly structured variations observed in Mercury's nightside magnetosphere are similar to variations in the corresponding regions of the Earth's magnetosphere during substorms. In order to compare substorms at Mercury and Earth, Siscoe et al. (1975) obtained the temporal and spatial scaling relationships between Mercury and Earth. The time scale of the convection cycle was estimated be about 30 times shorter than at Earth and the size of the magnetosphere relative to the planet was estimated to be about seven times smaller. Then the time scale of the observed magnetic variation was shown to be in the range of expected time-scale for substorms. Transient particle and magnetic field variations observed by Mariner 10 showed further supporting evidence for a substorm-like activity. These include transient magnetic field signatures similar to dipolarization events, enhancements in field-aligned currents suggesting the existence of a substorm current wedge (Slavin et al. 1997), and strong enhancements in the flux of >35 keV electrons analogous to energetic particle injection during substorms (Baker et al. 1986; Simpson et al. 1974; Eraker and Simpson 1986; Christon 1987).

Particles are accelerated by magnetic reconnection as they move in the direction of the induction electric field, which is the East–West direction. If the extent of the magnetosphere in this direction is limited, which is indeed very true for Mercury, particle acceleration requires stronger electric fields. That the reconnection rate is likely to be higher at Mercury may explain the fact that efficient particle acceleration was observed in such a small magnetosphere. On the other hand, some new acceleration mechanism that does not require a laterally elongated reconnection region may be operative. In this sense, it is curious to see how the (heavy) ions, which require more length for acceleration, behave in these energetic electron events.

The substorm interpretation was challenged by a simpler one in Luhmann et al. (1998), particularly with regard to the existence of energy storage and release phases in the Mercury's substorm cycle. The terrestrial magnetic field model T96 (Tsyganenko 1996) was modified and used in this study. What was done was as follows:

(1) Produce time varying magnetosphere by changing the input IMF parameters.
(2) IMF data used in (1) were taken from observations at 1 AU and enhanced in magnitude by a factor of 2.
(3) A virtual spacecraft was flown in the virtual magnetosphere.
(4) The spatial scaling factor of 7 was taken into account in mimicking the Mariner 10 orbit.

The good agreement shown in Fig. 4 implies that the major magnetic disturbances observed by Mariner 10 can be produced by direct influence of the changing IMF. The energetic electrons enhancement was interpreted to originate from either external sources, or a magnetospheric boundary layer which was encountered by the spacecraft because of a favorable reconfiguration of the magnetosphere due to solar wind changes.

Due to lack of simultaneous observations in the solar wind and in Mercury's magnetosphere the actual response of the magnetosphere to the solar wind input could not be confirmed. Hence, the existence of substorms—or more accurately magnetospheric disturbances caused by internal magnetotail current sheet instabilities—is still a debated issue. Global MHD simulations seem to suggest the presence of substorm-like activity at Mercury. Unless the kinetic corrections, which certainly exist and may be substantial at Mercury, enhance the instantaneous response to the external force and/or weaken the loading/unloading dynamics, it seems reasonable to expect substorms to occur.

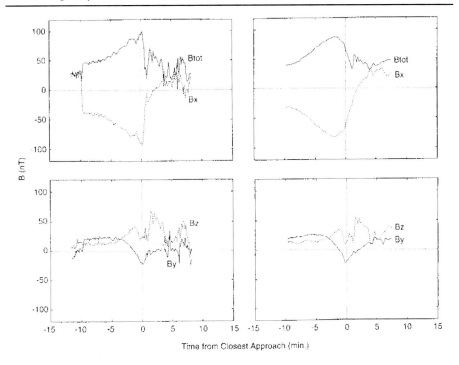

Fig. 4 *Left*: magnetic field measurements from Mariner 10. *Right*: a simulated time series produced by using a magnetospheric magnetic field model and varying IMF orientation (Luhmann et al. 1998)

While Mariner 10 flybys took place in the nightside inner magnetosphere (if scaled to Earth), upcoming missions like Messenger or BepiColombo-MMO will cross the magnetotail current sheet at distances where effects of a tail reconnection process are expected to be observed more clearly. Here let us emulate a possible MMO event using Geotail data. In the Earth's magnetotail, the storage-release (loading-unloading) cycle of substorms can be identified in the temporal variation of the total pressure, which shows gradual increase and then reduction (Caan et al. 1973). Figure 5 shows the results of the emulation. Here the expected total pressure variations in the Mercury's tail at the distance of $\sim 5.4 R_M$ is reproduced by scaling the empirical model results with the IMF input. Also shown are the scaled Geotail data obtained in the Earth's magnetotail. The empirical models are the statistical lobe field model of Fairfield and Jones (1996) and the modified T96 model as used by Luhmann et al. (1998). If Mercury's magnetospheric activity is driven only by the solar wind, the enhancement in the southward IMF and the solar wind pressure would cause the enhanced total pressure, which are the outputs from the two models (thin and dotted lines). In contrast, if the same process as in the Earth's magnetotail happens what will be observed are the increase before and the rapid decrease after substorm expansion onset (thick line). Ideally, Mercury's magnetotail response to solar wind changes should be studied in a way similar to Fig. 5, using both the solar wind and the magnetospheric data. Such simultaneous observations are available when MMO is located at the high latitude magnetosheath and MPO is in the inner magnetosphere.

The above arguments assume that there exists a similar magnetotail configuration based on the scaling relationships of Siscoe et al. (1975). Yet, one should note that from the limited observation of Mariner 10, quantitative characteristics of the magnetotail current sheet,

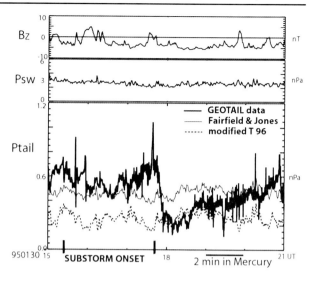

Fig. 5 Solar wind inputs (*two upper panels*) and the expected responses of Mercury's magnetotail according to the three models (*bottom panel*). The *thick line* shows a rescaled Geotail data plot showing how the magnetic field in the Mercury's tail would behave if there are substorms. If there is no substorm, the data would look similar to the plots from the lobe model of Fairfield and Jones (1996) or the modified T96 model of Luhmann et al. (1998)

such as the scale size and the stability, are still unknown. For example, the thickness of the current sheet crossed around the time of the closest approach in Fig. 4 was estimated to be 150 km. This thickness was obtained from the spacecraft motion, 3.7 km/s and the crossing duration of 40 s (Whang 1977). This is comparable to a 2 keV proton Larmor radius in the magnetic field of 40 nT and therefore suggests that it was a thin current sheet. On the other hand, this estimation has not taken into account any motion of the current sheet relative to the spacecraft. In the Earth's magnetosphere, such current sheet motion can have speeds of tens to hundreds of km/s (Runov et al. 2005). Corresponding current sheet motion at Mercury would then be about four times faster if we simply use the argument that the spatial scale is seven times shorter and the time scale is 30 times quicker. If such current sheet motion can exist, the actual current sheet obtained by Mariner 10 could be much thicker than the previous estimation (by a factor of 10 to 100). On the other hand, a 2-keV proton as a representative ion in the Mercury's plasma sheet is just an arbitrary choice since the plasma sheet conditions are not known at all. Simultaneous ion distribution and magnetic field observations by the future missions are therefore essential to determining the quantitative characteristics of the magnetotail current sheet.

The small size and likely substantial presence of the heavy ions make it very possible that the physics of the Mercury's magnetotail is governed more by kinetic effects than at Earth. Conditions for current sheet instabilities could be much easily fulfilled without waiting during the loading phase for the thin current sheet to develop. Then substorms at Mercury are inclined to possess a character that may be mistaken as directly driven by the IMF. There is even a possibility that the magnetotail is in an ever-unstable regime dominated by easily triggered "substorm"-type disturbances. In addition to the lack of the ionosphere, which is suggested to significantly modify the substorm field-aligned current system at Mercury (Glassmeier 2000), it is expected that this relatively "thin" current sheet of the magnetotail will result in an essentially different nature of the substorms at Mercury. If this is true, it would be one of the most interesting and exciting insights that Mercury's magnetosphere may provide to the comparative magnetosphere studies.

3 Heavily Loaded Hermean Magnetosphere?

Mercury has only a tenuous atmosphere and no ionosphere. While this seems to imply that its role as a plasma source for the magnetospheric is weak, this may not necessarily be so. In this section, we introduce an argument that claims that contribution by the ions of the planetary origin can be rather significant, so significant in fact that the mass density in the inner-magnetosphere is larger than that in the solar wind. The interaction of the solar wind with this heavily loaded magnetosphere can give rise to a possibly interesting magnetospheric dynamics.

3.1 Mercury as a Plasma Source

Mercury's ionized atmosphere is poorly known. The existence of neutral particles that are the source for the ions has been established for H, He, and O by Mariner 10 observations (Broadfoot et al. 1976), while Na, K, and Ca have been detected by ground-based measurements (Potter and Morgan, 1985, 1986; Bida et al. 2000). For example, the observed zenith column densities of Na and O are 2×10^{11} cm^{-2} and 3×10^{11} cm^{-2} with suggested number densities at the surface of 4×10^4 cm^{-3} and 4×10^4 cm^{-3}, respectively. Radio occultation measurements made during the first flyby of Mariner 10 indicated that electron number densities at the limb, both nightside and dayside, are less than 10^3 cm^{-3} (Fjeldbo et al. 1976). From these small number densities at the surface, one may conclude that the plasma supply from the planet to its magnetosphere is negligible.

The low density at the surface, however, does not necessarily imply a weak plasma source to the magnetosphere. Because of the proximity to the Sun, the exospheric neutrals are subject to efficient photoionization. The rates are, 7×10^{-5} to 2×10^{-4} s^{-1} for Na (Smyth and Marconi 1995), and 2×10^{-6} to 5×10^{-6} s^{-1} for O (Cravens et al. 1987). Note that Cremonese et al. (1997) suggested a value for Na that is three times smaller, but we use the value in Smyth and Marconi (1995) to get the upper limit. Integrating the ionization rates over the exosphere with the observed zenith column number densities, we obtain the total Na$^+$ and O$^+$ production rates of $\sim 2 \times 10^{25}$ and $\sim 8 \times 10^{23}$ ions/s, respectively.

While the scale-heights of the exospheric neutrals range from 30 km (for potassium) to 1,330 km (for hydrogen atoms), which are lower or comparable to the nominal magnetopause height of $H_{MP} = 1,000$ km, the ions produced are expected to quickly fill the entire inner magnetosphere. This is because significant heating associated with the ion pick-up process is expected and the scale heights for ions are considered to be much higher. The pick-up process turns the velocity of magnetospheric flux tubes ($E \times B$ drift velocity) to the thermal velocity of the newly born ions. The velocity can be more than several km/s near the planetary surface, leading to the ion thermal energy of ~ 10 eV just after ionization. The corresponding scale-height of the newly created ions is $\sim 10^4$ km. A note of caution, however, is that we do not know exactly how fast the magnetic flux tubes move relative to the surface because of uncertainties in the anchoring effect by the surface, which depends significantly on its electrical conductance (e.g., Coroniti and Kennel 1973; Hill et al. 1976; Slavin 2004).

Let us assume that produced ions are quickly spread over the inner magnetosphere and then are lost mainly due to magnetospheric convection of a timescale of ~ 100 s. A rough estimate of the volume of the inner magnetosphere is $H_{MP}S$ (where $H_{MP} \sim 1,000$ km is a nominal magnetopause altitude and $S = 7.5 \times 10^7$ km^2 the surface area of planet). From these we obtain the average Na$^+$ and O$^+$ number densities of 30 cm^{-3} and 1 cm^{-3}, respectively. These values are in good agreement with a previous estimate by Lundin et al.

(1997). That is, only Na$^+$ and O$^+$ are enough to make the plasma mass density in the inner magnetosphere to surpass the nominal solar wind mass density.

In the nightside neither photon-stimulated desorption of exospheric neutrals nor photo-ionization by solar photons is available. Meanwhile other processes such as thermal desorption, particle sputtering, and meteorite impact vaporization release neutral gases to form the nightside exosphere (e.g., Wurz and Lammer 2003; Marchi et al. 2005). Of the known six exospheric components, O may have a significant nightside enhancement, with densities of up to 50 times larger than on the dayside (Morgan and Shemansky 1991; Sprague et al. 1992). Even though the photo-ionization is not operative, some of the nightside gases are subject to ionization via electron impact and charge exchange processes. Assuming that the Hermean magnetosphere on the nightside is filled with high energy (>100 eV) electrons with the number density 1 cm^{-3}, the electron impact ionization rate is estimated to be 10^{-7} s^{-1} for O (Cravens et al. 1987), which is only an order of magnitude smaller than the dayside photo-ionization. In other words, even the night-side magnetosphere also possesses a substantial local plasma source.

While we have so far touched only on Na and O, there is a possibility that yet undetected gases are present in Mercury's atmosphere. The fact that the upper limit for the total atmospheric pressure constrained by the Mariner 10 radio occultation is two orders of magnitude greater than what the known exospheric components can carry (Fjeldbo et al. 1976; Hunten et al. 1988; Wurz and Lammer 2003) suggests that the known constituents are only a small fraction of the total (e.g., Morgan and Killen 1997). The enhancement of O$^+$(and/or O$^-$) caused by the surface's spatial inhomogeneity (Morgan and Shemansky 1991) may play a role in the mass-loading of the magnetosphere. Some of these species will be detected by UV and/or in situ instruments onboard the forthcoming missions.

3.2 Interaction of the Heavily Loaded Magnetosphere with the Solar Wind

The heavy mass-load of the Hermean magnetosphere has possible impacts on the magnetospheric dynamics. Here we discuss the Kelvin–Helmholtz (K–H) instability at the dayside magnetopause. The K–H instability is a convective instability that develops in a flow shear, whose propagation speed is approximately given by that of the center of mass frame. At the Earth's magnetopause, it usually evolves while propagating downstream along with the anti-sunward magnetosheath flow, which is much faster and denser than the dayside magnetospheric plasma. The instability has been studied well in theory (e.g., Miura 1984; Fujimoto and Terasawa 1994; Nakamura et al. 2004; Matsumoto and Hoshino 2006) as well as in observations (e.g., Fairfield et al. 2000; Hasegawa et al. 2004). Hasegawa et al. (2004) showed evidence that the instability grows into the non-linear stage to produce rolled-up vortices under northward IMF.

The likely presence of dense plasma in the Mercury's magnetosphere would change the situation. The center of mass frame of the magnetosheath and the magnetospheric plasmas can stagnate at some location (local-time) on the magnetopause, leading to an absolute growth of the K–H mode at the site. In the zero thickness shear layer and incompressible approximation, the linear dispersion relation for the K–H instability is written as (e.g., Nagano, 1978; 1979; Terada et al. 2002)

$$\omega = k\frac{\rho_1 U_1 + \rho_2 U_2}{\rho_1 + \rho_2} + k|k|\frac{v_{L1}\rho_1 - v_{L2}\rho_2}{\rho_1 + \rho_2} \pm i\gamma, \qquad (5)$$

where

$$\gamma = \sqrt{k^2 \frac{\rho_1 \rho_2 (U_1 - U_2)^2}{(\rho_1 + \rho_2)^2} + 2|k^3| \frac{(\nu_{L1} + \nu_{L2})\rho_1 \rho_2 (U_1 - U_2)}{(\rho_1 + \rho_2)^2} - k^4 \left(\frac{\nu_{L1}\rho_1 - \nu_{L2}\rho_2}{\rho_1 + \rho_2} \right)^2} \quad (6)$$

Here, subscripts 1 and 2 denote the values in the magnetosheath and the magnetosphere, respectively. ρ is the mass density, U the flow velocity tangential to the interface, k the wave number, ω the wave frequency, $\nu_L = \frac{1}{4} r_i^2 \Omega_i$ the gyro-viscous coefficient, r_i the ion gyro radius, and Ω_i the ion gyro frequency. In deriving (5), we assume that $k \parallel U \perp B$. The fastest growing wave number of the K–H instability is in general determined either by the finite Larmor radius effects or by the thickness of the boundary layer. As the parameters at the Mercury's magnetopause, we take $U_2 = -20$ km/s, $\rho_1 = m_{H^+} \times 100$ kg cm^{-3}. The sheath flow U_1 varies from 0 to 300 km/s with the increasing solar zenith angle (SZA) from 0 to 90 degrees according to a gas-dynamic model. If $\rho_2 = 6\rho_1$ is assumed, the group velocity of the fastest growing mode vanishes at SZA = 38 degrees with a growth rate of $\gamma = 0.15$ s^{-1}. This indicates that the K–H instability at Mercury's magnetopause indeed become absolutely unstable (rather than convectively unstable) at some local time.

Figure 6 illustrates non-linear simulation results of the solar wind interaction with a heavily loaded magnetosphere. A global MHD model for Earth is scaled to the Hermean parameters and a heavy plasma production term is included in the continuity equation. The plasma mass production rate is assumed to have a dependence of $P = P_0 \exp(-h/H_S)$ except for the dark side of the planet, where P_0 is the mass production rate at the planetary surface and h the altitude. For the less-loaded case (panel a) we use $P_0 = m_{H^+} \times 4$ kg cm^{-3} s^{-1} and $H_S 600$ km assuming a fictitious gas component whose total mass production rate above the surface is about half that of Na$^+$ discussed above. The production rate for the heavily loaded case (panel b) is ten times greater than that for panel a. The computational grid system is Cartesian and the grid spacing (150 km) is marginally sufficient to resolve the scale-height. We see the evolution of the K–H instability in the heavily loaded case (panel b), while not in the less-loaded case (panel a), consistent with the above argument based on the linear theory. The evolution of the K–H mode is affected significantly by the magnetospheric convection, which is dependent on the location of the near Mercury neutral line (NMNL) and hence on the electrical resistivity assumed in the MHD simulation. A uniform and constant resistivity is used here throughout the simulation box, which yields an NMNL location around 3.2 (2.2) R_M for panel a (b) downstream the planet.

The high mass density may set another interesting stage for the K–H instability. The higher mass density lowers the fast mode speed even if the total pressure is the same. This enlarges the possibility that the vortex propagation speed seen in the magnetospheric plasma frame is super-sonic while it remains sub-sonic on the magnetosheath side. When this super–sub sonic combination occurs, a recent study shows that a nicely rolled up vortex forms and at the same time it launches shock waves into the super-sonic (magnetospheric) side (Kobayashi et al. 2007). It has been known that

(1) in the sub–sub combination, which is the usual case that has been studied extensively, a nicely rolled up vortex forms (e.g., Miura 1984), and
(2) the super–super combination allows the instability to launch sound waves but not to produce a nice vortex (e.g., Miura 1990).

In the super–sub case, in contrast, the instability produces both a vortex and a shock wave. The setting at Mercury may allow us to study observationally this new regime of the K–H instability that does not seem to be available at any other planetary magnetosphere. The

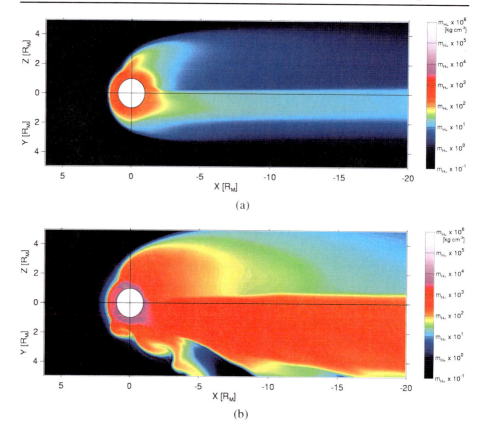

Fig. 6 Global MHD simulation of a weakly-loaded magnetosphere (**a**) and a heavily-loaded magnetosphere (**b**). The production rate of planetary plasma for **b** is ten 10 times greater than that for **a**. Upstream solar wind number density and velocity are $n = 35$ cm^{-3} and $U = 400$ km/s, respectively. IMF is set to point purely southward with an intensity of 5 nT

shock waves launched into the magnetosphere from the boundary may heat some of the heavy ions to supra-thermal energy altering the plasma environment of the planet.

4 Conclusion

We considered the physics of the Hermean magnetosphere based on what we know about Earth's magnetosphere. While the skeleton of the magnetospheric structure seems to be well predicted by the scaling arguments, the dynamic aspect seems to require substantial modification. The key questions are:

(1) How crucial is the kinetic correction due to the overall smallness of the magnetosphere?
(2) How substantial are the heavy ions of the planetary origin?
(3) How much does the unknown boundary condition at the planet's surface matter?

We should be ready for surprises in the data from the future missions.

References

D.N. Baker, J.A. Simpson, J.H. Eraker, J. Geophys. Res. **91**, 8742–8748 (1986)
A. Balogh, Planet. Space Sci. **45**, 1 (1997)
W. Baumjohann, R.A. Treumann, *Basic Space Plasma Physics* (Imperial College Press, London, 1996)
W. Baumjohann, A. Matsuoka, K.-H. Glassmeier, C.T. Russell, T. Nagai, M. Hoshino, T. Nakagawa, A. Balogh, J.A. Slavin, R. Nakamura, W. Magnes, Adv. Space Res. **38**, 604 (2006)
T.A. Bida, R.M. Killen, T.H. Morgan, Nature **404**, 159–161 (2000)
A.L. Broadfoot, D.E. Shemansky, S. Kumar, Geophys. Res. Lett. **3**, 577–580 (1976)
J. Büchner, L.M. Zelenyi, J. Geophys. Res. **94**, 11821–11842 (1989)
M.N. Caan, R.L. McPherron, C.T. Russell, J. Geophys. Res. **78**, 8087 (1973)
J. Chen, J. Geophys. Res. **97**, 15011–15050 (1992)
S.P. Christon, Icarus **71**, 448–471 (1987)
F.V. Coroniti, C.F. Kennel, J. Geophys. Res. **77**, 3361–3370 (1972)
F.V. Coroniti, C.F. Kennel, J. Geophys. Res. **78**, 2837–2851 (1973)
T.E. Cravens, J.U. Kozyra, A.F. Nagy, T.I. Gombosi, M. Kurtz, J. Geophys. Res. **92**, 7341–7353 (1987)
G. Cremonese, H. Boehnhardt, J. Crovisier, H. Rauer, A. Fitzsimmons, M. Fulle, J. Licandro, D. Pollacco, G.P. Tozzi, R.M. West, Astrophys. J. **490**, L199–L202 (1997)
H. DeSterck, S. Poedts, Phys. Rev. Lett. **84**, 5524–5527 (2000)
I.A. Daglis, in *Magnetic Storms*, ed. by B.T. Tsurutani et al., AGU Monograph, vol. 98 (AGU, Washington, 1997).
J.H. Eraker, J.A. Simpson, J. Geophys. Res. **91**, 9973–9993 (1986)
D.H. Fairfield, J. Jones, J. Geophys. Res. **101**, 9113–9119 (1996)
D.H. Fairfield, A. Otto, T. Mukai, S. Kokubun, R.P. Lepping, J.T. Steinberg, A.J. Lazarus, T. Yamamoto, J. Geophys. Res. **105**, 21159–21174 (2000)
G. Fjeldbo, A. Kliore, D. Sweetnam, P. Esposito, B. Seidel, T. Howard, Icarus **29**, 439–444 (1976)
M. Fujimoto, T. Terasawa, J. Geophys. Res. **99**, 8601 (1994)
G. Giampieri, A. Balogh, Planet. Space Sci. **49**, 1637–1642 (2001)
K.-H. Glassmeier, in *Magnetospheric Current Systems*. Geophys. Monograph, vol. 118 (AGU, Washington, 2000), pp. 371.
J. Grosser, K.-H. Glassmeier, S. Stadelmann, Planet. Space Sci. **52**, 1251–1260 (2004)
H. Hasegawa, M. Fujimoto et al., Nature **430**, 755 (2004)
T.W. Hill, A.J. Dessler, R.A. Wolf, Geophys. Res. Lett. **3**, 429–432 (1976)
R.E. Holzer, J.A. Slavin, J. Geophys. Res. **83**, 3831–3839 (1978)
L.L. Hood, G. Schubert, J. Geophys. Res. **84**, 2641–2647 (1979)
D.M. Hunten, T.M. Morgan, D.E. Shemansky, in *Mercury*, ed. by F. Vilas, C.R. Chapman, M.S. Matthews (University of Arizona Press, Tucson, 1988), pp. 562–612.
W.-H. Ip, A. Kopp, J. Geophys. Res. **107**, 1348 (2002). doi:10.1029/2001JA009171
W.-H. Ip, A. Kopp, Adv. Space Res. **33**, 2172–2175 (2004)
J.R. Johnson, C.Z. Cheng, Geophys. Res. Lett. **23**, 1423 (1997)
K. Kabin, T.I. Gombosi, D.L. DeZeeuw, K.G. Powell, Icarus **143**, 397–406 (2000)
K. Kabin, J. Plasma Phys. **66**, 259–274 (2001)
E. Kallio, P. Janhunen, Adv. Space Res. **33**, 2176–2181 (2004)
L.M. Kistler et al., J. Geophys. Res. **111**, A11222 (2006). doi:10.1029/2006JA011939
Y. Kobayashi et al., Adv. Space. Res. (2007, in press)
H. Korth et al., Planet. Space Sci. **54**, 733–746 (2004)
B. Lavraud, M.F. Thomsen, B. Lefebvre, S.J. Schwartz, K. Seki, T.D. Phan, Y.L. Wang, A. Fazakerley, H. Rème, A. Balogh, J. Geophys. Res. **111**, A05211 (2006). doi:10.1029/2005JA011266
W. Li, J. Raeder, J. Dorelli, M. Øieroset, T.D. Phan, Geophys. Res. Lett. **32**, L12S08 (2005). doi:10.1029/2004GL021524
R. Lundin, S. Barabash, P. Brandt, L. Eliasson, C.M.C. Naim, O. Norberg, I. Sandahl, Adv. Space Res. **19**, 1593–1607 (1997)
J.G. Luhmann, C.T. Russell, L.H. Brace, O.L. Vaisberg, H.H. Kiefer, in *Mars*, ed. by H.H. Kiefer et al., (University of Arizona Press, Tucscon, 1992)
J.G. Luhmann, C.T. Russell, N.A. Tsyganenko, J. Geophys. Res. **103**, 9113–9119 (1998)
S. Marchi, A. Morbidelli, G. Cremonese, Astron. Astrophys. **431**, 1123–1127 (2005)
Y. Matsumoto, M. Hoshino, J. Geophys. Res. **111**, A05213 (2006). doi:10.1029/2004JA010988
S.E. Milan, M. Lester, S.W.H. Cowley, K. Oksavik, M. Brittnacher, R.A. Greenwald, G. Sofko, J.-P. Villaian, Ann. Geophys. **21**, 1121–1140 (2003)
A. Miura, J. Geophys. Res. **89**, 9801 (1984)
A. Miura, Geophys. Res. Lett. **92**, 3195 (1990)

T.H. Morgan, R.M. Killen, Planet. Space Sci. **45**, 81–94 (1997)
T.H. Morgan, D.E. Shemansky, J. Geophys. Res. **96**, 1351–1367 (1991)
H. Nagano, J. Plasma Phys. **20**, 149–160 (1978)
H. Nagano, Planet. Space Sci. **27**, 881–884 (1979)
R. Nakamura, W. Baumjohann, A. Runov, M. Volwerk, T.L. Zhang, B. Klecker, Y. Bogdanova, A. Roux, A. Balogh, H. Réme, J.A. Sauvaud, H.U. Frey, Geophys. Res. Lett. **2**(9), 2140 (2002). doi:10.1029/2002GL016200
R. Nakamura, W. Baumjohann, Y. Asano, A. Runov, A. Balogh, C.J. Owen, A.N. Fazakerley, M. Fujimoto, B. Klecker, H. Rème, J. Geophys. Res. **111**, A11206 (2006). doi:10.1029/2006JA011706
T.K.M. Nakamura, D. Hayashi, M. Fujimoto, I. Shinohara, Phys. Rev. Lett. **92**, 145001 (2004)
N.F. Ness, K.W. Behannon, R.P. Lepping, Y.C. Whang, Nature **255**, 204 (1975)
A. Nishida et al., J. Geophys. Res. **100**, 23663–23675 (1995)
K.W. Ogilvie, J.D. Scudder, V.M. Vasyliunas, R.E. Hartle, G.L. Siscoe, J. Geophys. Res. **82**, 1807–1824 (1977)
M. Øieroset, J. Raeder, T.D. Phan, S. Wing, J.P. McFadden, W. Li, M. Fujimoto, H. Rème, A. Balogh, Geophys. Res. Lett. **32**, L12S07 (2005). doi:10.1029/2004GL021523
G. Paschmann, I. Papamastorakis, W. Baumjohann, N. Sckopke, C.W. Carlson, B.U.Ö. Sonnerup, H. Lühr, J. Geophys. Res. **91**, 11099 (1986)
A.E. Potter, T.H. Morgan, Science, **229**, 651–653 (1985)
A.E. Potter, T.H. Morgan, Icarus, **67**, 336–340 (1986)
A. Runov, V.A. Sergeev, W. Baumjohann, R. Nakamura, S. Apatenkov, Y. Asano, M. Volwerk, Z. Vörös, T.L. Zhang, A. Petrukovich, A. Balogh, J.-A. Sauvaud, B. Klecker, H. Rème, Ann. Geophys. **23**, 1391–1403 (2005)
A. Runov, V.A. Sergeev, R. Nakamura, W. Baumjohann, S. Apatenkov, Y. Asano, T. Takada, M. Volwerk, Z. Vörös, T.L. Zhang, J.-A. Sauvaud, H. Rème, A. Balogh, Ann. Geophys. **24**, 247–262 (2006)
C.T. Russell, D.N. Baker, J.A. Slavin, in *Mercury*, ed. by F. Vilas, C.R. Chapman, M.S. Matthews (University of Arizona Press, Tucscon, 1988), pp. 514–561.
M. Sarantos, P.H. Reiff, T.W. Hill, R.M. Killen, A.L. Urquhart, Planet. Space Sci. **49**, 1629–1635 (2001)
N. Sckopke, G. Paschmann, H. Rosenbauer, D.H. Fairfield, J. Geophys. Res. **81**, 2687–2691 (1976)
P. Song, C.T. Russell, J. Geophys. Res. **97**, 1411–1420 (1992)
J.A. Simpson, J.H. Eraker, J.E. Lamport, P.H. Walpole, Science **185**, 160–166 (1974)
D.G. Sibeck, R.E. Lopez, E.C. Roelof, J. Geophys. Res. **96**, 5489–5495 (1991)
G.L. Siscoe, N.F. Ness, C.M. Yeates, J. Geophys. Res. **80**, 4359–4363 (1975)
G.L. Siscoe, L. Christopher, Geophys. Res. Lett. **2**, 158–160 (1975)
J.A. Slavin, Adv. Space. Res. **33**, 1859–1874 (2004)
J.A. Slavin, C.J. Owen, J.E.P. Connerney, S.P. Christon, Planet. Space Sci. **45**, 133–141 (1997)
J.A. Slavin, R.E. Holzer, J. Geophys. Res. **84**, 2076–2082 (1979)
J.A. Slavin, R.E. Holzer, J. Geophys. Res. **86**, 11401–11418 (1981)
J.A. Slavin, B.T. Tsurutani, E.J. Smith, D.E. Jones, D.G. Sibeck, Geophys. Res. Lett. **10**(10), 973–976 (1983)
J.A. Slavin, E.J. Smith, D.G. Sibeck, D.N. Baker, R.D. Zwickl, S.-I. Akasofu, J. Geophys. Res. **90**, 10875–10895 (1985)
W.H. Smyth, M.L. Marconi, Astrophys. J. **441**, 839–864 (1995)
T. Spohn, F. Sohl, K. Wieczerkowski, V. Conzelmann, Planet. Space Sci. **49**, 1561–1570 (2001)
A.L. Sprague, R.W.H. Kozlowski, D.M. Hunten, W.K. Wells, F.A. Gross, Icarus **96**, 277–342 (1992)
J.R. Spreiter, A.L. Summers, A.Y. Alksne, Planet. Space Sci. **14**, 223–253 (1966)
S. Taguchi, R.A. Hoffman, J. Geophys. Res. **100**, 19313–19320 (1995)
N. Terada, S. Machida, H. Shinagawa, J. Geophys. Res. **107**, 1471 (2002). doi:10.1029/2001JA009224
T. Terasawa et al., Geophys. Res. Lett. **24**, 935–938 (1997)
N.A. Tsyganenko, in *Proc. 3rd Conf. Substorms (ICS3)*, ESA SP-389 (1996), pp. 181–185
Y.C. Whang, J. Geophys. Res. **82**, 1024–1030 (1977)
P. Wurz, H. Lammer, Icarus **164**, 1–13 (2003)
L.M. Zelenyi, H.V. Malova, V.Yu. Popov, D. Delcourt, A.S. Sharma, Nonlinear Process. Geophys. **11**, 1–9 (2004)
L.M. Zelenyi et al., Geophys. Res. Lett. (2007, submitted)

Magnetosphere–Exosphere–Surface Coupling at Mercury

S. Orsini · L.G. Blomberg · D. Delcourt · R. Grard · S. Massetti · K. Seki · J. Slavin

Originally published in the journal Space Science Reviews, Volume 132, Nos 2–4.
DOI: 10.1007/s11214-007-9222-2 © Springer Science+Business Media B.V. 2007

Abstract Mercury's environment is a complex system, resulting from the interaction between the solar wind, magnetosphere, exosphere and surface. A comprehensive description of its characteristics requires a detailed study of these four elements. This paper illustrates and discusses the key processes that are implicated in the strong coupling of the Hermean magnetosphere with the other elements. The magnetosphere of Mercury, frequently called "mini-magnetosphere", when compared to that of Earth, plays a significant role in controlling the planet source and loss processes, by means of both particle and field interactions. We review the status of our knowledge, and give possible interpretations of the still-limited data set presently available.

Keywords Planets and satellites: Mercury · Planetary environments · Planetary magnetospheres plasmas · Solarplanetary relatioships · Interplanetary medium

---

S. Orsini (✉) · S. Massetti
INAF-IFSI, via Fosso del Cavaliere 100, 00133 Roma, Italy
e-mail: stefano.orsini@ifsi-roma.inaf.it

L.G. Blomberg
Space and Plasma Physics, School of Electrical Engineering, Royal Institute of Technology, 10044 Stockholm, Sweden

D. Delcourt
CETP-CNRS-IPSL, 4 avenue de Neptune, Saint-Maur des Fosses, France

R. Grard
Research and Scientific Support Dept., ESA/ESTEC, Noordwijk, The Netherlands

K. Seki
Solar-Terrestrial Environment Lab., Nagoya University, Furocho, Chikusa-ku, Nagoya, Aichi 464-8601, Japan

J. Slavin
NASA/GSFC, Code 696, Greenbelt, MD 20771, USA

1 Introduction

Mariner 10 observations revealed that Mercury has a weak dipolar magnetic field, 5.0×10^{12} T/m^3 = 344 nT/R_M^3 (Ness et al. 1976) and possesses an Earth-like magnetosphere, though much smaller relative to the size of the planet. The interplanetary conditions at Mercury's orbit (0.29–0.44 AU) differ from those observed at 1 AU: the IMF is 3–6 times more intense, 15–30 nT (Burlaga 2001), and the nominal IMF Parker spiral angle is only 20°, less than half its value at 1 AU. The radial component of the interplanetary magnetic field (IMF B_X) is therefore expected to significantly contribute, or even dominate, the reconnection topology on the dayside magnetosphere.

Recent models of Mercury's magnetosphere (see, e.g., Kabin et al. 2000; Sarantos et al. 2001; Kallio and Janhunen 2003; Massetti et al. 2003, 2007; Scuffham and Balogh 2006) show that the planet should possess broad cusp regions, with a large fraction of the dayside surface connected to open field lines, and exposed to plasma precipitation form the magnetosheath. The magnetic merging associated with the IMF B_X component causes a north–south asymmetry, with the northern (southern) hemisphere directly connected with the upstream interplanetary field when B_X is positive (negative). The Hermean planetary surface, exosphere and magnetosphere, are highly affected by the solar wind and interstellar medium, and constitute a complex, strongly coupled system, dominated mostly by interactions of the neutral and ionized gas particles with the surface and with the magnetospheric plasma (see Milillo et al. 2005). The Hermean exosphere is continuously refilled and eroded, and the Hermean environment can be considered as a single magnetosphere–exosphere–surface system, since these three parts are strongly linked through several source and loss processes (see Fig. 1). In the following, we report about these processes and their impact in terms of balance and features. The source and dynamics of the magnetospheric particles are reviewed in Sect. 1. The exosphere–magnetosphere coupling is discussed in Sect. 2. Sections 3 and 4 are devoted to coupling effects in terms of electrodynamics and wave–particle interactions. The major issues are recapitulated in the last section.

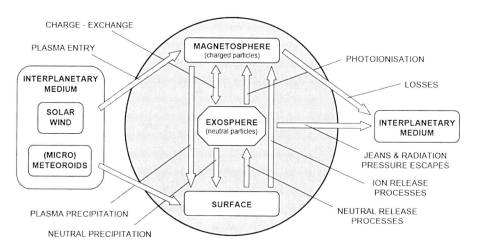

Fig. 1 The Hermean magnetosphere–exosphere–surface system (from Milillo et al. 2005)

2 Sources and Dynamics of the Magnetospheric Plasma

The magnetosphere of Mercury is strongly coupled to the interplanetary magnetic field with, on the dayside, large cusp regions directly exposed to the solar wind plasma penetration. Due to the effect of the dominating IMF B_X component, a strong north-south asymmetry is foreseen, with the northern/southern hemisphere directly interconnected to the IMF, for negative/positive IMF B_X. Even if the solar wind–magnetosphere interaction is very complex, the key features of the solar wind entry in the open dayside magnetosphere of Mercury can be analyzed with relatively simple models. Sarantos et al. (2001), and more recently Massetti et al. (2007), modeled Mercury's magnetosphere using an ad hoc adaptation of the semi-analytical TH93 magnetospheric model (Toffoletto and Hill 1989, 1993), which includes the effect of the IMF B_X in the magnetic merging topology. Sarantos et al. (2001) showed how wide portions of the dayside surface are directly exposed to the plasma precipitation, and also pointed out the role played by the IMF radial component in driving a net hemispheric asymmetry. Massetti et al. (2007) improved this analysis by constraining the model with the Mariner 10 data (Ness et al. 1976; Ogilvie et al. 1977), and the magnetic field geometry outlined by the Kallio and Janhunen (2003) hybrid model. In order to simulate the plasma entry at the dayside magnetopause, Massetti et al. (2007) included a first-order approximation of the magnetosheath plasma density, velocity and temperature along the magnetopause boundary, derived from the work of Spreiter et al. (1966), as a function of both the distance from the subsolar point of the magnetopause, and the free-stream Mach number. Using these models, we shall attempt to depict the flow pattern of the charged populations in the Hermean environment.

3 Populations of Solar Wind-Origin and Dayside Circulation

The cusps are the regions where most of the solar wind plasma entry takes place. Here the magnetosheath ions can cross the magnetopause discontinuity along the reconnected (open) field lines, and hence precipitate toward the planet. After their injection, the ions spread inside the magnetosphere, giving rise to a series of complex phenomena, such as proton-sputtering with the planetary surface (of the order of 10%) and charge-exchange with the neutral exosphere (about 1%), which strongly affect the environment of Mercury (e.g., Mura et al. 2005).

The reconnection process between the IMF and the magnetospheric field implies a transfer of energy from the field to the plasma, which flows across the magnetopause at the local Alfvénic speed. A fraction of the magnetosheath plasma population incident on the magnetopause is energized across the discontinuity. The key features of this process can be described on the basis of simple kinetic considerations, following the approach of Cowley and Owen (Cowley and Owen 1989; Cowley 1995). In order to approximate the magnetopause to a one-dimensional discontinuity (current layer), Massetti et al. (2003) used the *de Hoffmann–Teller* (HT) reference frame located at the rotational discontinuity produced by a newly reconnected field line moving along the magnetopause, away from the merging site. The fundamental points are that, in this new frame, the reconnected field line is at rest, the electric field vanishes, and the particles flow along the field lines with: (i) zero $\mathbf{E} \times \mathbf{B}$ drift, and (ii) no energy changes. With respect to the planet, the HT frame moves tailward along the magnetopause with a velocity (V_{HT}) that depends on both the flow speed in the magnetosheath and the magnetic tension of the kinked field line. The field tension generally dominates in the subsolar region where the magnetosheath flow is

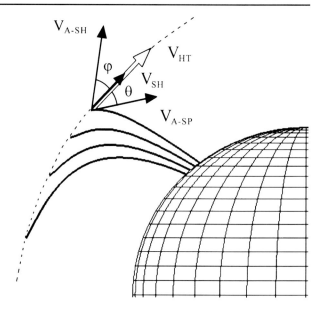

Fig. 2 Sketch of the reconnection geometry on the dayside magnetopause and related kinetic quantities derived by applying the stress balance condition to the HT frame

small (sub-Alfvénic regime), while the magnetosheath flow becomes dominant on the magnetospheric flanks, where its speed reaches super-Alfvénic values. Under the assumption that particles are not pitch-angle scattered while crossing the magnetopause, the bulk flow on either side of the discontinuity is field-aligned and moves at the local Alfvénic speed, the tangential stress balance condition implies that the change in the momentum of the plasma must balance the magnetic field tension, and that the field-aligned speed of the plasma in the HT frame must equal the local Alfvénic speed (Cowley and Owen 1989; Cowley 1995).

Figure 2 depicts the geometry of a reconnected field line on the XZ_{GSM} plane (dayside, i.e., $X_{GSM} > 0$), for the simple case of a southward pointing IMF ($B_Z < 0$): $V_{A\text{-}SH}$ and $V_{A\text{-}SP}$ are the Alfvén speed on the magnetosheath and on the magnetosphere sides, which are tangent to the local magnetic field; φ and θ are the angles formed by $V_{A\text{-}SH}$ and $V_{A\text{-}SP}$ with the magnetopause (dashed grey line); V_{SH} is the flow velocity in the magnetosheath (black arrow tangent to magnetopause); and V_{HT} is the HT reference frame velocity (white arrow).

$$V_{HT} = V_{SH} + V_{A\text{-}SH} \cos \varphi, \tag{1}$$

$$V_{min} = V_{HT} \cos \theta, \tag{2}$$

$$V_{peak} = V_{HT} \cos \theta + V_{A\text{-}SP}, \tag{3}$$

$$V_{max} = V_{HT} \cos \theta + V_{A\text{-}SP} + V_{th}. \tag{4}$$

Equations (1–4) summarize the stress-balance condition and the related kinetic quantities (e.g., Cowley and Owen 1989; Lockwood and Smith 1994; Cowley 1995; Lockwood 1995, 1997; Lockwood et al. 1996; Onsager et al. 1995). The minimum, peak and maximum field-aligned velocities (V_{min}, V_{peak} and V_{max}) are linked to the HT frame speed by mean of (2–4), where V_{th} is the plasma thermal speed. The ion energies E_{min}, E_{peak} and E_{max} can then be calculated from the corresponding velocities. A slightly more complex analysis, still based on the HT frame, includes the reconnection effect on the magnetosheath plasma population

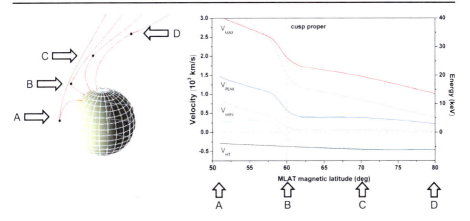

Fig. 3 Velocity distribution of proton injected across the dayside magnetopause, as a function of the latitude of the field line footprints (*right panel*). *Four reference points* are marked along the noon magnetic meridian (A–D, *left panel*) (from Massetti et al. 2007)

that is reflected back from the magnetopause, and on the magnetospheric plasma population that is transmitted across the magnetopause, into the magnetosheath (Lockwood 1997).

By applying the HT frame concept (1–4) to the Mercury's magnetospheric model described in the previous section, Massetti et al. (2007) analyzed the properties of the ion injection on the dayside. Figure 3 illustrates the results obtained for the simple case of a southward directed IMF, with the reconnected field line moving antisunward from A to D along the magnetic noon (left panel). The right panel shows the V_{min}, V_{peak} and V_{max} proton velocities (left axis), and the related energies (right axis), as functions of the magnetic latitude of the field line footprint. It must be emphasized that only the protons moving along the field line with positive field-aligned velocities can precipitate toward the planet. The protons are first accelerated as the newly reconnected field line moves from A to B; then, after crossing the cusp proper (that is: $\theta \sim 90°$), they decelerate progressively (B to D) as the field line moves antisunward ($\varphi + \theta > 90°$). This aspect can be better understood by means of the Poynting theorem, which describes the energy exchange between the electromagnetic field and the plasma. In the steady state, it is written in the simple form (e.g., Cowley 1991): $\text{div}\,\mathbf{S} + \mathbf{j}\cdot\mathbf{E} = 0$ where $\mathbf{S} = \mathbf{E}\times\mathbf{B}/\mu_0$ is the Poynting vector (the energy flux of the field), and $\mathbf{j}\cdot\mathbf{E}$ is the energy exchanged between the field and the plasma ($\mathbf{j}$ is the current density). When the second term is positive, the energy flows from the field to the plasma ($\text{div}\,\mathbf{S} < 0$), when it is negative, the plasma energy is transferred to the field ($\text{div}\,\mathbf{S} > 0$). In the subsolar region, equatorward of the cusp proper ($\theta \sim 90°$), the electric field and the dayside magnetopause current are both oriented in the dawn–dusk direction, $\mathbf{j}\cdot\mathbf{E} > 0$, and the field supplies energy to the plasma: the ions are accelerated by the magnetic tension of the (kinked) newly reconnected field lines. Poleward of the cusp, on the northern lobe, the tail current goes from dusk to dawn, so that $\mathbf{j}\cdot\mathbf{E} < 0$, and the plasma energy is supplied to the field: the magnetosheath flow stretches the magnetic field lines in the antisunward direction. Figure 4 shows the results obtained by Massetti et al. (2007): the quantities V_{min}, V_{peak} and V_{max} are mapped over the "open" dayside magnetopause of Mercury, by assuming IMF = [30, 10, −10] nT (i.e., nominal Parker's spiral direction, plus a negative B_Z component), $V_{SW} = 400$ km/s, and $D_{SW} = 60$ cm^{-3}. These panels show that the protons with higher velocities (i.e., energies) are injected at lower latitudes, in a belt that could be grossly identified with the cusp/LLBL (low latitude boundary layer) region in the Earth's magnetosphere. There is a relative minimum at the subsolar point, because of the minimum of the magnetosheath plasma speed

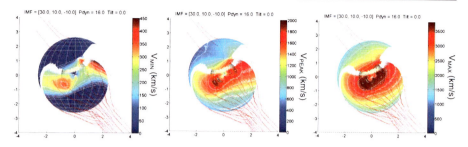

Fig. 4 Minimum, peak, and maximum, field-aligned velocities (V_{min}, V_{peak} and V_{max}) mapped on the open dayside magnetopause of Mercury. Exact values of V_{peak} and V_{max} depend upon the actual V_{A-SH} and V_{A-SP} Alfvénic speeds (from Massetti et al. 2007)

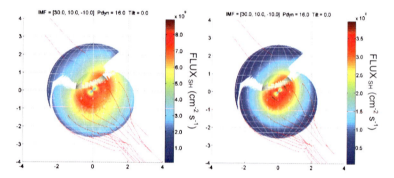

Fig. 5 Proton flux on the magnetosheath (*left panel*, $FLUX_{SH} = V_{A-SH} \cdot D_{SH}$), and magnetospheric (*right panel*, $FLUX_{SH} = t \cdot FLUX_{SP}$) sides of Mercury's magnetopause. The transmission factor $t = 1/2$ $(V_{T-SH}/V_{T-SH0})(1 - (V_{SH}/V_{SW0})^2)$ is a function of both the magnetosheath plasma temperature, normalized to its value at the subsolar point (V_{T-SH}/V_{T-SH0}), and the magnetosheath plasma velocity, normalized to the upstream solar wind speed (from Massetti et al. 2007)

($V_{SH} \sim 0$), and this affect the V_{HT} speed (see (1)). A significant north–south asymmetry is clearly visible; because of the strongly positive IMF B_X component, a wide portion of the southern hemisphere is connected to open field lines, directly to the upstream solar wind.

Massetti et al. (2007) estimated the fraction of the magnetosheath plasma that is expected to actually cross the discontinuity at the magnetopause, and hence enter the magnetosphere. They adopted a realistic transmission factor (t), which gradually decreases tailward, as the magnetosheath proton bulk velocity increases and the temperature decreases (e.g., Fedorov et al. 2000): the progressive antisunward "focusing" of the magnetosheath flow heavily hinders the plasma precipitation toward the planet. Figure 5 illustrates the proton flux on both the magnetosheath (left panel) and the magnetospheric (right panel) sides of the magnetopause, calculated with the model inputs used in Fig. 4: the proton flux crossing the magnetopause (right panel) is of the order of $1–4 \times 10^9$ cm^{-2} s^{-1}.

4 Non-adiabatic Effects and Particle Precipitation

Due to the relatively weak magnetic field of Mercury, the plasma precipitation throughout the open magnetospheric regions (cusps), and its subsequent circulation into the magnetosphere, is expected to strongly deviate from the adiabatic behavior

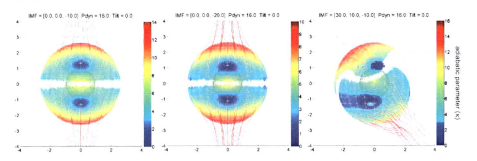

Fig. 6 The κ adiabatic parameter mapped over the dayside magnetopause of Mercury, for three different IMF configurations. The *blue areas* ($\kappa < 3$) are the regions affected by the non-adiabatic (from Massetti et al. 2007)

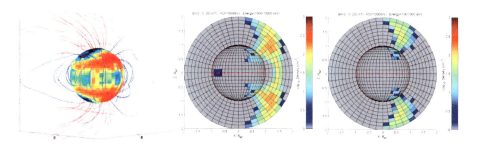

Fig. 7 Proton precipitation patterns on the dayside of Mercury, as modeled by Kallio and Janhunen (2003) (*left panel*), and Mura et al. (2005) (*right panels*), under different interplanetary conditions

(e.g., Delcourt et al. 2003). One way of estimating such non-adiabatic effects makes use of the κ parameter, which is defined as the square root of the minimum field line curvature to the maximum *Larmor* radius ratio (Büchner and Zelenyi 1989). For $\kappa > 3$ the particle motion is adiabatic (i.e., the magnetic moment is conserved), and the guide–centre approximation is applicable, while for $\kappa < 3$ the particles behave non-adiabatically. In the latter case, finite-gyroradius effects are expected to drive particle precipitation along closed field lines, where they bounce, drift (grad $\mathbf{B}$), and populate wide low latitude belts. For reference, Fig. 6 visualizes the κ adiabatic parameter, calculated for the peak value of the velocity distribution (4) and mapped over the dayside magnetopause of Mercury, for three different IMF configurations. Due to the reduced intensity of the magnetic field, the cusps (larger ion gyroradii) are the regions (blue-coded) where the non-adiabatic behaviour is more likely to take place. An increase of the magnetic merging causes a general widening of the non-adiabatic areas. Figure 7 shows the proton precipitation patterns on the dayside of Mercury, as calculated by Kallio and Janhunen (2003), and Mura et al. (2005). Their simulations show that, due to finite-gyroradius effects, a significant fraction of the proton precipitation across the magnetospheric cusps is actually scattered into closed field line regions. This scattered proton population tends to form two mid-latitude belts, whose extent depends upon the solar wind and IMF conditions, and upon the electric field intensity.

5 Heavy Ions of Planetary Origin: Sources and Nightside Circulation

5.1 Sources

Ions of planetary origin have not yet been observed in the Hermean environment. Nevertheless, it is expected that planetary sources of ions exist in the Hermean environment. First, photo-ionisation of exospheric atoms on the dayside should produce an ionised thermal population. For species, such as Li, Na, Mg and K, the photo-ionisation lifetimes are sufficiently small and there is a reasonable probability that they are ionised during their transit in the exosphere (Milillo et al. 2005). Lundin et al. (1997) derived the ion density from neutral density as being equal to $n_i = n_n T_c / \tau_{ph} (1/\beta)$, where n_n is the neutral density of the considered species, T_c its gyroperiod, β the escape fraction (maximum ion density is obtained for $\beta = 1$), and τ_{ph} the photo-ionisation life-time. They estimated densities of 10^{-3} cm^{-3} for He$^+$, 0.1 cm^{-3} for O$^+$, 10 cm^{-3} for Na$^+$ and 0.5 cm^{-3} for K$^+$ on the surface at the sub-solar point.

A second source of planetary ions should be generated by particle sputtering, since the impact of energetic ions of solar wind or magnetospheric origin releases not only neutral particles (e.g., Mura et al. 2005), but also ions from the planetary surface. The fraction of released ions is about 1% of that of the released neutrals (Sieveka and Johnson 1984). Sputtered ions have characteristic release energies of a few eV, similar to those of sputtered neutral atoms.

The ion populations, generated by the two above-mentioned mechanisms, can hardly constitute an ionosphere able to shield the magnetic and electric fields of the magnetosphere; hence, we may define it as exo-ionosphere because they are immediately energised and mixed with the magnetospheric ion populations.

Recently, Leblanc et al. (2004) considered many uncertainties that possibly could affect the ion density estimate, like the surface-binding energy (important for ion sputtering), the regolith composition, the porosity of the surface material and the mutual influence of surface-release processes. Moreover, Mercury's exosphere is far from being spherically homogeneous, being affected by both day/night and perihelion/aphelion conditions (Wurz and Lammer 2003; Leblanc and Johnson 2003). These features imply a wide range of variation for both neutral and ion densities. Leblanc et al. (2003) showed that the Na$^+$ density could spatially vary between 10^{-2} and 10^1 Na$^+$/cm^3 close to the planetary surface, whereas the Na neutral distribution is more uniform around Mercury at an altitude of about 400 km: at this altitude, the ion densities (e.g., H$^+$, He$^+$, O$^+$, Na$^+$, Si$^+$, S$^+$ and Mg$^+$) are expected to lie in the range 10^{-4}–10^2 (Leblanc et al. 2004).

5.2 Circulation

It is well established that ions from the topside ionosphere form a significant source of plasma for the Earth's magnetosphere (e.g., Chappell et al. 1987; Seki et al. 2001, 2003; Bouhram et al. 2005), leading to the "geopause" put forward by Moore and Delcourt (1995) which differentiates regions preferentially fed by the solar wind from those preferentially filled by ionospheric material (e.g., Winglee 1998). The question then arises whether ions of exospheric origin at Mercury may play a similar role for its miniature magnetosphere. In an attempt to estimate this contribution, Delcourt et al. (2003) showed that low-energy Na$^+$ ions sputtered from the planet surface may gain access to the magnetotail via convection over the polar cap, like the low-energy ions originating from the cleft region at Earth. This is illustrated in the left panels of Fig. 8 that show the trajectories of test Na$^+$

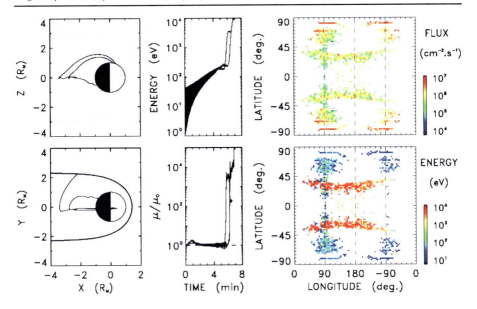

Fig. 8 (*left*) Model Na⁺ trajectories projected in the noon–midnight plane and in the equatorial plane. (*centre*) Corresponding energy and magnetic moment versus time. (*right*) Colour-coded flux and average energy of precipitating Na⁺ ions as a function of longitude and latitude (from Delcourt and Seki 2006)

launched from two distinct latitudes in the dayside sector (left panels), together with their energy and magnetic moment as a function of time (centre panels). It can be seen that, during their transport in the magnetospheric lobes, the ions are subjected to a prominent parallel energization, up to hundreds of eV. This is due to an enhanced $\mathbf{E} \times \mathbf{B}$ related centrifugal effect that is more pronounced than at Earth because of the larger angular speed of the convecting field lines at Mercury. Once they interact with the magnetotail current sheet, Na⁺ ions originating from the exosphere are accelerated in a non-adiabatic manner. They subsequently remain trapped in the equatorial vicinity or precipitate onto the planet surface, hence contributing to further regolith sputtering (e.g., Leblanc et al. 2003). This precipitation can be appreciated in the right panels of Fig. 8, which show the results of systematic Na⁺ trajectory computations, with colour-coded flux and energy of precipitating ions as a function of longitude and latitude. A model exosphere was considered in these calculations, assuming a variety of production processes such as photo-stimulated desorption, micro-meteoritic vaporization or sputtering due to the impinging solar wind (Leblanc and Johnson 2003). It is apparent from the right panels of Fig. 8 that Na⁺ precipitation occurs within two limited bands that extend over 10°–20° in latitude and a wide range of longitude. The average ion energy (in the keV range) within these bands increases from post-midnight to pre-midnight sectors, as expected from westward drift and associated acceleration by the large-scale convection electric field. These energetic ion bands contrast with the low-energy precipitation achieved at high (above ∼50°) latitudes which correspond to downflowing exospheric populations that do not interact with the magnetotail current sheet.

An intriguing feature in the right panels of Fig. 8 is the abrupt cut-off of the Na⁺ precipitation at latitudes of 35°–40°. In the Earth's auroral zone, this poleward boundary delineates the transition between polar cap and plasma sheet, that is, the transition between the open flux region and closed magnetic field lines that extend up to the reconnection site in the far tail. At Mercury, it can be seen in the bottom left panel of Fig. 8 that this poleward boundary does not map into the far tail but at distances of a few planetary radii. This results from the fact that ions intercept the dusk magnetopause in the course of their interaction with the current sheet. That is, in contrast to the situation prevailing at Earth, the poleward boundary of precipitation directly follows from the finite Larmor radius of plasma sheet ions. At large energies and/or large mass-to-charge ratios these Larmor radii become comparable to—or larger than—the characteristic width of the magnetotail, in which case ions do not execute a full Speiser orbit and are not reflected toward the planet. Note here that the width of the Hermean magnetotail is expected to be of the order of a few R_M, while 20 keV protons and 1 keV Na$^+$ ions have Larmor radii of the order of $1 R_M$ at a radial distance of $3 R_M$.

In terms of contribution to the magnetospheric populations, Delcourt et al. (2003) showed that exospheric Na$^+$ may form a quite substantial source of plasma for the magnetotail, with density levels up to several tens of ions/cm^3 (see, e.g., Fig. 5 of that paper). Still, the results in Delcourt et al. (2003) were obtained using a modified version of the Luhmann and Friesen (1979) model, together with a simple two-cell pattern of magnetospheric convection. Using MHD simulations, Seki et al. (2006) put forward that a far more complex plasma circulation may develop within the Hermean magnetotail, with the development of a Near-Mercury Neutral Line in the vicinity (2–3R_M) of the planet during active times. The contribution of planetary ions is in this case significantly altered, from about 15% of the total outflow from the exosphere (see, e.g., Leblanc et al. 2003) down to negligible recycling. This can be better appreciated in Fig. 9, which shows results of systematic trajectory tracings of Na$^+$ ions in the electric and magnetic fields obtained from a MHD simulation of the Mercury–solar wind interaction. The solar wind conditions (proton density: 35 cm^{-3}; flow speed: 400 km/s; and southward IMF field: 30 nT) similar to that used to construct a rescaled analytical model in Delcourt et al. (2003) was chosen in the MHD simulation. While the rescaled model assumes the existence of the distant neutral line (DNL) in Mercury's magnetotail, the MHD result shows the formation of the near-Mercury neutral line (NMNL) at

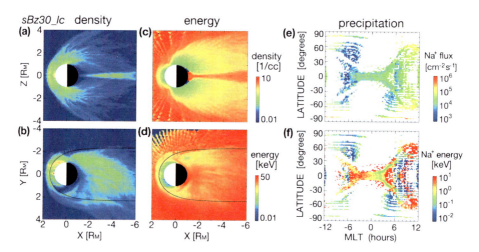

Fig. 9 **a, b** Density and **c, d** energy distributions of Na$^+$ ions in Mercury's magnetosphere obtained from systematic trajectory tracings in the MHD electric and magnetic field configuration using an exospheric Na model (Leblanc and Johnson 2003) as input. Panels **a** and **c** show cross sections in the noon–midnight meridian plane, while **b** and **d** in the equatorial plane. The density and energy is colour-coded according to the colour scales shown in the centre. **e** Flux and **f** average energy of precipitating Na$^+$ ions onto Mercury's surface are colour-coded (from Seki et al. 2006)

around $1.6R_M$ downtail, which is much closer to the planet than the rescaled location of DNL. In this case, a very low conductivity at the planetary surface was assumed as an inner boundary condition, which enables a fast magnetospheric convection without effects of the ionosphere, such as ion drag. Thus, the resultant Na^+ density (energy) in the magnetosphere is a little lower (higher) than that obtained in the rescaled analytical model, as shown in Figs. 9a, b (Figs. 8c, d).

The existence of the NMNL alters the trajectories of Na^+ ions and changes the fraction of ions that can precipitate back onto the planet surface. The resultant precipitation pattern of the Na^+ ions in the MHD fields (Figs. 8e and f) changes dramatically from that in an analytical model (right panels in Fig. 8). The ions reaching the region of the tailward flow in the plasma sheet do not return to the planet regardless of whether their motion is adiabatic or not. This causes the disappearance of the nightside precipitation bands of high-energy Na^+ ions at around $\pm 30°$ in latitude shown in Fig. 8, while a weak precipitation around the equator is seen in the MHD case, due to the strong sunward flow near the planet (Fig. 9). The other MHD simulation runs also show that the global convection pattern and location of the reconnection line in the magnetotail depend on the conductivity at the surface of the planet as well as on the strength of IMF B_Z. In Earth's magnetosphere, the near-Earth neutral line is often formed in association with substorm events. These results suggest that precipitation and recycling of planetary ions at Mercury critically depend upon the magnetic activity and the global convection pattern.

Because of the small spatial scales of Mercury's environment, ions have Larmor radii comparable to—or larger than—the field variation length scale and thus mostly behave in a non-adiabatic manner throughout the magnetosphere. This leads to magnetic moment scattering and prolonged trapping in the equatorial region or resonant (Speiser-like) interaction with the magnetotail field reversal, as illustrated in Fig. 8. As demonstrated by Delcourt et al. (2005), non-adiabatic transport may also affect energetic electrons with significant pitch angle diffusion inside the loss cone and subsequent precipitation, when we consider time-dependent magnetic and electric fields during the dipolarization.

Mariner 10 measurements during the 1974 pass displayed a rapid reconfiguration of the magnetotail with an abrupt increase of the Z component of the magnetic field (dipolarization) in conjunction with energetic particle injections (e.g., Slavin et al. 1997). In the Earth's magnetosphere, such injections in conjunction with magnetic field line dipolarization are well-established signatures of substorm expansion phase (e.g., Sauvaud and Winckler 1980; Moore et al. 1981; Lopez et al. 1989). Based on this interpretation, the Mariner 10 observations have thus been considered as evidence of substorm activity at Mercury (Slavin and Holzer 1979). The question then arises whether the electric field induced during such events significantly affects the particle dynamics in the Hermean magnetotail. In this regard, numerous observations in Earth's magnetosphere provide evidence of a prominent O^+ energization during substorm dipolarization. This energization appears to be mass selective since protons do not exhibit such large energy changes, as demonstrated by Mitchell et al. (2003). It is suspected that this preferential heating results from a resonance between the slowly gyrating O^+ and the surging electric field, a non-adiabatic process that contrasts with the rather adiabatic behavior of protons (e.g., Delcourt 2002). To investigate how this process translates to the miniature magnetosphere of Mercury, Delcourt et al. (2007) performed numerical simulations in a model dipolarization of Mercury's magnetotail. It appears from these simulations that the electric field induced by the magnetic field line relaxation does lead to a convection surge and rapid inward injections of plasma sheet populations (e.g., Mauk 1986). However, because of small temporal scales of field variations as compared to those at Earth, non-adiabatic "resonant" heating during substorms is here obtained for ions

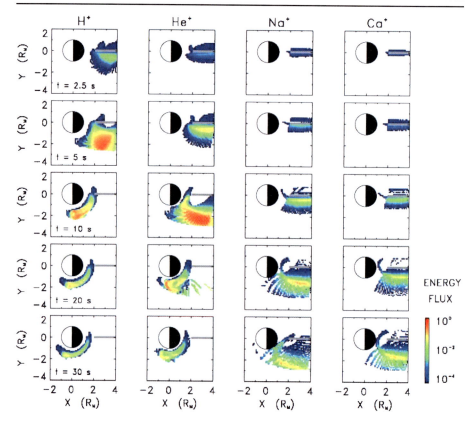

Fig. 10 (from *top* to *bottom*) Colour-coded energy flux (normalized to the maximum value) of equatorially trapped ions at distinct times of a model dipolarization. Distinct ion species are considered: (from *left* to *right*) H^+, He^+, Na^+, Ca^+. The initial ion distribution, launched from a line source extending from 2 to $4R_M$ along the tail axis, is assumed to be maxwellian with a temperature of 1 keV. The dipolarization time scale is set to 10 seconds

with small mass-to-charge ratios. This can be better appreciated in Fig. 10 that shows the results of systematic trajectory computations in a model dipolarization.

It is apparent from the left panels of Fig. 10 that light ions are rapidly energized in the course of the magnetic field line relaxation. Because of this energization and the large Larmor radii achieved, high cyclotron frequency H^+ are rapidly lost into the dusk magnetopause. These westward bursts of energetic protons are immediately followed by energetic He^+. Ultimately, only the innermost population that experiences moderate non-adiabatic heating during dipolarization is found to drift into the dayside sector. Also, the comparison of left and right panels in Fig. 10 clearly displays the weaker energization realized by heavy ions (Na^+, Ca^+). Here, a moderate energy flux gradually develops as the heavy ions drift westward, and are subsequently lost into the dusk magnetopause. In Earth's magnetosphere, it is well established that heavy ions of ionospheric origin play an important role in the development of the ring current (e.g., Hamilton et al. 1988; Daglis 1997). In view of Fig. 10, one would expect protons to play a similar role in the dynamics of the inner magnetosphere of Mercury, be they of solar wind or exospheric origin.

6 Magnetosphere–Exosphere Coupling

In 1974, the UVE experiment (Broadfoot et al. 1976) onboard Mariner 10 revealed the existence of a tiny exosphere around Mercury, and identified the presence of H, He, and, probably, O. In addition, ground-based observations have revealed the presence of Na (Potter and Morgan 1985), K (Potter and Morgan 1986) and Ca (Bida et al. 2000). So far, several processes have been proposed as exospheric sources, such as ion sputtering, thermal and photon-stimulated desorption, and micro-meteoroids vaporization. In particular, it has been hypothesized that ion sputtering could be responsible for the extended, and asymmetric, sodium population observed by (Potter and Morgan 1997).

Ion sputtering results from solar wind or planetary ion impact onto the surface of Mercury; multiple scattering processes can occur below the surface and a neutral can be extracted, with a probability (yield) of 1–10% per incoming ion, depending on projectile mass and energy (Lammer et al. 2003).

The peak of the energy distribution function of the ejected particles is usually of the order of few eV; however, there is a long, high-energy tail so that approximately 50% (on average) of the emitted particles have more than the escape energy (e.g. Grard 1997).

On Mercury, ion sputtering is likely to be caused by solar wind protons precipitating into the dayside cusps (Massetti et al. 2003; Kallio and Janhunen 2003); the H^+ flux onto the surface of Mercury may reach values of 10^9 cm^{-2} s^{-1}; the total H^+ flux onto the surface of Mercury can be estimated to be about 10^{25} s^{-1} (Mura et al. 2005); during solar energetic particle events, the high-energy proton flux may reach up to 10^{26}–10^{27} s^{-1} (Leblanc et al. 2003). Other solar wind components, like α particles, exhibit a smaller flux but produce a higher yield, so that they are expected to contribute to this process as well (Johnson and Baragiola 1991). The flux of the exospheric ions, such as sodium, is much lower, and can reach 10^6–10^7 cm^{-2} s^{-1} (Delcourt et al. 2003; Leblanc et al. 2004); those ions can precipitate and generate sputtering also on the nightside. The maximum sputtering flux due to this process has been estimated by Delcourt et al. (2003) and is about 10^4 cm^{-2} s^{-1}.

The intensity and the shape of the SW flux at the surface depend on the magnetospheric configuration, and can be simulated with MHD or quasi-neutral hybrid MHD models (Leblanc et al. 2003; Kallio and Janhunen 2003), loss-cone estimation (Massetti et al. 2003) or single particle models (Mura et al. 2005). As discussed in Sect. 1, when the IMF has either a negative B_Z or a strong B_X component, magnetic reconnection occurs between IMF and the Hermean magnetic field; SW particles from the magnetosheath can cross the magnetopause, enter the magnetosphere and precipitate along open field lines (see Fig. 11). During this motion, protons do not exactly follow magnetic lines: particles are convected northward by the $\mathbf{E} \times \mathbf{B}$ drift and westward by the grad $\mathbf{B}$ drift (Mura et al. 2005); non-adiabatic effects may become of crucial importance for most of the ion magnetospheric transport (Delcourt et al. 2003).

However, the shape of proton flux onto the surface mimics, approximately, the shape of the reconnection at the MP. In a purely IMF-B_Z case, reconnection occurs only if B_Z is negative; non-zero B_X values introduce a north–south asymmetry in the H^+ surface precipitation and, hence, of the related exospheric density. This feature is very important because it can easily explain north–south asymmetries in the ground-based observation of some exospheric components (e.g., sodium); Fig. 12 shows a numerical simulation of the sputtered sodium exosphere in a $B_X > 0$ case.

Charge-exchange is another relevant process in the frame of magnetosphere–exosphere interaction. Plasma particles may collide with exospheric neutrals and be neutralized; newly

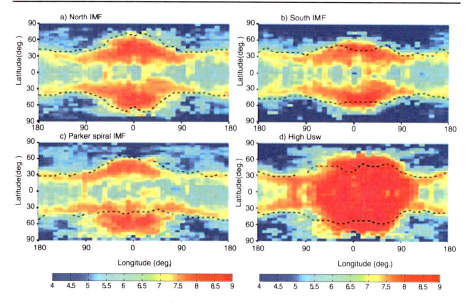

Fig. 11 Pseudo-colour maps of the H$^+$ flux onto the surface of Mercury under different external conditions. "Parker spiral" case (panel **c**) refers to IMF = (32, 10, 0) nT (from Kallio and Janhunnen 2003)

Fig. 12 Sodium density in a vertical (XZ) cut of the dayside exosphere [MTBW] (from Mura et al. 2006)

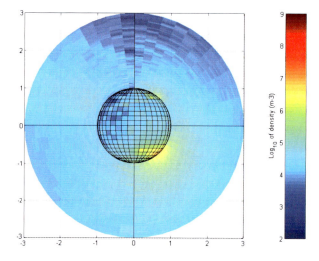

born neutrals (ENAs, Energetic Neutral Atoms) retain the energy and direction of the impacting ion, thus allowing a remote sensing of the process. The detection of ENA gives information about the plasma and exospheric distribution in the interaction region (Orsini et al. 2001; Barabash et al. 2001; Mura et al. 2005). The ENA flux at Mercury is mostly due to SW proton circulation and, hence, occurs on the dayside and in the dawn regions. Of the precipitating H$^+$, approximately 90% are bounced back due to the increasing magnetic field, about 10% reach the surface of Mercury and lead to ion-sputtering, and only 1% are neutralized due to charge–exchange, mostly on the dayside and in the dawn regions, and up to altitudes of hundreds of km (Mura et al. 2005, 2006). The charge–exchange process, hence, can be neglected as a plasma loss mechanism; however, the ENA production rate per

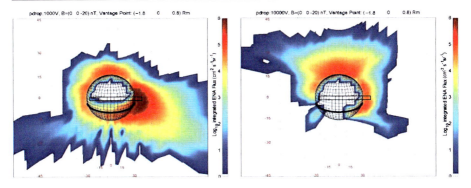

Fig. 13 Fish-eye projection of H-ENA simulated images, as seen from $(1.5, 0, 0.4) R_M$. The *two panels* refer to different magnetospheric conditions (see text for details). The colour is coded according to H-ENA flux, integrated over all energies (100 eV–10 keV) (from Mura et al. 2006)

unit length reaches values of up to 10 $(cm^{-2} s^{-1} sr^{-1} m^{-1})$, close to the dayside planetary surface. This value, if integrated over a long line-of-sight, could lead to H-ENA flux of up to $10^6 cm^{-1} s^{-1} sr^{-1}$. Such a flux is high enough to be remotely detected with proper instrumentation, such as the SERENA/ELENA sensor on board the ESA BepiColombo mission.

An example of simulated ENA flux at a realistic vantage point position is shown in Fig. 13. In principle, from this point it is possible to detect the H-ENA signal generated from two different H^+ populations. The first one consists of protons that are precipitating into the cusp regions or circulating over the North Pole. They produce an intense H-ENA signal in the upper part of the two pictures, which can be further enhanced by an intense electric field (due to the $\mathbf{E} \times \mathbf{B}$ drift). The second population is made of protons that are bouncing and drifting westward due to the grad $\mathbf{B}$ drift. These protons reach a low altitude at their mirroring points, where the exospheric densities are higher: even if the H^+ flux is generally lower there than on the dayside, they may produce an intense ENA signal that is visible in the right side of each image. The two panels of Fig. 13 refer to different magnetospheric configurations, when the electric field $\mathbf{E}$ is low (left) and high (right), and show how this affects the ENA signal that can be detected. In fact, the two above-mentioned drift effect ($\mathbf{E} \times \mathbf{B}$ and grad $\mathbf{B}$) and the related populations are, respectively, directly and inversely proportional to $\mathbf{E}$. Such simulations demonstrate that it is possible to obtain basic information about plasma circulation, as it happens at Earth (e.g. Roelof et al. 1985; Burch et al. 2001), and to study surface release processes (Milillo et al. 2005).

7 Magnetosphere–Exosphere–Regolith Electro-Dynamical Coupling at Mercury

7.1 Field-Aligned Currents

The stresses exerted on planetary magnetic fields by the solar wind and magnetospheric convection are transmitted down to the planet and its environs by Alfvén waves carrying field-aligned current. When the stresses are such that the current to be carried becomes continuous, these Alfvén waves add to produce "field-aligned currents" or "FACs." It is notable that all planetary magnetospheres visited thus far are field-aligned currents, including Mercury (Slavin et al. 1997; Kivelson 2005). At planets with electrically conductive ionospheres, such as the Earth, Jupiter, Saturn, Uranus, and Neptune, these current systems are the primary means for the transfer energy from the solar wind and magnetosphere to the ionosphere which, in turn, gives up energy and momentum to the surrounding neutral atmosphere via collisions. This coupling works in both directions and the ionosphere and neutral atmosphere operate in effect as a "brake" that limits the speed and rate of change of the plasma convection in the magnetosphere (Coroniti and Kennel 1973). Mercury's atmosphere, however, is very tenuous and collisionless, and the ionization of these neutrals by solar EUV cannot produce physically significant levels of electrical conductance (Lammer and Bauer 1997). Indeed, this lack of an ionosphere is often cited as the underlying reason for the sudden, very intense, but short-lived (i.e., ~1–2 min) substorm-like energetic particle events observed by Mariner 10 during its first traversal of Mercury's magnetic tail (Luhmann et al. 1998; Siscoe et al. 1975; Russell et al. 1988; Slavin 2004). When magnetospheric magnetic fields at the Earth reconnect with the interplanetary magnetic field and are pulled back into the tail, sheets of field-aligned current, termed "Region 1" currents, flow down into the high-latitude ionosphere on the dawn side of the polar cap and outward on the dusk side. These Region 1 currents are also expected to be present at Mercury, as schematically depicted in Fig. 14. However, their intensity and temporal evolution may be greatly modified depending upon the nature of current closure path and the electrical conductivity of the regolith or alternative closure path (Slavin et al. 1997; Blomberg 1997). When the magnetic flux tubes in the tail reconnect again and high-speed plasma jets are generated toward and away from the planet, another set of field-aligned currents are generated, termed the "substorm current wedge (SCW)" (McPherron et al. 1973; Hesse and Birn 1991; Shiokawa et al. 1998). The SCW field-aligned currents cause the magnetic field in the near-tail to become more dipolar and less tail-like or "dipolarize" (Nagai 1982; Baumjohann et al. 1999). These currents are also shown in Fig. 14.

The SCW currents connect the midnight region of the polar cap to the plasma sheet and transfer to the planet a significant fraction of the total energy being released in the tail (e.g., Fedder and Lyon 1987). Numerical simulations by Janhunen and Kallio (2004) and Ip and Kopp (2004) suggest that Region 1 and SCW field-aligned currents will have important consequences for Mercury's magnetosphere just as they do for that of Earth.

Figure 15 displays 1.2 s averages of the magnetic fields observed during the first Mariner 10 encounter in Mercury centred solar ecliptic coordinates. Vertical dashed lines mark the

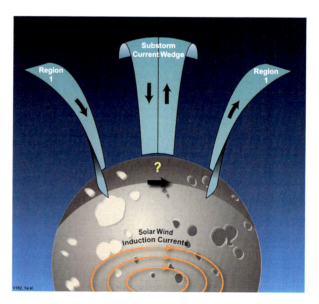

Fig. 14 Schematic image of possible Region 1 and substorm current wedge field-aligned currents at Mercury. Note: Planetary rotation and magnetic axes are approximately vertical in this depiction. Subsurface solar wind induction currents, flowing in the planetary interior, are shown at lower latitudes (from Slavin 2004)

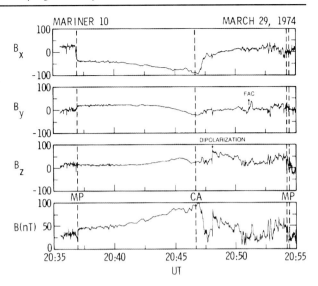

Fig. 15 Mariner 10 magnetic field observations (1.2 s averages) taken during the first encounter are displayed in Mercury centred solar ecliptic coordinates. *Vertical dashed lines* mark the inbound and outbound magnetopause crossings (MP) as well as the point of closest approach (CA). A dipolarization event (see B_Z panel) and field-aligned currents (FAC; see B_Y panel) are also identified (from Slavin and Holzer 1979)

inbound and outbound magnetopause crossings as well as the point of closest approach (CA) to the planet. The tail-like nature of the magnetic field during the inbound passage is very evident with B_X much greater than B_Y and B_Z. After closest approach the magnetic field variations become extremely dynamic. The large dip in field intensity just after closest approach coincided with the spacecraft becoming immersed in a hot, diamagnetic plasma sheet (Ogilvie et al. 1977; Christon 1987). Less than a minute after Mariner 10 entered the plasma sheet there is a sharp increase in the B_Z field component signifying a dipolarization event as shown in Fig. 14. This dipolarization event was nearly coincident with strong enhancements in the flux of >35 keV electrons observed by the cosmic ray telescopes (Simpson et al. 1974; Eraker and Simpson 1986; Christon 1987). This energetic particle signature, several weaker events observed later in the outbound pass, and the dipolarization signature were interpreted as evidence for substorm activity by a number of studies (Siscoe et al. 1975; Baker et al. 1986; Eraker and Simpson 1986; Christon 1987).

Two min after the dipolarization event, Mariner 10 was at a distance of $1.84 R_M$ from the centre of the planet at a local time of approximately 03:50. At this point several intense field-aligned current sheets are apparent in the magnetic field measurements (Fig. 15). As indicated, there is a strong bipolar variation in B_Y with a peak-to-peak amplitude of ~60 nT (Slavin and Holzer 1979). The main gradient in B_Y from negative to positive is indicative of an upward FAC at the spacecraft. This upward current sheet is largely balanced by two smaller downward current sheets before and after as is common above the Earth's auroral zones during substorms (Iijima and Potemra, 1978). These FACs could be associated with the Region 1 currents which flow into the poleward edge of the auroral zone in the dawn hemisphere. Alternatively, the Mariner 10 field-aligned currents could be associated with the east-most leg of the substorm current wedge (McPherron et al. 1973). This latter interpretation is of interest because of the substorm energetic particle injection observed a couple of minutes earlier.

A moderately conductive regolith is a likely candidate for FAC closure at Mercury. Hill et al. (1976) suggested a conductance value of 0.1 S, which is not unreasonable based upon the lunar measurements. If this value were indeed appropriate, however, the high rate of joule heating in the regolith would severely limit the duration of the current flow as the available magnetospheric energy would be quickly dissipated. Cheng et al. (1987) showed that sputtering is a possible means to generate the neutral sodium atmosphere of Mercury and a source population for magnetospheric ions. Cheng et al. (1987) pointed out that the new ions created by photo-ionization and impact ionization, and charge exchange are available to be "picked-up" by the convective motion of the magnetospheric flux tubes. In doing so, they would give rise to a "pick-up" or "mass loading" conductance (see Kivelson 2005) that might contribute to the generation and/or closure of FACs at Mercury at the 0.1–0.3 S level. Photoelectrons have also been suggested as a source of current carriers (Grard 1997; Grard et al. 1999; Grard and Balogh 2001). However, the pick-up of planetary ions and the photoelectron sheath over Mercury's sunlit surface provide conductances that are only slightly greater than the lunar values, that is, 0.1 S. The relative importance to the conductivity of a photoelectron sheath should be straightforward to determine by comparing measurement from different local time sectors from orbiting spacecraft (Grard et al. 1999). Interestingly, Glassmeier (2000) argued that field-aligned currents at Mercury may close as diamagnetic currents in regions of enhanced plasma density near the planet; perhaps associated with the sunward convection of magnetospheric plasma that might "pile-up" as stronger magnetic fields are encountered at low altitudes over the nightside of the planet. Finally, Janhunen and Kallio (2004) considered the range of possible surface materials and mineralogy and concluded that a wide range of effective height-integrated conductances are plausible.

7.2 Wave–Surface Interaction

As an example of wave–(exosphere)–surface interaction, we will consider standing waves along magnetic field lines that are reflected at or near the planetary surface. In the terrestrial magnetosphere (damped) standing Alfvén waves are quite frequently generated by instabilities that develop at the flanks of the magnetosphere by the streaming solar wind, which in turn generate broadband waves that propagate inward across the magnetic field. On field lines whose eigenfrequencies lie within the frequency range of the broadband waves, they may couple into a field-aligned wave mode. A similar effect is conceivable at Mercury and the Mariner 10 data indicates that it does indeed take place (Russell 1989); see Fig. 16. This

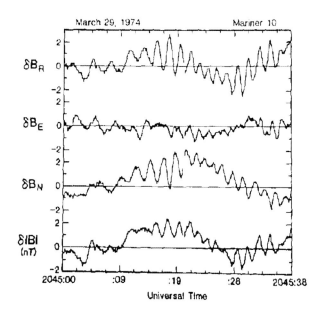

Fig. 16 Mariner 10 data showing a standing wave in Mercury's magnetosphere. This wave was interpreted as the fourth harmonic of a field–line resonance. After Russell (1989)

class of waves is often referred to as field line resonances, since it involves a large-scale fluctuating motion along the entire length of a set of magnetic field lines. However, because of the weakness of the magnetic field, the ion gyro radius is comparable to the radius of curvature of the magnetic field lines and it is not clear whether the field-line resonant waves are proper Alfvén waves at Mercury.

Depending on the conducting properties of the planet and of its immediate environment, the waves will be reflected either above, at, or below the surface. Also, depending on whether the conductance at the reflection boundary (Σ_R) is larger or smaller than the waveguide Alfvén conductance $\Sigma_A = (\mu_0 v_A)^{-1}$, where μ_0 is the permeability of vacuum and v_A is the Alfvén velocity, either the magnetic field or the electric field change phase when the wave is reflected. For a perfectly conducting reflection boundary the electric field is zero at the boundary whereas the magnetic field changes phase, and conversely for a perfectly insulating reflection boundary. In both cases no wave power is dissipated at the boundary. In the (unlikely) case where $\Sigma_R = \Sigma_A$ there would be perfect matching and therefore no reflection. Assuming a Σ_R in the range 0.1–0.3 S (as discussed earlier) and a surface magnetic field strength of 300 nT, matching would require a plasma mass density at low altitude in the range 10^{-21}–10^{-20} kg/m^3, corresponding to a number density of the order of 1 proton/cm^3. Such a low proton density at low altitude is not likely to occur and, thus, most realistically, there will be partial reflection and partial dissipation, resulting in a damped standing wave which eventually vanishes after the driver has switched off.

Studying these resonant waves by using electric and magnetic field data from orbiting spacecraft will reveal considerable information on the reflective, and thus also conductive, properties of the low-altitude region, as well as provide an additional constraint on the plasma distribution along the local magnetic field line.

7.3 Surface Charging and Charge Redistribution

Models of Mercury's magnetic field have been discussed by several authors (e.g., Ness et al. 1974, 1975, 1976; Whang 1977; Jackson and Beard 1977; Connerney and Ness 1988). The consensus is that a centred dipole is not a very good approximation. Following Blomberg and Cumnock (2004), we consider the case of a dipole offset along the axis by $0.2 R_M$ toward the north. This yields a simple but reasonable approximation of the average measured field.

Because of the dipole offset, the loss cones will be different as seen from the surfaces in the two hemispheres. As a consequence, all charged particles emitted from the planetary surface in the northern hemisphere impact the surface in the southern hemisphere in the absence of collisions and field-aligned potential gradients. However, not all particles leaving the southern hemisphere reach the northern surface even in the absence of collisions and potential drops. Instead, some of them are reflected well above the surface and return (approximately) to the point from where they were emitted. If there is an asymmetry in the net flux of escaping particles, between positive and negative charges, a weak field-aligned potential gradient builds up gradually. The effect of this will be to neutralize the charge imbalance by accelerating charged particles with the appropriate polarity toward the southern hemisphere. The net result is that plasma is transported from the northern to the southern hemisphere. Whether this inter-hemispheric plasma redistribution can be sustained depends on the conductivity at, or just above, the planetary surface.

If the photo-electron current emitted from Mercury's dayside surface exceeds the thermal electron current impacting the surface (e.g., Manka 1973; Willis et al. 1973) the dayside surface will develop a weak positive potential that limits the flow of photo-emitted electrons in order to maintain the current balance. Ip (1986) suggested that the nightside surface of Mercury might be negatively charged (up to perhaps a kilovolt) because of the difference in thermal velocity between ions and electrons. If this were indeed the case, a horizontal electric field would be set up on the dawn and dusk flanks, directed toward the nightside, potentially also contributing to a redistribution of charge, and possibly mapping into the magnetosphere along the magnetic field lines.

8 Summary

Mercury is the unique terrestrial planet possessing an Earth-like dipolar magnetic field, even if much weaker, which gives rise to a pocket-size magnetosphere. Due to the lack of a dense atmosphere, Mercury's environment is characterized by a tightly coupled magnetosphere–exosphere–surface system. In the following, we summarize the main issues discussed in this review paper.

- Mercury's magnetosphere is strongly coupled to the IMF and highly affected by the IMF B_X component, which dominates at that distance from the Sun. This results in a significant "opening" of the magnetosphere, even for small IMF B_Z values, and net north–south asymmetry. The characteristic time-scale of Mercury's magnetosphere is of the order of few tens of seconds, and about 1–2 min for substorm-like phenomena.
- The dayside magnetosphere, characterized by broad open magnetic field regions (cusps), is exposed to the direct penetration of the solar wind plasma. The magnetic merging process on the dayside magnetopause involves an energy exchange between the electromagnetic field and the plasma and is expected to accelerate solar wind protons up to a few tens of keV, depending on the local Alfvénic speed. Due to the absence of a dense atmosphere, the solar wind ions entering the cusp regions can impact the surface (about 10%) and contribute to the exospheric refilling, via the release of heavy ions produced by the ion-sputtering mechanism. This causes the exospheric abundances to be in part modulated by the upstream solar wind condition and IMF configuration.
- Due to the relative weakness of the Hermean magnetic field and the small size of its magnetosphere, the gyroradius of the most energetic ions is comparable to the magnetic field line curvature radius. In this case, charged particles do not move adiabatically and, due to magnetic moment scattering, can populate closed field line regions, where they bounce and drift (grad **B**).
- Low-energy heavy ions (Na$^+$) sputtered from the Mercury's surface can be transported into the magnetospheric lobes and then into the magnetotail, via convection over the polar cap. These ions are subjected to a significant parallel energization, up to hundreds of eV, due to an enhanced **E** × **B** centrifugal effect more pronounced than at Earth, because of the larger angular velocity of the convecting field lines at Mercury. Once in the magnetotail current sheet, the exospheric Na$^+$ ions are accelerated in a non-adiabatic way. They subsequently remain trapped in the vicinity of the equatorial plane, or precipitate onto the planet surface, contributing to further ion-sputtering.
- MHD simulation results suggest that precipitation and recycling of planetary ions at Mercury critically depend upon the magnetic activity and the global convection pattern. It appears that the electric field induced by the magnetic field line relaxation does lead to a convection surge and rapid inward injections of plasma sheet populations. However, because of small temporal scales of field variations as compared to those at Earth, nonadiabatic "resonant" heating takes place during substorms for ions with small mass-to-charge ratios.

- Among other processes proposed as exospheric sources (thermal- and photon-stimulated desorption, micro-meteoroids vaporization), ion-sputtering is expected to be the most effective in driving the observed Na$^+$ short-time variations in the exosphere of Mercury. Ion-sputtering is produced by solar wind H$^+$ and He^{++} precipitation through the magnetospheric cusps, and by planetary heavy ion circulation.
- Another relevant process, in the frame of magnetosphere–exosphere interaction, is the production of ENA (Energetic Neutral Atoms) via charge-exchange between the plasma particles and the local exospheric neutrals. The ENA flux at Mercury is mostly due to SW proton circulation, and, hence, occurs in the day- and dawn-side regions. Of the precipitating H$^+$ ions, approximately 90% are reflected by the increasing magnetic field, 10% reach the surface of Mercury and leads to ion-sputtering, and only 1% is neutralized due to charge-exchange, mostly in the dayside and dawn regions, and up to altitudes of hundreds of km. The charge-exchange process, hence slightly contributes to the plasma loss.
- The atmosphere of Mercury is very tenuous and collisionless, and the ionization induced by solar EUV likely cannot produce significant levels of electrical conductance, and the lack of a stable ionosphere could hinder the closure of the field-aligned currents (FACs), which couples the magnetosphere to the exosphere-surface system. A moderately conductive regolith is a likely candidate for FAC closure at Mercury, even if the high rate of Joule heating in the regolith severely limits the duration of the current flow and quickly dissipates the available magnetospheric energy. A "pickup" or "mass loading" conductance (Kivelson 2005) that might contribute to the generation and/or closure of FACs at Mercury, could be associated with the magnetospheric ions generated by the ion-sputtering mechanism, and the fresh ions created by photo-ionization, impact ionization, and charge exchange. Photoelectrons have also been suggested as a source of current carriers, however, the sum of both planetary ions and the photoelectron sheath over Mercury's sunlit surface provide conductances that are only slightly greater than the lunar values (i.e., 0.1 S). Another possibility is that field-aligned currents at Mercury may close as diamagnetic currents in regions of enhanced plasma density near the planet. The conductivity of the planetary surface, the low-altitude exosphere and, on the dayside, the photoelectron sheath, may however plays a significant role in the plasma circulation and re-distribution both along and across the magnetic field.

References

D.N. Baker, J.A. Simpson, J.H. Eraker, J. Geophys. Res. **91**, 8742 (1986)
S. Barabash, A.V. Lukyanov, P.C. Brandt, R. Lundin, Planet. Space Sci. **49**, 1685 (2001)
W. Baumjohann, R.A. Treumann, E. Georgescu, G. Haerendel, K.-H. Fornacon, U. Auster, Ann. Geophys. **17**, 1528 (1999)
T.A. Bida, R.M. Killen, T.H. Morgan, Nature **404**, 6774 (2000)
L.G. Blomberg, Planet. Space Sci. **45**, 143 (1997)
L.G. Blomberg, J.A. Cumnock, Adv. Space Res. **33**, 2161 (2004)
M. Bouhram, B. Klecker, G. Paschmann, S. Haaland, H. Hasegawa, A. Blagau, H. Rème, J.-A. Sauvaud, L.M. Kistler, A. Balogh, Ann. Geophys. **23**, 1281 (2005)
A.L. Broadfoot, D.E. Shemanky, S. Kumar, Geophys. Res. Lett. **3**, 577 (1976)
J. Büchner, L.M. Zelenyi, J. Geophys. Res. **94**, 11821 (1989)
J.L. Burch, S.B. Mende, D.G. Mitchell, T.E. Moore, C.J. Pollock, B.W. Reinisch, B.R. Sandel, S.A. Fuselier, D.L. Gallagher, J.L. Green, J.D. Perez, P.H. Reiff, Science **291**, 619 (2001)
L.F. Burlaga, Planet. Space Sci. **49**, 1619 (2001)
C.R. Chappell, T.E. Moore, J.H. Waite Jr., J. Geophys. Res. **92**, 5896 (1987)
A.F. Cheng, R.E. Johnson, S.M. Krimigis, L.J. Lanzerotti, Icarus **71**, 430 (1987)

S.P. Christon, Icarus **71**, 448 (1987)
J.E.P. Connerney, N.F. Ness, in *Mercury* (University of Arizona Press, Tucson, 1988), p. 494
F.V. Coroniti, C.F. Kennel, J. Geophys. Res. **78**, 2837 (1973)
S.W.H. Cowley, Ann. Geophys. **9**, 176 (1991)
S.W.H. Cowley, in *Physics of the Magnetopause*, ed. by P. Song, B.U.Ö. Sonnerup, M.F. Thomsen. AGU Geophysical Monograph, vol. 90 (American Geophysical Union, Washington, 1995), p. 29
S.W.H. Cowley, C.J. Owen, Planet. Space Sci. **37**, 1461 (1989)
I.A. Daglis, in *Magnetic Storms*, ed. by B.T. Tsurutani. Geophysical Monograph Series, vol. 98 (AGU, Washington, 1997), p. 107
D.C. Delcourt, J. Atmos. Sol. Terr. Phys. **64**, 551 (2002)
D.C. Delcourt, K. Seki, Adv. Geosci. **3**, 17 (2006)
D.C. Delcourt, S. Grimald, F. Leblanc, J.-J. Berthelier, A. Millilo, A. Mura, S. Orsini, T.E. Moore, Ann. Geophys. **21**, 1723 (2003)
D.C. Delcourt, K. Seki, N. Terada, Y. Miyoshi, Ann. Geophys. **23**, 3389 (2005)
D.C. Delcourt, F. Leblanc, K. Seki, N. Terada, T.E. Moore, M.-C. Fok, Planet. Space Sci. (2007, in press)
J.H. Eraker, J.A. Simpson, J. Geophys. Res. **91**, 9973 (1986)
J.A. Fedder, J.G. Lyon, Geophys. Res. Lett. **14**, 880 (1987)
A. Fedorov, E. Dubinin, P. Song, E. Budnick, P. Larson, J.-A. Sauvaud, J. Geophys. Res. **105**, 15945 (2000)
K.-H. Glassmeier, in *Magnetospheric Current Systems*, ed. by S.-I. Ohtani, R. Fujii, M. Hesse, R.L. Lysak. American Geophysical Union Mon., vol. 118 (2000), p. 371
R. Grard, Planet. Space Sci. **45**, 67–72 (1997)
R. Grard, A. Balogh, Planet. Space Sci. **49**, 1395 (2001)
R. Grard, H. Laakso, T.I. Pulkkinen, Planet. Space Sci. **47**, 1459 (1999)
D.C. Hamilton, G. Gloeckler, F.M. Ipavich, W. Studemann, B. Wilken, G. Kremser, J. Geophys. Res. **93**, 14343 (1988)
M. Hesse, J. Birn, J. Geophys. Res. **96**, 19417 (1991)
T.W. Hill, A.J. Dessler, R.A. Wolf, Geophys. Res. Lett. **3**, 429 (1976)
T. Iijima, T.A. Potemra, J. Geophys. Res. **83**, 599 (1978)
W.-H. Ip, Geophys. Res. Lett. **13**, 1133 (1986)
W.-H. Ip, A. Kopp, Adv. Space. Res. **33**, 2172 (2004)
D.J. Jackson, D.B. Beard, J. Geophys. Res. **82**, 2828 (1977)
P. Janhunen, E. Kallio, Ann. Geophys. **22**, 1829 (2004)
R.E. Johnson, R. Baragiola, Geophys. Res. Lett. **18**, 2169 (1991)
K. Kabin, T.I. Gombosi, D.L. DeZeeuw, K.G. Powell, Icarus **143**, 397 (2000)
E. Kallio, P. Janhunen, Geophys. Res. Lett. **30**, 1877 (2003)
M.G. Kivelson, Space Sci. Rev. **116**, 299 (2005)
H. Lammer, S.J. Bauer, Planet. Space Sci. **45**, 73 (1997)
H. Lammer, P. Wurz, M.R. Patel, R. Killen, C. Kolb, S. Massetti, S. Orsini, A. Milillo, Icarus **166**, 238 (2003)
F. Leblanc, R.E. Johnson, Icarus **164**, 261 (2003)
F. Leblanc, D.C. Delcourt, R.E. Johnson, J. Geophys. Res. **108**, 5136 (2003). doi 10.1029/2003JE002151
F. Leblanc, H. Lammer, K. Torkar, J.J. Berthelier, O. Vaisberg, J. Woch, *Notes du Pôle de Planétologie de l'IPSL*, No 5 (2004)
M. Lockwood, J. Geophys. Res. **100**, 21791 (1995)
M. Lockwood, Ann. Geophys. **15**, 1501 (1997)
M. Lockwood, M.F. Smith, J. Geophys. Res. **99**, 8531 (1994)
M. Lockwood, S.W.H. Cowley, T.G. Onsager, J. Geophys. Res. **101**, 21501 (1996)
R.E. Lopez, A.T.Y. Lui, D.G. Sibeck, K. Takahashi, R.W. McEntire, L.J. Zanetti, S.M. Krimigis, J. Geophys. Res. **94**, 17105 (1989)
J.G. Luhmann, L.M. Friesen, J. Geophys. Res. **84**, 4405 (1979)
G. Luhmann, C.T. Russell, N.A. Tsyganenko, J. Geophys. Res. **103**, 9113 (1998)
R. Lundin, S. Barabash, P.C. Brandt, L. Eliasson, C.M.C. Naim, O. Norberg, I. Sandahl, Ads. Space Res. **19**, 1593 (1997)
R.H. Manka, in *Photon and Particle Interactions with Surfaces*, ed. by R.J.L. Grard (Reidel, Dordrecht, 1973), p. 347
S. Massetti, S. Orsini, A. Milillo, A. Mura, E. De Angelis, H. Lammer, P. Wurz, Icarus **166**, 229 (2003)
S. Massetti, S. Orsini, A. Milillo, A. Mura, Planet. Space Sci. (2007, in press). doi: 10.1016/j.pss.2006.12.008
B. Mauk, J. Geophys. Res. **91**, 13423 (1986)
R.L. McPherron, C.T. Russell, M.P. Aubry, J. Geophys. Res. **78**, 3131 (1973)
A. Milillo, S. Orsini, D. Delcourt, E. Kallio, R.M. Killen, H. Lammer, F. Leblanc, S. Massetti, A. Mura, P. Wurz, S. Barabash, G. Cremonese, I.A. Daglis, E. De Angelis, A.M. Di Lellis, S. Livi, V. Mangano, Space Sci. Rev. **117**, 397 (2005)

D.G. Mitchell, P. C:son Brandt, E.C. Roelof, D.C. Hamilton, K. Retterer, S. Mende, Space Sci. Rev. **109**, 63 (2003)
T.E. Moore, D.C. Delcourt, Rev. Geophys. **33**, 175 (1995)
T.E. Moore, R.L. Arnoldy, J. Feynmann, D.A. Hardy, J. Geophys. Res. **86**, 6713 (1981)
A. Mura, S. Orsini, A. Milillo, D. Delcourt, S. Massetti, E. De Angelis, Icarus **177**, 305 (2005)
A. Mura, S. Orsini, A. Milillo, A.M. Di Lellis, E. De Angelis, Planet. Space Sci. **54**, 144 (2006)
T. Nagai, J. Geophys. Res. **87**, 4405 (1982)
N.F. Ness, K.W. Behannon, R.P. Lepping, Y.C. Whang, K.H. Schatten, Science **185**, 151 (1974)
N.F. Ness, K.W. Behannon, R.P. Lepping, Y.C. Whang, Nature **255**, 204 (1975)
N.F. Ness, K.W. Behannon, R.P. Lepping, Y.C. Whang, Icarus **28**, 479 (1976)
K.W. Ogilvie, J.D. Scudder, V.M. Vasyliunas, R.E. Hartle, G.L. Siscoe, J. Geophys. Res. **82**, 1807 (1977)
T.G. Onsager, S.-W. Chang, J.D. Perez, J.B. Austin, L.X. Janoo, J. Geophys. Res. **100**, 11831 (1995)
S. Orsini, A. Milillo, E. De Angelis, A.M. Di Lellis, V. Zanza, S. Livi, Planet. Space Sci. **49**, 1659 (2001)
A.E. Potter, T.H. Morgan, Science **229**, 651 (1985)
A.E. Potter, T.H. Morgan, Icarus **67**, 336 (1986)
A.E. Potter, T.H. Morgan, Adv. Space Res. **19**, 1571 (1997)
E.C. Roelof, D.G. Mitchell, D.J. Williams, J. Geophys. Res. **90**, 10991 (1985)
C.T. Russell, Geophys. Res. Lett. **16**, 1253 (1989)
C.T. Russell, D.N. Baker, J.A. Slavin, in *Mercury*, ed. by F. Vilas, C.R. Chapman, M.S. Matthews (University of Arizona Press, Tucscon, 1988), p. 514
M. Sarantos, P.H. Reiff, T.H. Hill, R.M. Killen, A.L. Urquhart, Planet. Space Sci. **49**, 1629 (2001)
J.-A. Sauvaud, J.R. Winckler, J. Geophys. Res. **85**, 2043 (1980)
J. Scuffham, A. Balogh, Adv. Space Res. **28**, 616 (2006)
K. Seki, R.C. Elphic, M. Hirahara, T. Terasawa, T. Mukai, Science **291**, 1939 (2001)
K. Seki, M. Hirahara, M. Hoshino, T. Terasawa, R.C. Elphic, Y. Saito, T. Mukai, H. Hayakawa, H. Kojima, H. Matsumoto, Nature **422**, 589 (2003)
K. Seki, N. Terada, D.C. Delcourt, T. Ogino, Dynamics of planetary ions in the Mercury's magnetosphere: comparison between ion trajectories in MHD fields and a rescaled analytical model, EGU General Assembly 2006, SRef-ID: 1607-7962/gra/EGU06-A-04012, April 2006 (abstract)
K. Shiokawa et al., J. Geophys. Res. **103**, 4491 (1998)
E.M. Sieveka, R.E. Johnson, Astrophys. J. **287**, 418 (1984)
J.A. Simpson, J.H. Eraker, J.E. Lamport, P.H. Walpole, Science **185**, 160 (1974)
G.L. Siscoe, N.F. Ness, C.M. Yeates, J. Geophys. Res. **80**, 4359 (1975)
J.A. Slavin, Adv. Space Res. **33**, 1587 (2004)
J.A. Slavin, R.E. Holzer, J. Geophys. Res. **84**, 2076 (1979)
J.A. Slavin, C.J. Owen, J.E.P. Connerney, S.P. Christon, Planet. Space Sci. **45**, 133 (1997)
J.R. Spreiter, A.L. Summers, A.Y. Alksne, Planet. Space Sci. **14**, 223 (1966)
F.R. Toffoletto, T.W. Hill, J. Geophys. Res. **94**, 329 (1989)
F.R. Toffoletto, T.W. Hill, J. Geophys. Res. **98**, 1339 (1993)
Y.C. Whang, J. Geophys. Res. **82**, 1024 (1977)
R.F. Willis, M. Anderegg, B. Feuerbacher, B. Fitton, in *Photon and Particle Interactions with Surfaces*, ed. by R.J.L. Grard (Reidel, Dordrecht, 1973), p. 389
R.M. Winglee, Geophys. Res. Lett. **25**, 4441 (1998)
P. Wurz, H. Lammer, Icarus **164**, 1 (2003)

Plasma Waves in the Hermean Magnetosphere

L.G. Blomberg · J.A. Cumnock · K.-H. Glassmeier ·
R.A. Treumann

Originally published in the journal Space Science Reviews, Volume 132, Nos 2–4.
DOI: 10.1007/s11214-007-9282-3 © Springer Science+Business Media B.V. 2007

Abstract The Hermean magnetosphere is likely to contain a number of wave phenomena. We briefly review what little is known so far about fields and waves around Mercury. We further discuss a number of possible phenomena, including ULF pulsations, acceleration-related radiation, bow shock waves, bremsstrahlung (or braking radiation), and synchrotron radiation. Finally, some predictions are made as to the likelihood that some of these types of wave emission exist.

Keywords Mercury · Magnetosphere · Plasma waves · Pulsations · Magnetospheric radiation

1 Introduction

Mariner 10 found, somewhat surprisingly, that Mercury possesses a sufficiently strong internal magnetic field for a proper magnetosphere to be set up as the solar wind flows by. The dipole moment of Mercury has been estimated in the range 200–400 nT R_H^3 where R_H is the Hermean radius, 2,440 km. The uncertainty in the estimate stems from the fact that the only available in situ magnetic data from Mercury were collected during Mariner 10's two near flybys, which is not sufficient to fully constrain the field. In addition, because of the relatively weak planetary field and the comparatively strong interplanetary magnetic field,

L.G. Blomberg (✉) · J.A. Cumnock
Space and Plasma Physics, School of Electrical Engineering, Royal Institute of Technology, Stockholm, Sweden
e-mail: lars.blomberg@ee.kth.se

K.-H. Glassmeier
Institute of Geophysics and Extraterrestrial Physics, Technical University of Braunschweig, Braunschweig, Germany

R.A. Treumann
Geophysics Section, Department of Geophysics and Environmental Physics, The University of Munich, Munich, Germany

Table 1 Typical solar wind parameters at Mercury

Velocity	300–800 km/s
Density	30–70 cm^{-3}
Magnetic field	20–40 nT
Ion temperature	11–16 eV
Electron temperature	17–21 eV

even with a larger data set the problem of separating the planetary and the interplanetary fields will remain.

Balance between the solar wind dynamic pressure and the planetary magnetic field typically results in a magnetopause stand off distance in the range 1.2–1.5 R_H, although it is believed that under extreme solar wind conditions the solar wind may impact directly on the dayside surface of the planet. Typical values of the solar wind parameters at the position of Mercury are given in Table 1.

The interaction of the solar wind with the planetary magnetic field also sets up a potential drop across the magnetosphere, typically in the range 5–25 kV, corresponding to a cross-tail electric field of 1–5 mV/m. These numbers assume an interaction similar to that at Earth, which is probably a valid assumption under normal solar wind conditions. However, under more extreme solar wind conditions, the cross-polar potential at Earth is known to saturate, an effect that has been attributed to the existence of a field-aligned current system on the dayside. The associated closure currents produce a magnetic field at the magnetopause which counteracts the magnetospheric field and, thus, weakens reconnection. However, because of the low conductivity of the Hermean (exo)ionosphere it is unclear whether a similar current system is set up at Mercury, which in turn means that a saturation effect similar to that at Earth may or may not exist at Mercury (Blomberg et al. 2006).

The interaction of the solar wind with the magnetosphere may also result in wave activity at a variety of frequencies. Figure 1 shows an overview of the wide variety of wave emissions that are expected at Mercury. In this paper, we discuss a selection of these different phenomena in more detail. The first comprehensive investigation of the wave spectrum around Mercury will be performed by the PWI instrument (Matsumoto et al. 2006) onboard the BepiColombo spacecraft. Until BepiColombo arrives at Mercury most of the topics discussed in this paper will remain speculative.

At the lower end of the spectrum we find ULF waves, which have been observed at Mercury (Russell 1989). Although ULF waves at Mercury may be conceptually similar to those at Earth, their physics are different. At Earth ULF waves have frequencies well below all relevant gyro frequencies in the system. However, at Mercury, the wave frequency is typically comparable to or even greater than the gyro frequencies of one or more particle species. This makes an MHD description irrelevant and leads to wave characteristics that depend on the plasma composition in an interesting way. Another interesting difference is found in the potential driver mechanisms for ULF waves, which we will return to later.

At the higher end of the spectrum we find radiation in the radio and X-ray bands. This is of particular interest at Mercury since its potential observation offers the possibility of remotely inferring the conditions in the radiation source regions. So far, however, observations of radiation have been unsuccessful. The reason for this might be that radiation has not been given enough attention; or that radiation is completely absent. We investigate the conditions under which radio radiation and natural X-rays can be expected to be emitted from Mercury. In principle, we can distinguish four types of radiation expected to be emitted from a strongly magnetized planet, some of them we would also expect to occur at Mercury. The first type is radiation in the radio band generated at the planetary bow shock or—more

Plasma Waves in the Hermean Magnetosphere

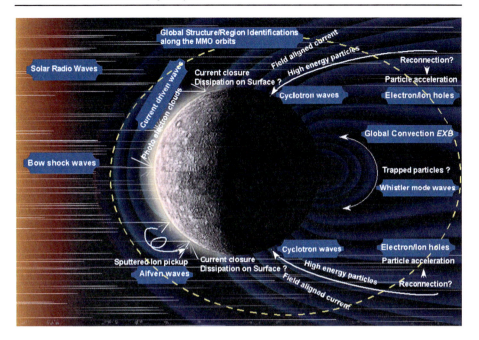

Fig. 1 Overview of expected wave activity in the Hermean magnetosphere. After Matsumoto et al. (2006)

precisely—in the electron foreshock of the planetary bow shock. The second type is the emission of synchrotron radiation. The third type is radio radiation from Mercury's auroral magnetosphere. Finally, the fourth type is X-ray emissions from the potential auroral regions on Mercury. A question of particular interest here is: Is there any equivalent at Mercury to the auroral kilometric radiation of Earth and the hectometric radiation from Jupiter? We will return to all four of these types of radiation in the following.

2 Ultra-Low Frequency Waves

Ultra-Low Frequency (ULF) waves are well known from the terrestrial magnetosphere. On the global scale we find the Pc5 type of waves, sometimes referred to as field-line resonances (Tamao 1965; Southwood 1974; Chen and Hasegawa 1974). These waves are standing oscillations on the magnetic field lines that bounce back and forth between the northern and southern ionosphere. In simple terms these waves arise when a fast mode wave is somehow generated and subsequently propagates across the magnetic field until it reaches a turning point where the fast mode and the shear Alfvén mode are in resonance, where part of the fast mode energy is coupled into the field-aligned mode and part of that energy is reflected. The fast mode wave may be generated by either a Kelvin–Helmholtz instability at the magnetospheric flanks, solar wind pressure pulses impinging on the magnetopause, or wave-particle interaction processes within the magnetosphere.

At Earth the ionosphere is a reasonably good conductor (compared to the effective conductance of the flux-tube waveguide) and therefore acts to more or less "short circuit" the wave electric field at the reflection points, resulting in a significant fraction of the wave energy being reflected back up along the field line. If the short circuiting is not perfect some wave energy will be dissipated into the ionosphere leading to damping of the wave unless energy is continuously being fed into it somewhere else. In the (unlikely) case of matching between the waveguide and reflection surface conductances, all of the wave energy would be transmitted into the ionosphere and thus there would be no reflected wave. In case the reflection surface is a poor conductor compared to the waveguide, the wave magnetic field rather than the wave electric field would be reduced (or short circuited, in the extreme case) still resulting in a reflected wave, although differently polarized from the case of a well-conducting reflecting surface. Although the latter case is not likely to occur at Earth, it cannot be ruled out at Mercury.

In the terrestrial magnetosphere, the bounce period between hemispheres for ULF waves is on the order of minutes, which corresponds to frequencies that are lower than the ion cyclotron frequencies of the ionospheric ions by several orders of magnitude. Thus, an MHD description is relevant and the waves are found to be shear Alfvén waves. At Mercury the situation is quite different. The bounce period is of the same order of magnitude as the proton cyclotron frequency and, thus, an MHD description is not immediately applicable. Othmer et al. (1999) and Glassmeier et al. (2004, 2003) addressed this problem with a multi-ion model and found what they called a crossover frequency at which the plasma supports a completely guided mode (corresponding to the transverse MHD mode). The crossover frequency depends on the ion composition. In the case of a two-ion plasma it lies in between the two ion cyclotron frequencies.

Using the multi-ion model the ion composition in the magnetosphere may be estimated. For a plasma with two ion species a direct estimate is obtained, whereas for a plasma with more than two ion species, further information or assumptions are needed. We exemplify this using Mariner 10's observations of a ULF wave at Mercury first published by Russell (1989); see Fig. 2. The waves are mostly transverse with a period of 2 s. Russell (1989) estimated the Alfvén wave bounce time between hemispheres to be 8 s and, thus, concluded that the observed wave could be the fourth harmonic of a standing wave. Othmer et al. (1999) applied their multi-ion model to the Mariner 10 data and derived a 14% (number density) sodium content, assuming protons and sodium as the only ion species. By including helium

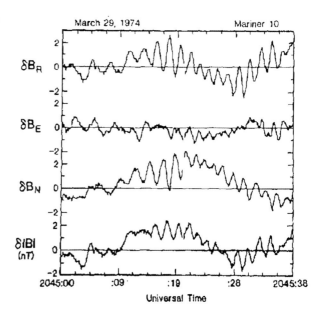

Fig. 2 ULF waves in Mercury's magnetosphere detected by the Mariner 10 magnetometer. After Russell (1989)

with a helium-to-proton ratio equal to that found in the solar wind, their model yielded an estimate of the sodium content of 20%.

ULF waves in Mercury's magnetosphere may also help understand the reflective and/or conductive properties of the planetary surface and/or low-altitude plasma, provided that both the electric and magnetic field are measured. From the phase difference between the electric and the magnetic field in a standing wave, the reflection coefficient at the boundary may be determined, which in turn gives information on the conductivity of the boundary relative to the waveguide conductivity. Thus, a properly instrumented satellite observing a ULF wave would remotely sense the conductivity of the planet at the magnetic footpoint.

Finally, a note on the driver mechanisms for ULF waves. In the terrestrial case, there are several instabilities within the magnetosphere that may drive waves via wave–particle interaction. However, some of these may be less likely to occur at Mercury since they are dependent on energetic particles bouncing and drifting in the planetary magnetic field. Mercury's small magnetosphere may not be able to contain sufficiently energetic ions in trapped orbits for the instabilities to develop.

3 Auroral Conditions at Mercury

All other magnetized planets in the solar system—Earth, Jupiter, Saturn, Uranus, and Neptune—are known to emit intense radio radiation from their auroral zones. Figure 3 shows a recent compilation of the known radio spectra of the planets. In addition, the auroral zones of the magnetized planets are subject to radiation of sporadic X-ray bursts that are related to electrons which have become accelerated in the auroral regions and precipitate into the planetary atmosphere.

The most intense planetary radiation is connected with the auroral activity of the planet, unless driven by the interaction of moons with the planet and the planetary magnetosphere. The latter is the case for the interaction of Jupiter with its satellite Io (Fig. 4).

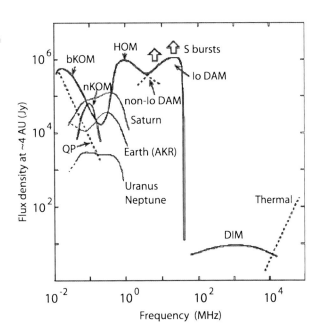

Fig. 3 Compilation of the known radio spectra emitted from the magnetized planets in the solar system (courtesy American Geophysical Union, after Zarka (1998)). The largest and most intense variety of radiation types is found for Jupiter covering the range from kilometric (KOM) through hectometric (HOM) to dekametric (DAM) wavelengths and includes Io-related radiation as well as quasiperiodic emissions (QP). Highest flux densities are found in Jupiter's S-shaped (in a frequency-time diagram) bursts. DIM in this figure indicates the decimetric solar radiation. At high frequencies the thermal galactic background radiation takes over

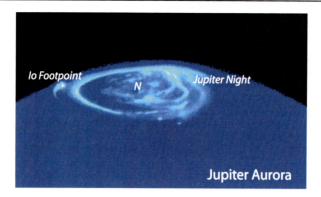

Fig. 4 Aurora on Jupiter, the largest and most strongly magnetized planet in the solar system. The aurora is seen here in a Hubble Space Telescope UV recording encircling the north pole of the planet (courtesy NASA/ESA/J. Clarke). The bright spot on the Jovian dayside is the footpoint of the magnetic flux tube that connects Jupiter's moon Io with the planet. The auroral oval maps the Jovian plasma sheet boundary region to the planetary atmosphere. Auroral substorm activity is seen from the evening through midnight to the morning sides in striking similarity to Earth, indicating substorm activity and thus reconnection in the Jovian tail plasma sheet even though the dynamics of the Jovian magnetosphere is different from that of the Earth and more resembles that of pulsars in that it is dominated by the planet rotation, indicating that substorms are a general phenomenon being insensitive to the nature of the gross magnetospheric dynamics and the cause of magnetospheric convection. For the slow rotator Mercury this implies that substorms will occur in the Hermean magnetosphere as well as long as a plasma sheet is formed in its magnetospheric tail

So far, no auroral radio emissions or auroral X-rays from Mercury have ever been reported. Therefore the intriguing question arises whether the Hermean conditions allow for generation of radiation. From in situ observations in the Earth's auroral zone and from remote observations of the Jovian radiation it has been established that the emissions are generated by the electron–cyclotron maser mechanism (for reviews see Zarka 1998; Treumann 2006) under the condition that the plasma density is very low and the electrons are at least weakly relativistic with bulk relativistic gamma factor γ_{rel}. These conditions can be written as

$$\frac{\omega_{pe}^2}{\omega_{ce}^2} \ll 1, \quad (1)$$

and

$$\gamma_{rel} - 1 > \frac{\omega_{pe}^2}{4\omega_{ce}^2}. \quad (2)$$

Hence, the electron plasma frequency ω_{pe} should be much less than the (nonrelativistic) electron cyclotron frequency ω_{ce} which implies that the Hermean magnetic field B_M must be correspondingly strong. $\gamma_{rel} - 1$ is the relativistic kinetic energy factor of the electrons that are responsible for the emission of the radiation. This condition is not very strong since by the first condition the right-hand side is small. If both conditions are satisfied, the radiation is emitted at frequency

$$\omega = \omega_{ce}(1 - \delta), \quad (3)$$

where $0 < \delta \ll 1$ so that the radiation frequency is just below but close to the electron cyclotron frequency ω_{ce}. In terms of the electron cyclotron frequency the maximum growth rate of the maser instability becomes

$$\frac{\Gamma_m}{\omega_{ce}} \approx \frac{\omega_{pe}}{\omega_{ce}} \sqrt{\frac{k_B T_e}{m_e c^2}} = \frac{\omega_{pe}}{\omega_{ce}} \sqrt{\gamma_{rel} - 1}, \qquad (4)$$

where $0 < \gamma_{rel} - 1 \ll 1$ and $k_B T_e = (\gamma_{rel} - 1) m_e c^2$ is the thermal kinetic energy of the radiating electrons. Since both factors on the right-hand side of this equation are small, the linear growth rate of the maser radiation is much less than the electron cyclotron frequency. Its lower limit is given by

$$\Gamma_m > \frac{1}{8} \left(\frac{\omega_{pe}}{\omega_{ce}} \right)^3 \omega_{ce}. \qquad (5)$$

From expressions (1–5) one immediately realizes that it is not easy to satisfy the conditions for any detectable auroral maser emission from Mercury unless the density in the radiation source region becomes very small. For the measured Hermean plasma sheet electron densities are $n_e \sim 3 \times 10^6$ m^{-3} (Ogilvie et al. 1977) and near-surface magnetic field strengths are $B_M \sim 300$ nT (Ness et al. 1975; Russell et al. 1988); the frequency ratio is $\omega_{pe}/\omega_{ce} \sim 2$, which violates the first condition. In addition, higher relativistic particle energies are required in order to obtain positive growth rates. There is also the possibility that the plasma sheet density varies significantly. Mukai et al. (2004) predicted plasma sheet densities in the range 0.1–32 cm^{-3} which, if correct, would mean that the radiation condition may sometimes be satisfied. The above conditions apply to maser radio-emission near the electron cyclotron frequency. At higher harmonics of the cyclotron frequency the conditions become less severe. Positive growth rates are obtained then even for plasma-to-cyclotron frequency ratios larger than one. Under Hermean conditions this implies emission frequencies >17 kHz. Radiation will then be emitted at oblique angles with respect to the magnetic field. Although this favors escape to free space, the expected growth rates are rather small for two reasons. First, the growth rates decrease with the square of the harmonic number. Second, the oblique resonance condition ceases to be circular. It becomes a displaced ellipse in momentum (velocity) space, only a small part of which matches the positive perpendicular phase space gradient of the electron velocity distribution. Thus, the phase space volume of particles actively contributing to radiation remains small only, while the phase space volume of absorbing particles increases. Both effects strongly reduce the growth rate to values of 10^{-5}–10^{-7} ω_{ce} such that one expects weak radiation only at higher harmonics. These values imply e-folding times in the range 1–100 s corresponding to e-folding lengths of 3×10^5–3×10^7 km which may be unrealistic.

The remaining possibility for intense radiation is that the radiation source is located very close to the Hermean surface in the strongest magnetic field available and, in addition, that the electron density is depleted by an order of magnitude or more, in which case emission will occur in the electromagnetic X-mode at about $f_{maser} \sim 8.4$ kHz. Near-planet plasma densities much higher than the one assumed earlier in the range from 10^7 m^{-3}–10^{10} m^{-3} have also been hypothesized (cf. the review by Hunten et al. 1988) in which case the plasma-to-cyclotron frequency ratio is much larger, requiring more violent density depletion factors for the generation of auroral radiation by the known maser mechanisms.

What might dilute the plasma close to Mercury? It is clear that there is no possibility of having low electron densities at low altitude on the illuminated dayside of Mercury. This side is covered by a layer of photoelectrons. However, during the long Mercury night in the shadow of the planet photoelectrons are about absent. Any electrons are either plasma sheet electrons or accompanying ion sputtering. These electrons are magnetically bound. In the absence of strong plasma diffusion processes (which are not completely excluded, however, because the Hermean magnetic field seems to be highly disturbed and, in addition, presence of high ion densities might contribute to a nonnegligible collisionality) the remaining possibility to deplete them locally on the small scale is by large magnetic-field-aligned potential drops, i.e., by parallel electric fields such as those known from Earth's auroral zone (e.g., Block 1988; Carlson et al. 1998).

Field-aligned potential drops on Earth occur in the presence of field-aligned current flow during substorms, and auroral radio emission correlates with auroral and substorm activity. The question of substorms on Mercury is addressed in a companion paper (Fujimoto et al. 2007).

Substorms are most probably driven by reconnection in the near-planet tail-current sheet. At Mercury the typical substorm time scale is 1–2 min (Baumjohann et al. 2006). Reconnection at Mercury must therefore be fast. This is possible only when the width of the tail current sheet, Δ, is on the order of or less than the ion inertial length $\lambda_i = c/\omega_{pi}$. For the Mariner 10 plasma sheet densities $n_e \sim 3 \times 10^6$ m^{-3}, assuming quasi-neutrality and a pure hydrogen plasma, the shortest ion skin depth is $\lambda_H = 130$ km. For observed heavier ion constituents like Na, K, and O it will be larger by a factor of 2 or 3. The transition time for this length at a convective inflow speed between a few 10 km s^{-1} and a few 100 km s^{-1} is between 10 s and 1 s, barely long enough for letting a fluid or ion instability grow enough to drive a substorm. Reconnection will thus proceed on electron inertial scales $\lambda_e = c/\omega_{pe} \approx 3$ km deep inside the "ion diffusion" Hall current region, which reduces the transition time to between 30 ms and 300 ms. The field-aligned currents are the closure currents of the Hall current system at the reconnection site (Nagai et al. 2001; Oieroset et al. 2001; Treumann et al. 2006). These field-aligned currents are probably carried by low-frequency electromagnetic plasma wave pulses. Since the transverse scale at the reconnection site is the ion inertial length, these waves are kinetic Alfvén waves with dispersion relation

$$\omega_{kA}^2 = k_\parallel^2 v_A^2 \left(1 + k_\perp^2 \rho_i^2\right), \tag{6}$$

where ρ_i is the thermal ion gyroradius. These waves travel along the magnetic field but propagate slowly transverse to the magnetic field thereby deviating from their original field line. Moreover, the waves are large-amplitude nonlinear waves. They also carry a transverse electric field that gives rise to shear motion along the field line where they propagate. In the presence of an ionosphere, current closure and shear motions would imply that the electric potential lines close to the planet deviate from being parallel to the magnetic field. This generates a field-aligned potential drop as large as the transverse electric field shear drop. It is the parallel drop that evaporates the plasma from the field region. By this mechanism a highly diluted plasma would result.

However, in the absence of an ionosphere it is highly uncertain how current closure is achieved and whether such a field-aligned potential drop is actually created. There are two possibilities for current closure in the absence of an ionosphere: either through transverse current diffusion in the thermospheric plasma or through current closure in the planetary body itself. Neither mechanism is understood so far. The former can proceed only via pressure-gradient drift currents which are carried by ions. The latter depends on the conductivity of the Hermean crust and uppermost mantle.

In addition to plasma depletion, in order to radiate, the weakly relativistic electron component must be in an "excited state," i.e., it must carry a substantial amount of free energy with practically no electrons in the lowest (thermal) energy states. Such electron phase space distributions—which in solid state physics are known as "inverted states"—can in a plasma

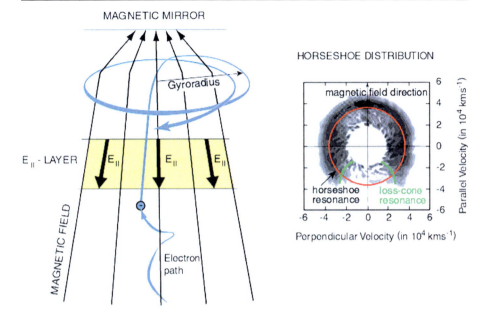

Fig. 5 Evolution of an electron ring-shell phase space distribution in the combined action of a mirror magnetic field and a magnetic field-aligned electrostatic potential drop. *Left*: Schematic of a converging magnetic mirror field geometry. The magnetic mirror is shown highly exaggerated. An electron spiralling along the magnetic field at small gyro-radius picks up the full energy of the electrostatic potential drop when passing the region of a parallel electric field directed away from the mirror point. Behind the electric field layer the parallel electron energy is converted by the magnetic mirror force under conservation of the electron's magnetic moment into perpendicular energy. This happens for all quasi-trapped electrons, i.e., the electrons outside the loss cone. *Right*: The resulting electron distribution in phase space (as has been observed in the terrestrial auroral region by the FAST spacecraft (Delory et al. 1998)) forms a ring that lacks the particles inside the loss cone and practically also lacks all thermal energy electrons. In practice such electrons will still be present due to photo emission from spacecraft. These photoelectrons would occupy the empty inner circle in the figure from where they have been artificially removed (from Treumann 2006)

most easily be generated in the presence of magnetic field-aligned electrostatic potential drops in a mirror magnetic field geometry. The "excited state" in a magnetic mirror without parallel electric field is a loss-cone distribution. This, however, is only weakly excited. Instead, the combined action of a field-aligned electric potential drop and a magnetic mirror produces a strongly excited "ring" or "shell distribution" in phase space by first accelerating the electrons along the field to energies on the order of the potential drop, and then mirroring the accelerated electrons by conserving their magnetic moments.

In this way the electric field energy that is fed into the electrons is transferred to the perpendicular velocity of the electrons, and almost all low-energy electrons are lifted into the excited ring state. This is schematically shown in Fig. 5. One should, however, note that in the weak Hermean magnetic field the mirror force is also weak. Most of the mirror points for the particles lie below the Hermean surface. The equatorial loss cone angle for particles having their mirror point right at the Hermean surface is

$$\sin\alpha_{eq} = R^{-3/2}, \qquad (7)$$

where $R = r/R_H$. In the small magnetosphere of Mercury the equatorial loss cone angles are large. For $R = 2-3$ the equatorial loss cone is about $20°$ wide and near surface occupies

60°. Therefore the loss cone might be much more important at Mercury for the generation of radiation than at Earth, where the loss-cone alone causes only a very small deviation from equilibrium and is thus very ineffective in generating radiation. At Mercury, however, filling almost all the upward half of the phase space the radiative importance of the loss cone is greater. Nevertheless, the conditions for emission in the absence of a magnetic field-aligned electric potential drop are unfavourable since the field-aligned electrons remain at low energy and the frequency ratio $\omega_{pe}/\omega_{ce} > 1$.

To our current knowledge the question whether or not Mercury emits auroral radio radiation is open. Detection, or lack thereof, of such radiation will shed light on the existence of magnetic field-aligned currents and electric potential drops, on the physics of substorms, the existence of reconnection, and on the way field-aligned currents are closed under severe conditions.

4 Bremsstrahlung X-Ray Emissions

Planetary aurorae are accompanied by intense particle precipitation and hence particle impact on the planetary atmosphere. If the atmosphere is sufficiently dense such that the precipitating particles experience frequent collisions with the atmospheric constituents, retardation of the electrons leads to direct emission of bremsstrahlung (braking radiation) X-rays. Figure 6 shows a typical example of PIXIE observations of X-ray emission during the interaction of a CME with Earth's magnetosphere on January 1, 2000, at 1945:00–2000:00 UT. PIXIE was at that time observing the northern hemisphere. The colour coded X-ray emission nicely covers part of the nighttime auroral oval circumventing the northern pole (southern magnetic pole) and consisting of several intense precipitation spots corresponding to the regions of strong substorm and auroral activities.

Observations from 120,000 km distance by the high-resolution X-ray camera (HRC) on Chandra at lower X-ray energies of 0.1–10 keV during aurorae have also been presented recently (Fig. 7a). Similar observations of aurora-related X-ray emissions have been reported for Jupiter. An example is shown in Fig. 7b.

X-rays from Mercury have so far not been observed either due to lack of any X-ray detector on the spacecraft flying by Mercury or to the lack of X-rays originating at Mercury. The Messenger spacecraft, which is currently on its way to Mercury, carries an X-ray spectrometer to measure the composition of the uppermost surface layer of the planet by analyzing solar X-ray photons reflected or scattered from the illuminated side of Mercury, so-called X-ray fluorescence. Seen from far away Mercury acts like a screen for solar X-rays emitted from the corona. Fluorescence measurements have in the past been made from the Moon by Rosat and with higher sensitivity by Chandra. They have also been reported recently from Saturn.

On the nightside of Mercury, any X-ray emission coming from the planet will, however, not be solar-related but should indicate precipitation of particles from the planetary magnetosphere. The Messenger instrument is not an X-ray imager and works at rather high X-ray energies. Since the accelerated electron and ion energies are expected to be lower on Mercury than on Earth it is thus uncertain whether Messenger will be able to detect X-rays from the acceleration process.

5 Bow Shock Radiation

Mercury possesses a solar wind generated bow shock which is a supercritical shock similar to Earth's bow shock though of much smaller extent and thus also smaller radius of cur-

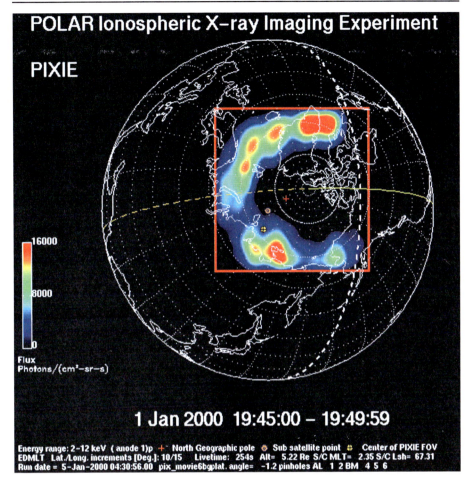

Fig. 6 PIXIE observations of intense auroral X-ray emission during the interaction of a CME with the Earth's magnetosphere (with permission of the PIXIE team, Principal Investigator Dr. Michael Schulz). The observed X-ray emission in the energy range of 2–12 keV nicely maps the nightside part of the auroral oval indicating its direct relation to auroral and substorm activities. The *dashed circle* indicates the position of the terminator at observation time. In addition the emission is spatially highly structured

vature. Because of the small radius of curvature this shock is predominantly parallel and because of the much weaker Hermean magnetic field also dynamically much more active than Earth's and Jupiter's bow shocks. They form and reform continuously, reflect and accelerate solar wind electrons and ions, and possess extended, highly turbulent electron and ion foreshocks. Ultimately, all reflected particles are picked up by the solar wind, i.e., they sense the solar wind electric field in their proper frame of reference and become accelerated up to a few times the solar wind streaming energy. For the electrons this is only a small fraction of their initial thermal energy. However, ions are effectively heated and isotropized by this process.

The mechanism of electron reflection from a supercritical shock is still under investigation. The simplest theories based on the specular reflection assumption refer to the strictly perpendicular region on the bent bow shock surface. At Mercury this region should be ex-

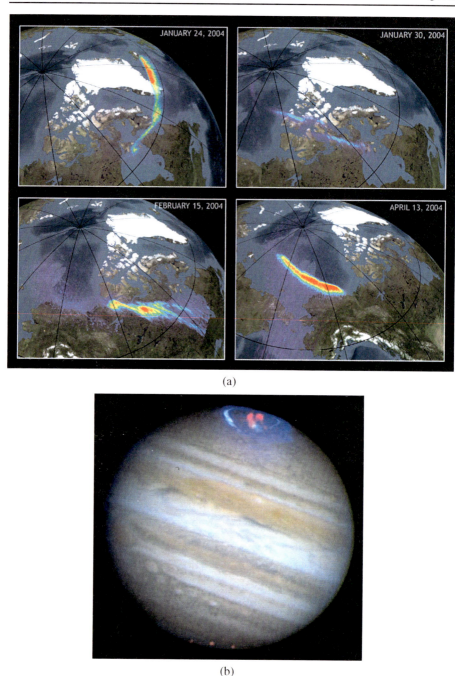

Fig. 7 Chandra HRC observations of low energy (0.1–10 keV) auroral X-ray emissions. (**a**) Earth's auroral X-rays detected from 120,000 km distance during early 2004 (courtesy NASA/MSFC/CXC/A.Bhardwaj & R. Elsner et al., *J. Atmos. Terrestr. Phys.*, in press). (**b**) A 45-minute Jovian auroral X-ray pulse (*magenta*) on December 18, 2000 (courtesy NASA/SWRI/R. Gladstone et al.), superimposed on the Hubble UV image (*blue*) of Jupiter's aurora. Surprisingly, the position of the pulse does not coincide with the Jovian auroral oval but is shifted closer to the pole, a fact that has not yet been understood properly

tremely narrow, forming a tiny speck only on the shock that is located where the instantaneous interplanetary magnetic field is tangential to the shock surface. Therefore, naively one does not expect large reflected electron fluxes in the foreshock region. This conclusion might, however, be completely wrong and will have to be corrected by observations and refinement of shock electron reflection/acceleration theory when taking into account the shock modification introduced by the reflected ion component.

Supercritical shocks reflect the flow in order to compensate for the excess energy of the inflowing plasma that cannot be transformed into heat and entropy during one shock crossing time. Since reflected electrons carry only a small fraction of the energy of the flow, curved shocks predominantly reflect ions back upstream into the inflowing plasma. Like the electrons these ions form fast beams counter-streaming to the supersonic/superAlfvénic plasma inflow. Since the velocity difference between the ion beams and the flow is far above the ion-acoustic speed, a variety of ion–ion and ion–electron beam instabilities are excited in the foreshock (e.g., Gary 1993). These waves are both electrostatic and electromagnetic.

In Earth's foreshock a number of wave modes excited in this way have been identified. The most interesting ones are the electromagnetic wave modes which propagate in the Alfvénic and magnetosonic ion-whistler branches at low frequency. The spectrum of turbulence to which these waves contribute has also been identified. This spectrum is intermittent since the waves evolve into large amplitude waves which themselves steepen and break. Breaking occurs because of the absence of dissipation and only partial cascading of the waves during the time of flow from their excitation site to the shock proper. But the steepened and breaking waves represent a whole system of small shocklets propagating in the foreshock. These shocklets have at least two effects. First, the turbulence to which they contribute retards the inflow in two ways by extracting flow energy and feeding it into the turbulence, and by trapping and scattering the ions on the ion-gyroscale. Both can be interpreted as collisionless dissipation, and hence the foreshock itself constitutes part of the shock transition. It represents the broadened shock transition region. Second, at the same time the turbulent wave spectrum is convected down to the shock ramp. Since it consists mainly of transverse waves in the Alfvénic and magnetosonic modes it leads to a turnover of the magnetic field direction at the shock front from parallel to perpendicular to the shock normal (in a statistical sense) such that on the ion scale the parallel shock remains quasi-parallel, while on the shorter electron scale the shock is practically quasi-perpendicular. The consequence is that the shock, almost over its entire surface, becomes quasi-perpendicular for electrons and thus is an electron reflector almost everywhere. In this way the electron and ion foreshock are mixed. For Mercury one expects that there is no distinction between electron and ion foreshocks because of the narrowness of the speck where the shock is genuinely perpendicular.

The electron foreshock is populated by electron beams escaping upstream along the solar wind magnetic field. Their effect is to generate Langmuir waves, ion acoustic waves, electron acoustic waves, Buneman modes, and possibly also electromagnetic electron oscillations in the whistler band. All these waves are of substantial interest in structuring the Mercury electron foreshock by formation of Langmuir solitons which may collapse whenever they become more than one-dimensional, ion-acoustic solitons which cause plasma density ripples, electron-acoustic solitons and electron holes, the latter being the consequence of the Buneman instability which also heats the electron plasma. The consequence is a coarse graining of the density in the electron foreshock on the scale of several Debye lengths.

Electron plasma waves have another interesting consequence that may serve as an identifier for electron reflection and particle acceleration. This is the generation of radio wave emission from the electron foreshock. The basic mechanism is very simple here: Electrostatic waves can be considered as particles. Hence, a three-wave process between waves must conserve energy and momentum. This can be symbolized as

$$L_1 + L_2 + L_3 (\text{or } T) = 0,$$

where L denotes longitudinal and T transverse (or electromagnetic). This equation describes either decay of a strong L-wave into other waves or the merging of two L-waves into another wave. In this way higher L-harmonics can be generated by merging, or an electromagnetic high-frequency wave is produced. Both cases are of interest in the electron foreshock. The most common one is the merging of two Langmuir waves into an electromagnetic wave which is described by the process

$$\omega_1 + \omega_2 = \omega, \qquad k_1 + k_2 = k. \tag{8}$$

The unindexed quantities belong to the radiation wave. Clearly, since this is long wavelength, $k \ll k_1, k_2$, and therefore the waves perform a head-on collision with $k_2 \approx k_1$ in order to merge into radiation. At the same time their energies add up, and the radiation is at frequency $\omega \approx 2\omega_1$, say, for Langmuir waves of similar frequencies. Such radiation is at the harmonic of the plasma frequency, $\omega \approx 2\omega_{pe}$, is not absorbed, and can escape from the plasma. This radiation is observed in Earth's foreshock and is expected for Mercury as well if only the Hermean bow shock is strong enough to reflect electrons. In turn, its properties can be used for probing the structure and strength of the bow shock. We should note that similar radiation is generated with other waves, described by similar conservation laws as those above. For instance, the merging of a Langmuir and an ion-acoustic wave produces radiation at the plasma frequency $\omega \approx \omega_{pe}$ since the ion acoustic frequency is much lower than the plasma frequency, and the merging of either a Langmuir wave with an electron acoustic wave produces radiation somewhere above the plasma frequency, $\omega_{pe} < \omega < 2\omega_{pe}$, while merging of two electron-acoustic waves generates just radiation at ω_{pe}. Moreover, bursts of radiation are generated when Langmuir solitons collapse, with highest radiation emission just before the collapse is completed, i.e., at the end of the collapse (Treumann and LaBelle 1992). In this process electrons become accelerated and hot plasma spots are generated in the course of the collapse.

6 Synchrotron Radiation

The elementary form of radiation in (sufficiently strong) magnetic fields is gyro-synchrotron radiation. In contrast to the electron cyclotron maser radiation this type of radiation is incoherent and therefore weak. Gyro-radiation consists of a sequence of emission lines at the electron cyclotron harmonics $\omega = l\omega_{ce}$, $l = 1, 2, \ldots$. These lines are of narrow width. The emissivity decreases with harmonic number. In addition, in hot plasma self-absorption damps the fundamental such that the "second harmonic," $l = 1$, is usually stronger than the fundamental. Gyro-synchrotron radiation depends on the availability of energetic electrons. The emission lines have a narrow but finite spectral width which from the uncertainty relation is proportional to the electron energy/temperature. The higher the temperature/energy the broader are the emission lines, and for relativistic electrons the lines overlap forming a continuous spectrum that decays towards higher frequency. For a power law electron distribution in energy $f(\varepsilon) = A\varepsilon^{-p}$ the total emitted power $P(\omega)$ per unit volume per unit frequency is

$$P(\omega) = C \frac{e^3 B \sin\alpha}{m_e c^2 (p+1)} \left(\frac{\omega}{\omega_c}\right)^{-(p-1)/2}, \qquad (9)$$

where C is a normalization factor that depends on A and p, B is the magnetic field strength, α is the pitch-angle (which can be put to 90° since perpendicular electrons are contributing most; otherwise for a more complicated pitch angle distribution one must average over pitch angle which contributes a numerical factor of order unity), $\omega_c = (3\gamma_{rel}^2 \, eB \sin\alpha)/2m_e$ is the critical frequency, and γ_{rel} the relativistic factor which is close to 1 for the expected electron energies in the Hermean magnetosphere. The critical frequency is roughly $\omega_c \approx 1.5\omega_{ce}$. The spectral maximum is then close to $\omega \approx 0.3\omega_c$. The stronger the field and the flatter the electron distribution the higher is the emitted power. We do, however, not expect high emissivities in the magnetosphere of Mercury since the magnetic field is very weak, and the expected electron energies are small such that the electron spectrum will decay steeply with energy. Even synchrotron radiation from Earth's radiation belts is weak, and substantial gyro-synchrotron emissivities in the solar system are measured only from the Sun and Jupiter's magnetosphere.

7 Conclusions and Predictions

ULF waves are known to exist in the Hermean magnetosphere. In addition to being an interesting physical phenomenon in themselves, they can also be used as a diagnostic tool. Their frequency may provide information on the plasma composition, and the phase difference between the electric and magnetic fields of the wave provides information on the reflection coefficient at low altitude, which in turn may be used to estimate the conductivity of the planetary surface or low-altitude (exo)ionosphere.

We are rather pessimistic about the likelihood of radiation from the polar regions of Mercury, a radiation analogue to Earth's AKR and Jupiter's S-bursts. If emitted, the radiation would be around or rather slightly above the local electron cyclotron frequency (≥ 8 kHz at ground level). This radiation should also be oblique because the condition that the plasma frequency is far below the local cyclotron frequency for fundamental emission is presumably not satisfied. Radiation above the local cyclotron frequency will be very weak, however, because the growth rates are small. On the other hand, intense emission at the fundamental depends on the presence of field-aligned electric potentials close to the planet. Such potentials may evacuate the plasma until $\omega_{pe} < \omega_{ce}$. Whether field-aligned potential drops occur on Mercury is a question of whether reconnection-driven substorms occur in the Hermean magnetosphere. Hence the possible detection of intense auroral radiation with wavelengths of some 100 km at a frequency below or very close to the local cyclotron frequency, transverse propagation in the X-mode and total power of the order of a few percent of the solar wind power transferred to the Hermean magnetosphere would illuminate a number of questions. First, it would suggest that substorms on Mercury exist and are reconnection-related. Second, it would prove that field-aligned electron currents flow from the magnetospheric tail region toward the planet. Third, it would indicate the existence of field-aligned electric potential drops generated in the Hermean magnetosphere, accelerating electrons and making possible electron-maser emission. On the other hand, when weak radiation of much lower power is detected at oblique angles, then it could still be generated by loss-cone distributions. These, however, would be the result from electrons hitting the body of the planet and not from atmospheric loss-cones since the expected atmospheric densities are too low to efficiently absorb electron fluxes. This is interesting in itself since it would confirm the presence of trapped particle fluxes in the Hermean magnetosphere and associated particle acceleration.

The other radiation source is the bow shock. Similar to Earth's and Jupiter's bow shocks, one expects that the Hermean shock is a strong shock which reflects electrons and ions. Radiation is then generated in the foreshock which for Mercury should fill almost all space upstream of the shock up to the distance that the reflected electron and ion beams can propagate until having dissipated their kinetic energy into feeding wave generation thereby pre-retarding the solar wind. Some of these waves, in particular Langmuir waves generated by electron beams, are capable of producing radiation at the solar wind plasma frequency $\omega = \omega_{pe}$ and at its second harmonic $\omega = 2\omega_{pe}$. However, other wave modes like electron acoustic and ion-sound waves may be involved as well. From their detection the state of the shock may be inferred.

Acknowledgements Support from the International Space Science Institute, Bern, Switzerland is gratefully acknowledged. Work at the Royal Institute of Technology was supported by the Swedish National Space Board and by the Alfvén Laboratory Centre for Space and Fusion Plasma Physics. The work by KHG was financially supported by the German Ministerium für Wirtschaft und Technologie and the German Zentrum für Luft- und Raumfahrt under contract 50 QW 0602.

References

W. Baumjohann, A. Matsuoka, K.-H. Glassmeier, C.T. Russell, T. Nagai, M. Hoshino, T. Nakagawa, A. Balogh, J.A. Slavin, R. Nakamura, W. Magnes, Adv. Space Res. **68** (2006). doi:10.1016/j.asr.2005.05.117
L.P. Block, Astrophys. Space Sci. **144**, 135 (1988)
L.G. Blomberg, J.A. Cumnock, Y. Kasaba, H. Matsumoto, H. Kojima, Y. Omura, M. Moncuquet, J.-E. Wahlund, Adv. Space Res. **38**, 627–631 (2006)
C.W. Carlson, R.F. Pfaff, J.G. Watzin, Geophys. Res. Lett. **25**, 2013–1016 (1998)
L. Chen, A. Hasegawa, J. Geophys. Res. **79**, 1024–1032 (1974)
G.T. Delory, R.E. Ergun, C.W. Carlson, L. Muschietti, C.C. Chaston, W. Peria, J.P. McFadden, R. Strangeway, Geophys. Res. Lett. **25** (1998). doi:10.1029/98GL00705
M. Fujimoto, W. Baumjohann, K. Kabin, R. Nakamura, J.A. Slavin, N. Terada, L. Zelenyi, Space Sci. Rev. (2007, this issue). doi:10.1007/s11214-007-9245-8
S.P. Gary, in *Theory of Space Plasma Microinstabilities* (Cambridge University Press, 1993), p. 193
K.-H. Glassmeier, P. Mager, D.Y. Klimushkin, Geophys. Res. Lett. **30** (2003). doi:10.1029/2003GL017175
K.-H. Glassmeier, D. Klimushkin, C. Othmer, P. Mager, Adv. Space Res. **33**, 1875–1883 (2004)
D.M. Hunten, T.H. Morgan, D.E. Shemansky, in *Mercury*, ed. by F. Vilas, C.R. Chapman, M.S. Matthews (Univ. Arizona Press, Tucson, 1988), pp. 562–312
H. Matsumoto, J.-L. Bougeret, L.G. Blomberg, H. Kojima, S. Yagitani, Y. Omura, M. Moncuquet, G. Chanteur, Y. Kasaba, J.-G. Trotignon, Y. Kasahara, BepiColombo MMO PWI Team, Adv. Geosci. **3**, 71–84 (2006)
T. Mukai, K. Ogasawara, Y. Saito, Adv. Space Res. (2004). doi:10.1016/S0273-1177(03)00443-5
T. Nagai, I. Shinohara, M. Fujimoto, M. Hoshino, Y. Saito et al., J. Geophys. Res. **106**, 25929–25950 (2001)
N.F. Ness, K.W. Behannon, R.P. Lepping, Y.C. Wang, Nature **255**, 204–205 (1975)
K.W. Ogilvie, J.D. Scudder, V.M. Vasyliunas, R.E. Hartle, G.L. Siccoe, J. Geophys. Res. **82**, 1807–1824 (1977)
C. Othmer, K.-H. Glassmeier, R. Cramm, J. Geophys. Res. **104**, 10,369–10,378 (1999)
M. Øieroset, T.D. Phan, M. Fujimoto, R.P. Lin, R.P. Lepping, Nature **412**, 414–417 (2001)
C.T. Russell, Geophys. Res. Lett. **16**, 1253–1256 (1989)
C.T. Russell, D.N. Baker, J.A. Slavin, in *Mercury*, ed. by F. Vilas, C.R. Chapman, M.S. Matthews (Univ. Arizona Press, Tucson, 1988), pp. 514–561
D.J. Southwood, Planet. Space Sci. **22**, 483–491 (1974)

T. Tamao, Science reports of Tohoku University, Series 5. Geophysics **43**, 43–72 (1965)
R.A. Treumann, Astron. Astrophys. Rev. **13**, 229–315 (2006)
R.A. Treumann, J. LaBelle, Astrophys. J. **399**, L167 (1992)
R.A. Treumann, C.H. Jaroschek, R. Nakamura, A. Runov, M. Scholer, Adv. Space Res. **67** (2006). doi:10.1016/j.asr.2004.11.045
P. Zarka, J. Geophys. Res. **103**, 20,159–20,194 (1998)

Particle Acceleration in Mercury's Magnetosphere

L. Zelenyi · M. Oka · H. Malova · M. Fujimoto ·
D. Delcourt · W. Baumjohann

Originally published in the journal Space Science Reviews, Volume 132, Nos 2–4.
DOI: 10.1007/s11214-007-9169-3 © Springer Science+Business Media, Inc. 2007

Abstract This paper is devoted to the problem of particle acceleration in the closest to the Sun Hermean magnetosphere. We discuss few available observations of energetic particles in Mercury environment made by Mariner-10 in 1974–1975 during Mercury flyby's and by Helios in 1979 upstream of the Hermean bow shock. Typically ions are non-adiabatic in a very dynamic and compact Mercury magnetosphere, so one may expect that particle acceleration will be very effective. However, it works perfectly for electrons, but for ions the scale of magnetosphere is so small that it allows their acceleration only up to 100 keV. We present comparative analysis of the efficiency of various acceleration mechanisms (inductive acceleration, acceleration by the centrifugal impulse force, stochastic acceleration in a turbulent magnetic fields, wave–particle interactions and bow shock energization) in the magnetospheres of the Earth and Mercury. Finally we discuss several points which need to be addressed in a future Hermean missions.

Keywords Hermean magnetosphere · Particle acceleration

L. Zelenyi · H. Malova (✉)
Space Research Institute, Russian Academy of Sciences, Moscow, Russia
e-mail: hmalova@mail.ru

M. Oka
Kwasan and Hida Observatory, Kyoto University, Kyoto, Japan

M. Fujimoto
Institute of Space and Astronautical Science, Sagamihara, Japan

D. Delcourt
Centre d'études des Environnements Terrestres et Planétaires, CNRS, Saint-Maur des Faussés, France

W. Baumjohann
Space Research Institute, Austrian Academy of Sciences, Graz, Austria

Observations of energetic particle fluxes in planetary magnetospheres are of general physical interest. Particles with large energies are observed both in the Earth's and in Mercury's magnetosphere. Mercury's case is of particular importance because of topological similarity and difference in scale as compared to the terrestrial case. Observations by Mariner 10 in 1974–1975 near Mercury showed that the size of the Hermean magnetosphere is only about 5% that of the Earth, although the planetary radii differ by less than a factor three. This means that Mercury occupies a much larger relative volume inside its magnetosphere. The difference between the scale sizes (with the subsolar magnetopause being at about 1.1 to 1.5 R_M from the planet's center) leads to significant differences in the dynamics of the terrestrial and Hermean magnetospheres and particle motion inside them. Since an ionosphere is absent at Mercury, this gives a good chance to estimate the importance of this plasma region for the dynamics of the terrestrial magnetosphere. Also charged particle motion in both magnetospheres might be quite different because of different magnetospheric scales and distances from the Sun. Due to the ratio of about 7 between the magnetospheric scales of Earth-to Mercury the Larmor radii of protons and especially of heavier ions are comparable with the size of the Hermean magnetosphere. Therefore one can easily anticipate that in the Hermean magnetosphere the particle motion should be strongly non-adiabatic. Mariner 10 observations have shown that relatively thick lobes of Mercury's magnetosphere are separated by a very thin current sheet. Particle motion typical for such configuration has the character of transient Speiser-like and "cucumber" orbits, which after crossing the neutral sheet plane are following nonadiabatic "serpentine" (meandering) orbits. The MHD approximation is severely violated for Mercury plasma environment.

1 Observations of Hermean Magnetospheric Particles

As in the Earth's case, the southward interplanetary magnetic field might also lead to the reconnection in the Hermean magnetosphere and consequently to the energy storage in the elongated Hermean magnetotail (Siscoe et al. 1975). Substorms at Mercury could, like in the terrestrial case, have explosive manifestations when this stored energy is dissipated. As a consequence large amounts of energy can be injected into the exosphere. A typical signature of a substorm is the so-called dipolarizations in the magnetotail region in which the stretched-out magnetic field lines suddenly snap back to a dipolar-like configuration (e.g. McPherron et al. 1973; Ip 1997, and references therein). Investigations at Mercury will give clues to the triggering of substorms, which are much more obscure in the Earth's magnetosphere because of its very large.

Due to the relatively weak dipole magnetic moment and intensive solar wind streams, the magnetosphere of Mercury is very dynamic. Mariner 10 data showed extremely strong changes of the Hermean magnetic field with a characteristic timescale of about a few minutes (Ness et al. 1974), which coincided with the sudden appearance of high energy particle fluxes (>35 keV) (Simpson et al. 1974). These data confirmed the suggestions that substorms should occur very often in the Hermean environment (Siscoe et al. 1975; Eraker and Simpson 1986). Ground spectroscopic investigations revealed that Mercury has a collisionless neutral exosphere with neutral particles Na, K, and Ca (Broadfoot et al. 1976; Goldstein et al. 1981; Potter and Morgan 1985; Bida et al. 2000). Various processes can populate the exosphere with neutrals including photo-sputtering, thermal desorption, particle sputtering and meteoric impact (Killen and Ip 1999; Milillo et al. 2005). Particle sputtering by direct impact of the solar wind with Mercury's surface might be an important source of neutrals. When a solar wind particle hits the surface, there is a probability that a neutral will be ejected and this neutral could then become ionized (e.g., Hunten et al. 1988). Once ionized the charged particle is then picked up by the magnetic field (e.g., Killen et al. 2001). Later it might be lost in the solar wind or return to the planetary surface (e.g., Potter et al. 2002; Delcourt et al. 2003) after transport in the magnetosphere. Thus the solar wind flow and magnetospheric magnetic field may play an important role in particle dynamics near the planet.

During the Mariner 10 Mercury flyby, observations of large fluxes of energetic electrons (energies in excess of 0.3 MeV) and protons (energies between 0.53 and 1.9 MeV) were reported by Simpson et al. (1974) at the time of a possible substorm onset identified later by Siscoe et al. (1975). These data were discussed in terms of particles acceleration by reconnection at a neutral line some 3–6 R_M behind the planet. The authors detected bursts of electrons with 5–10 s modulation which they attributed to repeated reconnection events. Baker et al. (1986) questioned this interpretation by Eraker and Simpson (1986). He did not doubt the original substorm scenario, but suggested that it is the drift of newly accelerated particles around the planet which gives rise to the repetitive observations. In other words, this means that at least temporarily a radiation belt could be formed around the planet.

Simultaneous strong enhancements of proton and electron fluxes in the magnetosphere of Mercury have raised many questions. After the pioneering measurements by Simpson et al. (1974) the critical point challenged by many authors was: did they really observe energetic ions? Armstrong et al. (1975) noticed that the response of the proton detector in the Mariner 10 experiment is most plausibly attributable to the pileup of low-energy electrons rather than the presence of protons in the vicinity of Mercury. They concluded that no 'new' acceleration mechanism has been identified at Mercury. Christon et al. (1979) investigated this question assuming some sensitivity of detectors to electrons with energies in excess of 35 keV due to pulse pile up.

Eraker and Simpson (1986) performed a detailed analysis of the measurements of the high-intensity bursts of electrons with energies of up to 600 keV, discovered in Mercury's magnetosphere (Simpson et al. 1974; Christon and Simpson 1979). The results provided strong evidence for particle acceleration during explosive magnetic field reconnection within Mercury's magnetotail and rapid release of magnetic free energy through instabilities during magnetic field reconnection in the planetary magnetotail. It was noted by Eraker and Simpson (1986) that "laboratory measurements with the Mariner instrumentation resolved questions regarding the electron measurements, but left open to question the proof of proton detection by LET". Later, Christon (1987) has done a comparison of the Hermean and terrestrial magnetospheres based on electron measurements and substorm time scales. His conclusion was quite certain: no proton fluxes associated with Mercury's magnetosphere were identified. This statement was directly confirmed by the theoretical analysis of Zelenyi et al. (1990) who proved that multiple sporadic reconnections (main drivers of inductive acceleration in the tail) could not provide an appreciable proton energization for $W \geq 50$–60 keV (we will discuss these results below). So both small-scale effects and large multiple scale modifications of magnetic topology do not provide strong ion acceleration at Mercury comparable to the terrestrial case. Later this assumption was also supported by simulations done by Ip (1997) who demonstrated that only keV acceleration may be achieved for ions during large-scale substorm reconfiguration events in the Hermean magnetosphere.

Over the years 1974–1980, the two Helios spacecraft located upstream of the magnetosphere of Mercury were, in principle, able to detect ion (greater than 80 keV) and electron (greater than 60 keV) fluxes coming from the direction of the Hermean magnetosphere, and propagating towards the sun. Kirsch and Richter (1985) gave an example of such event, when the fluxes of accelerated ions are statistically significant in the sense that they are 2–5

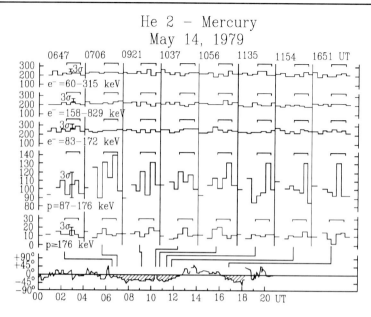

Fig. 1 Count rates from ion and electron channels obtained during the May, 15, 1979 event observed by the Helios 2 spacecraft. The *lower panel* shows the time profile of the out-of-ecliptic angle θ_b of the IMF, which was southward during these examples (from Kirsch and Richter 1985)

times larger than the statistical error. This is illustrated by Fig. 1, where some examples for sector plots of count rates from the ion and electron energy channels, obtained with higher-time resolution during the May 15, 1979, event, are shown. The lower panel demonstrates the time profile of the out-of-ecliptic angle of the IMF. It was concluded that the observed particles are of direct Hermean magnetospheric origin. The authors argued that solar wind particles reflected from or accelerated at the Hermean bow shock could be excluded from consideration.

It is interesting to compare this situation with the Earth's magnetosphere, where both ions and electrons might be accelerated to large energies about 1 MeV. In the Earth's magnetotail the maximum potential drop across the tail is at most 100–150 kV which is definitely not enough to support such strong acceleration. We will discuss below in this chapter, what are the relevant mechanisms for acceleration of electrons, protons and heavy planetary ions in the Hermean magnetosphere and will try to compare their operation with the terrestrial case

2 Mechanisms of Particle Acceleration

Due to its relatively close distance to the Sun and the weak magnetic moment of Mercury, its magnetosphere is very "open" in comparison with that of the Earth. As we have mentioned, its magnetotail contains relatively thick lobes and a "thin" current sheet, resembling a pre-substorm Earth's magnetotail. Therefore, by analogy with the Earth a few particle acceleration mechanisms in the Mercury magnetosphere might be discussed:

1. Magnetic reconnection between 3 to 6 R_M in the magnetotail may lead to the particle acceleration by inductive fields via substorm-like dipolarization (Delcourt et al. 2005).

2. Non-adiabatic scattering and acceleration may occur in weak magnetic field regions (centrifugal impulse force, Delcourt and Martin 1994) and in regions near X-lines (Zelenyi et al. 1990) with quasi-regular well-defined magnetic geometry.
3. Stochastic particle acceleration in the turbulent magnetic field varying with time, i.e., Fermi-type particle acceleration (Milovanov and Zelenyi 2002) could be another candidate mechanisms.
4. Interaction with ultra low frequency (ULF) waves, which have been observed at Mercury (Russell et al. 1988) and could play a role in ion acceleration (Glassmeier et al. 2003; Baumjohann et al. 2006).
5. Finally, non-adiabatic ion thermalization and acceleration could occur at shocks and at the magnetopause.

In general, the observations gave strong indications on particle acceleration during fast explosive magnetic field reconnection processes within Mercury's magnetotail and the release of magnetic free energy through instabilities in regions of magnetic field reconnection (Eraker and Simpson 1986). The electron fluxes with characteristic energies $W_e \leq 600$ keV have a time duration of about 10 sec and may be repeated in time every 2–3 min. One scenario relates these repetitions with multiple substorm onsets, which then could serve as the manifestation of strong variability of the Hermean magnetosphere.

As in the Earth's magnetosphere the maximum energy of electron bursts is close to 1 MeV. The value of the upper boundary of proton energy is not known. Let us estimate the inductive electric field

$$rot\, \vec{E}_{\text{ind}} = -\frac{\partial \vec{B}}{\partial t} \qquad (1)$$

at the time of a magnetic topology reconstruction during substorms. It is interesting that for the growth of magnetic islands in the magnetotail both for Earth and Mercury the value of the inductive electric field, E_{ind} is about 5–6 mV/m, taking into account the fact that the relative temporal scaling factor of both magnetospheres is of the order of 30.

3 Centrifugal Force Effects

The processes of particle acceleration and circulation were studied in a series of papers by Delcourt et al. (2002, 2003, 2005), Leblanc et al. (2003), with help of test particle simulations in a simple model of the Hermean magnetosphere based on a Luhmann and Friesen (1979) model. The authors focused on centrifugal effects associated with the large scale plasma convection. It was shown earlier that the influence of the centrifugal force during current sheet crossings results in particle acceleration in the Earth's magnetotail (Delcourt et al. 1996). Particles are accelerated by a "dawn-dusk" electric field, but centrifugal force effects are providing them mobility across the magnetic field. It was demonstrated that because the curvature radii of the $\vec{E} \times \vec{B}$ drift paths are much smaller at Mercury than at the Earth, centrifugal force effects are enhanced and lead to parallel energization during transport from high to low latitudes. For moderate convection rates, model trajectory calculations revealed that these $\vec{E} \times \vec{B}$ related centrifugal effects yield energization of heavy ions sputtered from Mercury's surface (e.g., Na^+) up to several tens or even a few hundreds of eV within minutes. These results suggest very substantial heating of planetary material within Hermean magnetosphere and thus contrast with that prevailing at the Earth where the centrifugal effects are relatively weak, yielding energization of ionospheric ions up to at most a few tens of eV on the time scale of hours.

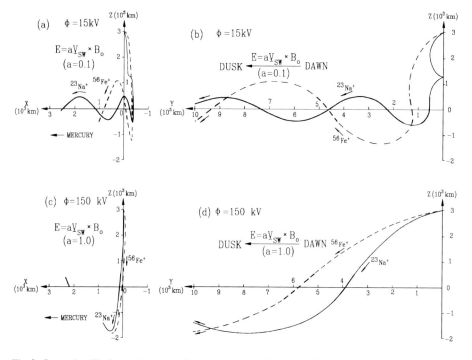

Fig. 2 Serpentine-like ion motion in the Hermean magnetosphere (according to Ip 1987)

A good contribution to the speculative modeling based on the Mariner 10 observations was done by Ip (1987). Several basic magnetospheric processes at Mercury have been investigated with simple models, including adiabatic acceleration and convection of equatorially mirroring charged particles, current sheet acceleration effects, and acceleration of Na$^+$ and other exospheric ions by the magnetospheric electric field near the planetary surface. Steady-state treatment of the magnetospheric drift and convection processes suggests that the region of the inner magnetosphere as explored by the Mariner 10 should be largely devoid of energetic (> 100 keV) electrons in equatorial mirroring motion. As for ion motion, the large gyro radii of the heavy ions permit surface re-impact as well as loss via intercepting the magnetopause. Because of the kinetic energy gained in the gyro motion, the first effect could lead to sputtering processes and hence generation of secondary ions and neutrals. The second effect could account for the loss of about 50% of Mercury's exospheric ions.

As a result of strong non-adiabatic ion motion, the relatively thick lobes of Mercury's magnetosphere should be separated by a very thin current sheet in the center plane, where the particle motion has a character of transient Speiser-like orbits, moving in a neutral sheet along nonadiabatic "serpentine-like" orbits (Ip 1987). This assumption is illustrated by simulation results in Fig. 2.

The actual structure of Mercury's magnetosphere and the existence of large-scale plasma cells (for example, lobe, plasma sheet, and boundary layers) remain to be elucidated. The hypotheses about a circulation of heavy ions of planetary origin within Mercury's magnetosphere could be qualitatively verified using a numerical model. Test particle simulations in three-dimensional electric and magnetic model fields that provide a first order description of Mercury's environment have been performed by Delcourt et al. (2003) to study the dynamics of sodium ions, ejected from the planetary surface to the magnetosphere. The numerical

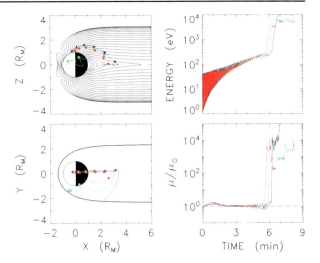

Fig. 3 Model of Na$^+$ trajectories during depolarization processes (from Delcourt et al. 2003). *Left panels* show the trajectory projections (*top*) in the noon-midnight meridian plane and (*bottom*) in the equatorial plane. *Right panels show* (*top*) the particle kinetic energy and (*bottom*) the magnetic moment (normalized to the initial value) versus time. The ions are launched from the planet's surface at different latitudes (color-coded in blue, green, and red) in the dayside sector. *Filled circles* in the *left panels* show the time of flight in steps of one minute

simulations revealed a significant sodium population in the Hermean night side sector, with energies about several keV.

The numerical simulations also display several features of interest that follow from the small spatial scales of Mercury's magnetosphere. First, in contrast to the situation prevailing at the Earth, ions in magnetospheric lobes are found to be relatively energetic (a few hundreds of eV), despite the low energy character of the exospheric source. This results from the enhanced centrifugal acceleration during $\vec{E} \times \vec{B}$ transport over the polar cap. Second, the large Larmor radii in the mid tail result in the loss of most Na$^+$ ions into the dusk flank at radial distances greater than a few planetary radii. Because gyro radii are comparable to, or larger than, the magnetic field variation length scale, the Na$^+$ motion is also found to be non-adiabatic throughout most of Mercury's equatorial magnetosphere, leading to chaotic scattering into the loss cone or meandering (Speiser-type) motion in the near tail. The nonadiabatic motion of Na$^+$ ions is illustrated in Fig. 3 from Delcourt et al. (2003). As a direct consequence, a localized region of energetic Na$^+$ precipitation develops at the planetary surface. In this region which extends over a wide range of longitudes at mid latitudes, one may expect additional sputtering of planetary material.

4 Inductive Acceleration of Electrons and Ions

High energy (up to several tens of keV) electron bursts are correlated with rapid variations in the orientation of the magnetic field. Unfortunately, particle measurements during Mariner 10 flybys at lower energies were not available. Existing particle experiments nevertheless indicated that the Hermean magnetosphere may exhibit substorm cycles, as the Earth's magnetosphere (e.g. Siscoe and Christopher 1975; Slavin 2004). Registered fluxes of energetic electrons experienced well pronounced temporal modulations with a period of ∼6 s, which are not yet explained (e.g. Christon et al. 1987). According to Eraker and Simpson (1986), these injections come from reconnection processes in Mercury's mid tail or from drift echoes of energetic electrons transported into the immediate vicinity of the planet (e.g. Baker et al. 1986). In contrast to these studies, Luhmann et al. (1998) suggested that the features observed by Mariner-10 may be directly driven by rapid changes in the solar wind.

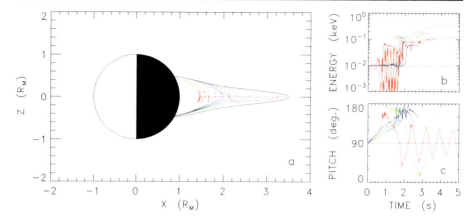

Fig. 4 Examples of electron orbits obtained in the model dipolarization displayed at the *left*. 10-eV test electrons were launched at 90° pitch angle from different (color-coded) locations on the field line intercepting the equator at $X = 3.5$ R$_M$ (from Delcourt et al. 2005)

Delcourt et al. (2005) investigated the dynamics of electrons during the expansion phase of Hermean substorms using test particle simulations in a simple model of magnetic field dipolarization. For this task a 3D time dependent test particle code previously developed for the Earth's magnetosphere (Delcourt et al. 2002) was rescaled to the environment of Mercury. The results of the simulations are shown in Fig. 4, which shows that electrons may be subjected to significant energization (right panel at the top) on the time scale of several seconds during reconfigurations of the magnetic field. As in the near Earth magnetosphere, electrons with energies up to several tens of eV may not conserve the second adiabatic invariant during dipolarization, which leads to the appearance of clusters of bouncing particles (left panel). On the other hand, it is found that, because of the stretching of the magnetic field lines, higher energy electrons (several keV and above) do not behave adiabatically and possibly experience meandering (Speiser-type) motion around the magnetospheric mid plane. It is shown that dipolarization of the magnetic field lines may be responsible for the significant and fast (few seconds) precipitation of these several keV electrons onto the planet's surface. Dipolarization also results in effective injections of energetic trapped electrons toward the planet.

Ion dynamics in the course of dipolarization processes was considered in the numerical model by Ip (1997). These results, shown in Fig. 5 demonstrate that the $\vec{E} \times \vec{B}$ related acceleration plays a much more important role in particle dynamics at Mercury than at the Earth. Even for moderate convection rates (e.g., 20 kV across the polar cap in the present calculations), the centrifugal effect is responsible for significant parallel energization. Luhmann et al. (1998) suggested the short-lived injections observed by Mariner-10 could be directly driven by abrupt IMF changes, whereas other studies (e.g., Christon et al. 1987) put forward these injections as evidences of magnetospheric substorms.

As a matter of fact, assuming dipolarization of the magnetic field lines on a time scale of 10 s, Ip (1997) demonstrated energization of H$^+$ and He^{2+} ions up to 10–20 keV. The results of these simulations are presented in Fig. 5. One certainly expects the electric field induced by rapid reconfiguration of the magnetospheric field lines to play a significant role in the net ion energization during transport (e.g., Delcourt and Sauvaud 1994).

The corresponding gains of energies by the charged particles, whose trajectories are presented in Fig. 5 are very much initial position dependent and even for the best scenario of

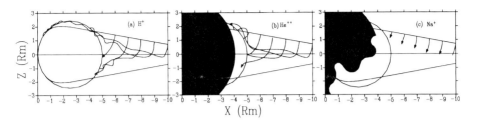

Fig. 5 Results of test particle simulations (Ip 1997) in a depolarization event (*from left to right*: p$^+$, He^{++}, Na$^+$)

acceleration both protons and α-particles could achieve about 10–12 keV gain in energy (Ip 1997).

5 Inductive Acceleration Near Reconnection Regions

A number of convincing experimental facts suggest acceleration processes are often related to spontaneous magnetic reconnection and the formation of a neutral regions in some parts of the magnetotail (see, e.g., classical papers by Krimigis and Sarris 1980; Hones 1984). Thus the generation of the energetic particle bursts can be related to the effective acceleration by inductive electric fields in the vicinity of the regions with small or zero magnetic fields when available magnetic energy is continuously (or intermittently) transferred to thermal and kinetic energy of plasma particles. Rather strong correlations between energetic particles bursts registered on board different spacecraft, and various signatures of neutral regions formation and motion, have been observed experimentally long time ago (Krimigis and Sarris 1980).

The energy source of the non-stationary impulsive processes such as substorms is the solar wind. Galeev et al. (1986) have shown that the energy transport from the dayside to the nightside of the magnetosphere usually is accomplished by separate "quanta" of magnetic flux. Such mode of reconnection is known in literature as flux transfer events and corresponds to the transfer of magnetic filaments of finite diameter from dayside to nightside of the magnetosphere. Later on, Kuznetsova and Zeleny (1986) have shown that the transfer process is much more effective for the compact magnetospheres of Earth and Mercury than for the giant magnetospheres of Jupiter or Saturn. For the terrestrial and Hermean magnetospheres the ratio of the magnetic flux in one flux tube transferred from dayside to nightside of the magnetosphere to the entire planetary tail magnetic flux is of the order of 2–3% (Kuznetsova and Zeleny 1986). Recently, some evidence of so-called "pulsating" dayside reconnection was found by Cluster. Relatively large value of these reconnection portions continuously added to the tail should induce pronounced transient effects in tail dynamics.

The most effective acceleration of particles during reconnection occurs in the vicinity of neutral lines of the magnetic field. Thus the reconnection process of this kind was very intensively discussed 2–3 decades ago in connection with the problem of charged particle acceleration in the course of solar flares (Friedman 1969; Syrovatskii 1981). The problem of spontaneous particle acceleration in the Earth's magnetosphere was investigated in detail by Galeev (1979), Zelenyi et al. (1984), and Terasawa (1981).

We discussed above that observations of energetic particle bursts in the magnetosphere of Mercury confirmed the existence of electron bursts with energies up to 600 keV at the night

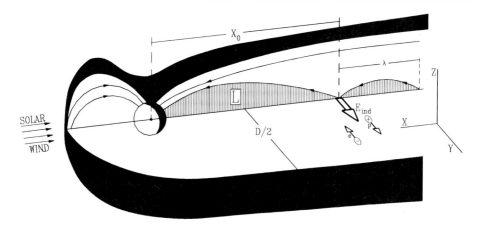

Fig. 6 The model of inductive acceleration after Zelenyi et al. (1990). Main energy gain occurs in the vicinity of X-line (AR – acceleration region)

side of the planetary magnetosphere. As for ions the authors of Mariner 10 papers finally argued that there are no definitive arguments in favor of the existence of ion bursts. The question whether experimental difficulties play the main role will be hopefully answered after Messenger observations.

The evaluation of the main parameters of particle acceleration (the maximum energy, their spectra and fluxes, burst duration, the time scale of the entire substorm processes) near magnetic X-line have been done by Zelenyi et al. (1990). The particles are accelerated by the inductive electric field (1) during the magnetic field topology reconstruction near the neutral regions of the magnetic field. The inductive electric field generated during reconnection in the vicinity of magnetotail X-lines is directed from the dawn to dusk flank for both terrestrial and Hermean magnetospheres.

For the beginning one could consider the simple quasi-2D model of Mercury's magnetotail

$$B = B_{0x} th(z/L) \vec{e}_x + B_z(t) \sin kx \cdot \vec{e}_z \qquad (2)$$

where the finite size of the reconnection domain in Y direction is also taken into account ($Y < D$). The evolution of the perturbed magnetic component $B_z(t)$ might have various forms. Zelenyi et al. (1984) have shown that the most effective acceleration is achieved for so-called explosive growth

$$B_z(t) = B_{0z}/(1 - t/\tau_r), \qquad B_{0z} = B_z(0), \qquad (3)$$

where τ_r is the characteristic time of this process. For the geometry represented in Fig. 6, protons are accelerated in the dawn to dusk direction, and electrons, naturally, to the opposite flank. The motion in X direction is unstable due to the influence of the Lorentz force $F_L \sim v_y \times B_z$, and particles are finally ejected out of it. In contrast, the motion in Z direction is stable, and particles accomplish rapid acceleration, being localized in the region of weak magnetic field.

Test particle experiments allowed estimating the maximum energies of ions in the Hermean and terrestrial magnetospheres after development of tearing-like perturbations in the

Table 1 Theoretical and observed acceleration parameters for the terrestrial and Hermean magnetosphere

Parameter	Observations theory			
	Earth		Mercury	
ε_e^{max}, MeV	≥ 1	≥ 1	0.600	1
ε_p^{max}, MeV	≥ 1	1.6	?	80
γ_e	3.4–8.1	4–10	7–9	4–11.8
γ_p	2–7	2–7	?	1.2–1.4
τ_B, s	10–10^2	50	1–2	0.2
τ_{ss}, min	10–60	20	1	0.4

tail magnetic field. Thus for protons at the Earth for $B_{0z}/B_0 = 0.1$ the maximum energy is $(E_p^{max})_{Earth} \approx 1.6$ MeV. For Mercury magnetosphere this energy is much smaller: $(E_p^{max})_{Mercury} \approx 60$ keV. For comparison, in the Helios observations proton and electron energies reach up to 80 keV and 60 keV. For sodium the maximum gained energy is about a few keV. The results of test simulations indicate that there is not enough room for acceleration of heavy particles in the compact Hermean magnetosphere. As it was shown, the electron acceleration is weakly impacted by the size of the Hermean magnetosphere. For example, the maximum energy of electrons in both cases $(E_e^{max})_{Mercury}$ and $(E_e^{max})_{Earth}$ could be as large as a few MeV for realistic tail parameters, but the flux of such particles could be smaller than the sensitivity of corresponding instruments. Estimates for the Earth give values of about 10 MeV.

The theory also allows estimating the shape of the energetic power spectra. Table 1 contains these results and a comparison with experimental data from Mariner 10 and few terrestrial magnetosphere spacecraft. The theoretical spectra, although obtained assuming explosive perturbation growth, conform rather well to experimental data both for Mercury and Earth. To have the complete picture, the table also includes characteristic times of energetic particle bursts τ_p. This confirms our assumption about the single common mechanism for the generation of energetic bursts of ions and electrons in both the terrestrial and Hermean magnetospheres. The chaotization processes near an X-line studied by Martin (1986) were neglected in such estimates because it was considered only a special group of particles whose motion is strongly controlled by electric field and is therefore regular. Thermal population otherwise may be scattered and isotropized due to chaotic effects.

Mariner 10 observed energetic particle bursts at Mercury at the same time when the magnetic field was changing rapidly. These observations have been interpreted as possible substorms analogous to substorms at the Earth (Siscoe et al. 1975; Ogilvie et al. 1977; Christon et al. 1987). Observed accelerated electron bursts in this interpretation are generated in the course of substorm dipolarization and non-adiabatic motion in the Hermean magnetosphere (Delcourt et al. 2005). At Earth ionospheric line-tying (Coroniti and Kennel 1973) and ionospheric–magnetospheric feedback (Baker et al. 1996) are thought to be very important. If substorms really are occurring at Mercury, the question is open as to how they operate in a spatial environment that has a minimal (if any) ionosphere.

6 Ergodic, Chaotic and Long-Range Correlation Effects

Simplistic early theories discussed above, mostly based on single-particle calculations gave a good insight into the basic physics of Hermean plasma acceleration missed few very important elements.

6.1 Chaotic Motion

Non-adiabatic motion on a longer time frame even in small κ limit (κ is the adiabaticity parameter) generally becomes chaotic (Büchner and Zelenyi 1989). Chaotic effects could also accompany X/O – line acceleration (Martin 1986). However strong inductive electric fields (few mV/m) usually produced during sporadic magnetic reconnection both in the terrestrial and the Hermean magnetosphere usually act to suppress chaotic motion. Later after leaving the acceleration region particles could get chaotized, but their motion during the short time interval of acquiring energy can be considered as regular.

6.2 Ergodicity

The estimates given in the previous section pertain to a single particle acceleration region (AR). In reality many such regions could independently (or almost independently, see below) operate in the tail. After escaping the AR, particles appear in different parts of the magnetotail at different times. During in situ measurements on a minute time scale one can observe the manifestations of the acceleration from an individual source. After averaging of the energetic particle data on significantly larger time scale (tens of minutes) we should observe the effect of the acceleration at multiple sources operating at various positions in the magnetotail. Energetic particles leaving the AR can "live" afterwards in the tail without additional acceleration (i.e. conserving their spectra) for quite a long time due to partial trapping near the current sheet (Savenkov and Zelenyi 1996). So the observer could also register particle accelerated at the reconnection sources well before the registration time. Thus, calculating the particle spectra by averaging the data over time, one can obtain time averages of the instant distributions provided by acceleration sources. In the analysis discussed above all particles are supposed to gain energy in corresponding acceleration region, irrespective of their further trajectories after the acceleration process is over. So, if one calculates their resulting accumulated spectra, this corresponds to spatial averaging of the particle distribution. The nontrivial, but reasonable "ergodic" assumption is that one could relate the spatially averaged theoretical distributions, formed at a single source, to ensemble averaged (equivalent to time averaged) experimental data. As usual, the validity of this assumption depends on the time scale of averaging. When averaging over a few minutes we will recover again "individual" very spiky spectra drastically different from the smooth averaged ones. One should also mention here that the influence of particle leakage from the tail (exits from the magnetospheric flanks and/or particle precipitation into the polar regions) on the spectra formation is not taken into account.

6.3 Long-Range Correlations of Multiple Acceleration Regions

More modern views on the structure of plasma sheet assume that it includes multi-scale magnetic perturbations operating at a variety of time scales. This is the generalization of the model of multiple ergodic acceleration regions described above. The multi-scale character of magnetic perturbations (very well demonstrated for the magnetospheric tail in the spectra of low-frequency magnetic fluctuations; see, e.g., Bauer et al. 1995; Vörös et al. 2003) allows the description of the tail as a fractal object (Milovanov and Zelenyi 1998). Figure 7, showing multiple ARs, illustrates this concept. Generally speaking, the reason for stochastic acceleration processes is the energy exchange between plasma particles wandering in the fractal set labyrinths (Milovanov and Zelenyi 2002) and the electromagnetic field fluctuations which scatter these particles. Such process may be considered as a generalization of stochastic Fermi acceleration.

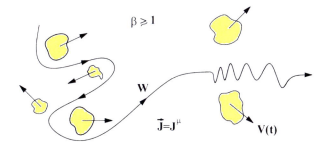

Fig. 7 The turbulent acceleration of plasma particles (Zelenyi and Milovanov 2004)

The mechanism of acceleration is actually the same, inductive electric fields, but now it operates in a much more complicated, entangled manner. In fact, dynamic relaxation processes assume an energy exchange between different subsystems comprising the turbulent ensemble. These subsystems usually include turbulent fields interacting both with themselves and with particles captured by them. The currents generated by these particles can in their turn become a source of the turbulent field. The system is reminiscent of a "boiling soup" of particles and fields. The transition of a turbulent system to the non-equilibrium stationary state is in many cases related to the occurrence of a population of energetic particles with the power-law tail in the velocity (energy) distribution function.

Planetary magnetotails represent space plasma structures with a large value of the plasma-to-field energy density ratio β (see, e.g., Fig. 1 of Baumjohann 2002). For such a condition the processes of self-organization in the system result in the turbulent magnetic field concentrating into magnetic clots, which then form fractal "mosaics" in the system configuration space. Mosaics are dynamical structures participating in the process of intrinsic variability resembling "self-pouring" of these magnetic structures. Inductive electric fields accompanying these variations are capable to accelerate particles up to high energies if the system exists for sufficiently long time (tens of hours for the Earth's magnetotail). These effects are especially important for the strongly solar wind-driven very dynamical and compact Hermean magnetosphere.

The interaction of particles with clots leads to a gradual heating of the plasma. The explanation is that for a chaotic velocity distribution of clots particles interact more often with the clots, moving in the opposite direction, which on average increases the particle kinetic energy. This mechanism was first proposed by Fermi under the assumption that particle scattering on clots has a random (Gaussian) character. For Gaussian scattering, the mean-square variation of the particle velocity is proportional to the time particles stay in the turbulent domain:

$$\langle \delta w^2(t) \rangle \propto t. \qquad (4)$$

This dependence leads to the standard diffusion equation for the probability density corresponding to the Gaussian diffusive acceleration:

$$\frac{\partial \psi}{\partial t} = \Delta_w \psi. \qquad (5)$$

Such process can be considered a random Brownian motion of a particle in the (three-dimensional) space of velocities. Diffusion coefficient Δ_w is the standard three-dimensional Laplacian, $\psi = \psi(t, \vec{w})$ is the particle velocity distribution.

New effects are appearing if one could not neglect the correlations between different clots. This naturally might happen for more compact systems when the influence of boundaries becomes sufficiently important. The size of the magnetic clots is usually controlled by particle Larmor radii and one may speculate that for the Hermean magnetosphere with its rather small size these correlation effects could become very important. Particle acceleration in the ensemble of non-Gaussian correlated fluctuations could have very different characteristics from the standard Fermi-like case (Milovanov and Zelenyi 2001). This acceleration is called "strange" Fermi acceleration and could be both faster and slower than the standard diffusion depending on the parameters of correlation function (4): $\langle \delta w^2(t) \rangle \sim t^\gamma$, $0 < \gamma < 2$. Most interesting are the cases of superdiffusion $\gamma \to 2$, when the motion of particles becomes almost ballistic. This happens when correlations line up fluctuations in special roads coherently accelerating particles interacting with them. This problem has a well developed mathematical tool to describe it (fractional derivatives) and its application to the problem of permanent acceleration in dynamic Hermean magnetotail could produce very interesting results. Figure 7 illustrates this concept of particle acceleration due to interaction with correlated "clots" of magnetic turbulences in the magnetotail.

7 Shock Particle Acceleration in the Hermean Environment

Shock acceleration of charged particles is the one of the most prominent problems in the space plasma physics. The observations in the Hermean environment, that is, observations of the interplanetary shocks (IPSs) near Mercury's orbit, provide us with the opportunity to look further into this process on the basis of in situ measurements. The interplanetary conditions around Mercury have been intensively studied by the solar-orbiting Helios spacecraft in the 1970s. According to a statistical study by Russell et al. (1988), the solar wind average speed, density, and interplanetary magnetic field magnitude are 430 km/s, 30–70 cm^{-3}, and 20–45 nT, respectively. These numbers yield an average Alfven speed of 80–120 km/s. Hence, the Alfven Mach numbers of strong IPSs would be as high as $M_A \sim 40$ because their expected propagation speeds range between 1000–4000 km/s (Smart and Shea 1985; the speed of a shock remains high up to the Mercury orbit but decreases as it propagates further outward). This makes such high Mach numbers, in contrast with those observed at 1 AU by Earth-orbiting spacecraft. The IPSs at 1 AU have at most $M_A \sim 10$ even in super-flare associated events.

Electron acceleration is one of the most outstanding problems of collisionless shock physics and in situ measurements of high M_A shocks are indeed crucial to advance our understanding of this issue. While it is widely accepted that the terrestrial bow shock ($M_A \sim 6$) accelerates electrons (Gosling et al. 1989; Oka et al. 2006), IPSs at 1 AU ($M_A \sim 10$) rarely shows electron acceleration (Treumann and Terasawa 2001). On the other hand, X-ray emission from supernova remnants is evidence for electron acceleration at those extremely high Mach number shocks ($M_A \sim 100$–1000). Since the Mach number of IPSs at the Mercury orbit can be as high as 40, close to the supernova shock regime, observations there could shed some light on this issue, i.e., on how the electron acceleration efficiency depends on the shock Mach number.

Recently, the understanding of non-stationarity, or reformation, of shock fronts has advanced considerably owing to a number of self-consistent particle-in-cell simulation studies. However, the condition for reformation to occur is still unclear. While an analytical treatment with help of computer simulation predicts reformation to occur above the so-called non-linear whistler critical Mach number (Krasnoselskikh et al. 2002), another study discusses the possibility that reformation may not exist at higher Mach numbers (Shimada and Hoshino 2005). Observational tests in the low Mach number regime by the data obtained at the Earth's bow shock are being conducted, but observations at the Mercury orbit of high

Mach number shocks are also required to cross-check the above theoretical results. Since electron dynamics are sensitive to the temporal behavior of the shock front and to the waves radiated from it, understanding the physics of the shock front is an indispensable step towards the understanding of electron shock acceleration.

The shocks, especially high Mach number shocks, produce non-thermal component of particles and non-linear effects emerge when the contribution from non-thermal energy density can no longer be neglected. This effect is believed to be quite effective in astrophysical shocks, such as supernova remnant shocks, and has been studied intensively with the name "cosmic ray mediated shock" (CRMS). A possible scenario in this line is that some fraction of non-thermal energy have dynamic effects on the macroscopic shock structure and thus on the efficiency of acceleration (Eichler 1979). Another possible scenario is that the accelerated particles stream toward upstream to produce magnetic field turbulence in the background plasma (Lucek and Bell 2000). Such turbulence should further scatter particles and increases acceleration efficiency. However, direct observational evidence of CRMS has been limited only to the cases of the Earth's bow shock (Zhang et al. 1995) and interplanetary shocks at 1 AU (Terasawa et al. 2006). Higher Mach number shocks will show more profound features of CRMS and observations at the Mercury orbit provide us with opportunity to study the details of the CRMS and its effect on particle acceleration efficiency.

As mentioned, plasma measurements at the Mercury orbit will make a crucial contribution for our understanding of the Plasma Universe. The relatively high Mach numbers of the IPSs in the Hermean environment, which is not realized at 1 AU, makes observations at the Mercury position necessary to fill the parameter gap between the low Mach number regime at 1 AU ($M_A < 10$) and the high Mach number regime ($M_A > 100$). In other words, they will bridge between our knowledge from numerous Earth's bow shock crossings and the astrophysical situations that can be sensed only remotely and indirectly. Especially, the BepiColombo MMO is equipped with fairly good plasma instruments (as good as the ones onboard Earth-orbiting spacecraft) and will arrive at Mercury during the next solar maximum. A number of super-solar-flare events will occur and we can expect high-quality in situ data of high Mach number shocks. It is noted that, besides Alfven Mach number, shock angle is also an important parameter which characterizes shocks. It is the cone angle between shock normal and magnetic field direction in the upstream side of a shock, and since the interplanetary magnetic field (IMF) is nominally in the Parker spiral configuration, the average shock angle of IPSs at the Mercury orbit is $\theta_{Bn} \sim 30°$ ($\sim 45°$ at 1 AU). Therefore a considerable amount of studies on high Mach number and quasi-parallel shocks can be anticipated. High-resolution X-ray images of a supernova remnant (SNR1006) by ASCA and more recently Chandra show clearly a non-spherical symmetric feature, which revitalizes interest in the shock angle control on the particle acceleration efficiency.

Studies on particle acceleration by shocks have close connection with space weather research. The Solar Energetic Particles (SEPs) are one of the main topics of space weather prediction and it is important to estimate quantitatively the particle flux associated with IPSs. Understanding the physics of particle acceleration is a way and the observations at the Mercury orbit can contribute in the sense as described above. The observations can also contribute from another aspect and that is to realize multi-point observations over the heliospheric-scale. Although the basic morphology of SEPs at 1 AU, including time profile and particle compositions, is now known, its evolution over the heliospheric scale is still unclear. The next decade, including the BepiColombo MMO period, will be the best time to perform this study to advance our understanding of particle acceleration in the inner heliosphere. A number of spacecraft, that are now under development or already launched, will constitute the fleet deployed over the heliospheric scale. These multi-point measurements allow us to study not only the dynamical structures of flux-ropes or Coronal Mass Ejections (CMEs) that generate IPSs but also long-time and large-scale evolution of micro-physics relevant to particle acceleration, such as the initial injection or non-linear effects.

In the study of SEPs, the contributions from the pickup ions of interstellar origin (PUIs) should not be neglected. Because of their peculiar position in the velocity space (Moebius et al. 1985; Oka et al. 2002a), these ions are expected to be accelerated efficiently. Indeed, there are evidences of PUI acceleration by shocks (Gloeckler et al. 1994; Oka et al. 2002b; Kucharek et al. 2003) and these ions must be taken into account to predict quantitatively the SEP flux. Furthermore, the studies of PUI acceleration by shocks can be extrapolated to the physics of the heliospheric termination shocks where incoming upstream plasma is expected to be dominated by the PUIs.

Acknowledgements The work by L.Z. and H.M. is supported by the RFBR grants 05-02-17003, 07-02-00319, 06-05-90631, INTAS grant 06-1000017-8943 and grant of Scientific schools HIII-5359.2006.2.

References

T.P. Armstrong, S.M. Krimigis, L.J. Lanzerotti, J. Geophys. Res. **80**, 4015 (1975)
D.N. Baker, J.A. Simpson, J.H. Eraker, J. Geophys. Res. **91**, 8742 (1986)
D.N. Baker, T.I. Pulkkinen, V. Angelopoulos, W. Baumjohann, R.L. McPherron, J. Geophys. Res. **101**, 12975–13010 (1996)
T.M. Bauer, W. Baumjohann, R.A. Treumann, N. Sckopke, H. Lühr, J. Geophys. Res. **100**, 9605–9617 (1995)
W. Baumjohann, Phys. Plasmas **9**, 3665–3667 (2002)
W. Baumjohann, A. Matsuoka, K.-H. Glassmeier, C.T. Russell, T. Nagai, M. Hoshino, T. Nakagawa, A. Balogh, J.A. Slavin, R. Nakamura, W. Magnes, Adv. Space Res. **38**, 604–609 (2006)
T.A. Bida, R.M. Killen, T.H. Morgan, Nature **404**, 159 (2000)
A.L. Broadfoot, D.E. Shemanky, S. Kumar, Geophys. Res. Lett. **3**, 577 (1976)
J. Büchner, L.M. Zelenyi, J. Geophys. Res. **94**, 11821 (1989)
S.P. Christon, S.F. Daly, J.H. Eraker, M.A. Perkins, J.A. Simpson, A.J. Tuzzolino, J. Geophys. Res. **84**, 4277–4288 (1979)
S.P. Christon, J.A. Simpson, Astrophys. J. **227**, L49–L53 (1979)
S.P. Christon, Icarus **71**, 448 (1987)
S.P. Christon, J. Feynman, J.A. Slavin, in *Magnetotail Physics*, ed. by A.T.Y. Lui (John Hopkins Univ. Press, Baltimore, 1987), p. 393
F.V. Coroniti, C.F. Kennel, J. Geophys. Res. **78**, 2837 (1973)
D.C. Delcourt, J.A. Sauvaud, J. Geophys. Res. **99**, 97–108 (1994)
D.C. Delcourt, R.F. Martin, J. Geophys. Res. **99**, 23583 (1994)
D. Delcourt, R.F. Martin, F. Alem, Adv. Space Res. **18**, 295–298 (1996)
D.C. Delcourt, T.E. Moore, S. Orsini, A. Millilo, J.A. Sauvaud, Geophys. Res. Lett. **29**, 1591 (2002). doi:10.1029/2001GL013829
D.C. Delcourt, S. Grimald, F. Leblanc, J.-J. Berthelier, A. Millilo, A. Mura, S. Orsini, T.E. Moore, Ann. Geophys. **21**, 1723 (2003)
D.C. Delcourt, K. Seki, N. Terada, Y. Miyoshi, Ann. Geophys. **23**, 3389 (2005)
D. Eichler, Astrophys. J. **229**, 419–423 (1979)
J.H. Eraker, J.A. Simpson, J. Geophys. Res. **91**, 9973 (1986)
M. Friedman, Phys. Rev. **182**(5), 1408–1414 (1969)
A.A. Galeev, Space Sci. Rev. **23**, 411–425 (1979)
A.A. Galeev, M.M. Kuznetsova, L.M. Zelenyi, Space Sci. Rev. **44**, 1–41 (1986)
K.-H. Glassmeier, N.P. Mager, D.Y. Klimushkin, Geophys. Res. Lett. **30**, 1928 (2003). doi:10.1029/2003GL017175
G. Gloeckler, J. Geiss, E.C. Roelof, L.A. Fisk, F.M. Ipavich, K.W. Ogilvie, L.J. Lanzerotti, R. von Steiger, B. Wilken, J. Geophys. Res. **99**, 17637–17643 (1994)

B.E. Goldstein, S.T. Suess, R.J. Walker, J. Geophys. Res. **86**, 5485 (1981)
J.T. Gosling, M.F. Thomsen, S.J. Bame, C.T. Russell, J. Geophys. Res. **94**, 10011–10025 (1989)
E.W. Hones Jr., *Magnetic Reconnection in Space and Laboratory Plasmas*. Geophysical Monograph, vol. 30, (American Geophysical Union, Washington, DC, 1984)
D.M. Hunten, D.E. Shemansky, T.H. Morgan, in: *Mercury (A89-43751, 19-91)* (University of Arizona Press, Tucson, 1988), pp. 562–612
W.-H. Ip, Icarus **71**, 441 (1987)
W.-H. Ip, Adv. Space Res. **19**(10), 1615 (1997)
R.M. Killen, W.-H. Ip, Rev. Geophys. **37**, 361 (1999)
R.M. Killen et al., J. Geophys. Res. **106**, 20509 (2001)
E. Kirsch, A.K. Richter, Ann. Geophys. **3**, 13–18 (1985)
V.V. Krasnoselskikh, B. Lembege, P. Savoini, V.V. Lobzin, Phys. Plasmas **9**, 1192–1209 (2002)
S.M. Krimigis, E.T. Sarris, in: *Dynamics of the Magnetosphere* (Reidel, Dordrecht, 1980), pp. 599–630
H. Kucharek, E. Möbius, W. Li, C.J. Farrugia, M.A. Popecki, A.B. Galvin, B. Klecker, M. Hilchenbach, P.A. Bochsler, J. Geophys. Res. **108**, LIS 15-1 (2003). doi:10.1029/2003JA009938
M.M. Kuznetsova, L.M. Zeleny, in: *ESA Proceedings of the Joint Varenna-Abastumani International School and Workshop on Plasma Astrophysics* (1986), pp. 137–146
F. Leblanc, D. Delcourt, R.E. Johnson, J. Geophys. Res. **108**, 5136 (2003). doi:10.1029/2003JE00215112
S.G. Lucek, A.R. Bell, Mon. Not. Roy. Astron. Soc. **314**, 65–74 (2000).
J.G. Luhmann, L.M. Friesen, J. Geophys. Res. **84**, 4405–4408 (1979)
J.G. Luhmann, C.T. Russell, N.A. Tsyganenko, J. Geophys. Res. **103**, 9113–9120 (1998)
R.F. Martin Jr., J. Geophys. Res. **91**, 11985–11992 (1986)
R.L. McPherron, C.T. Russell, M.P. Aubry, J. Geophys. Res. **78**, 3131 (1973)
A.V. Milovanov, L.M. Zelenyi, Astrophys. Space Sci. **264**, 317–345 (1998)
A.V. Milovanov, L.M. Zelenyi, Phys. Rev. E **64**, 052101 (2001)
A.V. Milovanov, L.M. Zelenyi, Adv. Space Res. **30**, 2667–2674 (2002)
A. Milillo, P. Wurz, S. Orsini, D. Delcourt, E. Kallio, R.M. Killen, H. Lammer, S. Massetti, A. Mura, S. Barabash, G. Cremonese, I.A. Daglis, E. De Angelis, A.M. Di Lellis, S. Livi, V. Mangano, K. Torkar, Space Sci. Rev. **117**, 397 (2005)
E. Moebius, D. Hovestadt, B. Klecker, M. Scholer, G. Gloeckler, Nature **318**, 426–429 (1985)
N.F. Ness, K.W. Behannon, R.P. Lepping, Y.C. Whang, K.H. Schatten, Science **185**, 151 (1974)
K.W. Ogilvie, J.D. Scudder, V. Vasyliunas, R.E. Hartle, G.L. Siscoe, J. Geophys. Res. **82**, 1807 (1977)
M. Oka, T. Tersaawa, H. Noda, Y. Saito, T. Mukai, Geophys. Res. Lett. **29** (2002a). doi:10.1029/2002GL015111
M. Oka, T. Terasawa, H. Noda, Y. Saito, T. Mukai, Geophys. Res. Lett. **29** (2002b). doi:10.1029/2001GL014150
M. Oka, T. Terasawa, Y. Seki, M. Fujimoto, Y. Kasaba, H. Kojima, I. Shinohara, H. Matsui, H. Matsumoto, Y. Saito, T. Mukai, Geophys. Res. Lett. **33**, L24104 (2006). doi:10.1029/2006GL028156
A.E. Potter, T.H. Morgan, Science **229**, 651 (1985)
A.E. Potter, R.M. Killen, T.H. Morgan, Meteorit. Planet. Sci. **37**, 1165–1172 (2002)
C.T. Russell, D.N. Baker, J.A. Slavin, in *Mercury*, ed. by F. Vilas, C.R. Chapman, M.S. Matthews (University of Arizona Press, 1988), pp. 514
B.V. Savenkov, L.M. Zelenyi, Geophys. Res. Lett. **23**, 3255 (1996)
N. Shimada, M. Hoshino, J. Geophys. Res. **110**, A02105 (2005)
J.A. Simpson, J.H. Eraker, J.E. Lamport, P.H. Walpole, Science **185**, 160 (1974)
G.L. Siscoe, L. Christopher, Geophys. Res. Lett. **2**, 158 (1975)
G.L. Siscoe, N.F. Ness, C.M. Yeates, J. Geophys. Res. **80**, 4359 (1975)
J.A. Slavin, Adv. Space Res. **33**, 1859 (2004)
D.F. Smart, M.A. Shea, J. Geophys. Res. **90**, 183–190 (1985)
S.I. Syrovatskii, Ann. Rev. Astron. Astrophys. **19**, 163–229 (1981)
T. Terasawa, J. Geophys. Res. **86**, 9007 (1981)
T. Terasawa, M. Oka, K. Nakata, K. Keika, M. Nose, R.W. McEntire, Y. Saito, T. Mukai, Adv. Space Res. **37**, 1408–1412 (2006)
R.A. Treumann, T. Terasawa, Space Sci. Rev. **99**, 135–150 (2001)
Z. Vörös, W. Baumjohann, R. Nakamura, A. Runov, T.L. Zhang, M. Volwerk, H.U. Eichelberger, A. Balogh, T.S. Horbury, K.-H. Glassmeier, B. Klecker, H. Réme, Ann. Geophys. **21**, 1955–1964 (2003)
L.M. Zelenyi, A.S. Lipatov, J.G. Lominadze, A.A. Taktakishvili, Planet. Space. Sci. **32**, 312–324 (1984)
L.M. Zelenyi, J.G. Lominadze, A.L. Taktakishvili, J. Geophys. Res. **95**, 3883 (1990)
L.M. Zelenyi, A.V. Milovanov, Uspekhi Fizicheskikh Nauk (Transl. from Russian) **47**(8), 749–788 (2004)
T.-L. Zhang, K. Schwingenschuh, C.T. Russell, Adv. Space Res. **15**, 137–140 (1995)

Missions to Mercury

André Balogh · Réjean Grard · Sean C. Solomon ·
Rita Schulz · Yves Langevin · Yasumasa Kasaba ·
Masaki Fujimoto

Originally published in the journal Space Science Reviews, Volume 132, Nos 2–4.
DOI: 10.1007/s11214-007-9212-4 © Springer Science+Business Media B.V. 2007

Abstract Mercury is a very difficult planet to observe from the Earth, and space missions that target Mercury are essential for a comprehensive understanding of the planet. At the same time, it is also difficult to orbit because it is deep inside the Sun's gravitational well. Only one mission has visited Mercury; that was Mariner 10 in the 1970s. This paper provides a brief history of Mariner 10 and the numerous imaginative but unsuccessful mission proposals since the 1970s for another Mercury mission. In the late 1990s, two missions—MESSENGER and BepiColombo—received the go-ahead; MESSENGER is on its way to its first encounter with Mercury in January 2008. The history, scientific objectives, mission designs, and payloads of both these missions are described in detail.

Keywords Mercury · Mariner 10 · MESSENGER · BepiColombo

1 Introduction

Mercury is the innermost planet, the terrestrial planet closest to the Sun. It is very difficult to observe from the Earth because it can be viewed in visible light only just before sunrise

A. Balogh (✉)
International Space Science Institute, Bern, Switzerland
e-mail: balogh@issibern.ch

R. Grard · R. Schulz
Research and Scientific Support Department, ESA, Noordwijk, The Netherlands

S.C. Solomon
Department of Terrestrial Magnetism, Carnegie Institution of Washington, Washington, DC 20015, USA

Y. Langevin
Institut d'Astrophysique Spatiale, Orsay, France

Y. Kasaba · M. Fujimoto
Japan Aerospace Exploration Agency (JAXA), Tokyo, Japan

or just before sunset, low above the horizon. The elongation of Mercury is always less than 30° from the Sun. Despite this handicap, it was already well known in the world of classical Greece and was named Hermes, the winged messenger of the gods; its Roman name was Mercury. (Of course, seeing with the naked eye was considerably better then, without the atmospheric and light pollution that we have to contend with today.)

The precise orbital period of Mercury around the Sun has been known for a long time. However, the difficulties of seeing features on Mercury to help with determining its rotation period (the Mercury day) delayed the recognition of the 3 : 2 resonance between Mercury's spin rate and orbital mean motion until the mid-1960s (Colombo 1965). The reason for this difficulty was another near-resonance: the synodic period of Mercury (the orbital period when seen from the Earth) is in an almost 4 : 3 resonance with Mercury's orbital period (e.g., Balogh and Giampieri 2002), so that the same face of Mercury is seen repeatedly from the Earth. This led to the earlier, erroneous conclusion that Mercury was in a 1 : 1 spin–orbit resonance state, as is the Moon with respect to the Earth.

Because of the proximity of Mercury to the Sun, there have been several difficulties in gaining better information about the planet. The first of these difficulties is to observe it from Earth, although significant progress has been made in radar, visual, and infrared (IR) observations (see articles in this volume: Harmon 2007; Ksanfomality et al. 2007; ...). The second challenge is to orbit Mercury by spacecraft, as the planet is deep inside the gravitational potential well of the Sun. The third obstacle is the very hostile thermal environment that awaits any spacecraft in Mercury orbit; this environment consists of increased solar irradiance (up to a factor 10) as well as the thermal radiation from the sunlit side of the planet. As a result, to date only one spacecraft, Mariner 10, reached Mercury more than 30 years ago. Another spacecraft, MESSENGER, is on its way at present to a first flyby in 2008 and insertion into Mercury orbit in 2011. Both spacecraft have been flown by the U.S. National Aeronautics and Space Administration (NASA). A third, more ambitious two-spacecraft mission, BepiColombo, is a joint undertaking by the European Space Agency (ESA) and the Japan Aerospace Exploration Agency (JAXA). Its construction will start in 2007 for a launch in 2013 and insertion into Mercury orbit in 2019.

This paper traces the history of the space missions and mission plans to Mercury, from Mariner 10 through the many proposals to space agencies that were never realised, to the present when, at last, two missions, MESSENGER and BepiColombo, will be targeting Mercury. These missions, their objectives, and their scientific payloads are described in some detail. Much is expected from these missions (Grard and Balogh 2001; Solomon 2003; McNutt et al. 2004; Solomon et al. 2007). It is clear that such a concentrated effort is required to resolve the many outstanding questions regarding this important planet, the end member of the terrestrial planets.

2 Mariner 10: Brief History and Achievements

The difficulties in observing Mercury from the ground, and even with space-based telescopes in Earth-orbit, have meant that little could be known of the planet without a close-up look with a space probe that actually travelled to it. Remarkably, it was only about 11 years after the launch of Sputnik 1 inaugurated the space era that NASA first considered launching a spacecraft to Mercury. The Space Science Board of the National Academy of Sciences, as part of a planetary exploration program developed in 1968, proposed a mission to Mercury via Venus for a 1973 launch opportunity. This was to be Mariner 10, a remarkable and still-unique mission that provided much of what we know, even now, about Mercury.

This was an era of unequalled activity in space, led by the United States, when not only the Apollo missions to the Moon became almost commonplace, but there were also numerous unmanned programs. Many scientific satellites, with a wide range of objectives, were orbiting the Earth, and several spacecraft, both American and Soviet, were sent to explore the two nearest planets, Venus and Mars. The main elements of NASA's planetary exploration program were the Mariner space probes. The objective of six of the 10 Mariner spacecraft built and launched between 1962 and 1973 was Mars; four of these reached their objective and successfully returned data from the red planet. Three were targeted to Venus; two of these successfully returned data from Venus and one of these, Mariner 2, also confirmed the existence of the solar wind during its interplanetary cruise in 1962. The tenth and last in the series, originally called the Mariner Earth–Venus–Mercury mission, acquired the Mariner 10 name after launch.

The mission to Venus and Mercury was made possible by the technique of gravity-assist flybys. Although some orbital calculations provided a likely basis for such a mission, the specific Venus gravity-assist opportunity that enabled it was worked out only in the late 1960s at the Jet Propulsion Laboratory (JPL) in Pasadena, CA. In the simple scenario first adopted for the Mariner Earth–Venus–Mercury mission, a single gravity-assist flyby of Venus was to modify both the velocity of the spacecraft and its orbital plane around the Sun to bring it to the orbital inclination of Mercury. As both Venus and Mercury had to be in specific points in their orbit for the two encounters, opportunities for the launch were identified in 1970 and 1973. The latter was in fact the launch date recommended by the Space Science Board.

Instruments for the scientific payload for what became known as Mariner 10 were selected in mid-1970, and a Project Management Team was formed at JPL. A contract to build the spacecraft was placed in mid-1971. It was a remarkable achievement that the spacecraft was ready and tested for launch on November 3, 1973, from Cape Kennedy onboard an Atlas-Centaur launcher. A number of "firsts" was achieved by Mariner 10. It performed the first gravity-assist flyby of a planet (Venus) on the way to another (Mercury), and it was the first to reach so close to the Sun, with all the challenges that represented for the thermal design of the payload and spacecraft.

It is interesting to note the scientific objectives as represented by the instruments selected for the payload: imaging (using dual television cameras and telescopes), IR radiometry, extreme ultraviolet (EUV) spectroscopy, magnetometry, plasma and charged particle characteristics, and radio wave propagation. The Mariner 10 spacecraft and its payload are illustrated in Fig. 1. The objective was to learn as much as possible about Mercury. This involved primarily the television (TV) cameras, as no reliable images of the planet existed. An interesting aspect of the payload was the inclusion of a magnetometer, as Mercury was thought not to possess a planetary magnetic field; it is fortunate that this instrument was included, as it led to the discovery of perhaps the most puzzling aspect of Mercury.

The Mariner 10 spacecraft weighed 503 kg; it was 140 cm diagonally and 46 cm high. It was three-axis stabilised using cold-gas thrusters. The two solar panels measured 2.7×1.0 m each, which represented in total an area of 5.1 m^2 of solar cells. The maximum power delivered by the solar arrays was 540 W. The parabolic antenna, for communication with the Earth through NASA's Deep Space Network, had a diameter of 137 cm.

From a technical point of view, there were a number of serious challenges during the construction and testing of the spacecraft, but these were overcome prior to the launch. One of these problems related to the tape recorder that was to be used during mission, in particular during the flybys, to buffer the data acquired by the instruments prior to transmission to the ground. This was clearly a critical subsystem; extensive testing led to the identification of

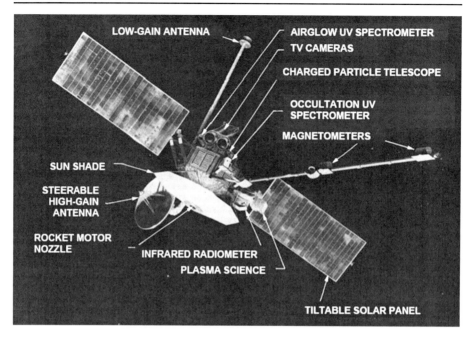

Fig. 1 The Mariner 10 spacecraft and its scientific payload

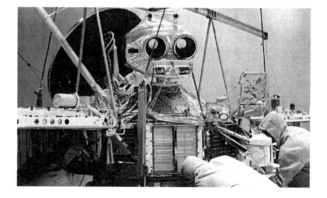

Fig. 2 The final assembly stage of the Mariner 10 spacecraft. Its most prominent feature is the dual vidicon stereo camera sent back images from the planet

the problem (the relative humidity levels of the tape and the internal atmosphere of the tape transport system) and its solution. In the end, the tape recorder worked very well throughout the mission, ensuring that all the flyby data were transmitted to the ground. A photo of the spacecraft during its final assembly stage is shown in Fig. 2.

There were also a number of difficulties and near-failures that occurred during the mission, such as the problem with the heaters which kept the telescopes and the TV cameras (vidicon tubes) warm. In fact, the heaters failed, but by changing the operation modes (keeping the vidicon tubes switched on) and by optimising the attitude of the spacecraft to prevent the cameras from cooling too much, the mission objectives were fully achieved. Other problems affected the gyros that controlled the attitude of the spacecraft; these problems, also potentially fatal to the mission, were overcome by using the star tracker, a risky strategy that paid off throughout the mission. There were several other moments of concern during the

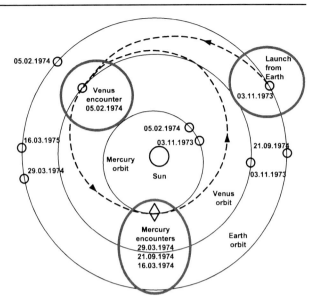

Fig. 3 The ecliptic projection of the trajectory of Mariner 10 and the orbits of Earth, Venus and Mercury, indicating the launch, the Venus flyby and the three Mercury flybys

mission, but thanks to the resilience of the spacecraft and the ingenuity of the mission team all these were overcome to ensure the success of the mission.

Following its launch in November 1973, Mariner 10 flew by Venus three months later on February 5, 1974, with a closest approach at an altitude of 5,768 km. Other than the gravitational kick that was necessary to place the spacecraft on its flyby trajectory towards Mercury, the scientific instruments on Mariner 10 targeted Venus and its atmosphere, returning some previously unknown data about the cloud cover of the planet. Following the Venus flyby, Mariner 10 followed a direct trajectory to Mercury for the first flyby. Originally only a single flyby had been foreseen, but Giuseppe (Bepi) Colombo—a professor from Padua, Italy, visiting JPL in the early 1970s—pointed out to NASA the possibility of a transfer, after the first flyby, to a resonant, multiple flyby orbit. This meant that Mariner 10 orbited the Sun in an eccentric orbit in a time (0.48 years) exactly twice the period of Mercury's orbit around the Sun. The mission trajectory is shown in Fig. 3.

The first encounter with Mercury, on March 29, 1973 (less than 5 months after the launch!), was targeted behind the planet. The spacecraft was able to take a remarkable set of pictures both before and after closest approach (at a height of 705 km). A bare, cratered surface, not unlike that of the Moon, was seen by the TV cameras. A typical illustration of the imaging of Mercury achieved by Mariner 10 is shown in Fig. 4. It shows the very large Caloris impact basin, about half of which was documented by Mariner 10. The major surprise was the detection of a magnetosphere around Mercury, implying the existence of an internal magnetic field that could form a protective shield preventing the solar wind from directly impacting the surface. The implied magnetic field was much weaker than the Earth's, so the magnetosphere was much smaller in relative terms than at the Earth. The existence of an internal magnetic field led to a fundamental revision of ideas regarding the internal evolution of the planet since its formation; the debate about the origin of this magnetic field continues to this day.

The second encounter, on September 21, 1974, had a trajectory on the sunlit side of the planet, with a more distant "closest" approach distance of 48,069 km. This flyby was

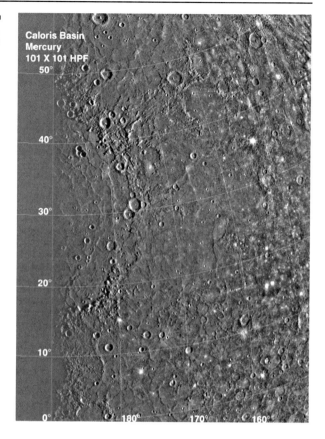

Fig. 4 The Caloris basin, shown here in a 1-km-per-pixel mosaic, is one of the largest basins in the solar system. Its diameter exceeds 1,300 km, and in many ways it is similar to the great Imbrium basin on the Moon (diameter >1,100 km). To enhance landforms a high-pass filter was used in processing. (Photo courtesy NASA)

dedicated to taking further images of Mercury, to ensure as much coverage of the planet as possible; in total, more than 750 images were taken during this encounter.

The dramatic nature of this first mission to Mercury is illustrated by the way the third encounter was achieved (Dunne and Burgess 1978). After the second flyby, the spacecraft was placed in a cruise mode in which the high-gain antenna and solar panels were oriented to use solar radiation pressure to maintain the orientation of the spacecraft to save attitude control gas for a third encounter. However, the Canopus star tracker lost its lock on the reference star, and the spacecraft went into an uncontrolled roll. Attempts to reacquire Canopus, however, depleted the gas supply below that required to achieve a third encounter. It was decided to abandon roll-axis stabilisation, and the spacecraft was allowed to roll slowly, the rate being controlled by differentially tilting the solar panels. The roll rates had to be maintained quite low to prevent excessive use of the pitch and yaw jets, and also to allow gyro turn-on for trajectory correction manoeuvres and pre-encounter reacquisition without inducing an oscillation. However, due to failures in the star tracker and its electronics, it was very difficult to reconstruct the roll position. Modulation in the intensity of the signal from the low-gain antenna dependent on roll position and roll rate was used for this. This emergency procedure was successful, and just enough fuel remained on board to implement the third flyby.

An additional, related problem was that the modulation of the signal from the spacecraft due to the roll introduced a modulation in the Doppler signal and therefore made the reconstruction of Mariner 10's orbit considerably more difficult. However, the three trajectory-correction manoeuvres needed for the third flyby could still be achieved, thanks to an ad hoc modification of the orbit-determination software. Another problem that arose in the approach to the third flyby was that the spacecraft rolled into a position that was a null in the low-gain antenna pattern, so that the spacecraft could no longer be tracked. Through some extraordinary effort and the use of the then-largest ground-tracking stations of NASA, reacquisition was achieved only hours before the encounter.

The objective of the third encounter was to investigate Mercury's magnetic field, one of the major discoveries made during the first flyby. For this, the encounter trajectory aimed at a closest approach point at an altitude of 327 km, the closest of all three flybys, and at higher latitude to see a stronger magnetic field. This tactic was completely successful; the resulting observations eliminated any doubt about the existence of the relatively weak, but still significant, magnetic field of planetary origin. In addition, many more images were returned from the vantage point along this orbit.

Until the new generation of spacecraft arrive at Mercury, Mariner 10 remains the most abundant source of our knowledge about that planet. For a summary, see Murray (1975); for a comprehensive description of our understanding of Mercury after Mariner 10, see the Mariner 10 special issues: *Science*, 185, No. 4146, July 12, 1974; *J. Geophys. Res.*, 80, 2341–2514, 1975; and the chapters in *Mercury*, edited by Villas et al. (1988). For more recent assessments, see Strom and Sprague (2003); and Shirley (2003). Although other planets (Mars, Venus, Jupiter, and now Saturn) have been quite thoroughly explored in comparison since the pioneering era of the 1960s and 1970s, Mercury waits for another space mission, more than 30 years after Mariner 10's achievements. Much is expected even from the first flyby of Mercury by MESSENGER in January 2008.

3 Plans to Follow up Mariner 10

The results of Mariner 10 justified the expectation of planetary scientists that another mission would soon follow the initial success. Further missions were sought to complete the imaging of Mercury, to map its internal magnetic field in order to determine its origin, and to provide the necessary data for better understanding this key terrestrial planet. The Space Science Board (later the Space Studies Board) of the U.S. National Academy of Sciences carried out in-depth studies of the priorities and objectives for exploring the inner planets, first in 1978, then in 1990. These reports, while noting the significance of Mercury, placed much greater emphasis on the detailed exploration of Mars and Venus. For Mercury itself, it was recognised that the difficulty of getting to Mercury probably implied the use of low-thrust propulsion systems. In the context of development studies for both solar sail and solar electric propulsion (SEP) missions, Mercury served as a potential target. However, such propulsion systems had not been developed at that time. In any case, the scientific arguments called for ambitious objectives for an eventual mission to Mercury, to complete the imaging began by Mariner 10 at higher resolution and to investigate the internal evolution and state of the planet. To meet this objective, clearly the full magnetic mapping of the planet was essential.

Even then, in fact, there was no shortage of mission proposals to Mercury in the 1980s and early 1990s, to both NASA and ESA. One of the first was called Messenger, submitted to ESA in 1983 by A.K. Richter and A. Balogh; this mission was to be a multiple flyby, somewhat similar to Mariner 10, but with space physics objectives. Other than flying by Mercury, the mission was also aimed at the further exploration of the inner heliosphere, following up on the successful German/American Helios probes. However, it was clear from the start that an orbiter was needed rather than just another flyby mission. The first such fully worked out proposal for a Mercury polar orbiter was made to ESA in 1985 by a consortium of scientists led by G. Neukum (1985). This proposal took into account the necessary propulsion requirements by including a gravity assist at Venus and the use of SEP. It was to be a large spacecraft, with a mass of 2,900 kg, carrying the SEP module as well as the spacecraft bus and a comprehensive payload for both planetary and space physics objectives. At that time, NASA was developing SEP for such missions (although these were not followed up at the time), but there was no prospect for a similar development in Europe, so Neukum's proposal was not considered further by ESA.

A major development for improving the prospect of a mission to Mercury was the discovery of a new class of gravity-assist missions by Chen-Wan Yen of JPL (Yen 1985, 1989) who identified ballistic trajectories using multiple-braking gravity-assist flybys denoted E-VVMM-M or E-VV-MMM-M that had, after a ballistic transfer from Earth (E) to Venus (V), repeated encounters with Venus and Mercury (M) before finally being placed in orbit around Mercury. In the first case, E-VVMM-M, there are two Venus flybys and two Mercury flybys before the arrival velocity at Mercury on the third approach is sufficiently low for the orbit insertion manoeuvre. The E-VV-MMM-M trajectory has an additional Mercury flyby before the final, orbit-insertion encounter. The braking gravity assists would decrease the final arrival velocity at Mercury down to 2 to 3 km/s that could be handled with an onboard chemical propulsion module for orbit insertion.

Taking advantage of Yen's mission concept, J. Belcher and J. Slavin led a very detailed study of a dual Mercury orbiter mission for NASA in 1988–1990 (Belcher et al. 1991; Rideourne et al. 1990). Using a Titan IV (with a solid booster added and the Centaur upper stage) the mission design, with the multiple Venus and Mercury flybys, was due to take between 3 and 7 years, depending on the selected launch date. Only rare launch opportunities provide an optimal alignment of the Earth, Venus, and Mercury to minimise the trip time to Mercury orbit insertion. Under conditions of optimal alignment, an E-VVMM-M trip takes 3 years, while a more realistic launch (for which there are more frequent opportunities) takes 4.9 years. Such a mission design was considered as the baseline for this mission, scheduled for launch in 1997. For comparison, an E-VVMMM-M trip, with a launch in 1999, was calculated to take almost 7 years.

Given the budgetary limits set for the mission ($500 million in 1988), the science team involved in the study put a higher priority on the space physics (magnetospheric) objectives, while accommodating as best they could some planetary objectives. Even then, the best way to implement the mission was through a dual spacecraft design (designated SC-1 and SC-2). In order to accommodate the space physics payloads, the two spacecraft were designed to be identical (including the payload) and spin-stabilised; this choice was also to help in the thermal design of the spacecraft in Mercury orbit. With the envisaged launcher, each spacecraft could have a dry mass of about 800 kg with a fuel load of 1,600 kg.

The two-spacecraft approach recognised that the short timescales and small spatial scales of Mercury's magnetosphere required at least two measurement points simultaneously for the interpretation of the observations. In particular, for determining the internal magnetic field of Mercury, the magnetic fields due to the magnetosphere need to be subtracted from the measurements along the orbit. This is possible only if the magnetosphere is well understood and well modelled, and if the state of the variable magnetosphere is monitored simultaneously.

The design of the Mercury orbiting phase of the mission was to be complex but ingenious, in order to reconcile the different mission objectives. Spacecraft 1 was to have an eccentric polar orbit of 12-hour period, with a periapsis of 200 km over Mercury's north pole and an

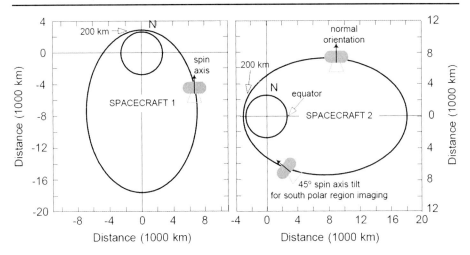

Fig. 5 NASA's proposed dual Mercury orbiter mission. *On the left*, the polar orbit of spacecraft 1 is shown, with the spin axis pointing north. *On the right*, the orbit of spacecraft 2, showing the 45° orientation of the spin axis optimised for imaging the south pole region. (From NASA Technical Memorandum 4255)

apoapsis at a height of 6.2 R_M. Spacecraft 2 was to have three different orbit phases. It was to have first a highly eccentric equatorial orbit, reaching out to over 80 R_M in the anti-Sun direction from the planet, deep into its magnetotail. This was to be a single orbit, almost a month in duration; following the completion of this orbit, the apoapsis of the elliptical, equatorial orbit was to be reduced to 32 R_M, with a periapsis altitude of 200 km. Spacecraft 2 was to remain in this orbit for more than a Mercury year (thus precessing around Mercury by over 360°). The final phase was to follow a change in the orbit plane to polar inclination and a 12-hour period, to engage in the closer observation of the planet, in particular to achieve imaging of the surface with a best resolution of about 100 m. This phase of the proposed mission is shown in Fig. 5.

This was an ambitious mission study, seriously addressing the complexities and challenges of a Mercury orbiter mission. In particular, the dual spacecraft approach, taken up later by the BepiColombo mission, pioneered the concept of simultaneous observations to reduce the uncertainties that arise from the complex and rapidly changing Mercury environment. However, the dual Mercury orbiter concept was not selected by NASA for implementation, so other proposals continued to be submitted.

The new proposals were submitted to NASA's Discovery Program, the framework for a range of medium-scale missions that was initiated in 1992. (For a brief history of the mission proposals to NASA at this time, see McNutt et al. 2006.) One such proposal, submitted in 1993 by JPL for selection as a Discovery-class mission, was a considerably more modest one than the dual orbiter. This mission, called Hermes (Nelson et al. 1994; Cruz and Bell 1995), was to carry only three scientific instruments: an imaging system that also included a dual wavelength laser altimeter, an ultraviolet (UV) spectrometer, and a boom-mounted magnetometer. The spacecraft was to be three-axis stabilised, but quite light, with a mass of only 320 kg. The mission was to follow a Yen design of the E-VVMM-M type with a flight time to orbit insertion of about 3 years. Launch opportunities were identified in 1999, 2000, 2004, and 2005. Another proposal was based on the successful Clementine mission to the Moon (Ely et al. 1995). The payload was to comprise UV/visible/IR imagers and a laser altimeter, but no magnetometer. It was an ambitious concept, with the spacecraft in a

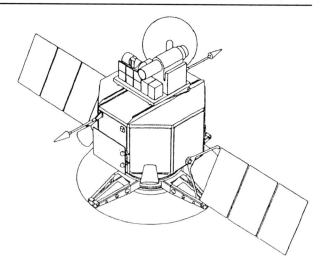

Fig. 6 Conceptual design of the spacecraft proposed to NASA as the Discovery Mercury Polar Flyby mission. (From Spudis et al. 1994)

300-km-altitude circular polar orbit around Mercury. All but one of the proposed missions were to use a Yen-type transfer to Mercury.

The only flyby mission proposed in this framework was the Mercury Polar Flyby mission (Spudis et al. 1994). The conceptual design of this spacecraft is shown in Fig. 6. It was to be launched into a similar transfer trajectory to that of Mariner 10, with a transfer to Venus and gravity-assist flyby, then targeted to Mercury in a solar resonant orbit for nominally three Mercury flybys, the first of which was to be over Mercury's north pole, the second one equatorial, and the third one over the south pole. The imaging was to be multi-spectral from 200 to 1,000 nm, with a coverage complementary to that of Mariner 10. Great emphasis was placed on investigating the polar regions with a neutron spectrometer, a radar scatterometer, and a thermal emission spectrometer. In particular, the nature and origin of polar deposits in the permanently shadowed craters discovered by Earth-based radar were to be studied. On the equatorial flyby, an X-ray fluorescence spectrometer was to study the rock-forming elements; use would also be made of data obtained by the thermal emission spectrometer.

Another proposal was made to both NASA (as a Pathfinder mission) and ESA (as a Flexi mission) quite late, effectively after the MESSENGER selection; this was LUGH (Low-cost Unified Geophysics at Hermes). It was to consist of three elements, a main spacecraft that would perform an equatorial flyby, with two polar nanoprobes released from the main spacecraft (Clark et al. 2003). While imaginative, given that MESSENGER had been approved, the LUGH proposal could not cover the detailed objectives made possible by an orbiter.

None of these proposals, or other, similar Mercury missions proposed at this time, was selected by NASA. However, it was in this framework that, after the second attempt a few years later, MESSENGER was selected. It was also at about this time, in 1993, that a mission proposal was made to ESA for a Mercury orbiter that eventually became the BepiColombo mission.

4 The MESSENGER Mission

4.1 Mission Origin and Design

In 1999, while ESA was studying what was to become the BepiColombo mission, NASA selected MESSENGER (MErcury Surface, Space ENvironment, GEochemistry, and Rang-

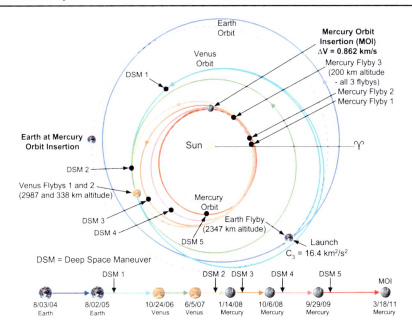

Fig. 7 The trajectory of MESSENGER with the mission timeline

ing) as a Discovery-class mission for launch in 2004. MESSENGER had originally been proposed in 1996 but was not successful until its second proposal in 1998 (McNutt et al. 2006). Three launch opportunities were considered, one in March, one in May and one in August 2004. The three launch opportunities were not equivalent: if launch were in March or May 2004, Mercury orbit would be reached in 2009; if launch were in August 2004, orbit insertion would not occur until 2011. In the end, it was the third launch date that was used (see the timeline of the MESSENGER mission in Fig. 7 and Table 1). MESSENGER had been proposed to NASA by a team led by S.C. Solomon and is described in detail by Solomon et al. (2001), Santo et al. (2001) and Gold et al. (2001, 2003). A single spacecraft will be placed in orbit around Mercury. The spacecraft is three-axis stabilised and carries a range of instruments that combines planetary and magnetospheric objectives.

The mission design is based on the multiple gravity-assist principles of Chen-Wan Yen (McAdams et al. 1998, 2006). Following the launch on August 3, 2004, MESSENGER spent a year in an orbit close to that of the Earth prior to a gravity-assist flyby on August 2, 2005, that placed it on a trajectory to Venus for two gravity-assist flybys in 2006 and 2007. The first Mercury flyby will be take place January 14, 2008. The geometry of this flyby is shown in Fig. 8. All three scheduled gravity-assist flybys of Mercury (used for slowing down the spacecraft relative to Mercury) have a similar geometry. However, contrary to the Mariner 10 flybys, successive flybys by MESSENGER will have their closest approaches at different longitudes of the planet, thereby extending the coverage of the imaging initiated by Mariner 10 (so that most of the planet will be imaged during the flybys) and providing a more global coverage of the magnetic field. Aspects of the first two flybys are illustrated in Figs. 8 and 9. In this way, these first two flybys will bring a very early harvest of important scientific data, some three years before orbit insertion and the more comprehensive coverage that will then be undertaken.

Table 1 The milestones of the MESSENGER mission

Events	Date
Selection as a Discovery mission	July 1999
Phase B (detailed design)	January 2000–June 2001
Phase C/D (fabrication, assembly and test)	July 2001–July 2004
Launch	3 August 2004
Earth flyby	2 August 2005
Venus flybys	24 October 2006, 5 June 2007
Mercury flybys	14 January 2008, 6 October 2008, 29 September 2009
Mercury orbit insertion	18 March 2011
Mercury orbit	March 2011–March 2012

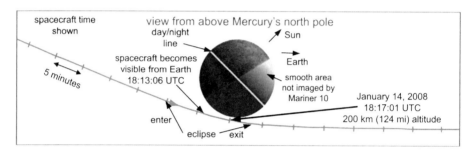

Fig. 8 The trajectory of MESSENGER around Mercury, seen from above the north pole during the first flyby on January 14, 2008. During flyby 1, approximately half of the hemisphere unseen by Mariner 10 will be imaged. It is also expected that the close flyby will give an early indication of the nature of the magnetic field of Mercury

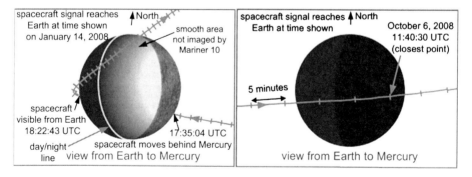

Fig. 9 The trajectory of MESSENGER around Mercury, seen from the Earth, during the first and second flybys of the planet on January 14, 2008, and October 6, 2008. The second flyby will almost complete the coverage of the imaging of the planet's surface not seen by Mariner 10. The two flybys are also complementary for mapping the planetary field of Mercury, with the first one covering a higher range in latitudes than the second, which is nearly equatorial

The planetary flybys make the mission possible by reducing the cumulative change in spacecraft velocity (ΔV) required for the mission, but MESSENGER was launched with a total ΔV capability of more than 2.2 km s^{-1}; the fuel necessary for this corresponds to

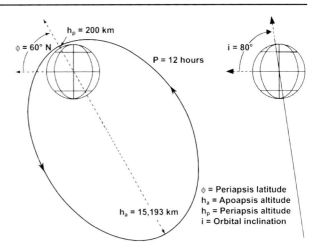

Fig. 10 The operational orbit of MESSENGER

about 54% of the launch mass of MESSENGER. About a third of this fuel is required for Mercury orbit insertion; the rest is used for trajectory manoeuvres and orbit maintenance around Mercury.

MESSENGER's operational orbit, shown in Fig. 10, has an altitude of 200 km at periapsis and 15,193 km at apoapsis and a period of 12 hours. The plane of the orbit is nearly polar, with an initial inclination of 80°; periapsis is initially at 60°N and increases to 72°N by the end of one Earth year. This geometry allows a very close survey of the northern hemisphere. Part of the fuel carried is used to correct the perturbations that tend to raise the periapsis altitude. There is sufficient fuel to carry out these maintenance manoeuvres during the operational phase of the mission.

4.2 The MESSENGER Scientific Objectives and Payload

The questions that MESSENGER has set out to answer all relate to the unique properties of the planet. Although a member of the family of terrestrial planets, Mercury—just like Venus, the Earth, and Mars—has its own specific characteristic features that, in the case of Mercury, are poorly understood because of the lack of detailed observational evidence. The questions relate to Mercury's unusually large density, its geologic history (and the specific surface features discovered by Mariner 10), the structure of the planet's core (solid or partly liquid), the origin of its magnetic field, the nature of the radar-bright deposits in craters close to the poles, and the characteristics of Mercury's dynamic exosphere. The payload of MESSENGER was selected in the light of these objectives, which require a specific range of observations.

There are seven scientific instruments on board MESSENGER, as shown in Table 2. An artist's sketch of the spacecraft with the payload instruments is shown in Fig. 11. These instruments have been described in considerable detail by Gold et al. (2001), including their scientific objectives, placed in the general context of the scientific exploration of Mercury by Solomon et al. (2001). After the earlier papers, some of the instruments were redesigned prior to launch; an up-to-date, if brief, account is given by Gold et al. (2003). In the following is a summary of the instruments and their capabilities; more detailed descriptions of the individual instruments can be found in the set of recent papers referenced.

Mercury Dual Imaging System (MDIS) contains a reflective narrow-angle (NA) camera and a refractive wide-angle (WA) camera (Hawkins et al. 2007). MDIS will map the entire

Table 2 The scientific payload of MESSENGER

Instrument		Mass (kg)	Power* (W)
MDIS	Dual imagers, narrow and wide angle FOV	8	7.6
GRNS	Gamma-Ray and Neutron Spectrometer	13.1	22.5
XRS	X-ray spectrometer, 1–10 keV	3.4	6.9
MAG	Fluxgate magnetometer + 3.6 m boom	4.4	4.2
MLA	Laser altimeter, 1,200 km range	7.4	16.4
MASCS	UV/Visible spectrometer, visible/IR spectrograph	3.1	6.7
EPPS	Energetic particle spectrometer, fast imaging plasma spectrometer	3.1	7.8
DPU	Integrated electronics, power processing for all instruments, MDIS electronics	3.1	12.3
	Payload harness, purge system, magnetic shielding etc.	1.7	
Payload totals:		47.2	84.4

*Orbit average

planet in monochrome to an average resolution of 250 m per pixel or better, global colour images will be obtained at an average resolution of 2 km or better, and high-resolution images will be obtained of selected features at a resolution of 20–50 m per pixel. Because of the highly elliptical orbit at Mercury, MDIS has been constructed using on-board pixel summing to provide images of uniform resolution throughout the orbit. There is a common scan platform on which the two imagers are mounted to allow pointing the instruments with some independence from the attitude of the spacecraft, to optimise the coverage and to assist with navigation. The NA and WA imagers have fields of view, respectively, of 1.5° and 10.5°. The CCD detector is 1,024 × 1,024 pixels in size, with 14 µm/pixel. Spectral coverage is provided by the WA imager using 11 colour filters from 415 to 1,020 nm; there is also a clear filter (centred on 700 nm) for navigation.

The Gamma-Ray and Neutron Spectrometer (GRNS) consists of two sensors (Goldsten et al. 2007), the Gamma-Ray Spectrometer (GRS) and the Neutron Spectrometer (NS). The GRS is based on a cryogenically cooled, high purity germanium detector which is actively shielded by a surrounding plastic scintillator. The combination, using a Stirling-cycle active cooler that keeps the germanium sensor at 90 K, optimises, for the mass and power available, the signal-to-background ratio. This is needed for measuring the surface elemental abundances of O, Si, S, Fe, H, K, Th, and U. This instrument was developed late in the programme, to replace a scintillator-based instrument that had been originally proposed but could not meet the scientific requirements. The separate NS sensor consists of a slab of plastic scintillator (to measure fast neutrons) sandwiched between lithium glass scintillators to measure the thermal neutrons, taking advantage of the orbital velocity from the ratio of the fluxes in the ram and the wake directions.

The X-Ray Spectrometer (XRS) measures the surface abundances of the elements Mg, Al, Si, Ca, Ti, and Fe from the fluorescence induced by solar X-rays (Schlemm et al. 2007). The instrument uses an assembly of three copper honeycomb collimators with three proportional counters that have been developed from previously used components. The field of view is 12°. The instrument also includes a silicon PIN detector that looks at the Sun through an opening of 0.03 mm^2 in the spacecraft's sunshade to monitor the solar X-ray flux. This detector is protected by two foils from the heat of the solar radiation, ensuring its operation at a temperature at −45°C.

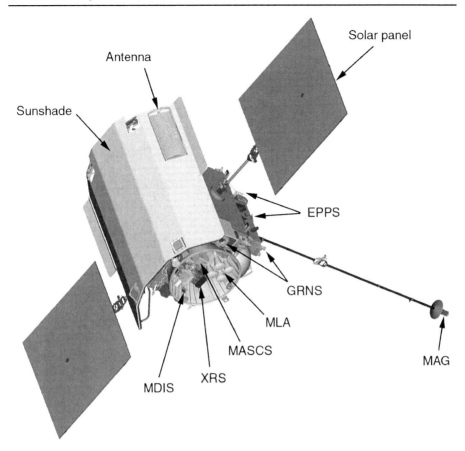

Fig. 11 The MESSENGER spacecraft showing the location of the scientific instruments (see text for explanation of the acronyms), as well as the sunshade, the solar panels, and one of the two phased-array antennas

The Mercury Laser Altimeter (MLA) uses a diode-pumped, Q-switched, Nd:Cr:YAG laser transmitter operating at 1,064 nm and four receiver telescopes with sapphire lenses (Cavanaugh et al. 2007). In orbit around Mercury, the MLA will measure at altitudes up to 1,200 km with 30-cm precision. Together with the exact positioning of the spacecraft using the tracking information from NASA's Deep Space Network and the imaging from MDIS, the MLA will deliver topographical information of very high accuracy. The elliptical orbit of MESSENGER means that the MLA will operate for about 30 minutes around the periapsis of each orbit. The performance of the MLA was tested in space in 2005 by exchanging laser pulses between MESSENGER, at the time at a distance of about 25 million km, and a ground station on Earth (Smith et al. 2006).

The Mercury Atmospheric and Surface Composition Spectrometer (MASCS) consists of two sensors, one covering the ultraviolet and visible wavelengths from 115 to 600 nm, the other the visible and infrared wavelengths in two bands, from 300 to 1,050 nm and 0.85 to 1.45 μm (McClintock and Lankton 2007). The two instruments share a common front-end Cassegrain telescope but have separate detection and signal-processing units. The UV/visible sensor will be used to study Mercury's very tenuous exosphere by scanning over the limb of the planet to determine its composition, while the visible/IR sensor will observe Mercury's surface composition.

Plasma and high-energy particles in the Mercury environment will be measured by two complementary sensors in the Energetic Particle and Plasma Spectrometer (EPPS) instrument (Andrews et al. 2007). For high-energy particles, an Energetic Particle Spectrometer (EPS) sensor measures the fluxes of ions from 10 keV/nuc to ~3 MeV and electrons up to about 400 keV, using time-of-flight and residual energy techniques. The directional measurements are performed over a field of view of 160° by 12° by a 24-pixel silicon detector array that is divided into six segments of 25° each. The Fast Imaging Plasma Spectrometer (FIPS) is a new design, using a complex electrostatic deflection geometry followed by a position-sensing time-of-flight measurement element for determining ion species in the plasma up to 20 keV per electronic charge. This instrument has a solid angle viewing coverage of almost 2π.

The Magnetometer (MAG) will map the magnetic field along the orbit, to study the magnetosphere and the origin of the planetary field (Anderson et al. 2007). MAG is a three-axis fluxgate instrument, following a very long line of similar instruments in space. In order to minimise the effect of the spacecraft-generated magnetic field on the measurements, the MAG sensor is mounted on a 3.6-m boom. The measurement rate will be 20 vector samples/s, but the transmitted rate will vary according to the telemetry capability; at lower transmission rates, the measurements will be averaged onboard prior to transmission. The normal range is $\pm 1,530$ nT, with a 16-bit or a 0.05-nT resolution.

5 BepiColombo

5.1 Origins of The BepiColombo Mission

5.1.1 Concepts of a Mercury Orbiter Mission in Europe

In the 1980s, the ESA received two proposals for a Mercury orbiter mission. Neither were selected by the European Space Agency (ESA), but European scientists interested in Mercury were optimistic because the science case for such a mission remained strong. Once a first study established the technical feasibility, it was expected that the scientific arguments would convince ESA and the planetary science community to undertake a mission to Mercury. In response to ESA's call for mission proposals in 1992, a Mercury Orbiter proposal was submitted in May 1993 by a team of European scientists led by A. Balogh (1993). The proposed mission, using a Chen-Wan Yen design of multiple Venus and Mercury flybys prior to insertion into orbit around Mercury, combined space plasma (magnetospheric) and planetary objectives. The prime motivation for the proposal was the study of the very intimate interdependence of the planetary interior and its magnetic field with the magnetosphere formed around Mercury by the interaction with the solar wind. The link provided by the magnetic field observations that needed interpretation in terms of the comparable external and internal contributions appealed to both planetary and magnetospheric scientists in ESA's science advisory committees.

A study was duly carried out within ESA in 1993–1994; the conceptual spacecraft design is shown in Fig. 12. However, at the conclusion of the study, the mission design was regretfully deemed to be beyond the budgetary limit of the "medium" scale missions in ESA's Horizon 2000 framework. However, by 1996, another round of strategic planning was carried out to define the Horizon 2000+ follow-up programme; in the context of this

Fig. 12 Concept of the Mercury Orbiter spacecraft from the first study by ESA in 1993–1994. It was to be a spinning spacecraft, with a despun antenna to ensure a high data rate to Earth. Several of the instruments considered for the model payload were similar to those now selected for JAXA's Mercury Magnetospheric Orbiter

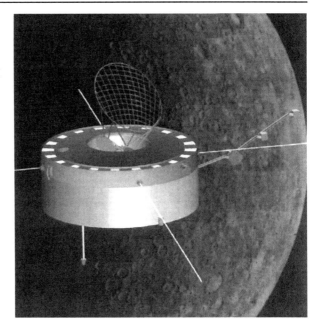

programme, a Mercury mission was to be the planetary "cornerstone" of the new programme phase, enabling the likely costs of the mission to be included in ESA's future budget.

One of the foundations of the new Mercury Orbiter cornerstone programme was the new mission design, based on solar electric propulsion (SEP, Racca 1997). This technology was to be tested by ESA through the first Small Mission for Advanced Research in Technology (SMART-1) to the Moon. This mission, launched in 2003 and very successfully implemented, used SEP for transfer to lunar orbit and concluded in 2006 when the spacecraft was crashed into the Moon, following a successful technological and scientific programme (Foing et al. 2006).

An industrial study of the new design for BepiColombo was undertaken under ESA direction in 1997. The design consisted of two orbiters, one a large, three-axis stabilized platform with large planetary instruments, the other a small spinning subsatellite with a modest space plasma payload. However, this study ran into considerable problems when the Mercury approach, orbit insertion, and operational phases were studied. The mission design was very reminiscent of the earlier proposal by Neukum (1985). Reconciling the conflicting requirements of instrument pointing, propulsion, and thermal design proved to be very difficult and could not be satisfactorily resolved within the mass budget.

At that point a new mission design was proposed by Y. Langevin, a member of ESA's scientific advisory team for the Mercury cornerstone mission (Langevin 2000, 2005). This scheme involved a single Ariane 5 launch with the composite spacecraft shown in Fig. 13. The mission plan consisted of a ballistic orbit to Venus followed by a gravity assist at Venus, solar electric propulsion for transfer to Mercury, followed by the jettisoning of the SEP module and orbit insertion using chemical propulsion. This plan provided a realistic basis for a new study that involved a two-spacecraft approach, one for planetary and one for magnetospheric objectives. At that point a lander, or surface element, was also included. The problem for a lander at Mercury is that due to the absence of an atmosphere, landing involves active retrorockets and flight control as on the Moon, but with the added difficulty of a very challenging thermal environment. During the study, even a "hard" lander was investigated,

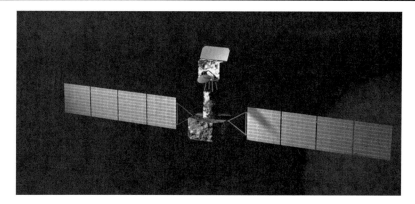

Fig. 13 The configuration of the Colombo (as it was then called) composite on its way to Mercury, comprising the planetary and magnetospheric orbiters, together with the large solar panels of the SEP subsystem. This concept was the outcome of the second industrial study in 1999

one that once released from the orbiter would free-fall onto Mercury's surface. Through the use of some exotic technologies, this option was found to be marginally feasible, but in the final version of the study the "soft" landing approach was adopted (Novara 2001). Although the lander was a popular element of the planned mission, it became a cost driver and was later dropped. At about the same time, in 2000, an agreement was reached with the Japanese Institute of Space and Astronautical Science (ISAS, now JAXA/ISAS) for the construction of the Mercury Magnetospheric Orbiter with ESA providing the other elements (Mercury Planetary Orbiter, SEP, and chemical propulsion modules and the launch of the composite). In the late 1990s, ISAS had in fact undertaken a study for a Mercury Orbiter mission (Yamakawa et al. 1999), so a cooperative mission with ESA could increase the scope of the science objectives beyond what could be achieved independently by the two agencies.

5.1.2 Japan's Plans for a Mercury Mission

In Japan, as a result of the successful technical developments that had been required for the first planetary mission, 'Planet-B' (Nozomi, the spacecraft intended to orbit Mars and launched in 1998), ISAS initiated consultations with the scientific community for missions to other planetary targets in the mid-1990s. Venus (Planet-C, to be launched in 2010) and Mercury were the strongest candidates. The latter target was supported by two groups, magnetospheric scientists motivated by the great success of Akebono (1989–) and Geotail (1992–), and the lunar science community as a follow-up to the Lunar-A (postponed) and Selene (to be launched in 2007) missions. The Mercury Exploration Working Group (MEWG) was formed in July 1997 to carry out detailed feasibility studies and an official proposal was submitted in November 1998 to the Steering Committee for Space Science (SCSS) of ISAS in November 1998 (Yamakawa et al. 1996, 1999).

ISAS M-V and NASDA H-IIA were considered as possible launch vehicles. The former was derived from the M-III-S2, which launched the first Japanese interplanetary missions Sakigake and Suisei to P/Comet Halley in 1985. However, its capability was deemed not sufficient for a realistic payload mass for a Mercury mission, so the larger H-IIA was assumed for system design. Two mission profiles were studied. The classical one was a spacecraft

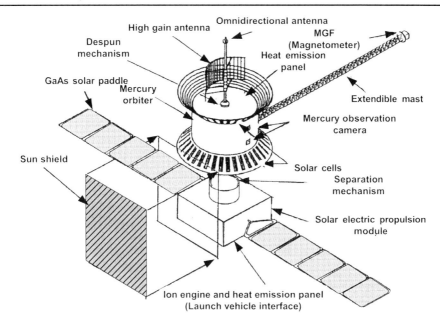

Fig. 14 The ISAS Mercury Orbiter design with SEP module in 1998 (Yamakawa et al. 1999)

with chemical propulsion that used a Chen-Wan Yen trajectory. For launch in 2005 with two Venus and two Mercury flybys, the overall transit time was 4.2 years. The launch mass was 1,650 kg including 950 kg fuel. A more ambitious proposal used SEP, based on the Hayabusa mission to asteroid Itokawa (2003–). For a launch window in 2005 with a single Venus flyby, the total flight time was to be 2.3 years. Wet mass at launch was to be 1,500 kg with an SEP module of 1,100 kg. To address simultaneously magnetospheric, exospheric, and planetological investigations, a 0.5-day-period polar orbit (300 km × 6 Mercury radii altitude, inclination = 90°) was selected. The argument of periapsis was proposed to be 30° with respect to the ecliptic plane to avoid long shadow periods.

The proposed orbiter was to be a Helios-type spinning spacecraft (6–10 rpm). A schematic view with the SEP module is shown in Fig. 14. The chemical propulsion was identical to the Mercury Orbiter in Fig. 12. The main objectives of the payload were the interior, surface, atmosphere, and magnetosphere of Mercury. The model payload (total mass 70 kg) consisted of multiband cameras, X- and gamma-ray spectrometers, magnetometers, plasma and energetic particle analyzers, plasma wave analyzer with radar sounder, dust detector, laser altimeter, and radio science instrument. The feasibility and required capabilities for a lander (300 kg), a penetrator (90 kg), and a small orbiter (20–30 kg) were also investigated. At the end of 1998, the remaining problems were cost, thermal design, and kick-motor development. The SCSS encouraged the MEWG to study those issues in more detail before a decision on the proposal could be made. This was to come to fruition in the context of the collaboration between ESA and (then) ISAS.

5.1.3 The BepiColombo Mission

In the meantime, ESA's Mercury Cornerstone mission was renamed BepiColombo, in memory of Giuseppe Colombo (see Fig. 15), who had played a crucial role in the design of

Fig. 15 Giuseppe Colombo (1920–1984), who discovered the orbital resonance of Mercury and who advised NASA to achieve three flybys of Mercury by Mariner 10. ESA and JAXA's joint, two-spacecraft mission is named in his memory

the Mariner 10 mission and in the determination of Mercury's orbital resonance (Colombo 1965). The discussion about the possibility of a collaboration with ESA was initiated in ISAS following the Inter-Agency Consultative Group (IACG) meeting of November 1999. The official request was sent in a letter from the Director-General of ISAS to the Directorate of Scientific Programme of ESA, dated July 31, 2000. Following the approval of BepiColombo as the fifth Cornerstone of ESA, the MEWG was re-convened to evaluate the role of the Japanese orbiter as one element of the BepiColombo mission. On the basis of this study, 'the international Mercury exploration mission BepiColombo' was approved by the SCSS of ISAS in January 2002, and by the Space Activities Commission of Japan in June 2003. In October 2003 JAXA was formed through the merger of ISAS, the National Space Development Agency (NASDA) and the National Aerospace Laboratory (NAL). ISAS is now part of JAXA.

The status of the BepiColombo project studies at the conclusion of these preliminary, but in-depth, technical and scientific studies is described by Grard and Balogh (2001), Anselmi and Scoon (2001), and Novara (2001, 2002).

The two-spacecraft approach, together with a possible a lander (called the Mercury Surface Element, or MSE), and both an SEP and a chemical propulsion subsystems, was quite a complex assemblage of elements for launch and delivery to Mercury. Two launch possibilities were envisaged: a launch of the whole assembly on a single Ariane 5, or two launches with Soyuz-Fregat, one for the Magnetospheric Orbiter and the Surface Element, the other for the Planetary Orbiter. A serious disadvantage of this second scenario was that two propulsion elements (SEP and chemical) were needed, as well as two service modules to support the two composites during the launch and the cruise to Mercury.

A more comprehensive study indicated that the Surface Element was too heavy and would need a separate (third) launcher, which finally led to the abandonment of the lander.

Although ESA approved BepiColombo as a cornerstone-class mission in late 2000, it became necessary to reassess the mission in 2002, due to budget restrictions within ESA. The approval was made subject to a satisfactory outcome of the reassessment studies that were initiated to refine the definition of the mission and its cost. The study was carried out from 2002 to 2003, with the help of both internal ESA support and further industrial contracts. A number of technical solutions were proposed at this stage, particularly for optimising the accommodation of the payload. In addition, the prospect of a more powerful Soyuz-Fregat launch vehicle (an upgraded Soyuz-2B with Fregat M upper stage, to be launched from Kourou) led to a single launch for the MPO, MMO, and propulsion modules assembly.

In February 2004, ESA's Science Programme Committee approved the Science Management Plan; this was followed by an Announcement of Opportunity to the scientific community for proposals for the payload of the European orbiter, and signalled the start of the industrial-definition Phase 2 studies. The instrument proposals were reviewed by ESA through 2004 and a payload was selected for the Planetary Orbiter in November 2004. At the same time, ISAS/JAXA successfully carried out a similar approval and payload selection procedure.

For ESA, it remained to determine that the payload instruments would be adequately supported by the national agencies; this was completed in the course of 2005, but not without some difficulties in some of ESA's member states. This ended the approval procedure of the MPO payload. The stage was thus set for ESA to issue an invitation to tender for the mission elements and the evaluation of the tenders and selection of the industrial Prime Contractor for the European elements of BepiColombo.

5.2 Mission Objectives and Mission Design

The objectives of BepiColombo have been formulated as a comprehensive set of questions that relate to all aspects of the planet Mercury and its environment (Grard et al. 2000). The objectives and the requirements on the mission take into account the capabilities and expectations of MESSENGER, which will bring answers to some of the key questions. The emphasis for BepiColombo is not so much the discovery of new features of Mercury (although this cannot be excluded), but rather the collection of a comprehensive set of observations that will bring knowledge of the planet on a par with the other terrestrial planets and describe satisfactorily its origin and evolution. The following summary of the mission objectives takes into account the two BepiColombo spacecraft, the MPO and MMO, each with its specific emphasis. It is anticipated, however, that for most objectives, the joint analysis of the observations will bring more substantial results than if the data collected by each spacecraft are analysed separately.

The questions follow the structure of the planet and its environment as a system of interacting parts. Starting with Mercury's internal structure (targeted more specifically for the MPO), the objective is to determine precisely the sizes and masses of the major chemical reservoirs, the crust, mantle, and core. The state of the core and the existence of an outer liquid layer need to be determined to account for the origin of the planetary magnetic field. Is a classic dynamo possible within Mercury's core? Observations are also required to detect radial and lateral heterogeneities in the crust and mantle structure and, in addition, topographic variations in the core–mantle boundary. The observations will be used to constrain models of the rheology and of the tectonic, volcanic, chemical, and thermal evolution of Mercury. Such modelling will also constrain hypotheses for the formation of the planet and even of the terrestrial planets as a family.

The requirement is to obtain accurate measurements of the gravity field, the topography, the amplitude of forced libration, and the obliquity. The observations will be solved for the long wavelength gravity field with a high relative accuracy (10^{-4}). Local gravity anomalies down to a resolution of 400 km will be determined. The Love number k_2 needs to be determined and the dissipation factor Q sufficiently constrained. In addition, the mean planetary moment of inertia and the ratio between the moments of inertia of the solid upper layer and of the entire planet need to be determined.

The measurement of the planetary magnetic field, up to high-order terms, will help constrain the interior structure. The existence of magnetic anomalies and/or short-spatial-wavelength magnetic structures needs to be investigated. This will be a difficult task, as the measurements will give the sum of internal and external contributions, possibly of comparable magnitude. This objective is intimately related to the magnetospheric objectives described in the following, and observations by both the MPO and MMO will contribute to meeting these, in particular by the joint analysis of both data sets.

The resolution with which Mariner 10 imaged somewhat less than half Mercury's surface will be considerably improved by MESSENGER, which will cover the entire surface, with a considerably improved resolution especially on features of interest. The objectives include (1) the global characterisation of tectonic and volcanic features (lineaments, scarps, and, possibly, domes); (2) the assessment of the roles of global cooling, major basin-formation events (e.g., Caloris), viscous relaxation and tidal stresses for the endogenic modification of the surface; and (3) the study of the crustal rheology from the relaxation of surface features (e.g., multi-ring basins). Similarly, the altimetry of surface features and crustal structure from geodesy at a lateral scale of 500 km or less are important related objectives.

The elemental and mineralogical compositions of the surface need to be determined on large, regional, and small scales. Of particular importance is the determination of key element ratios, such as potassium to thorium, iron to silicon, again on a range of scales. An important objective is the determination of the nature of volatiles in the permanently shadowed craters revealed by Earth-based radar imaging. There will still be a need to confirm and complement MESSENGER's mapping of craters and crater sizes, in particular over the southern hemisphere where BepiColombo will have a more uniform coverage.

The exosphere of Mercury is both tenuous and highly variable. Composition and height distributions of the constituent species of the exosphere will be determined, as will the dependence of the density vs. height profiles of the constituent species on local time (especially the differences between the nightside, terminator and dayside differences) and latitude. A particularly important topic is exospheric dynamics in response to solar wind variations and magnetospheric processes. There are intimate links between the surface and the exosphere on the one hand, and the exosphere and the magnetosphere on the other. Important to unravelling these links is a determination of the sources, production mechanisms, and loss processes of the different constituents of Mercury's exosphere. With observations from both the MPO and MMO of the exosphere and its dynamics, spatial ambiguities can be resolved and the data can be used to refine three-dimensional exospheric models.

Mercury's magnetosphere has significant differences from that of the Earth: it is considerably smaller, it is variable on very short timescales, there is no ionosphere, and the planetary surface is acting as an absorber. Furthermore, solar wind in the inner heliosphere has a higher density and stronger magnetic fields that strongly affect the structure, dynamics, and physical processes in the Hermean magnetosphere. The objectives therefore include the determination of the structure and temporal variability of the magnetosphere in response to the strongly variable solar wind and interplanetary magnetic field. In terms of structures, the dayside cusp, the location of the bow shock and the magnetopause will be studied, as well as the dynamics of the magnetotail (for instance, are there substorms on Mercury?). Magnetospheric current systems pose a particularly important question; although their existence cannot be in doubt, their topology is fundamentally unknown. Given the very different parameter regime, the observation of Mercury's magnetosphere will help to assess the importance of key parameters that control the structure, dynamics, and physical processes of other planetary magnetospheres. The MMO is optimised to make the necessary comprehensive observations, but the MPO will also contribute significantly to meeting these objectives.

Table 3 The milestones of the BepiColombo mission

Events	Date
Agreement ESA–ISAS (later JAXA)	2000
Confirmation as a Cornerstone mission	September 2000
Reassessment	2002 to 2003
Payload selection, detailed approval	2004 to 2005
Phase C/D (fabrication, assembly and test)	January 2007 to mid 2013
Launch	August–October 1013
Moon flyby	31 October 2013
Earth flyby	1 February 2015
Venus flybys	February and September 2016
Mercury flybys	August and October 2018
Capture into Mercury orbit	March 2019
Mercury Magnetospheric Orbiter in place	June 2019
Mercury Planetary Orbiter in place	July 2019
Mercury Orbit	July 2019–July 2020

In addition to the objectives described above, the MPO and MMO each have additional objectives that go beyond planetology. The MPO, as part of its capabilities for mapping the gravitational potential of Mercury, can also address general relativity, testing it to a level better than 10^{-5} by measuring the time delay and Doppler shift of radio waves and the precession of Mercury's perihelion. Furthermore, the MPO can test the very strong equivalence principle to a level better of 4×10^{-5}, determine the oblateness (J_2) of the Sun to better than 10^{-8}, and set upper limits to the time variation of the gravitational "constant" G.

The MMO also provides an opportunity to revisit the inner heliosphere, which was only once explored previously 30 years ago by Helios 1 and 2. The MMO equipped with modern instrumentation will provide measurements of the solar electromagnetic radiation, solar wind and interplanetary dust at 0.3–0.5 AU. Given its orbit, the MMO will spend a significant fraction of time well in front of the magnetosphere, in the undisturbed solar wind, when its apoapsis is on the sunward side of Mercury. This will occur, due to the thermal design of the mission, around the perihelion of Mercury. Depending on the scheduling of the missions, collaboration with the planned ESA Solar Orbiter mission can also be envisaged.

The mission design of BepiColombo has been thoroughly analysed during the successive studies. The milestones of the mission are listed in Table 3. Essentially, once the cornerstone mission was selected as a dual spacecraft mission launched by a single Soyuz-Fregat, a ballistic-chemical propellant mission design was no longer possible and SEP became a requirement. The mission, as indicated above, combines chemical propulsion, repeated planetary flybys, and SEP.

The launch configuration consists of four main elements in the BepiColombo stack. There are two propulsion modules, the Chemical and the Solar Electric Propulsion Modules that constitute the Mercury Transfer Module, and the two spacecraft, the planetary and the magnetospheric orbiters. This configuration is shown in Fig. 16 after launch with the solar panels used for powering the SEP module deployed (Förstner et al. 2006).

Launch will take place from Kourou on board a Soyuz 2-1B/Fregat M in August 2013. Following injection into a Geostationary Transfer Orbit (GTO), chemical propulsion is used to raise the apogee to the Moon's orbit. A Moon flyby is foreseen to place the BepiColombo

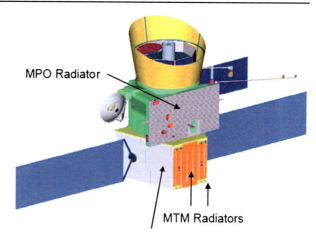

Fig. 16 The BepiColombo Mercury Composite Spacecraft (MCO) in the cruise configuration, with the Mercury Transfer Module (MTM) *at the bottom*, the Mercury Planetary Orbiter (MPO) *in the middle*, and the Mercury Magnetospheric Orbiter (MMO) *at the top* of the stack protected by a heat shield (Förstner et al. 2006)

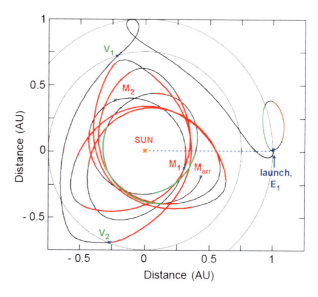

Fig. 17 The cruise trajectory of BepiColombo in an ecliptic projection, in a *coordinate system* with the *Sun–Earth line* fixed. The gravity-assist flyby encounters with the Earth (E_1), Venus (V_1 and V_2), and Mercury (M_1 and M_2) are indicated as is the final arrival to Mercury (M_{arr}). SEP thrust arcs are shown in red and green; coasting arcs are shown in black

composite into an interplanetary escape trajectory. The planned transfer to Mercury is shown in Fig. 17 (Förstner et al. 2006). There are gravity-assist flybys at Earth, at Venus (twice), and at Mercury (also twice) before a gravity-capture manoeuvre at Mercury. The strategy for planetary capture makes use of the weak stability boundary technique (see, e.g., Belbruno and Carrico 2000; Circi and Teofilatto 2001) to eliminate the risk of a critical injection burn. By approaching the planet slowly enough in the vicinity of one of the two libration points, L1 and L2, it is possible to weakly capture the spacecraft around Mercury without any propulsive manoeuvres (except for small trajectory corrections). The initial orbit (400 × 180,000 km) is stable for about one Mercury year (88 days) and can be readily stabilised with a very small manoeuvre. The apoapsis is then lowered to reach the operational orbit (400 × 11,824 km) of the MMO (Förstner et al. 2006).

Here the MMO as well as the Sun shield, which had been protecting the MMO during the cruise phase, will be separated from the MPO. Subsequently, the apoapsis of the MPO

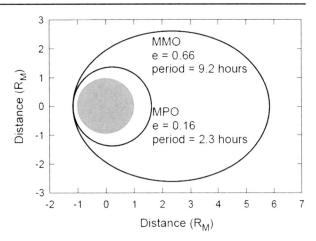

Fig. 18 The operational orbits of the Mercury Planetary and Magnetospheric Orbiters

Table 4 The scientific investigations of the BepiColombo Mercury Planetary Orbiter

Instrument		Mass (kg)	Power (W)
BELA	Laser Altimeter	12.5	52
ISA	Radio Science: Accelerometer	5.9	7
MERMAG	Magnetometer	2.0	3
MERTIS	IR Spectrometer	2.9	8.5
MGNS	Gamma Ray and Neutron Spectrometer	5.2	4
MIXS / SIXS	X-ray Spectrometer and Solar Monitor	5.5	15
		1.5	4
MORE	Radio Science: Ka-band Transponder	3.3	16
PHEBUS	UV Spectrometer	4.6	6
SERENA	Neutral Particle Analyser / Ion Spectrometers	5.0	21
SIMBIO-SYS	High Res.+ Stereo Cameras / Visual and NIR Spectrometer	7.2	23
Totals:		55.6	159.5

is further lowered until it reaches its operational orbit of 400 × 1,508 km in July 2019. The mission, in its operational orbit, is expected to last one Earth year.

5.3 The BepiColombo Mercury Planetary Orbiter (MPO)

The Mercury Planetary Orbiter is a three-axis-stabilised and nadir-pointing spacecraft. Its prime objective is to carry out remote sensing of the planet. It will be placed into a low-eccentricity polar orbit (400 × 1,500 km) that will provide an excellent spatial resolution over the entire surface of the planet. The MPO payload as selected and confirmed by ESA is listed in Table 4. The instruments on the MPO will concentrate on the investigation of Mercury's interior, surface, and exosphere (Schulz and Benkhoff 2006).

The imaging instruments of MPO will map the complete surface of Mercury with a resolution better than ∼100 m globally in the stereo mode and ∼5 m for selected areas. For meeting these measurement objectives, the composite instrument, SIMBIO-SYS consists of two imagers, one a stereo channel (STC) with a field of view of 4° and 50 m/pixel resolution, the other is the High Spatial Resolution Imaging Channel (HRIC) with a field of view of 1.47° and a resolution of 5 m/pixel. In both cases, the resolution is given from an altitude of 400 km. Both imagers have a spectral range of 400 to 900 nm and four spectral channels. Another instrument in this package is the Visible Infrared Hyperspectral Imager Channel (VIHI) targeting the mineral composition of the surface in the spectral range 400 to 2000 nm. The objective is to cross-correlate the mineralogical maps with the morphological maps produced by the visible/near infrared imagers.

The Mercury Radiometer and Thermal Infrared Spectrometer (MERTIS) is an IR-imaging spectrometer that will also provide information on the mineralogical composition of Mercury's surface by mapping its spectral emittance in the wavelength range of 7 to 14 μm.

The global abundance of rock-forming elements on Mercury's surface (in the uppermost one or two microns) will be measured by the X-ray spectrometer (MIXS). This instrument uses the activation of elements on the surface by solar X-rays; it will measure fluorescence line emission in the 0.5 to 7.5 keV energy range that corresponds to the emission energy of some important elements such as magnesium, aluminium, silicon, sulphur, calcium, titanium, and iron. Because the excitation of fluorescence emission lines depends on the variable solar X-ray and energetic particle flux, the latter will also be measured directly by another instrument, the Solar Intensity X-ray and particle Spectrometer (SIXS), to allow the absolute calibration of the MIXS measurements.

The composition of the surface layers deeper than those accessible to the X-ray measurements will be measured by the Mercury Gamma and Neutron Spectrometer (MGNS). The gamma-ray line spectra activated by the cosmic-ray flux are characteristic of elements that are within about 1 to 2 m from the surface. The instrument uses newly developed scintillation detectors which provide adequate energy resolution for the identification of elements. The same instrument also contains a detector for measuring the neutron flux (proportional counters).

The collective objective of these instruments is to characterise—morphologically and compositionally—Mercury's surface features to identify compositional variations. This will help determine whether specific landmarks have been produced by endogenic processes (e.g., volcanism) or exogenic processes (e.g., impacting objects). Knowledge of Mercury's surface composition will provide a key test of competing models for the formation and evolution of Mercury and the other terrestrial planets. The neutron spectrometer will additionally take measurements of the radar-bright spots observed from ground in the polar regions to identify their composition. If these spots, initially thought to reflect the presence of water ice, are covered with sulphur, this finding would support the idea that the planet's core is composed of Fe–FeS alloys which, compared with pure iron, remain liquid at lower temperatures.

The interior structure of Mercury will be investigated by measuring the planet's gravitational potential using the radio science experiment in combination with a laser altimeter, the high-resolution camera, an accelerometer, and (indirectly) by measuring the planet's magnetic field.

The radio science experiment (MORE) is a very sophisticated combination of data generated, gathered, and collectively analysed to determine with extreme precision the range and range rate of the spacecraft. These will be determined with an accuracy of 15 cm for range and 1.5 μm/s for the range rate at 1,000-s integration time. The key component of the "instrument" is an advanced Ka-band transponder carried on board for the very precise determination, in combination with the ground tracking and telemetry station, of the Doppler signal in the up-down (coherent two-way mode) radio link. The non-gravitational acceleration will be determined using an accelerometer (ISA). The data will be used to generate orbital solutions from which the gravitational terms can be derived. The surface gravitational potential will be linked to landmarks with high precision using the high-resolution

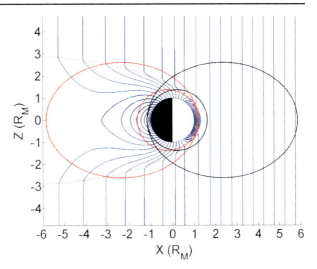

Fig. 19 The orbits of the BepiColombo Planetary and Magnetospheric Orbiters in the context of a model Mercury magnetosphere. Two sets of orbits are shown: at perihelion (with apoapsis in the solar wind, *black lines*) and at aphelion (with apoapsis in the tail of the magnetosphere, *red lines*)

camera and the topographic information from the laser altimeter. Thus, the joint analysis of observations from these three instruments will provide unprecedented quality of data on the geodesic properties of the surface of Mercury and the planet's gravitational potential.

The laser altimeter (BELA) provides absolute topographic height and position with respect to a Mercury-centred coordinate system. It uses a passive Q-switched Nd:YAG (neodymium-doped yttrium aluminium garnet; Nd:$Y_3Al_5O_{12}$) laser at 1,064 nm, generating high-energy pulses (50 mJ) at a rate of 10 Hz. The beam width is 50 microrads and results in a spot size on Mercury of 20 to 50 m. The time resolution of processing the return pulse is 2 ns, corresponding to a range resolution of 30 cm (comparable to the expected knowledge of the position of the spacecraft). This performance will be maintained up to an altitude of 1,000 km, above which it is expected that the laser will not be operated. Other than contributing to the joint analysis of the data, the laser altimeter will also provide information on the tidal deformation of the surface, surface roughness and local slopes, and albedo variations.

The determination of the planet's internal magnetic field will also contribute to the determination of its internal structure. The questions concerning the core, and the likely existence of an outer liquid layer that decouples the core from the mantle, will be answered from a combination of gravitational, orbital, and magnetic field measurements. The magnetometer (MERMAG) is due to map the magnetic field along the orbit of the MPO that is due in part to the internally generated field and in part to fields generated by the interaction of the internal field with the solar wind in Mercury's magnetosphere. In fact, estimates indicate that even along the low-altitude orbit of the MPO the contribution of externally generated magnetic field to the total field measured can be 20% to 50%. As described in more detail in Sect 5.4, Mercury's magnetosphere is not only small but is also very variable, so the external contributions (together with induction effects in the core due to the externally variable currents) make the separation of internal and external terms very difficult. It is expected that MESSENGER's magnetic survey of Mercury will determine the origin (dynamo, crustal magnetism, or other) of the internal field (Korth et al. 2004). But the detailed data that are relevant for better understanding the dynamo (such as the higher-order terms in the scalar potential of the magnetic field) will be more securely obtained by combining the data from the two BepiColombo orbiters, so that by modelling the variable external field its contribution to the MPO data can be removed. The possibility that Mercury's internal field displays secular variations can also be assessed from a comparison of MESSENGER and BepiColombo observations. The orbits of the MPO and MMO are shown in Fig. 17, in the context of Mercury's magnetosphere, for two epochs, at perihelion and aphelion. The overlap of the orbits with the magnetosphere (when the planetary magnetic field can be measured, i.e. when the spacecraft is not in the solar wind) shows how much the measurements will be affected by the magnetosphere. The synchronism of the orbits will be used for correlating the measurements.

In the absence of a stable atmosphere, Mercury has a tenuous and highly variable exosphere. Mariner 10 and ground-based observations have established the presence of Ca, Na, K, H, He, and O in the exosphere, and other elements are also likely to be present. The key questions relate to the sources and sinks of the variable exospheric populations and to their dynamics as a function of external forcing (solar and solar wind effects). The global state of the exosphere will be observed by a UV spectrometer (PHEBUS) and by the visible spectrometer component of SIMBIO-SYS. The UV spectrometer has two spectrographs, one covering the wavelength range 55 to 155 nm, the other from 145 to 315 nm. A scanning mirror is aimed at the planet's limb (one degree of freedom), and the instrument operates in the push-broom mode (viewing a swath of the limb in the direction of travel). The objectives include the discovery of new species in the exosphere and the temporal variations due to external conditions and dynamical effects.

A complex package of instruments (SERENA) will carry out in situ observations of the exosphere, and its objective is the investigation of the surface-exosphere-magnetosphere coupling. There are four spectrometers in the SERENA group of instruments. One is a neutral particle camera (ELENA) that will study the escape of, and dynamic processes in populations of neutral gases from the planet's surface. A neutral particle spectrometer (STROFIO) will measure the composition of exospheric gases directly with high sensitivity. Precipitating plasma will be measured by an ion monitor (MIPA) using an electrostatic deflector, followed by an electrostatic analyser and a time-of-flight section for determining the velocity, energy, electrical charge, and mass of the incident ions. An all-sky camera for charged particles (PICAM), acting as an ion mass spectrometer, will make observations of the generation of neutral particles from the planet's surface, their subsequent ionisation and their transport through the exosphere and magnetosphere.

The objective of the remote and in situ observations of the exosphere and its interactions with the surface and the magnetosphere is to understand the surface release processes and the source–sink balance of the exosphere. Together these measurements will help to explain the cycling of volatile elements between Mercury's interior, surface, and exosphere, and the contribution of meteoritic–cometary material and solar wind plasma to Mercury's near-surface volatile budget.

5.4 The BepiColombo Mercury Magnetospheric Orbiter (MMO)

The MMO's goal is to study Mercury's magnetic field, exosphere, and magnetosphere as well as the inner heliosphere. The spacecraft will accommodate instruments mostly dedicated to the study of the magnetic field, waves, and particles in the Mercury's environment. Four main scientific targets have been set for the MMO spacecraft on the basis of BepiColombo mission objectives. Achieving these objectives will significantly advance comparative studies of the magnetic fields and magnetospheres of the terrestrial planets (Hayakawa et al. 2004a, 2004b).

To achieve those objectives, the MMO spacecraft ($\sim$250 kg, octagonal shape with 180-cm diameter and 90-cm height, Fig. 20) is spin stabilized at 15 rpm, which facilitates the

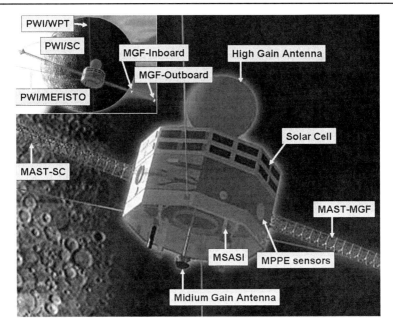

Fig. 20 The Mercury Magnetospheric Orbiter (MMO) in Mercury orbit. (Figure courtesy of JAXA and Kyoto University)

azimuthal scan of the particle detectors and the deployment of four wire antennas and two masts. Its spin axis is nearly perpendicular to the Mercury equator. The orbit is polar and highly elliptic, and its major axis lies in the equatorial plane to permit a global exploration of the exosphere, magnetic field, and magnetosphere up to an altitude of nearly six planetary radii, as well as the inner heliosphere. Details have been given by Yamakawa et al. (2002, 2004).

The MMO payload selected by JAXA in 2005 consists of five instruments or instrument packages (see Fig. 21): (1) wide range of capabilities for observing charged particles and energetic neutral atoms, (2) magnetic field, (3) electric field/plasma waves/radio waves, (4) dust, and (5) exospheric constituents (Mukai et al. 2006; Kasaba et al. 2007). Table 5 shows the list of MMO instruments. The Magnetic Field Investigation (MGF) consists of two sub-instruments, and the Mercury Plasma Particle Experiment (MPPE) for plasma and neutral particle observations consists of seven sub-instruments. The Plasma Wave Investigation (PWI) for electric field, plasma wave, and radio wave measurements with seven sub-instruments will be provided by large consortia from Japan, Europe, and elsewhere. Those payload packages will perform in-situ measurements of particles and fields in the magnetosphere of Mercury and its solar wind environment. The Mercury Sodium Atmosphere Spectral Imager (MSASI) is an imaging system included to map the sodium exosphere. The Mercury Dust Monitor (MDM) will characterise dust information around Mercury and the inner heliosphere. These scientific payload groups are coordinated by the Mission Data Processor (MDP) provided by JAXA which will ensure that the scientific objectives of the mission are fulfilled.

The MGF is designed to measure the magnetic field with an accuracy of about 10 pT, a dynamic range of $\pm 2,048$ nT and a time resolution of up to 128 Hz. Magnetic field measurements are essential to the fulfilment of the MMO scientific objectives; hence, two sets

Table 5 The scientific investigations on the BepiColombo Mercury Magnetospheric Orbiter

Instrument		Mass* (kg)	Power* (W)
MGF	Magnetic Field Investigation	1.4	17.2
	[2 sensors: MGF-Outboard, MGF-Inboard]		
MPPE	Mercury Plasma Particle Investigation	13.4	2.6
	[7 sensors: MEA1, MEA2, MIA, MSA,		
	HEP-ion, HEP-ele, ENA]		
PWI	Plasma Wave Investigation	5.4	8.5
	[4 sensors: WPT, MEFISTO, SC-LF, SC-DB]		
	[3 receivers: EWO, SORBET, AM2P]		
MSASI	Mercury Sodium Atmosphere Spectral Imager	3.6	18.6
MDM	Mercury Dust Monitor	0.5	2.4
Totals**:		24.2	48.5

*Assigned values in Apr. 2005

**Mission Data Processor (MDP: for all) and mast (for MGF and PWI) are not included

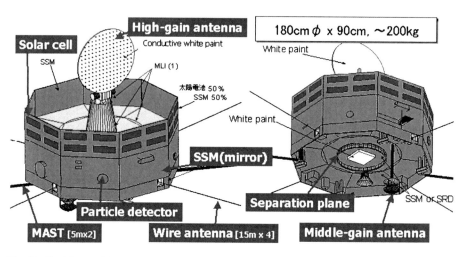

Fig. 21 The Mercury Magnetospheric Orbiter. Two views are shown to illustrate key features of the spacecraft. (Figure courtesy of JAXA)

of three-axis fluxgate sensors are installed in the middle and at the tip of a 5-m extendable mast (MAST-MGF) for redundancy, as well as for estimation of the residual field of the spacecraft. The outboard and inboard sensors are of different designs. The outboard sensor is a digital type developed in Europe, while the inboard one is an analogue type developed in Japan.

The MPPE is a composite of particle instruments: two Mercury Electron Analyzer (MEA) instruments, a Mercury Ion Analyzer (MIA), a Mercury Mass Spectrum Analyzer (MSA) for ions, High Energy Particle detectors for electrons and ions (HEP-ele and HEP-ion), and an Energetic Neutral Analyzer (ENA). These instruments are mounted on the side panels. Most of the plasma sensors have a 180° field of view (FOV) to yield full three-dimensional (3D) distribution within half a spin. Two electron sensors are looking in orthogonal direction and the full 3D distribution of electrons is measured with 1/4 spin, i.e. 1 s. MEA (3 eV ~ 0 keV) employs two separate sensors, with variable geometrical factors, in order to cover the wide dynamic range (~10^6) required for accurate measurements in the low-density plasma of Mercury's magnetotail on the one hand, and the dense plasma in the solar wind and the magnetosheath, cusp and boundary layers on the other hand. The two sensors point 90° apart in order to cover a 4π steradian solid angle in a quarter of the spin period when both sensors are working. The MIA (5 eV ~ 30 keV) and MSA (5 eV ~ 40 keV) also point 90° apart and have variable geometrical factors, but information about the ion species can be obtained only with the MSA. Measurements of low-energy electrons and ions, below 30 keV/q, have functional redundancies. Both the HEP-ele (30 keV ~ 700 keV) and HEP-ion (30 keV ~ 1.5 MeV) sensors use solid-state detectors (SSDs), and the HEP-ion sensor employs a time-of-flight (TOF) technique as well. The velocity analysis using the TOF technique is promising for high-energy ion measurements, even in the severe thermal Mercury environment. The combination of the SSD energy analysis and the TOF velocity analysis also provides information about ion species and a redundant estimation of electron fluxes. The ENA (25 eV ~ 3.3 keV) is designed to measure neutral atoms produced by charge exchange between magnetospheric ions and exospheric particles, as well as neutrals sputtered from the planetary surface by impinging magnetospheric and solar wind ions. Hence the ENA provides imaging complementary to the MSASI observation of the exosphere.

The PWI is another composite instrument to study various plasma processes associated with radio and plasma waves (electric and magnetic components) and DC electric fields in the magnetosphere of Mercury and the solar wind environment. The instrument consists of three receivers. The first is the Electric Field Detector, Waveform Capture and Onboard Frequency Analyser (EWO) for frequencies from DC to 120 kHz for electric fields and from a few Hz to 20 kHz for magnetic fields. The second, the Spectroscopie Ondes Radio et Bruit Electrostatique Thermique (SORBET) covers the frequency range from 2.5 kHz to 10 MHz for electric fields and from 2.5 kHz to 640 kHz for magnetic fields. The third is the Active Measurement of Mercury's Plasma (AM2P) with a signal output in the range 0.7 to 120 kHz. These receivers are connected to two electric field sensors, the Wire Probe Antenna (WPT) and the Mercury Electric Field In-Situ Tool (MEFISTO) and also to two magnetic field sensors, the Low-Frequency Search Coil (LF-SC) and the Dual-Band Search Coil (DB-SC) magnetometers. Observations of solar radio activity will also be useful as background information on the solar activity level at the heliocentric longitude of Mercury. Each of the WPT and MEFISTO antennas are extended orthogonally to measure electric fields and consists of a pair of wire sensors, with an overall length of 32 m, tip-to-tip, but their configurations and frequency characteristics are different. LF-SC and DB-SC constitute a complementary set of three-axis search coil sensors.

The MSASI is located on the lower deck. It is a high-dispersion spectrometer working in the visible spectral range, around the sodium D2 emission line (589 nm), and is devoted to the study of Mercury's exosphere. A tandem Fabry–Perot etalon is used to achieve a compact design, and a one degree-of-freedom scanning mirror is employed to obtain full-disk images of the planet and selected regions of interest, such as the polar regions, Caloris basin, and the magnetosphere. The MSASI will provide information on the regolith-exosphere-magnetosphere interaction as well as of the dynamics governing the surface-bounded exosphere. The MSASI data are complementary to those obtained in situ by ENA, plasma and dust measurements, as well as surface composition investigations by the X-ray and gamma-ray instruments aboard the MPO.

The MDM is located on the side panel. The purpose of the MDM is to study the near-Mercury dust environment in terms of its interaction with the planetary surface and to improve our knowledge of the interplanetary meteoroid population obtained with the Helios spacecraft in the inner heliosphere (0.31–0.47 AU). The MDM employs four 5 cm × 5 cm light-weight, heat-resistant (up to about 300°C) piezoelectric ceramic (PZT) sensors, a combined area of 100 cm^2. Those sensors will have the capabilities for counting the number and determining (roughly) the direction and momentum of dust particles. The mass and velocity, if possible, might also be separated based on recent experiments carried out by the MDM team.

The two orbiters, MPO and MMO, have different orbits and attitudes around Mercury, optimized for each orbiter's set of main observational objectives directed at the planet and the magnetosphere, respectively. The MMO and MPO instrument teams are working in close collaboration. Simultaneous measurements of the magnetic field on the MMO and MPO will enable the separation of internal and external sources, thus resolving ambiguities in the higher orders of the internal magnetic field. This investigation is linked with the planetary interior studies and the gravitational field and composition observations of the MPO. Both the exospheric and magnetospheric objectives will benefit from simultaneous measurements on the MPO and MMO. Combining in-situ particles, fields and remote sensing measurements at different altitudes will resolve spatial and temporal ambiguities that would arise from single point observations. Simultaneous MMO and MPO measurements will elucidate the physical processes and interactions that take place in the magnetosphere–exosphere–surface system of Mercury.

6 Summary and Prospects

The exploration of the planets of the solar system has progressed, if occasionally slowly, from its early beginnings in the first decade of the space age to the sophistication of the current generation of missions to Venus (Venus Express), Mars (Mars Global Surveyor, Mars Odyssey, Mars Express, the Spirit and Opportunity rovers, Mars Reconnaissance Orbiter, with others to follow), and Saturn (Cassini-Huygens). The Galileo Jupiter Orbiter explored extensively that giant planet in the 1990s. Mercury is the only reasonably accessible planet for which the only completed mission remains Mariner 10, carried out in the first phase of planetary exploration.

Looking back at the remarkable achievements of Mariner 10, its rapid development is striking: the mission was undertaken in 1970, launched in late 1973, and by early 1975 had performed its three flybys and gathered all the vital data that have been the basis of much of what we know, even now, of the planet. In addition, like many other space missions from that early phase of space exploration, Mariner 10 remains, from the vantage point of the early twenty-first century, a source of wonder at the technological and scientific achievements of a bygone age.

The reasons for not undertaking another mission before the present have been described earlier and relate both to the technical difficulties (and therefore costs) of placing and operating an orbiter around Mercury and to the attraction of other planets, in particular Mars, to the planetary community. While clear priorities can be drawn up in planetary exploration that take into account costs and technical difficulties, as well as questions which have wider than scientific implications (e.g., the origin of life), Mercury nevertheless remains a key to understanding the formation and evolution of the family of terrestrial planets, to the same extent as the other three.

MESSENGER and BepiColombo are complementary missions (Grard and Balogh 2001; McNutt et al. 2004). Clearly, MESSENGER will be able to provide answers to many of the questions left unanswered by Mariner 10 and ground-based observations, by completing the imaging of the entire surface and determining its composition, mapping the magnetic field, and resolving the nature of the radar-bright spots in the permanently shaded craters. BepiColombo will arrive at Mercury eight years after MESSENGER; as a two-spacecraft mission, it can carry a more diverse payload, more specific to planetary and magnetospheric objectives. The two orbiters can be more specifically targeted, with the Planetary Orbiter in a low-altitude polar orbit, and the Magnetospheric Orbiter in a more eccentric, but synchronised orbit. The questions that will be addressed by BepiColombo will be evidently refined by the study of the earlier MESSENGER observations. At the same time, the two-spacecraft approach will provide coordinated measurements to investigate in detail the intimate relationship between the planet and its environment. The sophisticated gravitational field investigation in the low-altitude, low-eccentricity orbit of the BepiColombo MPO can provide more detailed answers related to the planet's interior, and the combination of measurements on the MPO and MMO will discriminate between external and internal magnetic fields to a greater extent than is possible by a single orbiter. Again, the orbit of the MPO will allow a more uniform, generally higher resolution coverage for imaging Mercury's surface that will be guided by, and complement, the MESSENGER observations.

The current opportunity with the two missions, MESSENGER and BepiColombo, represents a once-in-a-generation opportunity not only to carry out a detailed survey of Mercury and its environment, but also, as a result, to integrate our knowledge of the terrestrial planets as we face the prospect of having to study and explain similar, Earth-like planets around other stars.

References

B.J. Anderson, M.H. Acuña, D.A. Lohr, J. Scheifele, A. Raval, H. Korth, J.A. Slavin, Space Sci. Rev. (2007, in press). doi:10.1007/s11214-007-9246-7

G.B. Andrews, T.H. Zurbuchen, B.H. Mauk, H. Malcom, L.A. Fisk, G. Gloeckler, G.C. Ho, J.S. Kelley, P.L. Koehn, T.W. LeFevere, S.S. Livi, R.A. Lundgren, J.M. Raines, Space Sci. Rev. (2007, in press). doi:10.1007/s11214-007-9272-5

A. Anselmi, G.E.N. Scoon, Planet. Space Sci. 49, 1409–1420 (2001)

A. Balogh (Coordinator), Mercury Orbiter, A mission proposal to ESA to be considered for the M3 mission (1993)

A. Balogh, G. Giampieri, Rep. Prog. Phys. 65, 529–560 (2002)

E.A. Belbruno, J.P. Carrico, Calculation of weak stability boundary ballistic lunar transfer trajectories, paper AIAA 2000-4142, AIAA/AAS Astrodynamics Specialist Conference, Denver, CO (2000)

J.W. Belcher et al., Mercury Orbiter: Report of the Science Working Team, Technical Memorandum 4255, NASA, Washington, D.C. 1991, p. 132

J.F. Cavanaugh, J.C. Smith, X. Sun, A.E. Bartels, L. Ramos-Izquierdo, D.J. Krebs, A.M. Novo-Gradac, J.F. McGarry, R. Trunzo, J.L. Britt, J. Karsh, R.B. Katz, A. Lukemire, R. Szymkiewicz, D.L. Berry, J.P. Swinski, G.A. Neumann, M.T. Zuber, D.E. Smith, Space Sci. Rev. (2007, in press). doi:10.1007/s11214-007-9273-4

C. Circi, P. Teofilatto, Celest. Mech. Dyn. Astron. 79, 41–72 (2001)

P.E. Clark, S.A. Curtis, G. Marr, D. Reuter, S. McKenna-Lawlor, Acta Astronautica 52, 181 (2003)

G. Colombo, Nature 208, 575 (1965)

M.I. Cruz, G.J. Bell, Acta Astronautica 35(Suppl.), 427–433 (1995)

J.A. Dunne, E. Burgess, The voyage of Mariner 10, Special Publication 424, NASA History Office, Washington, D.C. 1978, xx pp. http://history.nasa.gov/SP-424/sp424.htm

K.J. Ely, W.T. Fowler, B.D. Tapley, Acta Astronautica 35(Suppl.), 445–454 (1995)

B.H. Foing, G.D. Racca, A. Marini, E. Evrard, L. Stagnaro, M. Almeida, D. Koschny, D. Frew, J. Zender, J. Heather, M. Grande, J. Huovelin, H.U. Keller, A. Nathues, J.L. Josset, A. Malkki, W. Schmidt, G. Noci, R. Birkl, L. Iess, Z. Sodnik, P. McManamon, Adv. Space Res. 37, 6–13 (2006)

R. Förstner, R. Best, M. Steckling, BepiColombo—Mission to Mercury, International Astronautical Congress, paper IAC-06-A3.2.7, 2006

R.E. Gold, S.C. Solomon, R.L. McNutt Jr., A.G. Santo, J.B. Abshire, M.H. Acuña, R.S. Afzal, B.J. Anderson, G.B. Andrews, P.D. Bedini, J. Cain, A.F. Cheng, L.G. Evans, R.B. Follas, G. Gloeckler, J.O. Goldsten, S.E. Hawkins III, N.R. Izenberg, S.E. Jaskulek, E.A. Ketchum, M.R. Lankton, D.A. Lohr, B.H. Mauk, W.E. McClintock, S.L. Murchie, C.E. Schlemm II, D.E. Smith, R.D. Starr, T.H. Zurbuchen, Planet. Space Sci. **49**, 1467–1479 (2001)

R.E. Gold, R.L. McNutt Jr., S.C. Solomon, in *Proceedings of the 5th American Institute of Aeronautics and Astronautics International Conference on Low-Cost Planetary Missions*. Special Publication, vol. 542 (European Space Agency, Noordwijk, 2003), pp. 399–405

J.O. Goldsten, E.A. Rhodes, W.V. Boynton, W.C. Feldman, D.J. Lawrence, J.I. Trombka, D.M. Smith, L.G. Evans, J. White, N.W. Madden, P.C. Berg, G.A. Murphy, R.S. Gurnee, K. Strohbehn, B.D. Williams, E.D. Schaefer, C.A. Monaco, C.P. Cork, J.D. Eckels, W.O. Miller, M.T. Burks, L.B. Hagler, S.J. De teresa, M.C. Witte, Space Sci. Rev. (2007, in press). doi:10.1007/s11214-007-9262-7

R. Grard et al., BepiColombo: An interdisciplinary cornerstone mission to the planet Mercury. System and Technology Study Report, ESA-SCI (2000), 1

R. Grard, A. Balogh, Planet. Space Sci. **41**, 1395–1407 (2001)

J. Harmon, Space Sci. Rev. (2007, this issue). doi:10.1007/s11214-007-9234-y

S.E. Hawkins III, J. Boldt, E.H. Darlington, R. Espiritu, R.E. Gold, B. Gotwols, M. Grey, C. Hash, J. Hayes, S. Jaskulek, C. Kardian, M. Keller, E. Malaret, S.L. Murchie, P. Murphy, K. Peacock, L. Prockter, A. Reiter, M.S. Robinson., E. Schaefer, R. Shelton, R. Sterner, H. Taylor, B. Williams, T. Watters, Space Sci. Rev. (2007, in press). doi:10.1007/s11214-007-9266-3

H. Hayakawa, Y. Kasaba, H. Yamakawa, H. Ogawa, T. Mukai, Adv. Space Res. **33**, 2142–2146 (2004a)

H. Hayakawa, JAXA/BepiColombo Project Office, Science Requirement Document, JX-MMO-0003, Ver.2.0, ISAS/JAXA, 2004b

Y. Kasaba, T. Takashima, M. Fujimoto, H. Hayakawa, T. Mukai, the BepiColombo Science Working Team, in *Proc. International Symposium Space Technology and Science (ISTS)*, paper 2006-k-21 (2007, in press)

L. Ksanfomality, J. Harmon, J. Warell, I. Veselovsky, A. Sprague, N. Thomas, J. Helbert, Space Sci. Rev. (2007, this issue)

Y. Langevin, Acta Astronautica **47**, 443–452 (2000)

Y. Langevin, in *Payload and Mission Definition in Space Sciences*, ed. by V. Martinez-Pillet, A. Aparicio, F. Sanchez (Cambridge University Press, Cambridge, 2005), pp. 17–87

H.J. Korth, B.J. Anderson, M.H. Acuña, J.A. Slavin, N.A. Tsyganenko, S.C. Solomon, R.L. McNutt, Planet. Space Sci. **52**, 733–746 (2004)

J.V. McAdams, J.L. Horsewood, C.L. Yen, Discovery-class Mercury orbiter trajectory design for the 2005 launch opportunity, AIAA/AAS Astrodynamics Specialist Conference and Exhibit, Boston, MA, paper AIAA-98-4283, 1998, pp. 109–115

J.V. McAdams, D.W. Dunham, R.W. Farquhar, A.H. Taylor, B.G. Williams, J. Spacecr. Rockets **43**, 1054–1064 (2006)

W.E. McClintock, M.R. Lankton, Space Sci. Rev. (2007, in press). doi:10.1007/s11214-007-9264-5

R.L. McNutt Jr., S.C. Solomon, R. Grard, M. Novara, T. Mukai, Adv. Space Res. **33**, 2126–2132 (2004)

R.L. McNutt Jr., S.C. Solomon, R.E. Gold, J.C. Leary, the MESSENGER Team, Adv. Space Res. **38**, 564–571 (2006)

T. Mukai, H. Yamakawa, H. Hayakawa, Y. Kasaba, H. Ogawa, Adv. Space Res. **38**, 578–582 (2006)

B.C. Murray, Sci. Am. **233**, 58–68 (1975)

R.M. Nelson, L.J. Horn, J.R. Weiss, W.D. Smythe, Lunar Planet. Sci. **25**, 985–986 (1994)

G. Neukum (Coordinator), Mercury Polar Orbiter. Mission proposal to ESA, 1985

M. Novara, Planet. Space Sci **49**, 1421–1435 (2001)

M. Novara, Acta Astronautica **51**, 387–395 (2002)

G. Racca, Mercury orbiter mission with solar electric propulsion, ESA/PF/1440.97/GR, 1997

R.W. Rideourne, C.L. Yen, D.H. Collins, Mercury Dual Orbiter: Mission and flight system definition report. Publication D-7443, Jet Propulsion Laboratory, Pasadena, CA, 1990

A.G. Santo, R.E. Gold, R.L. McNutt Jr., S.C. Solomon, C.J. Ercol, R.W. Farquhar, T.J. Hartka, J.E. Jenkins, J.V. McAdams, L.E. Mosher, D.F. Persons, D.A. Artis, R.S. Bokulic, R.F. Conde, G. Dakermanji, M.E. Gross Jr., D.R. Haley, K.J. Heeres, R.H. Maurer, R.C. Moore, E.H. Rodberg, T.G. Stern, S.R. Wiley, B.G. Williams, C.L. Yen, M.R. Peterson, Planet. Space Sci. **49**, 1481–1500 (2001)

C.E. Schlemm II, R.D. Starr, G.C. Ho, K.E. Bechtold, S.A. Benedict, J.D. Boldt, W.V. Boynton, W. Bradley, M.E. Fraeman, R.E. Gold, J.O. Goldsten, J.R. Hayes, S.E. Jaskulek, E. Rossano, R.A. Rumpf, E.D. Schaefer, K. Strohbehn, R.G. Shelton, R.E. Thompson, J.I. Trombka, B.D. Williams, Space Sci. Rev. (2007, in press). doi:10.1007/s11214-007-9248-5

R. Schulz, J. Benkhoff, Adv. Space Res. **38**, 572–577 (2006)

D.L. Shirley, Acta Astronautica **53**, 375–385 (2003)

D.E. Smith, M.T. Zuber, X. Sun, G.A. Neumann, J.F. Cavanaugh, J.F. McGarry, T.W. Zagwodski, Science **311**, 53 (2006)

S.C. Solomon, R.L. McNutt Jr., R.E. Gold, M.H. Acuña, D.N. Baker, W.V. Boynton, C.R. Chapman, A.F. Cheng, G. Gloeckler, J.W. Head III, S.M. Krimigis, W.E. McClintock, S.J. Peale, S.L. Murchie, R.J. Phillips, M.S. Robinson, J.A. Slavin, D.E. Smith, R.G. Strom, J.I. Trombka, M.T. Zuber, Planet. Space Sci. **49**, 1445–1465 (2001)

S.C. Solomon, Earth Planet. Sci. Lett. **216**, 441–455 (2003)

S.C. Solomon, R.L. McNutt Jr., R.E. Gold, D.L. Domingue, Space Sci. Rev. (2007, in press). doi:10.1007/s11214-007-9247-6

P.D. Spudis, J.B. Plescia, A.D. Stewart, Lunar Planet. Sci. **25**, 1323–1324 (1994)

R.G. Strom, A.L. Sprague, *Exploring Mercury: The Iron Planet* (Springer/Praxis Publishing, Chichester, 2003)

F. Villas, C.R. Chapman, M.S. Matthews (eds.), *Mercury* (University of Arizona Press, Tucson, 1988)

H. Yamakawa, J. Kawagushi, K. Uesugi, H. Matsuo, Acta Astronautica **39**, 133–142 (1996)

H. Yamakawa, H. Saito, J. Kawaguchi, Y. Kobayashi, H. Hayakawa, T. Mukai, Acta Astronautica **45**, 187–195 (1999)

H. Yamakawa, H. Ogawa, Y. Kasaba, H. Hayakawa, T. Mukai, M. Adachi, Acta Astronautica **51**, 397–404 (2002)

H. Yamakawa, H. Ogawa, Y. Kasaba, H. Hayakawa, T. Mukai, M. Adachi, Adv. Space Res. **33**, 2133–2141 (2004)

C.-W. Yen, Proc. American Aeronautical Society/American Institute of Aeronautics and Astronautics (AAS/AIAA) Astrodynamics Specialist Conference, paper AIAA No. 85-346, San Diego, CA, 1985

C.-W. Yen, J. Astron. Sci. **37**, 417–432 (1989)

Space Science Series of ISSI

1. R. von Steiger, R. Lallement and M.A. Lee (eds.): *The Heliosphere in the Local Interstellar Medium.* 1996　　ISBN 0-7923-4320-4
2. B. Hultqvist and M. Øieroset (eds.): *Transport Across the Boundaries of the Magnetosphere.* 1997　　ISBN 0-7923-4788-9
3. L.A. Fisk, J.R. Jokipii, G.M. Simnett, R. von Steiger and K.-P. Wenzel (eds.): *Cosmic Rays in the Heliosphere.* 1998　　ISBN 0-7923-5069-3
4. N. Prantzos, M. Tosi and R. von Steiger (eds.): *Primordial Nuclei and Their Galactic Evolution.* 1998　　ISBN 0-7923-5114-2
5. C. Fröhlich, M.C.E. Huber, S.K. Solanki and R. von Steiger (eds.): *Solar Composition and its Evolution – From Core to Corona.* 1998　　ISBN 0-7923-5496-6
6. B. Hultqvist, M. Øieroset, Goetz Paschmann and R. Treumann (eds.): *Magnetospheric Plasma Sources and Losses.* 1999　　ISBN 0-7923-5846-5
7. A. Balogh, J.T. Gosling, J.R. Jokipii, R. Kallenbach and H. Kunow (eds.): *Co-rotating Interaction Regions.* 1999　　ISBN 0-7923-6080-X
8. K. Altwegg, P. Ehrenfreund, J. Geiss and W. Huebner (eds.): *Composition and Origin of Cometary Materials.* 1999　　ISBN 0-7923-6154-7
9. W. Benz, R. Kallenbach and G.W. Lugmair (eds.): *From Dust to Terrestrial Planets.* 2000　　ISBN 0-7923-6467-8
10. J.W. Bieber, E. Eroshenko, P. Evenson, E.O. Flückiger and R. Kallenbach (eds.): *Cosmic Rays and Earth.* 2000　　ISBN 0-7923-6712-X
11. E. Friis-Christensen, C. Fröhlich, J.D. Haigh, M. Schüssler and R. von Steiger (eds.): *Solar Variability and Climate.* 2000　　ISBN 0-7923-6741-3
12. R. Kallenbach, J. Geiss and W.K. Hartmann (eds.): *Chronology and Evolution of Mars.* 2001　　ISBN 0-7923-7051-1
13. R. Diehl, E. Parizot, R. Kallenbach and R. von Steiger (eds.): *The Astrophysics of Galactic Cosmic Rays.* 2001　　ISBN 0-7923-7051-1
14. Ph. Jetzer, K. Pretzl and R. von Steiger (eds.): *Matter in the Universe.* 2001　　ISBN 1-4020-0666-7
15. G. Paschmann, S. Haaland and R. Treumann (eds.): *Auroral Plasma Physics.* 2002　　ISBN 1-4020-0963-1
16. R. Kallenbach, T. Encrenaz, J. Geiss, K. Mauersberger, T.C. Owen and F. Robert (eds.): *Solar System History from Isotopic Signatures of Volatile Elements.* 2003　　ISBN 1-4020-1177-6
17. G. Beutler, M.R. Drinkwater, R. Rummel and R. Von Steiger (eds.): *Earth Gravity Field from Space – from Sensors to Earth Sciences.* 2003　　ISBN 1-4020-1408-2
18. D. Winterhalter, M. Acuña and A. Zakharov (eds.): *"Mars" Magnetism and its Interaction with the Solar Wind.* 2004　　ISBN 1-4020-2048-1
19. T. Encrenaz, R. Kallenbach, T.C. Owen and C. Sotin: *The Outer Planets and their Moons*　　ISBN 1-4020-3362-1
20. G. Paschmann, S.J. Schwartz, C.P. Escoubet and S. Haaland (eds.): *Outer Magnetospheric Boundaries: Cluster Results*　　ISBN 1-4020-3488-1
21. H. Kunow, N.U. Crooker, J.A. Linker, R. Schwenn and R. von Steiger (eds.): *Coronal Mass Ejections*　　ISBN 978-0-387-45086-5

22. D.N. Baker, B. Klecker, S.J. Schwartz, R. Schwenn and R. von Steiger (eds.): *Solar Dynamics and its Effects on the Heliosphere and Earth*　　ISBN 978-0-387-69531-0
23. Y. Calisesi, R.-M. Bonnet, L. Gray, J. Langen and M. Lockwood (eds.): *Solar Variability and Planetary Climates*　　ISBN 978-0-387-48339-9
24. K.E. Fishbaugh, P. Lognonné, F. Raulin, D.J. Des Marais, O. Korablev (eds.): *Geology and Habitability of Terrestrial Planets*　　ISBN 978-0-387-74287-8
26. A. Balogh, L. Ksanfomality, R. von Steiger (eds.): *Mercury*
　　ISBN 978-0-387-77538-8
27. R. von Steiger, G. Gloeckler, G.M. Mason (eds.): *The Composition of Matter*
　　ISBN 978-0-387-74183-3

Springer – Dordrecht / Boston / London

Printed by Publishers' Graphics LLC